Springer Proceedings in Mathematics & Statistics

Volume 179

Springer Proceedings in Mathematics & Statistics

This book series features volumes composed of selected contributions from workshops and conferences in all areas of current research in mathematics and statistics, including operation research and optimization. In addition to an overall evaluation of the interest, scientific quality, and timeliness of each proposal at the hands of the publisher, individual contributions are all refereed to the high quality standards of leading journals in the field. Thus, this series provides the research community with well-edited, authoritative reports on developments in the most exciting areas of mathematical and statistical research today.

More information about this series at http://www.springer.com/series/10533

Sergei Silvestrov · Milica Rančić
Editors

Engineering Mathematics II

Algebraic, Stochastic and Analysis Structures
for Networks, Data Classification
and Optimization

 Springer

Editors
Sergei Silvestrov
Division of Applied Mathematics,
  School of Education, Culture and
  Communication
Mälardalen University
Västerås
Sweden

Milica Rančić
Division of Applied Mathematics,
  School of Education, Culture and
  Communication
Mälardalen University
Västerås
Sweden

ISSN 2194-1009   ISSN 2194-1017 (electronic)
Springer Proceedings in Mathematics & Statistics
ISBN 978-3-319-82499-4   ISBN 978-3-319-42105-6 (eBook)
DOI 10.1007/978-3-319-42105-6

Mathematics Subject Classification (2010): 00A69, 08Axx, 16-XX, 37-XX, 46-XX, 46E30, 47H10, 60Jxx, 60K15, 92Bxx, 92Cxx, 94Cxx, 00B15

This Springer imprint is published by Springer Nature
The registered company is Springer International Publishing AG
The registered company address is: Gewerbestrasse 11, 6330 Cham, Switzerland

Preface

This book highlights the latest advances in engineering mathematics with a main focus on the mathematical models, structures, concepts, problems and computational methods and algorithms most relevant for applications in modern technologies and engineering. It addresses mathematical methods of noncommutative algebra, applied matrix analysis, operator analysis, probability theory and stochastic processes, geometry, computational mathematics, optimization and operations research with applications in network analysis, ranking in networks, networks in bioinformatics, genetic analysis and cancer research, data mining and classification, production logistics optimization.

The individual chapters cover both theory and applications, and include a wealth of figures, schemes, algorithms, tables and results of data analysis and simulation. Presenting new methods and results, reviews of cutting-edge research, and open problems for future research, they equip readers to develop new mathematical methods and concepts of their own, and to further compare and analyze the methods and results discussed.

Chapter "Classification of Low Dimensional 3-Lie Superalgebras" by Viktor Abramov and Priit Lätt is concerned with extension of a notion of n-Lie algebra to $\mathbb{Z}_2$-graded structures by means of a graded Filippov identity giving a notion of n-Lie superalgebra. Classification of low dimensional 3-Lie superalgebras is proposed, and it is shown that given an n-Lie superalgebra equipped with a supertrace one can construct the $(n+1)$-Lie superalgebra which is referred to as the induced $(n+1)$-Lie superalgebra. Based on Clifford algebra which, when endowed with a $\mathbb{Z}_2$-graded structure and a graded commutator, can be viewed as the Lie superalgebra and supertrace defined via its matrix representation, the 3-Lie superalgebras are constructed and explicitly described by their ternary commutators. In Chap. "Semi-Commutative Galois Extension and Reduced Quantum Plane" by Viktor Abramov and Md. Raknuzzaman, it is shown that a semi-commutative Galois extension of associative unital algebra by means of an element τ, which satisfies $\tau^N = 1$ (1 is the identity element of an algebra and $N \geq 2$ is an integer) induces a structure of graded q-differential algebra, where q is a primitive Nth root of unity. The graded q-differential algebra is constructed and its first order noncommutative differential calculus is studied. Moreover, the higher order noncommutative differential calculus induced by a semi-commutative Galois extension of associative unital algebra is studied, and it is shown that a reduced quantum plane can be viewed as a semi-commutative Galois extension of a fractional one-dimensional space. Chapter "Valued Custom Skew Fields with Generalised PBW Property from Power Series Construction" describes an interesting construction of associative algebras with a number of useful properties. The construction is basically that of a power series algebra with given commutation relation. The constructed algebras have a Poincaré–Birkhoff–Witt type basis, are equipped with a norm (actually an ultranorm) that is trivial to compute for basis elements, are topologically complete, and satisfy their given commutation relation. In addition, parameters can be chosen so that the algebras will in fact turn out to be skew fields and the norms become valuations. Chapter "Computing Burchnall-Chaundy Polynomials with Determinants" by Johan Richter and Sergei Silvestrov concerned with generalization of a method of computing the Burchnall–Chaundy polynomial of two commuting differential operators based on Burchnall–Chaundy eliminant determinant construction to the class of rings known as Ore extensions. It is shown that the eliminant construction partially generalizes and also counterexamples showing that these generalizations do not always retain all desired properties are provided. In Chap. "Centralizers and Pseudo-Degree Functions" by Johan Richter, a generalization of a proof of certain results by Hellström and Silvestrov on centralizers in graded algebras is presented, centralizers in certain algebras with valuations are considered and a proof that the centralizer of an element in these algebras is a free module over a certain ring is given. Under further assumptions it is also shown that the centralizer is also commutative. In Chap. "Crossed Product Algebras for Piece-Wise Constant Functions" by Johan Richter, Sergei Silvestrov, Vincent Ssembatya and Alex Behakanira Tumwesigye, algebras of functions that are constant on the sets of a partition are considered together with their crossed product algebras with the group of integers and the commutant of the function algebra in the crossed product algebra. In Chap. "Commutants in Crossed Product Algebras for Piece-Wise Constant Functions" by Johan Richter, Sergei Silvestrov and Alex Behakanira Tumwesigye, crossed product algebras of algebras of piece-wise constant functions on the real line with the group of integers are considered, and for an increasing sequence of algebras the set difference between the corresponding commutants is described.

Chapter "Asymptotic Expansions for Moment Functionals of Perturbed Discrete Time Semi-Markov Processes" by Mikael Petersson is devoted to the study of moment functionals of mixed power-exponential type for nonlinearly perturbed semi-Markov processes in discrete time. Conditions under which the moment functionals of interest can be expanded in asymptotic power series with respect to the perturbation parameter are given and it is shown how the coefficients in these expansions can be computed from explicit recursive formulas. The results of this chapter have applications for studies of quasi-stationary distributions. In Chap. "Asymptotics for Quasi-Stationary Distributions of Perturbed Discrete

Time Semi-Markov Processes" by Mikael Petersson, quasi-stationary distributions of nonlinearly perturbed semi-Markov processes in discrete time are studied. This type of distributions is of interest for analysis of stochastic systems which have finite lifetimes but are expected to persist for a long time. Asymptotic power series expansions for quasi-stationary distributions are obtained, it is shown how the coefficients in these expansions can be computed from a recursive algorithm, and a numerical example for a discrete time Markov chain is presented as an illustration of this algorithm. Chapter "Asymptotic Expansions for Stationary Distributions of Perturbed Semi-Markov Processes" by Dmitrii Silvestrov and Sergei Silvestrov presents new algorithms for computing asymptotic expansions for stationary distributions of nonlinearly perturbed semi-Markov processes based on special techniques of sequential phase space reduction, which can be applied to processes with asymptotically coupled and uncoupled finite phase spaces. Chapter "PageRank, a Look at Small Changes in a Line of Nodes and the Complete Graph" is about the PageRank algorithm used as part of the ranking process of different Internet pages in search engines, ranking in citation networks as well as other information, communication and big data networks. The chapter focuses on the behavior of PageRank as the system dynamically changes either by contracting or expanding such as when subtracting or adding nodes or links or groups of nodes or links. PageRank is considered as the solution of a linear system of equations and examined in both the ordinary normalized version of PageRank as well as the non-normalized version, and explicit formulas for the PageRank of some simple link structures are obtained. Chapter "PageRank, Connecting a Line of Nodes with a Complete Graph" is focused on the PageRank algorithm following original definition of PageRank by Sergey Brin and Larry Page as the stationary distribution of a certain random walk on a graph used to rank homepages on the Internet. Specifically, this chapter is concerned with PageRank changes after adding or removing edge between otherwise disjoint subgraphs, for example link structures consisting of a line of nodes or a complete graph and different ways to combine the two. Both the ordinary normalized version of PageRank as well as a non-normalized version of PageRank can be found by solving corresponding linear system, and it is demonstrated that it is possible to find moreover explicit formulas for the PageRank in some simple link structures and using these formulas take a more in-depth look at the behavior of the ranking as the system changes. Chapter "Graph Centrality Based Prediction of Cancer Genes" by Holger Weishaupt, Patrik Johansson, Christopher Engström, Sven Nelander, Sergei Silvestrov and Fredrik J. Swartling focuses on how graph centralities obtained from biological networks have been used to predict cancer genes. As current cancer therapies including surgery, radiotherapy and chemotherapy are often plagued by high failure rates, designing more targeted and personalized treatment strategies requires a detailed understanding of druggable tumor driver genes. Specifically, the chapter begins with describing the current problems in cancer therapy and the reasoning behind using network based cancer gene prediction, followed by an outline of biological networks, their generation and properties, and finely by a review of major concepts, recent results as well as future challenges regarding the use of graph centralities in cancer gene prediction. Chapter "Output Rate Variation Problem: Some Heuristic Paradigms and Dynamic Programming" by Gyan Bahadur Thapa and Sergei Silvestrov is concerned with the output rate variation problem, which is one of the important research directions in the area of multi-level just-in-time production systems. A short survey of the mathematical models of this problem is provided together with consideration of its NP-hardness, a brief review of heuristic approaches to the problem, the discussion on the dynamic programming approach and pegging assumption reducing the multi-level problem to weighted single-level problem as well as some open problems.

In Chap. "L^p-Boundedness of Two Singular Integral Operators of Convolution Type" by Sten Kaijser and John Musonda, boundedness properties investigated for two singular integral operators defined on L^p-spaces $(1 < p < \infty)$ on the real line, both as convolution operators on $L^p(\mathbb{R})$ and on the weighted spaces $L^p(\omega)$, where $\omega(x) = 1/(2\cosh\frac{\pi}{2}x)$. In the Chap. "Fractional-Wavelet Analysis of Positive definite Distributions and Wavelets on $\mathscr{D}'(\mathbb{C})$" by Emanuel Guariglia and Sergei Silvestrov, a wavelet expansion theory for positive definite distributions over the real line is considered and a fractional derivative operator for complex functions in the distribution sense is defined. The Ortigueira–Caputo fractional derivative operator is rewritten as a convolution according to the fractional calculus of real distributions, and the fractional derivatives of the complex Shannon wavelet and Gabor–Morlet wavelet are computed together with their plots and main properties. Chapters "Linear Classification of Data with Support Vector Machines and Generalized Support Vector Machines" and "Linear and Nonlinear Classifiers of Data with Support Vector Machines and Generalized Support Vector Machines" by Talat Nazir, Xiaomin Qi and Sergei Silvestrov are devoted to support vector machine for linear and nonlinear classification of data. Generalized support vector machine for classification of data is introduced, and it is shown that the problem of generalized support vector machine is equivalent to the problem of generalized variational inequality. Various results for the existence of solutions are established and several examples are constructed. In Chaps. "Common Fixed Points of Weakly Commuting Multivalued Mappings on a Domain of Sets Endowed with Directed Graph" and "Common Fixed Point Results for Family of Generalized Multivalued F-contraction Mappings in Ordered Metric Spaces" by Talat Nazir and Sergei Silvestrov, the existence of coincidence points and common fixed points for multivalued mappings satisfying certain graphic ψ-contraction contractive conditions with set-valued domain endowed with a graph, without appealing to continuity, is established, the existence of common fixed points of family of multivalued mappings satisfying generalized F-contractive conditions in ordered metric spaces is also investigated.

The book consists of carefully selected and refereed contributed chapters covering research developed as a result of a focused international seminar series on mathematics and applied mathematics and a series of three focused international research workshops on engineering mathematics organized by the Research Environment in Mathematics and Applied Mathematics at Mälardalen University from autumn 2014 to autumn 2015: the International Workshop on Engineering Mathematics for Electromagnetics and Health Technology; the International Workshop on Engineering Mathematics, Algebra, Analysis and Electromagnetics; and the 1st Swedish-Estonian International Workshop on Engineering Mathematics, Algebra, Analysis and Applications.

This book project has been realized, thanks to the strategic support by Mälardalen University to the research and research education in Mathematics, which is conducted by the research environment Mathematics and Applied Mathematics (MAM), in the established research area of Educational Sciences and Mathematics at the School of Education, Culture and Communication at Mälardalen University. We are grateful also to the EU Erasmus Mundus projects FUSION, EUROWEB and IDEAS, the Swedish International Development Cooperation Agency (Sida) and International Science Programme in Mathematical Sciences, Swedish Mathematical Society, Linda Peetre Memorial Foundation, as well as other national and international funding organizations and the research and education environments and institutions of the individual researchers and research teams that contributed to this book.

We hope that this book will serve as a source of inspiration for a broad spectrum of researchers and research students in mathematics and applied mathematics, as well as in the areas of applications of mathematics considered in the book.

Västerås, Sweden   Sergei Silvestrov
July 2016   Milica Rančić

Contents

Contributors

Viktor Abramov Institute of Mathematics and Statistics, University of Tartu, Tartu, Estonia

Alex Behakanira Tumwesigye Department of Mathematics, College of Natural Sciences, Makerere University, Kampala, Uganda

Christopher Engström Division of Applied Mathematics, School of Education, Culture and Communication, Mälardalen University, Västerås, Sweden

Emanuel Guariglia Department of Physics "E. R. Caianiello", University of Salerno, Fisciano, Italy; Division of Applied Mathematics, School of Education, Culture and Communication, Mälardalen University, Västerås, Sweden

Lars Hellström Division of Applied Mathematics, School of Education, Culture and Communication, Mälardalen University, Västerås, Sweden

Patrik Johansson Department of Immunology, Genetics and Pathology, Science for Life Laboratory, Uppsala University, Uppsala, Sweden

Sten Kaijser Department of Mathematics, Uppsala University, Uppsala, Sweden

Priit Lätt Institute of Mathematics and Statistics, University of Tartu, Tartu, Estonia

John Musonda Division of Applied Mathematics, School of Education, Culture and Communication, Mälardalen University, Västerås, Sweden

Talat Nazir Division of Applied Mathematics, School of Education, Culture and Communication, Mälardalen University, Västerås, Sweden; Department of Mathematics, COMSATS Institute of Information Technology, Abbottabad, Pakistan

Sven Nelander Department of Immunology, Genetics and Pathology, Uppsala University, Uppsala, Sweden

Mikael Petersson Department of Mathematics, Stockholm University, Stockholm, Sweden

Xiaomin Qi Division of Applied Mathematics, School of Education, Culture and Communication, Mälardalen University, Västerås, Sweden

Md. Raknuzzaman Institute of Mathematics, University of Tartu, Tartu, Estonia

Johan Richter Division of Applied Mathematics, School of Education, Culture and Communication, Mälardalen University, Västerås, Sweden

Dmitrii Silvestrov Department of Mathematics, Stockholm University, Stockholm, Sweden

Sergei Silvestrov Division of Applied Mathematics, School of Education, Culture and Communication, Mälardalen University, Västerås, Sweden

Vincent Ssembatya Department of Mathematics, College of Natural Sciences, Makerere University, Kampala, Uganda

Fredrik J. Swartling Department of Immunology, Genetics and Pathology, Uppsala University, Uppsala, Sweden

Gyan Bahadur Thapa Pulchowk Campus, Institute of Engineering, Tribhuvan University, Kathmandu, Nepal

Holger Weishaupt Department of Immunology, Genetics and Pathology, Science for Life Laboratory, Uppsala University, Uppsala, Sweden

Classification of Low Dimensional 3-Lie Superalgebras

Viktor Abramov and Priit Lätt

Abstract A notion of n-Lie algebra introduced by V.T. Filippov can be viewed as a generalization of a concept of binary Lie algebra to the algebras with n-ary multiplication law. A notion of Lie algebra can be extended to $\mathbb{Z}_2$-graded structures giving a notion of Lie superalgebra. Analogously a notion of n-Lie algebra can be extended to $\mathbb{Z}_2$-graded structures by means of a graded Filippov identity giving a notion of n-Lie superalgebra. We propose a classification of low dimensional 3-Lie superalgebras. We show that given an n-Lie superalgebra equipped with a supertrace one can construct the $(n+1)$-Lie superalgebra which is referred to as the induced $(n+1)$-Lie superalgebra. A Clifford algebra endowed with a $\mathbb{Z}_2$-graded structure and a graded commutator can be viewed as the Lie superalgebra. It is well known that this Lie superalgebra has a matrix representation which allows to introduce a supertrace. We apply the method of induced Lie superalgebras to a Clifford algebra to construct the 3-Lie superalgebras and give their explicit description by ternary commutators.

Keywords n-Lie algebras · n-Lie superalgebras · Clifford algebras · Induced n-Lie superalgebras

1 Introduction

Recently, there was markedly increased interest of theoretical physics towards the algebras with n-ary multiplication law. Due to the fact that the Lie algebras play a crucial role in theoretical physics, it seems that development of n-ary analog of a concept of Lie algebra is especially important. In [5] V.T. Filippov proposed a notion of n-Lie algebra which can be considered as a possible generalization of a concept

V. Abramov (✉) · P. Lätt
Institute of Mathematics and Statistics, University of Tartu, Liivi 2–602, 50409 Tartu, Estonia
e-mail: viktor.abramov@ut.ee

P. Lätt
e-mail: priit_lt@ut.ee

© Springer International Publishing Switzerland 2016
S. Silvestrov and M. Rančić (eds.), *Engineering Mathematics II*,
Springer Proceedings in Mathematics & Statistics 179,
DOI 10.1007/978-3-319-42105-6_1

of Lie algebra to structures with n-ary multiplication law. In approach proposed by V.T. Filippov an n-ary commutator of n-Lie algebra is skew-symmetric and satisfies an n-ary analog of Jacobi identity which is now called Filippov identity. It is worth to mention that there is an approach different from the one proposed by V.T. Filippov, where a ternary commutator is not skew-symmetric but it obeys a symmetry based on a representation of the group of cyclic permutations $\mathbb{Z}_3$ by cubic roots of unity [2]. It is well known that a concept of Lie algebra can be extended to $\mathbb{Z}_2$-graded structures with the help of graded commutator and graded Jacoby identity, and a corresponding structure is known under the name of Lie superalgebra.

In the present paper we show that a notion of n-Lie algebra proposed by V.T. Filippov can be extended to $\mathbb{Z}_2$-graded structures by means of graded n-commutator and a graded analog of Filippov identity. This $\mathbb{Z}_2$-graded n-Lie algebra will be referred to as a n-Lie superalgebra. We show that a method of induced n-Lie algebras proposed in [3] and based on an analog of a trace can be applied to n-Lie superalgebras if instead of a trace we will be using a supertrace. We introduce the notions such as an ideal of n-Lie superalgebra, subalgebra of n-Lie superalgebra, descending series and prove several results analogous to the results proved in [3] for n-Lie algebras. We propose a classification of low dimensional 3-Lie superalgebras and find their commutation relations. A Clifford algebra can be used to construct a Lie superalgebra if one equips it with a graded commutator. This Lie superalgebra has a matrix representation called supermodule of spinors and this representation can be endowed with a supertrace. Thus we have all basic components of a method of induced n-Lie superalgebras and applying this method we construct a series of 3-Lie superalgebras.

2 Supertrace and Induced n-Lie Superalgebras

A notion of Lie algebra can be extended from binary algebras to algebras with n-ary multiplication law with the help of a notion of n-Lie algebra, where n is any integer greater or equal to 2. This approach was proposed by V. T. Filippov in [5], and it is based on n-ary analog of Jacobi identity which is now called the Filippov identity. It is well known that a concept of binary Lie algebra can be extended to $\mathbb{Z}_2$-graded structures giving a notion of Lie superalgebra. Similarly a notion of n-Lie algebra can be extended to $\mathbb{Z}_2$-graded structures giving a structure which we call an n-Lie superalgebra [1, 4]. In this section we give the definitions of n-Lie algebra, n-Lie superalgebra and show that a structure of induced n-Lie algebra based on an analog of trace [3] can be extended to n-Lie superalgebras with the help of supertrace.

Definition 1 Vector space $\mathfrak{g}$ endowed with a mapping $[\cdot, \ldots, \cdot] : \mathfrak{g}^n \to \mathfrak{g}$ is said to be a n-*Lie algebra*, if $[\cdot, \ldots, \cdot]$ is n-linear, skew-symmetric and satisfies the identity

$$[x_1, \ldots, x_{n-1}, [y_1, \ldots, y_n]] = \sum_{i=1}^{n} [y_1, \ldots, [x_1, \ldots, x_{n-1}, y_i], \ldots, y_n], \quad (1)$$

where $x_1, \ldots, x_{n-1}, y_1, \ldots, y_n \in \mathfrak{g}$.

In the definition of n-Lie algebra the identity (1) is called the Filippov identity [5]. It is clear that for $n = 2$ Filippov identity yields the classical Jacobi identity of a binary Lie algebra.

Definition 2 Let $\phi : \mathfrak{g}^n \to \mathfrak{g}$. A linear map $\tau : \mathfrak{g} \to \mathbb{K}$ will be referred to as a ϕ-trace if $\tau (\phi(x_1, \ldots, x_n)) = 0$ for all $x_1, \ldots, x_n \in \mathfrak{g}$.

In [3] the authors proposed a method based on ϕ-trace which can be applied to an n-Lie algebra to construct the $(n + 1)$-Lie algebra.

Theorem 1 Let $(\mathfrak{g}, [\cdot, \ldots, \cdot])$ be an n-Lie algebra and τ be a $[\cdot, \ldots, \cdot]$-trace. Define $[\cdot, \ldots, \cdot]_\tau : \mathfrak{g}^{n+1} \to \mathfrak{g}$ by

$$[x_1, \ldots, x_{n+1}]_\tau = \sum_{i=1}^{n+1} (-1)^i \tau(x_i)[x_1, \ldots, x_{i-1}, x_{i+1}, \ldots, x_{n+1}]. \tag{2}$$

Then $(\mathfrak{g}, [\cdot, \ldots, \cdot]_\tau)$ is the $(n + 1)$-Lie algebra.

It was shown in [1] that a similar method based on a notion of a supertrace can be used in the case of n-Lie superagebras, and given an n-Lie superalgebra one can apply this method to induce the $(n + 1)$-Lie superalgebra. Let us remind that super vector space is a direct sum of two vector spaces, i.e. $V = V_{\bar{0}} \oplus V_{\bar{1}}$. The dimension of finite dimensional super vector space is denoted as $m|n$ if $dim V_{\bar{0}} = m$ and $dim V_{\bar{1}} = n$. Element $x \in V \setminus \{0\}$ is said to be homogeneous if either $x \in V_{\bar{0}}$ or $x \in V_{\bar{1}}$. For homogeneous elements we can define parity by

$$|x| = \begin{cases} \bar{0}, & x \in V_{\bar{0}}, \\ \bar{1}, & x \in V_{\bar{1}}. \end{cases} \tag{3}$$

In what follows we will assume that element x is homogeneous whenever $|x|$ is used.

Definition 3 We say that super vector space $\mathfrak{g}$ endowed with n-linear map $[\cdot, \ldots, \cdot] : \mathfrak{g}^n \to \mathfrak{g}$ is n-Lie superalgebra if for all $x_1, \ldots, x_n, y_1, \ldots, y_{n-1} \in \mathfrak{g}$

1. $|[x_1, \ldots, x_n]| = \sum_{i=1}^n |x_i|$,
2. $[x_1, \ldots, x_i, x_{i+1}, \ldots, x_n] = -(-1)^{|x_i||x_{i+1}|}[x_1, \ldots, x_{i+1}, x_i, \ldots, x_n]$,
3. $[y_1, \ldots, y_{n-1}, [x_1, \ldots, x_n]] =$
   $= \sum_{i=1}^n (-1)^{|\mathsf{x}|_{i-1}|\mathsf{y}|_{n-1}}[x_1, \ldots, x_{i-1}, [y_1, \ldots, y_{n-1}, x_i], x_{i+1}, \ldots, x_n]$,

where $\mathsf{x} = (x_1, \ldots, x_n)$, $\mathsf{y} = (y_1, \ldots, y_{n-1})$, and $|\mathsf{x}_i| = \sum_{j=1}^i |x_j|$.

Definition 4 Let $V = V_{\bar{0}} \oplus V_{\bar{1}}$ be a super vector space and let $\phi : V^n \to V$. We say that linear map $S : V \to \mathbb{K}$ is a ϕ-*supertrace* if

1. $S\left(\phi(x_1, \ldots, x_n)\right) = 0$ for all $x_1, \ldots, x_n \in V$,
2. $S(x) = 0$ for all $x \in V_{\bar{1}}$.

Given an n-Lie superalgebra endowed with a supertrace (which satisfies the conditions of the previous definition with respect to a graded commutator of this algebra) we can construct the $(n + 1)$-Lie superalgebra by means of a method described in [1].

Theorem 2 *Let* $(\mathfrak{g}, [\cdot, \ldots, \cdot])$ *be a n-Lie superalgebra and let* $S : \mathfrak{g} \to \mathbb{K}$ *be a* $[\cdot, \ldots, \cdot]$-*supertrace. Define* $[\cdot, \ldots, \cdot]_S : \mathfrak{g}^{n+1} \to \mathfrak{g}$ *by*

$$[x_1, \ldots, x_{n+1}]_S = \sum_{i=1}^{n+1} (-1)^{i-1} (-1)^{|x_i||x|_{i-1}} S(x_i)[x_1, \ldots, x_{i-1}, x_{i+1}, \ldots, x_{n+1}].$$

(4)

Then $(\mathfrak{g}, [\cdot, \ldots, \cdot]_S)$ *is a $(n + 1)$-Lie superalgebra.*

3 Properties of Induced n-Lie Superalgebras

In this section we study a structure of induced n-Lie superalgebra, and introducing the notions such as ideal of an n-Lie superalgebra, derived series, subalgebra of n-Lie superalgebra we prove several results which are analogous to the results proved in [3] in the case of n-Lie algebras.

Definition 5 Let $(\mathfrak{g}, [\cdot, \ldots, \cdot])$ be a n-Lie superalgebra and let $\mathfrak{h}$ be a subspace of $\mathfrak{g}$. We say that $\mathfrak{h}$ is an *ideal* of $\mathfrak{g}$, if for all $h \in \mathfrak{h}$ and for all $x_1, \ldots, x_{n-1} \in \mathfrak{g}$, it holds that $[h, x_1, \ldots, x_{n-1}] \in \mathfrak{h}$.

Definition 6 Let $(\mathfrak{g}, [\cdot, \ldots, \cdot])$ be a n-Lie superalgebra and let $\mathfrak{h}$ be an ideal of $\mathfrak{g}$. *Derived series* of $\mathfrak{h}$ is defined as

$$D^0(\mathfrak{h}) = \mathfrak{h} \quad \text{and} \quad D^{p+1}(\mathfrak{h}) = [D^p(\mathfrak{h}), \ldots, D^p(\mathfrak{h})], \quad p \in \mathbb{N},$$

and the *descending central series* of $\mathfrak{h}$ as

$$C^0(\mathfrak{h}) = \mathfrak{h} \quad \text{and} \quad C^{p+1}(\mathfrak{h}) = [C^p(\mathfrak{h}), \mathfrak{h}, \ldots, \mathfrak{h}], \quad p \in \mathbb{N}.$$

An ideal $\mathfrak{h}$ of n-Lie superalgebra $\mathfrak{g}$ is said to be *solvable* if there exists $p \in \mathbb{N}$ such that $D^p(\mathfrak{h}) = \{0\}$, and we call $\mathfrak{h}$ *nilpotent* if $C^p(\mathfrak{h}) = \{0\}$ for some $p \in \mathbb{N}$.

Proposition 1 *Let* $(\mathfrak{g}, [\cdot, \ldots, \cdot])$ *be a n-Lie superalgebra and let* $\mathfrak{h} \subset \mathfrak{g}$ *be a subalgebra. If S is supertrace of* $[\cdot, \ldots, \cdot]$, *then $\mathfrak{h}$ is also subalgebra of* $(\mathfrak{g}, [\cdot, \ldots, \cdot]_S)$.

Proof Let $\mathfrak{h}$ be subalgebra of n-Lie superalgebra $(\mathfrak{g}, [\cdot, \ldots, \cdot])$, $x_1, \ldots, x_{n+1} \in \mathfrak{h}$ and assume S is a supertrace of $[\cdot, \ldots, \cdot]$. Then $[x_1, \ldots, x_{n+1}]_S$ is a linear combination of elements of $\mathfrak{h}$ as desired. $\square$

Proposition 2 *Let $\mathfrak{h}$ be an ideal of $(\mathfrak{g}, [\cdot, \ldots, \cdot])$ and assume S is supertrace of $[\cdot, \ldots, \cdot]$. Then $\mathfrak{h}$ is ideal of $(\mathfrak{g}, [\cdot, \ldots, \cdot]_S)$ if and only if $[\mathfrak{g}, \mathfrak{g}, \ldots, \mathfrak{g}] \subseteq \mathfrak{h}$ or $\mathfrak{h} \subseteq \ker S$.*

Proof Let $h \in \mathfrak{h}$ and $x_1, \ldots, x_n \in \mathfrak{g}$. Then

$$[x_1, \ldots, x_n, h]_S = \sum_{i=1}^{n} (-1)^{i-1} (-1)^{|x_i||x|_{i-1}} S(x_i)[x_1, \ldots, x_{i-1}, x_{i+1}, \ldots, x_n, h] + \quad (5)$$

$$(-1)^n (-1)^{|h||x|_n} S(h)[x_1, \ldots, x_n]. \quad (6)$$

Since $\mathfrak{h}$ is ideal we have $[x_1, \ldots, x_{i-1}, x_{i+1}, \ldots, x_n, h] \in \mathfrak{h}$ for all $i = 1, \ldots, n$. Thus $[x_1, \ldots, x_n, h]_S \in \mathfrak{h}$ is equivalent to

$$(-1)^n (-1)^{|h||x|_n} S(h)[x_1, \ldots, x_n] \in \mathfrak{h},$$

which clearly holds when $S(h) = 0$ or $[x_1, \ldots, x_n] \in \mathfrak{h}$. $\qquad\square$

Proposition 3 *Let $(\mathfrak{g}, [\cdot, \ldots, \cdot])$ be n-Lie superalgebra and let S be supertrace of $[\cdot, \ldots, \cdot]$. Then induced $(n+1)$-Lie superalgebra $(\mathfrak{g}_S, [\cdot, \ldots, \cdot]_S)$ is solvable.*

Proof Assume $(\mathfrak{g}, [\cdot, \ldots, \cdot])$ is a n-Lie superalgebra and S is supertrace of $[\cdot, \ldots, \cdot]$, and let $x_1, \ldots, x_{n+1} \in D^1(\mathfrak{g}_S)$.

Then for every $i = 1, \ldots, n+1$ we have $x_i^1, \ldots, x_i^{n+1} \in \mathfrak{g}$ such that $x_i = [x_i^1, \ldots, x_i^{n+1}]_S$, in which case

$$[x_1, \ldots, x_{n+1}]_S =$$
$$\sum_{i=1}^{n+1} (-1)^{i-1} (-1)^{|x_i||x|_{i-1}} S([x_i^1, \ldots, x_i^{n+1}]_S)[x_1, \ldots, x_{i-1}, x_{i+1}, \ldots, x_{n+1}] = 0.$$

$\qquad\square$

In the light of the last proposition we can immediately see that if $(\mathfrak{g}, [\cdot, \ldots, \cdot])$ is an n-Lie superalgebra, then for the induced $(n+1)$-Lie superalgebra it holds $D^p(\mathfrak{g}_S) = \{0\}$, whenever $p \geq 2$.

Proposition 4 *Let $(\mathfrak{g}, [\cdot, \ldots, \cdot])$ be n-Lie superalgebra, S supertrace of $[\cdot, \ldots, \cdot]$ and assume $(n+1)$-Lie superalgebra $(\mathfrak{g}_S, [\cdot, \ldots, \cdot]_S)$ is induced by S. Denote descending central series of $\mathfrak{g}$ by $(C^p(\mathfrak{g}))_{p=0}^{\infty}$ and denote descending central series of $\mathfrak{g}_S$ by $(C^p(\mathfrak{g}_S))_{p=0}^{\infty}$. Then*

$$C^p(\mathfrak{g}_S) \subseteq C^p(\mathfrak{g}) \quad \text{for all} \quad p \in \mathbb{N}.$$

If there exists $g \in \mathfrak{g}$ such that $[g, x_1, \ldots, x_n]_S = [x_1, \ldots, x_n]$ holds for all $x_1, x_2, \ldots, x_n \in \mathfrak{g}$, then

$$C^p(\mathfrak{g}_S) = C^p(\mathfrak{g}) \quad \text{for all} \quad p \in \mathbb{N}.$$

Proof Case $p = 0$ is trivial. Note that for $p = 1$ any $x = [x_1, \ldots, x_{n+1}]_S \in C^1(\mathfrak{g}_S)$ can be expressed as

$$x = \sum_{i=1}^{n+1} (-1)^{i-1} (-1)^{|x_i||X|_{i-1}} S(x_i)[x_1, \ldots, x_{i-1}, x_{i+1}, \ldots, x_{n+1}],$$

meaning x is a linear combination in $C^1(\mathfrak{g})$.

Assume now that there exists $g \in \mathfrak{g}$ such that for all $y_1, \ldots, y_n \in \mathfrak{g}$ it holds that $[g, y_1, \ldots, y_n]_S = [y_1, \ldots, y_n]$. Then $x = [x_1, \ldots, x_n] \in C^1(\mathfrak{g})$ can be written as $[g, x_1, \ldots, x_n]_S$ and thus $x \in C^1(\mathfrak{g}_S)$.

Next assume that the statement holds for some $p \in \mathbb{N}$ and let $x \in C^{p+1}(\mathfrak{g}_S)$. Then there are $x_1, \ldots, x_n \in \mathfrak{g}$ and $g \in C^p(\mathfrak{g}_S)$ such that

$$x = [g, x_1, \ldots, x_n]_S = (-1)^{n+|g||x|_n} [x_1, \ldots, x_n, g]_S =$$

$$= (-1)^{n+|g||x|_n} \sum_{i=1}^{n} (-1)^{i-1} (-1)^{|x_i||X|_{i-1}} S(x_i)[x_1, \ldots, x_{i-1}, x_{i+1}, \ldots, x_n, g],$$

since g can be expressed as a bracket of some elements, and hence $S(g) = 0$. On the other hand, as $g \in C^p(\mathfrak{g}_S)$, by our inductive assumption $g \in C^p(\mathfrak{g})$, and thus $x \in C^{p+1}(\mathfrak{g})$.

To complete the proof, assume that there exists $g \in \mathfrak{g}$ such that for all $y_1, \ldots, y_n \in \mathfrak{g}$ equality $[g, y_1, \ldots, y_n]_S = [y_1, \ldots, y_n]$ holds. If $x \in C^{p+1}(\mathfrak{g})$, then $x = [h, x_1, \ldots, x_{n-1}]$, where $x_1, \ldots, x_{n-1} \in \mathfrak{g}$ and $h \in C^p(\mathfrak{g})$. Altogether we have

$$x = [h, x_1, \ldots, x_{n-1}] = [g, h, x_1, \ldots, x_{n-1}]_S = -(-1)^{|g||h|} [h, g, x_1, \ldots, x_{n-1}]_S.$$

At the same time $h \in C^p(\mathfrak{g}) = C^p(\mathfrak{g}_S)$, which gives us $[h, g, x_1, \ldots, x_{n-1}]_S \in C^{p+1}(\mathfrak{g}_S)$, meaning $x \in C^{p+1}(\mathfrak{g}_S)$, as desired. □

4   Low Dimensional Ternary Lie Superalgebras

In this section we propose a classification of low dimensional ternary Lie superalgebras.

First of all we find the number of different (non-isomorphic) 3-Lie superalgebras over $\mathbb{C}$ of dimension $m|n$ where $m + n < 5$. We also find the explicit commutation relations of these 3-Lie superalgebras. We use a method which is based on the structure constants of an n-Lie superalgebra.

Definition 7 Let $\mathfrak{g} = \mathfrak{g}_{\bar{0}} \oplus \mathfrak{g}_{\bar{1}}$ be n-Lie superalgebra, denote

$$\mathcal{B} = \{e_1, \ldots, e_m, f_1, \ldots, f_n\}$$

and assume $\{e_1, \ldots, e_m\}$ spans $\mathfrak{g}_{\bar{0}}$ and $\{f_1, \ldots, f_n\}$ spans $\mathfrak{g}_{\bar{1}}$. Elements $K^B_{A_1 \ldots A_n}$ defined by

$$[z_{A_1}, \ldots, z_{A_n}] = K^B_{A_1 \ldots A_n} z_B,$$

where $z_{A_1}, \ldots, z_{A_n}, z_B \in \mathcal{B}$, are said to be *structure constants* of $\mathfrak{g}$ with respect to $\mathcal{B}$.

Assume we have a 3-Lie superalgebra $(\mathfrak{g}, [\cdot, \cdot, \cdot])$ of dimension $m|n$ over $\mathbb{C}$. Denote

$$\mathcal{B} = \{e_1, \ldots, e_m, f_1, \ldots, f_n\} = \{z_1, \ldots, z_{m+n}\}$$

and assume e_α, $1 \le \alpha \le m$, spans the even part of $\mathfrak{g}$ and f_i, $1 \le i \le n$, spans the odd part of $\mathfrak{g}$. Additionaly, let $z_A = e_A$, when $1 \le A \le m$, and $z_A = f_{A-m}$, when $m < A \le m + n$. Since $|[z_1, z_2, z_3]| = |z_1| + |z_2| + |z_3|$ we can express the values of commutator on generators using structure constants in the following form:

$$\begin{aligned}
[e_\alpha, e_\beta, e_\gamma] &= K^\lambda_{\alpha\beta\gamma} e_\lambda, \\
[e_\alpha, e_\beta, f_i] &= K^j_{\alpha\beta i} f_j, \\
[e_\alpha, f_i, f_j] &= K^\beta_{\alpha i j} e_\beta, \\
[f_i, f_j, f_k] &= K^l_{ijk} f_l,
\end{aligned}$$

where $\alpha \le \beta \le \gamma$ and $i \le j \le k$. As all other possible orderings and combinations of generators can be transformed into one of these four forms by graded skew-symmetry of $[\cdot, \cdot, \cdot]$, we will not consider them.

As a next step we can eliminate the combinations that are trivial. To find such brackets we can observe different permutations of arguments. If some permutation yields the initial ordering without preserving the sign, then this bracket must be zero, as in

$$[e_1, e_1, f_i] = -(-1)^{|e_1||e_1|}[e_1, e_1, f_1] = -[e_1, e_1, f_i].$$

Finally we can use the graded Filippov identity. Observe $[z_A, z_B, z_C] = K^D_{ABC} z_D \ne 0$, where $1 \le A \le B \le C \le m + n$, and calculate

$$[z_E, z_F, [z_A, z_B, z_C]]$$

using two different paths. Firstly use what is known and write

$$[z_E, z_F, [z_A, z_B, z_C]] = K^D_{ABC}[z_E, z_F, z_D].$$

Then transform bracket $[z_E, z_F, z_D]$ to $(-1)^{\circlearrowleft_{DEF}}[z_{D'}, z_{E'}, z_{F'}]$, where $\{D, E, F\} = \{D', E', F'\}$, but $D' \le E' \le F'$, and $(-1)^{\circlearrowleft_{DEF}}$ gives the sign that comes from graded skew-symmetry. Note that $[z_{D'}, z_{E'}, z_{F'}]$ can be expressed using structure constants and generators as well and thus we have $[z_{D'}, z_{E'}, z_{F'}] = K^H_{D'E'F'} z_H$, which means that on the one hand

$$[z_E, z_F, [z_A, z_B, z_C]] = (-1)^{\circlearrowleft_{DEF}} K^D_{ABC} K^H_{D'E'F'} z_H.$$

On the other hand we can use Filippov identity to calculate $[z_E, z_F, [z_A, z_B, z_C]]$:

$$[z_E, z_F, [z_A, z_B, z_C]] = [[z_E, z_F, z_A], z_B, z_C] + (-1)^{|z_A|(|z_E|+|z_F|)} [z_A, [z_E, z_F, z_B], z_C] +$$
$$(-1)^{(|z_A|+|z_B|)(|z_E|+|z_F|)} [z_A, z_B, [z_E, z_F, z_C].]$$

In every summand we can apply the same construction as described above. To do that, let us denote $z_{AEF} = |z_A|(|z_E| + |z_F|)$ and $z_{ABEF} = (|z_A| + |z_B|)(|z_E| + |z_F|)$. Now reorder the arguments in increasing order and replace the result with linear combination of generators and structure constants. By doing so we end up having

$$[z_E, z_F, [z_A, z_B, z_C]] = (-1)^{\circlearrowleft_{AEF} + \circlearrowleft_{B'C'G'}} K^G_{A'E'F'} K^H_{B'C'G'} z_H +$$
$$(-1)^{z_{AEF} + \circlearrowleft_{BEF} + \circlearrowleft_{A'C'G'}} K^G_{B'E'F'} K^H_{A'C'G'} z_H +$$
$$(-1)^{z_{ABEF} + \circlearrowleft_{CEF} + \circlearrowleft_{A'B'G'}} K^G_{C'E'F'} K^H_{A'B'G'} z_H.$$

In other words the following system of quadratic equations emerges:

$$(-1)^{\circlearrowleft_{DEF}} K^D_{ABC} K^H_{D'E'F'} z_H = (-1)^{\circlearrowleft_{AEF} + \circlearrowleft_{B'C'G'}} K^G_{A'E'F'} K^H_{B'C'G'} z_H +$$
$$(-1)^{z_{AEF} + \circlearrowleft_{BEF} + \circlearrowleft_{A'C'G'}} K^G_{B'E'F'} K^H_{A'C'G'} z_H +$$
$$(-1)^{z_{ABEF} + \circlearrowleft_{CEF} + \circlearrowleft_{A'B'G'}} K^G_{C'E'F'} K^H_{A'B'G'} z_H,$$

where generators z_H are known and structure constants K^D_{ABC} are unknown. Furthermore, for every $H \in \{1, 2, \ldots, m + n\}$ we have

$$(-1)^{\circlearrowleft_{DEF}} K^D_{ABC} K^H_{D'E'F'} = (-1)^{\circlearrowleft_{AEF} + \circlearrowleft_{B'C'G'}} K^G_{A'E'F'} K^H_{B'C'G'} +$$
$$(-1)^{z_{AEF} + \circlearrowleft_{BEF} + \circlearrowleft_{A'C'G'}} K^G_{B'E'F'} K^H_{A'C'G'} +$$
$$(-1)^{z_{ABEF} + \circlearrowleft_{CEF} + \circlearrowleft_{A'B'G'}} K^G_{C'E'F'} K^H_{A'B'G'},$$

In summary, we have a system of quadratic equations whose solutions are possible structure constants for $m|n$-dimensional 3-Lie superalgebra. We note however, that the structure constants are depending on the choice of basis for the super vector space and thus invariant solutions have to be removed case by case.

Applying the described algorithm to concrete cases gives us the following theorems.

Theorem 3 *3-Lie superalgebras over* $\mathbb{C}$, *whose super vector space dimension is* $0|1$ *or* $1|1$, *is Abelian.*

Theorem 4 *3-Lie superalgebras over* $\mathbb{C}$, *whose super vector space dimension is* $0|2$ *or* $1|2$, *are either Abelian or isomorphic to 3-Lie superalgebra* $\mathfrak{h}$ *whose non-trivial commutation relations are*

$$
\begin{cases}
[f_1, f_1, f_1] = -f_1 + f_2, \\
[f_1, f_1, f_2] = -f_1 + f_2, \\
[f_1, f_2, f_2] = -f_1 + f_2, \\
[f_2, f_2, f_2] = -f_1 + f_2,
\end{cases}
\quad or \quad [f_1, f_1, f_1] = f_2,
$$

where f_1, f_2 are odd generators of $\mathfrak{h}$.

Theorem 5 3-*Lie superalgebras over* $\mathbb{C}$, *whose super vector space dimension is* $2|1$, *are either Abelian or isomorphic to* 3-*Lie superalgebra* $\mathfrak{h}$ *whose non-trivial commutation relations are*

$$
\begin{cases}
[e_1, f_1, f_1] = e_1 + e_2, \\
[e_2, f_1, f_1] = -e_1 - e_2,
\end{cases}
\quad [e_1, e_2, f_1] = f_1, \quad or \quad [f_1, f_1, f_1] = f_1,
$$

where e_1, e_2 are even generators of $\mathfrak{h}$ and f_1 is odd generator of $\mathfrak{h}$.

5 Supermodule Over Clifford Algebra

In this section we apply the method described in Sect. 2 to a Clifford algebra. It is well known that a Clifford algebra can be equipped with the structure of superalgebra if one associates degree 1 to each generator of Clifford algebra and defines the degree of product of generators as the sum of degrees of its factors. Then making use of a graded commutator we can consider a Clifford algebra as the Lie superalgebra. A Clifford algebra has a matrix representation and this allows to introduce a supertrace. Hence we have a Lie superalgebra endowed with a supertrace, and we can apply the method described in Theorem 2 to construct the 3-Lie superalgebra. In this section we will give an explicit description of the structure of this constructed 3-Lie superalgebra.

A Clifford algebra C_n is the unital associative algebra over $\mathbb{C}$ generated by $\gamma_1, \gamma_2, \ldots, \gamma_n$ which obey the relations

$$
\gamma_i \gamma_j + \gamma_j \gamma_i = 2 \delta_{ij} e, \quad i, j = 1, 2, \ldots, n, \tag{7}
$$

where e is the unit element of Clifford algebra. Let $\mathcal{N} = \{1, 2, \ldots, n\}$ be the set of integers from 1 to n. If I is a subset of $\mathcal{N}$, i.e. $I = \{i_1, i_2, \ldots, i_k\}$ where $1 \le i_1 < i_2 < \cdots < i_k \le n$, then one can associate to this subset I the monomial $\gamma_I = \gamma_{i_1} \gamma_{i_2} \cdots \gamma_{i_k}$. If $I = \emptyset$ one defines $\gamma_\emptyset = e$. The number of elements of a subset I will be denoted by $|I|$. It is obvious that the vector space of Clifford algebra C_n is spanned by the monomials γ_I, where $I \subseteq \mathcal{N}$. Hence the dimension of this vector space is 2^n and any element $x \in C_n$ can be expressed in terms of these monomials as

$$
x = \sum_{I \subseteq \mathcal{N}} a_I \gamma_I,
$$

where $a_I = a_{i_1 i_2 ... i_k}$ is a complex number. It is easy to see that one can endow a Clifford algebra C_n with the $\mathbb{Z}_2$-graded structure by assigning the degree $|\gamma_I| = |I| \pmod 2$ to monomial γ_I. Then a Clifford algebra C_n can be considered as the superalgebra since for any two monomials it holds $|\gamma_I \gamma_J| = |\gamma_I| + |\gamma_J|$.

Another way to construct this superalgebra which does not contain explicit reference to Clifford algebra is given by the following theorem.

Theorem 6 *Let I be a subset of $\mathcal{N} = \{1, 2, \ldots, n\}$, and γ_I be a symbol associated to I. Let C_n be the vector space spanned by the symbols γ_I. Define the degree of γ_I by $|\gamma_I| = |I| (\text{mod}\, 2)$, where $|I|$ is the number of elements of I, and the product of γ_I, γ_J by*

$$\gamma_I \, \gamma_J = (-1)^{\sigma(I,J)} \gamma_{I \Delta J}, \tag{8}$$

where $\sigma(I, J) = \sum_{j \in J} \sigma(I, j)$, $\sigma(I, j)$ is the number of elements of I which are greater than $j \in J$, and $I \Delta J$ is the symmetric difference of two subsets. Then C_n is the unital associative superalgebra, where the unit element e is $\gamma_\emptyset$.

This theorem can be proved by means of the properties of symmetric difference of two subsets. We remind a reader that the symmetric difference is commutative $I \oplus J = J \oplus I$, associative $(I \Delta J) \Delta K = I \Delta (J \Delta K)$ and $I \Delta \emptyset = \emptyset \Delta I$. The latter shows that $\gamma_\emptyset$ is the unit element of this superalgebra. The symmetric difference also satisfies $|I \Delta J| = |I| + |J| \pmod 2$. Hence C_n is the superalgebra.

The superalgebra C_n can be considered as the super Lie algebra if for any two homogeneous elements x, y of this superalgebra one introduces the graded commutator $[x, y] = xy - (-1)^{|x||y|} yx$ and extends it by linearity to a whole superalgebra C_n. We will denote this super Lie algebra by $\mathfrak{C}_n$. Then $\{\gamma_I\}_{I \subseteq \mathcal{N}}$ are the generators of this super Lie algebra $\mathfrak{C}_n$, and its structure is entirely determined by the graded commutators of γ_I. Then for any two generators γ_I, γ_J we have

$$[\gamma_I, \gamma_J] = f(I, J) \, \gamma_{I \Delta J}, \tag{9}$$

where $f(I, J)$ is the integer-valued function of two subsets of $\mathcal{N}$ defined by

$$f(I, J) = (-1)^{\sigma(I,J)} \left(1 - (-1)^{|I \cap J|}\right),$$

It is easy to verify that the degree of graded commutator is consistent with the degrees of generators, i.e. $[\gamma_I, \gamma_J] = |\gamma_I| + |\gamma_J|$. Indeed the function $\sigma(I, J)$ satisfies

$$\sigma(I, J) = |I||J| - |I \cap J| - \sigma(I, J),$$

and

$$\begin{aligned}
f(J, I) &= (-1)^{\sigma(J,I)} \left(1 - (-1)^{|I \cap J|}\right) \\
&= (-1)^{|I||J| - |I \cap J| - \sigma(I,J)} \left(1 - (-1)^{|I \cap J|}\right) \\
&= (-1)^{|I||J|} (-1)^{\sigma(I,J)} \left((-1)^{|I \cap J|} - 1\right) = -(-1)^{|I||J|} f(I, J).
\end{aligned}$$

Hence $[\gamma_I, \gamma_J] = -(-1)^{|I||J|}[\gamma_J, \gamma_I]$ which shows that the relation (9) is consistent with the symmetries of graded commutator. It is obvious that if the intersection of subsets I, J contains an even number of elements then $f(I, J) = 0$, and the graded commutator of γ_I, γ_J is trivial. Particularly if at least one of two subsets I, J is the empty set then $f(I, J) = 0$. Thus any graded commutator (9) containing e is trivial.

As an example, consider the super Lie algebra $\mathfrak{C}_2$. Its underlying vector space is 4-dimensional and $\mathfrak{C}_2$ is generated by two even degree generators e, γ_{12} and two odd degree generators γ_1, γ_2. The non-trivial relations of this Lie superalgebra are given by

$$[\gamma_1, \gamma_1] = [\gamma_2, \gamma_2] = 2\,e, \ [\gamma_1, \gamma_{12}] = 2\,\gamma_2, \ [\gamma_2, \gamma_{12}] = -2\,\gamma_1. \tag{10}$$

Now we assume that $n = 2m, m \geq 1$ is an even integer. The Lie superalgebra $\mathfrak{C}_n$ has a matrix representation which can be described as follows. Fix $n = 2$ and identify the generators γ_1, γ_2 with the Pauli matrices σ_1, σ_2, i.e.

$$\gamma_1 = \begin{pmatrix} 0 & 1 \\ 1 & 0 \end{pmatrix}, \quad \gamma_2 = \begin{pmatrix} 0 & -i \\ i & 0 \end{pmatrix}. \tag{11}$$

Then $\gamma_{12} = \gamma_1\gamma_2 = i\,\sigma_3$ where

$$\sigma_3 = \begin{pmatrix} 1 & 0 \\ 0 & -1 \end{pmatrix}.$$

Let S^2 be the 2-dimensional complex super vector space $\mathbb{C}^2$ with the odd degree operators (11), where the $\mathbb{Z}_2$-graded structure of S^2 is determined by $\sigma_3 = i^{-1}\gamma_{12}$. Then $C_2 \simeq \mathrm{End}\,(S^2)$, and S^2 can be considered as a supermodule over the superalgebra C_2. Let $S^n = S^2 \otimes S^2 \otimes \ldots \otimes S^2 (m - \text{times})$. Then S^n can be viewed as a supermodule over the m-fold tensor product of C_2, which can be identified with C_n by identifying γ_1, γ_2 in the jth factor with $\gamma_{2j-1}, \gamma_{2j}$ in C_n. This C_n-supermodule S^n is called the supermodule of spinors [6]. Hence we have the matrix representation for the Clifford algebra C_n, and this matrix representation or supermodule of spinors allows one to consider the supertrace, and it can be proved [6] that

$$\mathrm{Str}(\gamma_I) = \begin{cases} 0 & \text{if } I < \mathbb{N}, \\ (2i)^m & \text{if } I = \mathbb{N}. \end{cases} \tag{12}$$

Now we have the Lie superalgebra $\mathfrak{C}_n$ with the graded commutator defined in (9) and its matrix representation based on the supermodule of spinors. Hence we can construct a 3-Lie superalgebra by making use of graded ternary commutator (4). Applying the formula (4) we define the graded ternary commutator for any triple $\gamma_I, \gamma_J, \gamma_K$ of elements of basis for $\mathfrak{C}_n$ by

$$[\gamma_I, \gamma_J, \gamma_K] = \mathrm{Str}(\gamma_I)\,[\gamma_J, \gamma_K] - (-1)^{|I||J|}\mathrm{Str}(\gamma_J)\,[\gamma_I, \gamma_K]$$
$$+(-1)^{|K|(|I|+|J|)}\mathrm{Str}(\gamma_K)\,[\gamma_I, \gamma_J], \quad (13)$$

where the binary graded commutator at the right-hand side of this formula is defined by (9). According to Theorem 2 the vector space spanned by γ_I, $I \subset \mathcal{N}$ and equipped with the ternary graded commutator (13) is the 3-Lie superalgebra which will be denoted by $\mathfrak{C}_n^{(3)}$. Making use of (9) we can write the expression at the right-hand side of the above formula in the form

$$[\gamma_I, \gamma_J, \gamma_K] = f(J, K)\mathrm{Str}(\gamma_I)\,\gamma_{J\Delta K} - (-1)^{|I||J|}f(I, K)\mathrm{Str}(\gamma_J)\,\gamma_{I\Delta K}$$
$$+(-1)^{|K|(|I|+|J|)}f(I, J)\mathrm{Str}(\gamma_K)\,\gamma_{I\Delta J}.$$

From the formula for supertrace (12) it follows immediately that the above graded ternary commutator is trivial if none of subsets γ_i, γ_J, γ_K is equal to $\mathcal{N}$. Similarly this graded ternary commutator is also trivial if all three subsets I, J, K are equal to $\mathcal{N}$, i.e. $I = J = K = \mathcal{N}$, or two of them are equal to $\mathcal{N}$.

Proposition 5 *The graded ternary commutators of the generators γ_I, $I \subseteq \mathcal{N}$ of the 3-Lie superalgebra $\mathfrak{C}_n^{(3)}$ are given by*

$$[\gamma_I, \gamma_J, \gamma_K] = \begin{cases} (2i)^m f(I, J)\gamma_{I\Delta J} & \text{if } I \neq \mathcal{N}, J \neq \mathcal{N}, K = \mathcal{N}, \\ 0 & \text{in all other cases}. \end{cases} \quad (14)$$

Acknowledgement The authors is gratefully acknowledge the Estonian Science Foundation for financial support of this work under the Research Grant No. ETF9328. This research was also supported by institutional research funding IUT20-57 of the Estonian Ministry of Education and Research. The authors are also grateful for partial support from Linda Peetres Foundation for cooperation between Sweden and Estonia provided by Swedish Mathematical Society.

References

1. Abramov, V.: Super 3-Lie algebras induced by super Lie algebras. (to appear in Advances in Applied Clifford Algebras)
2. Abramov, V., Kerner, R., Le Roy, B.: Hypersymmetry: A $\mathbb{Z}_3$-graded generalization of super-symmetry. J. Math. Phys. **38**, 1650–1669 (1997)
3. Arnlind, J., Kitouni, A., Makhlouf, A., Silvestrov, S.: Structure and cohomology of 3-Lie algebras induced by Lie algebras. In: Makhlouf, A., Paal, E., Silvestrov, S.D., Stolin, A. (eds.) Algebra, Geometry and Mathematical Physics, pp. 123–144. Springer Proceedings in Mathematics & Statistics, Mulhouse, France (2014)
4. Daletskii, Yu. L., Kushnirevitch,V. A.: Formal differential geometry and Nambu-Takhtajan algebra. In: Budzynski, R., Pusz, W., Zakrzewski, S. (eds.) Quantum Groups and Quantum Spaces, **40**, pp. 293–302. Banach Center Publications (1997)
5. Filippov, V.T.: n-Lie algebras. Sib. Math. J. **26**, 879–891 (1985)
6. Mathai, V., Quillen, D.: Superconnections, Thom classes, and equivariant differential forms. Topology **25**(1), 85–110 (1986)

Semi-commutative Galois Extension and Reduced Quantum Plane

Viktor Abramov and Md. Raknuzzaman

Abstract In this paper we show that a semi-commutative Galois extension of associative unital algebra by means of an element τ, which satisfies $\tau^N = \mathbb{1}$ ($\mathbb{1}$ is the identity element of an algebra and $N \geq 2$ is an integer) induces a structure of graded q-differential algebra, where q is a primitive Nth root of unity. A graded q-differential algebra with differential d, which satisfies $d^N = 0, N \geq 2$, can be viewed as a generalization of graded differential algebra. The subalgebra of elements of degree zero and the subspace of elements of degree one of a graded q-differential algebra together with a differential d can be considered as a first order noncommutative differential calculus. In this paper we assume that we are given a semi-commutative Galois extension of associative unital algebra, then we show how one can construct the graded q-differential algebra and when this algebra is constructed we study its first order noncommutative differential calculus. We also study the subspaces of graded q-differential algebra of degree greater than one which we call the higher order noncommutative differential calculus induced by a semi-commutative Galois extension of associative unital algebra. We also study the subspaces of graded q-differential algebra of degree greater than one which we call the higher order noncommutative differential calculus induced by a semi-commutative Galois extension of associative unital algebra. Finally we show that a reduced quantum plane can be viewed as a semi-commutative Galois extension of a fractional one-dimensional space and we apply the noncommutative differential calculus developed in the previous sections to a reduced quantum plane.

Keywords Noncommutative differential calculus · Galois extension · Reduced quantum plane

V. Abramov (✉) · Md. Raknuzzaman
Institute of Mathematics, University of Tartu, Liivi 2–602, Tartu 50409, Estonia
e-mail: viktor.abramov@ut.ee

Md. Raknuzzaman
e-mail: raknuzza@ut.ee

© Springer International Publishing Switzerland 2016
S. Silvestrov and M. Rančić (eds.), *Engineering Mathematics II*,
Springer Proceedings in Mathematics & Statistics 179,
DOI 10.1007/978-3-319-42105-6_2

13

1 Introduction

Let us briefly remind a definition of noncommutative Galois extension [12–15]. Suppose $\tilde{\mathscr{A}}$ is an associative unital $\mathbb{C}$-algebra, $\mathscr{A} \subset \tilde{\mathscr{A}}$ is its subalgebra, and there is an element $\tau \in \tilde{\mathscr{A}}$ which satisfies $\tau \notin \mathscr{A}$, $\tau^N = \mathbb{1}$, where $N \geq 2$ is an integer and $\mathbb{1}$ is the identity element of $\tilde{\mathscr{A}}$. A noncommutative Galois extension of $\mathscr{A}$ by means of τ is the smallest subalgebra $\mathscr{A}[\tau] \subset \tilde{\mathscr{A}}$ such that $\mathscr{A} \subset \mathscr{A}[\tau]$, and $\tau \in \mathscr{A}[\tau]$. It should be pointed out that a concept of noncommutative Galois extension can be applied not only to associative unital algebra with a binary multiplication law but as well as to the algebra with a ternary multiplication law, for instant to a ternary analog of Grassmann and Clifford algebra [6, 14, 15], and this approach can be used in particle physics to construct an elegant algebraic model for quarks.

A graded q-differential algebra can be viewed as a generalization of a notion of graded differential algebra if we use a more general equation $d^N = 0$, $N \geq 2$ than the basic equation $d^2 = 0$ of a graded differential algebra. This idea was proposed and developed within the framework of noncommutative geometry [10], where the author introduced the notions of N-complex, generalized cohomologies of N-complex and making use of an Nth primitive root of unity constructed an analog of an algebra of differential forms in n-dimensional space with exterior differential satisfying the relation $d^N = 0$. Later this idea was developed in the paper [9], where the authors introduced and studied a notion of graded q-differential algebra. It was shown [1, 2, 4, 5] that a notion of graded q-differential algebra can be applied in noncommutative geometry in order to construct a noncommutative generalization of differential forms and a concept of connection.

In this paper we will study a special case of noncommutative Galois extension which is called a semi-commutative Galois extension. A noncommutative Galois extension is referred to as a semi-commutative Galois extension [15] if for any element $x \in \mathscr{A}$ there exists an element $x' \in \mathscr{A}$ such that $x\tau = \tau x'$. In this paper we show that a semi-commutative Galois extension can be endowed with a structure of a graded algebra if we assign degree zero to elements of subalgebra $\mathscr{A}$ and degree one to τ. This is the first step on a way to construct the graded q-differential algebra if we are given a semi-commutative Galois extension. The second step is the theorem which states that if there exists an element v of graded associative unital $\mathbb{C}$-algebra which satisfies the relation $v^N = \mathbb{1}$ then this algebra can be endowed with the structure of graded q-differential algebra. We can apply this theorem to a semi-commutative Galois extension because we have an element τ with the property $\tau^N = \mathbb{1}$, and this allows us to equip a semi-commutative Galois extension with the structure of graded q-differential algebra. Then we study the first and higher order noncommutative differential calculus induced by the N-differential of graded q-differential algebra. We introduce a derivative and differential with the help of first order noncommutative differential calculus developed in the papers [3, 7]. We also study the higher order noncommutative differential calculus and in this case we consider a differential d as an analog of exterior differential and the elements of higher order differential calculus as analogs of differential forms. Finally we apply our calculus to reduced quantum plane [8].

2 Graded q-Differential Algebra Structure of Noncommutative Galois Extension

In this section we remind a definition of noncommutative Galois extension, semi-commutative Galois extension, and show that given a semi-commutative Galois extension we can construct the graded q-differential algebra.

First of all we remind a notion of a noncommutative Galois extension [12–15].

Definition 1 Let $\tilde{\mathscr{A}}$ be an associative unital $\mathbb{C}$-algebra and $\mathscr{A} \subset \tilde{\mathscr{A}}$ be its subalgebra. If there exist an element $\tau \in \tilde{\mathscr{A}}$ and an integer $N \geq 2$ such that

(i) $\tau^N = \pm \mathbb{1}$,
(ii) $\tau^k \notin \mathscr{A}$ for any integer $1 \leq k \leq N - 1$,

then the smallest subalgebra $\mathscr{A}[\tau]$ of $\tilde{\mathscr{A}}$ which satisfies

(iii) $\mathscr{A} \subset \mathscr{A}[\tau]$,
(iv) $\tau \in \mathscr{A}[\tau]$,

is called the noncommutative Galois extension of $\mathscr{A}$ by means of τ.

In this paper we will study a particular case of a noncommutative Galois extension which is called a semi-commutative Galois extension [15]. A noncommutative Galois extension is referred to as a semi-commutative Galois extension if for any element $x \in \mathscr{A}$ there exists an element $x' \in \mathscr{A}$ such that $x \tau = \tau x'$. We will give this definition in terms of left and right $\mathscr{A}$-modules generated by τ. Let $\mathscr{A}_l^1[\tau]$ and $\mathscr{A}_r^1[\tau]$ be respectively the left and right $\mathscr{A}$-modules generated by τ. Obviously we have

$$\mathscr{A}_l^1[\tau] \subset \mathscr{A}[\tau], \quad \mathscr{A}_r^1[\tau] \subset \mathscr{A}[\tau].$$

Definition 2 A noncommutative Galois extension $\mathscr{A}[\tau]$ is said to be a right (left) semi-commutative Galois extension if $\mathscr{A}_r^1[\tau] \subset \mathscr{A}_l^1[\tau]$ ($\mathscr{A}_l^1[\tau] \subset \mathscr{A}_r^1[\tau]$). If $\mathscr{A}_r^1[\tau] \equiv \mathscr{A}_l^1[\tau]$ then a noncommutative Galois extension will be referred to as a semi-commutative Galois extension, and in this case $\mathscr{A}^1[\tau] = \mathscr{A}_r^1[\tau] = \mathscr{A}_l^1[\tau]$ is the $\mathscr{A}$-bimodule.

It is well known that a bimodule over an associative unital algebra $\mathscr{A}$ freely generated by elements of its basis induces the endomorphism from an algebra $\mathscr{A}$ to the algebra of square matrices over $\mathscr{A}$. In the case of semi-commutative Galois extension we have only one generator τ and it induces the endomorphism of an algebra $\mathscr{A}$. Indeed let $\mathscr{A}[\tau]$ be a semi-commutative Galois extension and $\mathscr{A}^1[\tau]$ be its $\mathscr{A}$-bimodule generated by $[\tau]$. Any element of the right $\mathscr{A}$-module $\mathscr{A}_r^1[\tau]$ can be written as τx, where $x \in \mathscr{A}$. On the other hand $\mathscr{A}[\tau]$ is a semi-commutative Galois extension which means $\mathscr{A}_r^1[\tau] \equiv \mathscr{A}_l^1[\tau]$, and hence each element $x \tau$ of the left $\mathscr{A}$-module can be expressed as $\tau \phi_\tau(x)$, where $\phi_\tau(x) \in \mathscr{A}$. It is easy to verify that the linear mapping $\phi : x \to \phi_\tau(x)$ is the endomorphism of subalgebra $\mathscr{A}$, i.e. for any elements $x, y \in \mathfrak{A}$ we have $\phi_\tau(xy) = \phi_\tau(x)\phi_\tau(y)$. This endomorphism will play an important role in our differential calculus, and in what follows we will also use the notation $\phi_\tau(x) = x_\tau$. Thus

$$u\,\tau = \tau\,\phi_\tau(x), \quad u\,\tau = \tau\,u_\tau.$$

It is clear that

$$\phi_\tau^N = \mathrm{id}_\mathscr{A}, \quad u_{\tau^N} = u,$$

because for any $u \in \mathscr{A}$ it holds $u\,\tau^N = \tau^N\,\phi^N(u)$ and taking into account that $\tau^N = \mathbb{1}$ we get $\phi_\tau^N(u) = u$.

Proposition 1 *Let $\mathscr{A}[\tau]$ be a semi-commutative Galois extension of $\mathscr{A}$ by means of τ, and $\mathscr{A}_l^k[\tau]$, $\mathscr{A}_r^k[\tau]$ be respectively the left and right $\mathscr{A}$-modules generated by τ^k, where $k = 1, 2, \ldots, N - 1$. Then $\mathscr{A}_l^k[\tau] \equiv \mathscr{A}_r^k[\tau] = \mathscr{A}^k[\tau]$ is the $\mathscr{A}$-bimodule, and*

$$\mathscr{A}[\tau] = \oplus_{k=0}^{N-1}\mathscr{A}^k[\tau] = \mathscr{A}^0[\tau] \oplus \mathscr{A}^1[\tau] \oplus \cdots \oplus \mathscr{A}^{N-1}[\tau],$$

where $\mathscr{A}^0[\tau] \equiv \mathscr{A}$.

Evidently the endomorphism of $\mathscr{A}$ induced by the $\mathscr{A}$-bimodule structure of $A^k[\tau]$ is ϕ^k, where $\phi : \mathscr{A} \to \mathscr{A}$ is the endomorphism induced by the $\mathscr{A}$-bimodule $\mathscr{A}^1[\tau]$. We will also use the notation $\phi^k(x) = x_{\tau^k}$.

It follows from Proposition 1 that a semi-commutative Galois extension $\mathscr{A}[\tau]$ has a natural $\mathbb{Z}_N$-graded structure which can be defined as follows: we assign degree zero to each element of subalgebra $\mathscr{A}$, degree 1 to τ and extend this graded structure to a semi-commutative Galois extension $\mathscr{A}[\tau]$ by determining the degree of a product of two elements as the sum of degree of its factors. The degree of a homogeneous element of $\mathscr{A}[\tau]$ will be denoted by $|\ |$. Hence $|u| = 0$ for any $u \in \mathscr{A}$ and $|\tau| = 1$.

Now our aim is to show that given a noncommutative Galois extension we can construct a graded q-differential algebra, where q is a primitive Nth root of unity. First of all we remind some basic notions, structures and theorems of theory of graded q-differential algebras.

Let $\mathscr{A} = \oplus_{k \in \mathbb{Z}_N}\mathscr{A}^k = \mathscr{A}^0 \oplus \mathscr{A}^1 \oplus \cdots \oplus \mathscr{A}^{N-1}$ be a $\mathbb{Z}_N$-graded associative unital $\mathbb{C}$-algebra with identity element denoted by $\mathbb{1}$. Obviously the subspace $\mathscr{A}^0$ of elements of degree 0 is the subalgebra of a graded algebra $\mathscr{A}$. Every subspace $\mathscr{A}^k$ of homogeneous elements of degree $k \geq 0$ can be viewed as the $\mathscr{A}^0$-bimodule. The graded q-commutator of two homogeneous elements $u, v \in \mathscr{A}$ is defined by

$$[v, u]_q = v\,u - q^{|v||u|}u\,v.$$

A graded q-derivation of degree m of a graded algebra $\mathscr{A}$ is a linear mapping $d : \mathscr{A} \to \mathscr{A}$ of degree m, i.e. $d : \mathscr{A}^i \to \mathscr{A}^{i+m}$, which satisfies the graded q-Leibniz rule

$$d(u\,v) = d(u)\,v + q^{ml}u\,d(v), \tag{1}$$

where u is a homogeneous element of degree l, i.e. $u \in \mathscr{A}^l$. A graded q-derivation d of degree m is called an inner graded q-derivation of degree m induced by an element $v \in \mathscr{A}^m$ if

$$d(u) = [v, u]_q = v\,u - q^{ml}u\,v, \tag{2}$$

where $u \in \mathscr{A}^l$.

Now let q be a primitive Nth root of unity, for instant $q = e^{2\pi i/N}$. Then

$$q^N = 1, \quad 1 + q + \cdots + q^{N-1} = 0.$$

A graded q-differential algebra is a graded associative unital algebra $\mathscr{A}$ endowed with a graded q-derivation d of degree one which satisfies $d^N = 0$. In what follows a graded q-derivation d of a graded q-differential algebra $\mathscr{A}$ will be referred to as a graded N-differential. Thus a graded N-differential d of a graded q-differential algebra is a linear mapping of degree one which satisfies a graded q-Leibniz rule and $d^N = 0$. It is useful to remind that a graded differential algebra is a graded associative unital algebra equipped with a differential d which satisfies the graded Leibniz rule and $d^2 = 0$. Hence it is easy to see that a graded differential algebra is a particular case of a graded q-differential algebra when $N = 2, q = -1$, and in this sense we can consider a graded q-differential algebra as a generalization of a concept of graded differential algebra. Given a graded associative algebra $\mathscr{A}$ we can consider the vector space of inner graded q-derivations of degree one of this algebra and put the question: under what conditions an inner graded q-derivation of degree one is a graded N-differential? The following theorem gives answer to this question.

Theorem 1 *Let $\mathscr{A}$ be a $\mathbb{Z}_N$-graded associative unital $\mathbb{C}$-algebra and $d(u) = [v, u]_q$ be its inner graded q-derivation induced by an element $v \in \mathscr{A}^1$. The inner graded q-derivation d is the N-differential, i.e. it satisfies $d^N = 0$, if and only if $v^N = \pm \mathbb{1}$.*

Now our goal is apply this theorem to a semi-commutative Galois extension to construct a graded q-differential algebra with N-differential satisfying $d^N = 0$.

Proposition 2 *Let q be a primitive Nth root of unity. A semi-commutative Galois extension $\mathscr{A}[\tau]$, equipped with the $\mathbb{Z}_N$-graded structure described above and with the inner graded q-derivation $d = [\tau, \,]_q$ induced by τ, is the graded q-differential algebra, and d is its N-differential. For any element ξ of semi-commutative Galois extension $\mathscr{A}[\tau]$ written as a sum of elements of right $\mathscr{A}$-modules $\mathscr{A}^k[\tau]$*

$$\xi = \sum_{k=0}^{N-1} \tau^k u_k = \mathbb{1}\,u_0 + \tau\,u_1 + \tau^2\,u_2 + \cdots \tau^{N-1}\,u_{N-1}, \quad u_k \in \mathscr{A},$$

it holds

$$d\xi = \sum_{k=0}^{N-1} \tau^{k+1}(u_k - q^k(u_k)_\tau),$$ (3)

where $u_k \rightarrow (u_k)_\tau$ *is the endomorphism of* $\mathscr{A}$ *induced by the bimodule structure of* $\mathscr{A}^1[\tau]$.

3   First Order Differential Calculus over Associative Unital Algebra

In this section we describe a first order differential calculus over associative unital algebra [7]. If an associative unital algebra is generated by a family of variables, which obey commutation relations, then one can construct a coordinate first order differential calculus over this algebra. A coordinate first differential calculus induces the partial derivatives with respect to generators of algebra and these partial derivatives satisfy the twisted Leibniz rule.

A first order differential calculus is a triple $(\mathscr{A}, \mathscr{M}, d)$ where $\mathscr{A}$ is an associative unital algebra, $\mathscr{M}$ is an $\mathscr{A}$-bimodule, and d, which is called a differential of first order differential calculus, is a linear mapping $d : \mathscr{A} \rightarrow \mathscr{M}$ satisfying the Leibniz rule $d(fh) = dfh + fdh$, where $f, h \in \mathscr{A}$. A first order differential calculus $(\mathscr{A}, \mathscr{M}, d)$ is referred to as a coordinate first order differential calculus if an algebra $\mathscr{A}$ is generated by the variables $x^1, x^2, \ldots, x^n$ which satisfy the commutation relations, and an $\mathscr{A}$-bimodule $\mathscr{M}$, considered as a right $\mathscr{A}$-module, is freely generated by $dx^1, dx^2, \ldots, dx^n$. It is worth to mention that a first order differential calculus was developed within the framework of noncommutative geometry, and an algebra $\mathscr{A}$ is usually considered as the algebra of functions of a noncommutative space, the generators $x^1, x^2, \ldots, x^n$ of this algebra are usually interpreted as coordinates of this noncommutative space, and an $\mathscr{A}$-bimodule $\mathscr{M}$ plays the role of space of differential forms of degree one. In this paper we will use the corresponding terminology in order to stress a relation with noncommutative geometry.

Let us consider a structure of coordinate first order differential calculus. This differential calculus induces the differentials $dx^1, dx^2, \ldots, dx^n$ of the generators $x^1, x^2, \ldots, x^n$. Evidently $dx^1, dx^2, \ldots, dx^n \in \mathscr{M}$. $\mathscr{M}$ is a bimodule, i.e. it has a structure of left $\mathscr{A}$-module and right $\mathscr{A}$-module. Hence for any two elements $f, h \in \mathscr{A}$ and $\omega \in \mathscr{M}$ it holds $(f\omega)h = f(\omega h)$. According to the definition of a coordinate first order differential calculus the right $\mathscr{A}$-module $\mathscr{M}$ is freely generated by the differentials of generators $dx^1, dx^2, \ldots, dx^n$. Thus for any $\omega \in \mathscr{M}$ we have $\omega = dx^1 f_1 + dx^2 f_2 + \ldots + dx^n f_n$ where $f_1, f_2, \ldots, f_n \in \mathscr{A}$. A coordinate first order differential calculus $(\mathscr{A}, \mathscr{M}, d)$ is an algebraic structure, which extends to noncommutative case the classical differential structure of a manifold. From the point of view of noncommutative geometry $\mathscr{A}$ can be viewed as an algebra of smooth functions, d is the exterior differential, and $\mathscr{M}$ is the bimodule of differential 1-forms. In order to stress this analogy we will call the elements of algebra $\mathscr{A}$ "functions" and the elements of $\mathscr{A}$-bimodule $\mathscr{M}$ "1-forms".

Because $\mathscr{M}$ is $\mathscr{A}$-bimodule, for any function $f \in \mathscr{A}$ we have two products $f\,dx^i$ and $dx^i f$. Since $dx^1, dx^2, \ldots, dx^n$ is the basis for the right $\mathscr{A}$-module $\mathscr{M}$, each element of $\mathscr{M}$ can be expressed as linear combination of $dx^1, dx^2, \ldots, dx^n$ multiplied by the functions from the right. Hence the element $f\,dx^i \in \mathscr{M}$ can be expressed in this way, i.e.

$$f\,dx^i = dx^1 r_1^i(f) + dx^2 r_2^i(f) + \cdots + dx^n r_n^i(f) = dx^j r_j^i(f), \tag{4}$$

where $r_1^i(f), r_2^i(f), \ldots, r_n^i(f) \in \mathscr{A}$ are the functions. Making use of these functions we can compose the square matrix

$$R(f) = (r_j^i(f)) = \begin{pmatrix} r_1^1(f) & r_1^2(f) & \cdots & r_1^n(f) \\ \vdots & \vdots & \vdots & \vdots \\ r_n^1(f) & r_n^2(f) & \cdots & r_n^n(f) \end{pmatrix}.$$

It is worth to point out that an entry $r_j^i(f)$ stands on intersection of i-th column and j-th row. This square matrix determines the mapping $R : \mathscr{A} \to \mathrm{Mat}_n(\mathscr{A})$ where $\mathrm{Mat}_n(\mathscr{A})$ is the algebra of n order square matrices over an algebra $\mathscr{A}$. It can be proved

Proposition 3 $R : \mathscr{A} \to Mat_n(\mathscr{A})$ *is the homomorphism of algebras.*

Proof We need to prove that for any $f, g \in \mathscr{A}$ it holds $R(fg) = R(f)R(g)$. Now according to the Eq. (4) we have

$$(fg)dx^i = dx^j r_j^i(fg).$$

The left hand side of the above relation can be written as

$$f(g\,dx^i) = f(dx^j r_j^i(g)) = (f\,dx^j)r_j^i(g) = (dx^k r_k^j(f))r_j^i(g) = dx^k (r_k^j(f)r_j^i(g)).$$

Now we can write

$$dx^j r_j^i(fg) = dx^k (r_k^j(f)r_j^i(g)) \Rightarrow r_k^i(fg) = r_k^j(f)r_j^i(g),$$

or in matrix form $R(fg) = R(f)R(g)$, which ends the proof. $\qquad\square$

Let $\mathscr{A}, \mathscr{M}, d$ be a coordinate first order differential calculus such that right $\mathscr{A}$-module $\mathscr{M}$ is a finite freely generated by the differentials of coordinates $\{dx_i\}_{i=1}^n$. The mappings $\partial_k : \mathscr{A} \to \mathscr{A}$, where $k \in \{1, 2, \ldots, n\}$, uniquely defined by

$$df = dx^k \, \partial_k(f), \quad f \in \mathscr{A}, \tag{5}$$

are called the right partial derivatives of a coordinate first order differential calculus. It can be proved

Proposition 4 *If $\mathscr{A}$, $\mathscr{M}$, d is a coordinate first order differential calculus over an algebra $\mathscr{A}$ such that $\mathscr{M}$ is a finite freely generated right $\mathscr{A}$-module with a basis $\{dx_i\}_{i=1}^n$ then the right partial derivatives $\partial_k : \mathscr{A} \to \mathscr{A}$ of this differential calculus satisfy*

$$\partial_k(fg) = \partial_k(f)\,g + r(f)_k^i\,\partial_i(g). \tag{6}$$

The property (6) is called the twisted (with homomorphism R) Leibniz rule for partial derivatives.

If $\mathscr{A}$ is a graded q-differential algebra with differential d then evidently the subspace of elements of degree zero $\mathscr{A}^0$ is the subalgebra of $\mathscr{A}$, the subspace of elements of degree one $\mathscr{A}^1$ is the $\mathscr{A}^0$-bimodule, a differential $d : \mathscr{A}^0 \to \mathscr{A}^1$ satisfies the Leibniz rule. Consequently we have the first order differential calculus $(\mathscr{A}^0, \mathscr{A}^1, d)$ of a graded q-differential algebra $\mathscr{A}$. If $\mathscr{A}^0$ is generated by some set of variables then we can construct a coordinate first order differential calculus with corresponding right partial derivatives.

4 First Order Differential Calculus of Semi-commutative Galois Extension

It is shown in Sect. 2 that given a semi-commutative Galois extension we can construct a graded q-differential algebra. In the previous section we described the structure of a coordinate first order differential calculus over an associative unital algebra, and at the end of this section we also mentioned that the subspaces $\mathscr{A}^0$, $\mathscr{A}^1$ of a graded q-differential algebra together with differential d of this algebra can be viewed as a first order differential calculus over $\mathscr{A}^0$. In this section we apply an approach of first order differential calculus to a graded q-differential algebra of a semi-commutative Galois extension.

Let $\mathscr{A}[\tau]$ be a semi-commutative Galois extension of an algebra $\mathscr{A}$ by means of τ. Thus we have an algebra $\mathscr{A}$ and $\mathscr{A}$-bimodule $\mathscr{A}^1[\tau]$. Next we have the N-differential $d : \mathscr{A}[\tau] \to \mathscr{A}[\tau]$ induced by τ, and if we restrict this N-differential to the subalgebra $\mathscr{A}$ of Galois extension $\mathscr{A}[\tau]$ then $d : \mathscr{A} \to \mathscr{A}^1[\tau]$ satisfies the Leibniz rule. Consequently we have the first order differential calculus which can be written as the triple $(\mathscr{A}, d, \mathscr{A}^1[\tau])$. In order to describe the structure of this first order differential calculus we will need the vector space endomorphism $\Delta : \mathscr{A} \to \mathscr{A}$ defined by

$$\Delta u = u - u_\tau, \quad u \in \mathscr{A}.$$

For any elements $u, v \in \mathscr{A}$ this endomorphism satisfies

$$\Delta(u\,v) = \Delta(u)\,v + u_\tau\,\Delta(v).$$

Let us assume that there exists an element $x \in \mathscr{A}$ such that the element $\Delta x \in \mathscr{A}$ is invertible, and the inverse element will be denoted by Δx^{-1}. The differential dx of an element x can be written in the form $dx = \tau \, \Delta x$ which clearly shows that dx has degree one, i.e. $dx \in \mathscr{A}^1[\tau]$, and hence dx can be used as generator for the right $\mathscr{A}$-module $\mathscr{A}^1[\tau]$. Let us denote by $\phi_{dx} : u \to \phi_{dx}(u) = u_{dx}$ the endomorphism of $\mathscr{A}$ induced by bimodule structure of $\mathscr{A}^1[\tau]$ in the basis dx. Then

$$u_{dx} = \Delta x^{-1} u_\tau \, \Delta x = \mathrm{Ad}_{\Delta x} u_\tau. \tag{7}$$

Definition 3 For any element $u \in \mathscr{A}$ we define the right derivative $\frac{du}{dx} \in \mathscr{A}$ (with respect to x) by the formula

$$du = dx \, \frac{du}{dx}. \tag{8}$$

Analogously one can define the left derivative with respect to x by means of the left $\mathscr{A}$-module structure of $\mathscr{A}^1[\tau]$. Further we will only use the right derivative which will be referred to as the derivative and often will be denoted by u_x'. Thus we have the linear mapping

$$\frac{d}{dx} : \mathscr{A} \to \mathscr{A}, \quad \frac{d}{dx} : u \mapsto u_x'.$$

Proposition 5 *For any element $u \in \mathscr{A}$ we have*

$$\frac{du}{dx} = \Delta x^{-1} \, \Delta u. \tag{9}$$

The derivative (8) satisfies the twisted Leibniz rule, i.e. for any two elements $u, v \in \mathscr{A}$ it holds

$$\frac{d}{dx}(u\,v) = \frac{du}{dx} v + \phi_{dx}(u) \frac{dv}{dx} = \frac{du}{dx} v + \mathrm{Ad}_{\Delta x} u_\tau \frac{dv}{dx}.$$

We have constructed the first order differential calculus with one variable x, and it is natural to study a transformation rule of the derivative of this calculus if we choose another variable. From the point of view of differential geometry we will study a change of coordinate in one dimensional space. Let $y \in \mathscr{A}$ be an element of $\mathscr{A}$ such that $\Delta y = y - y_\tau$ is invertible.

Proposition 6 *Let x, y be elements of $\mathscr{A}$ such that $\Delta x, \Delta y$ are invertible elements of $\mathscr{A}$. Then*

$$dy = dx \, y_x', \quad \frac{d}{dx} = y_x' \frac{d}{dy}, \quad dx = dy \, x_y', \quad \frac{d}{dy} = x_y' \frac{d}{dx},$$

where $x_y' = (y_x')^{-1}$.

Indeed we have $dy = \tau \, \Delta \, y$, $dx = \tau \, \Delta \, x$. Hence $\tau = dx \, \Delta \, x^{-1}$ and

$$dy = dx \, (\Delta \, x^{-1} \Delta \, y) = dx \, y'_x.$$

If u is any element of $\mathscr{A}$ the for the derivatives we have

$$\frac{du}{dx} = \Delta \, x^{-1} \, \Delta \, u = (\Delta \, x^{-1} \, \Delta \, y)(\Delta \, y^{-1} \, \Delta \, u) = y'_x \, \frac{du}{dy}.$$

As an example of the structure of graded q-differential algebra induced by d_τ on a semi-commutative Galois extension we can consider the quaternion algebra $\mathbb{H}$. The quaternion algebra $\mathbb{H}$ is associative unital algebra generated over $\mathbb{R}$ by i, j, k which are subjected to the relations

$$i^2 = j^2 = k^2 = -\mathbb{1}, \; ij = -ji = k, \; jk = -kj = i, \; ki = -ik = j,$$

where $\mathbb{1}$ is the unity element of $\mathbb{H}$. Given a quaternion

$$\mathfrak{q} = a_0 \, \mathbb{1} + a_1 \, i + a_2 j + a_3 \, k$$

we can write it in the form $\mathfrak{q} = (a_0 \, \mathbb{1} + a_2 j) + i \, (a_1 + a_3 j)$. Hence if we consider the coefficients of the previous expression $z_0 = a_0 \, \mathbb{1} + a_2 j, z_1 = a_1 + a_3 j$ as complex numbers then $\mathfrak{q} = z_0 \, \mathbb{1} + i \, z_1$ which clearly shows that the quaternion algebra $\mathbb{H}$ can be viewed as the semi-commutative Galois extension $\mathbb{C}[i]$. Evidently in this case we have $N = 2, q = -1$, and $\mathbb{Z}_2$-graded structure defined by $|\mathbb{1}| = 0, |i| = 1$. Hence we can use the terminology of superalgebras. It is easy to see that the subspace of odd elements (degree 1) can be considered as the bimodule over the subalgebra of even elements $a \, \mathbb{1} + bj$ and this bimodule induces the endomorphism $\phi : \mathbb{C} \to \mathbb{C}$, where $\phi(z) = \bar{z}$. Let d be the differential of degree one (odd degree operator) induced by i. Then making use of (3) for any quaternion $\mathfrak{q}$ we have

$$d\mathfrak{q} = d(z_0 \, \mathbb{1} + i \, z_1) = -(\bar{z}_1 + z_1) \, \mathbb{1}.$$

Obviously $d^2\mathfrak{q} = 0$.

5   Higher Order Differential Calculus of Semi-commutative Galois Extension

Our aim in this section is to develop a higher order differential calculus of a semi-commutative Galois extension $\mathscr{A}[\tau]$. This higher order differential calculus is induced by the graded q-differential algebra structure. In Sect. 2 it is mentioned that a graded q-differential algebra can be viewed as a generalization of a concept of graded differential algebra if we take $N = 2, q = -1$. It is well known that one of the most important realizations of graded differential algebra is the algebra of differential forms on a smooth manifold. Hence we can consider the elements of the graded q-differential algebra constructed by means of a semi-commutative Galois extension $\mathscr{A}[\tau]$ and expressed in terms of differential dx as noncommutative analogs of differential forms with exterior differential d which satisfies $d^N = 0$. In order to stress this analogy we will consider an element $x \in \mathscr{A}$ as analog of coordinate, the elements of degree zero as analogs of functions, elements of degree k as analogs of k-forms, and we will use the corresponding terminology. It should be pointed out that because of the equation $d^N = 0$ there are higher order differentials $dx, d^2x, \ldots, d^{N-1}x$ in this algebra of differential forms.

Before we describe the structure of higher order differentials forms it is useful to introduce the polynomials $P_k(x), Q_k(x)$, where $k = 1, 2, \ldots, N$. Let us remind that $\Delta x = x - x_\tau \in \mathscr{A}$. Applying the endomorphism τ we can generate the sequence of elements

$$\Delta x_\tau = x_\tau - x_{\tau^2}, \Delta x_{\tau^2} = x_{\tau^2} - x_{\tau^3}, \ldots, \Delta x_{\tau^{N-1}} = x_{\tau^{N-1}} - x.$$

Obviously each element of this sequence is invertible. Now we define the sequence of polynomials $Q_1(x), Q_2(x), \ldots, Q_N(x)$, where

$$Q_k(x) = \Delta x_{\tau^{k-1}} \Delta x_{\tau^{k-2}} \ldots \Delta x_\tau \Delta x.$$

These polynomials can be defined by means of the recurrent relation

$$Q_{k+1}(x) = (Q_k(x))_\tau \Delta x.$$

It should be mentioned that $Q_k(x)$ is the invertible element and

$$(Q_k(x))^{-1} = \Delta x^{-1} \Delta x_\tau^{-1} \ldots \Delta x_{\tau^{k-1}}^{-1}.$$

We define the sequence of elements $P_1(x), P_2(x), \ldots, P_N(x) \in \mathscr{A}$ by the recurrent formula

$$P_{k+1}(x) = P_k(x) - q^k (P_k(x))_\tau, \quad k = 1, 2, \ldots, N - 1,$$

and $P_1(x) = \Delta x$. Clearly $P_1(x) = Q_(x)$ and for the $k = 2, 3$ a straightforward calculation gives

$$P_2(x) = x - (1 + q) x_\tau + q x_{\tau^2},$$
$$P_3(x) = x - (1 + q + q^2) x_\tau + (q + q^2 + q^3) x_{\tau^2} - q^3 x_{\tau^3}.$$

Proposition 7 *If q is a primitive Nth root of unity then there are the identities*

$$P_{N-1}(x) + (P_{N-1}(x))_\tau + \cdots + (P_{N-1}(x))_{\tau^{N-1}} \equiv 0, \quad P_N(x) \equiv 0.$$

Now we will describe the structure of higher order differential forms. It follows from the previous section that any 1-form ω, i.e. an element of $\mathscr{A}^1[\tau]$, can be written in the form $\omega = dx\, u$, where $u \in \mathscr{A}$. Evidently $d : \mathscr{A} \to \mathscr{A}^1[\tau]$, $d\omega = dx\, u'_x$. The elements of $\mathscr{A}^2[\tau]$ will be referred to as 2-forms. In this case there are two choices for a basis for the right $\mathscr{A}$-module $\mathscr{A}^2[\tau]$. We can take either τ^2 or $(dx)^2$ as a basis for $\mathscr{A}^2[\tau]$. Indeed we have

$$(dx)^2 = \tau^2\, Q_2(x).$$

It is worth mentioning that the second order differential d^2x can be used as the basis for $\mathscr{A}^2[\tau]$ only in the case when $P_2(x)$ is invertible. Indeed we have

$$d^2x = \tau^2\, P_2(x), \quad d^2x = (dx)^2\, Q_2^{-1}(x)P_2(x).$$

If we choose $(dx)^2$ as the basis for the module of 2-forms $\mathscr{A}^2[\tau]$ then any 2-form ω can be written as $\omega = (dx)^2\, u$, where $u \in \mathscr{A}$. Now the differential of any 1-form $\omega = dx\, u$, where $u \in \mathscr{A}$, can be expressed as follows

$$d\omega = (dx)^2 \left(q\, u'_x + Q_2^{-1}(x)P_2(x)\, u\right). \tag{10}$$

It should be pointed out that the second factor of the right-hand side of the above formula resembles a covariant derivative in classical differential geometry. Hence we can introduce the linear operator $D : \mathscr{A} \to \mathscr{A}$ by the formula

$$Du = q\, u'_x + Q_2^{-1}(x)P_2(x)\, u, \quad u \in \mathscr{A}. \tag{11}$$

If $\omega = dv, v \in \mathscr{A}$, i.e. ω is an exact form, then

$$d\omega = d^2v = (dx)^2\, Dv'_x = (dx)^2 \left(q\, v''_x + Q_2^{-1}(x)P_2(x)\, v'_x\right).$$

If we consider the simplest case $N = 2, q = -1$ then

$$d^2v = 0, \quad P_2(x) \equiv 0, \quad (dx)^2 \neq 0,$$

and from the above formula it follows that $v''_x = 0$.

Proposition 8 *Let $\mathscr{A}[\tau]$ be a semi-commutative Galois extension of algebra $\mathscr{A}$ by means of τ, which satisfies $\tau^2 = \mathbb{1}$, and d be the differential of the graded differential algebra induced by an element τ as it is shown in Proposition 2. Let $x \in \mathscr{A}$ be an element such that Δx is invertible. Then for any element $u \in \mathscr{A}$ it holds $u''_x = 0$, where u'_x is the derivative (8) induced by d. Hence any element of an algebra $\mathscr{A}$ is linear with respect to x.*

The quaternions considered as the noncommutative Galois extension of complex numbers (Sect. 3) provides a simple example for the above proposition. Indeed in this case $\tau = i, \mathscr{A} \equiv \mathbb{C}$, where the imaginary unit is identified with j, $(a\, \mathbb{1} + b\, j)_\tau =$

$a\,\mathbb{1} - bj$. Hence we can choose $x = a\,\mathbb{1} + bj$ iff $b \neq 0$. Indeed in this case $\Delta x = x - x_\tau = a\,\mathbb{1} + bj - a\,\mathbb{1} + bj = 2bj$, and Δx is invertible iff $b \neq 0$. Now any $z = c\,\mathbb{1} + dj \in \mathscr{A}$ can be uniquely written in the form $z = \tilde{c}\,\mathbb{1} + \tilde{d}\,x$ iff

$$\begin{vmatrix} 1 & a \\ 0 & b \end{vmatrix} = b \neq 0.$$

Thus any $z \in \mathscr{A}$ is linear with respect to x.

Now we will describe the structure of module of k-forms $\mathscr{A}^k[\tau]$. We choose $(dx)^k$ as the basis for the right $\mathscr{A}$-module $\mathscr{A}^k[\tau]$, then any k-form ω can be written $\omega = (dx)^k u, \ u \in \mathscr{A}$. We have the following relations

$$(dx)^k = \tau^k Q_k(x), \quad d^k x = \tau^k P_k(x).$$

In order to get a formula for the exterior differential of a k-form ω we need the polynomials $\Phi_1(x), \Phi_2(x), \ldots, \Phi_{N-1}(x)$ which can be defined by the recurrent relation

$$\Phi_{k+1}(x) = \mathrm{Ad}_{\Delta x}(\Phi_k) + q^{k-1}\Phi_1(x), \quad k = 1, 2, \ldots, N - 1, \tag{12}$$

where $\Phi_1(x) = Q_2^{-1}(x)P_2(x)$. These polynomials satisfy the relations $d(dx)^k = (dx)^{k+1}\Phi_k(x)$ and given a k-form $\omega = (dx)^k u, \ u \in \mathscr{A}$ we find its exterior differential as

$$d\omega = (dx)^{k+1}\left(q^k u_x' + \Phi_k(x)\,u\right) = (dx)^{k+1} D^{(k)}u.$$

The linear operator $D^{(k)} : \mathscr{A} \to \mathscr{A}, k = 1, 2, \ldots, N - 1$ introduced in the previous formula has the form

$$D^{(k)}u = q^k u_x' + \Phi_k(x)\,u, \tag{13}$$

and, as it was mentioned before, this operator resembles a covariant derivative of classical differential geometry. It is easy to see that the operator (11) is the particular case of (13), i.e. $D^{(1)} \equiv D$.

6 Semi-commutative Galois Extension Approach to Reduced Quantum Plane

In this section we show that a reduced quantum plane can be considered as a semi-commutative Galois extension. We study a first order and higher order differential calculus of a semi-commutative Galois extension in the particular case of a reduced quantum plane.

Let x, y be two variables which obey the commutation relation

$$x\,y = q\,y\,x, \tag{14}$$

where $q \neq 0, 1$ is a complex number. These two variables generate the algebra of polynomials over the complex numbers. This algebra is an associative algebra of polynomials over $\mathbb{C}$ and the identity element of this algebra will be denoted by $\mathbb{1}$. In noncommutative geometry and theoretical physics a polynomial of this algebra is interpreted as a function of a quantum plane with two noncommuting coordinate functions x, y and the algebra of polynomials is interpreted as the algebra of (polynomial) functions of a quantum plane. If we fix an integer $N \geq 2$ and impose the additional condition

$$x^N = y^N = \mathbb{1}, \tag{15}$$

then a quantum plane is referred to as a reduced quantum plane and this polynomial algebra will be denoted by $\mathcal{A}_q[x, y]$.

Let us mention that from an algebraic point of view an algebra of functions on a reduced quantum plane may be identified with the generalized Clifford algebra $\mathfrak{C}_2^N$ with two generators x, y. Indeed a generalized Clifford algebra is an associative unital algebra generated by variables $x_1, x_2, \ldots, x_p$ obeying the relations $x_i x_j = q^{\text{sg}(j-i)} x_j x_i$, $x_i^N = 1$, where sg is the sign function.

It is well known that the generalized Clifford algebras have matrix representations, and, in the particular case of the algebra $\mathcal{A}_q[x, y]$, the generators of this algebra x, y can be identified with the square matrices of order N

$$x = \begin{pmatrix} 1 & 0 & 0 & \ldots & 0 & 0 \\ 0 & q^{-1} & 0 & \ldots & 0 & 0 \\ 0 & 0 & q^{-2} & \ldots & 0 & 0 \\ \vdots & \vdots & \vdots & \ddots & \vdots & \vdots \\ 0 & 0 & 0 & \ldots & q^{-(N-2)} & 0 \\ 0 & 0 & 0 & \ldots & 0 & q^{-(N-1)} \end{pmatrix}, \quad y = \begin{pmatrix} 0 & 1 & 0 & \ldots & 0 & 0 \\ 0 & 0 & 1 & \ldots & 0 & 0 \\ 0 & 0 & 0 & \ldots & 0 & 0 \\ \vdots & \vdots & \vdots & \ddots & \vdots & \vdots \\ 0 & 0 & 0 & \ldots & 0 & 1 \\ 1 & 0 & 0 & \ldots & 0 & 0 \end{pmatrix}, \tag{16}$$

where q is a primitive Nth root of unity. As the matrices (16) generate the algebra $\text{Mat}_N(\mathbb{C})$ of square matrices of order N we can identify the algebra of functions on a reduced quantum plane with the algebra of matrices $\text{Mat}_N(\mathbb{C})$.

The set of monomials $B = \{\mathbf{1}, y, x, x^2, yx, y^2, \ldots, y^k x^l, \ldots, y^{N-1} x^{N-1}\}$ can be taken as the basis for the vector space of the algebra $\mathcal{A}_q[x, y]$. We can endow this vector space with an $\mathbb{Z}_N$-graded structure if we assign degree zero to the identity element $\mathbb{1}$ and variable x and we assign degree one to the variable y. As usual we define the degree of a product of two variables x, y as the sum of degrees of factors. Then a polynomial

$$w = \sum_{l=0}^{N-1} \beta_l y^k x^l, \quad \beta_l \in \mathbb{C}, \tag{17}$$

will be a homogeneous polynomial with degree k. Let us denote the degree of a homogeneous polynomial w by $|w|$ and the subspace of the homogeneous polynomials of degree k by $\mathscr{A}_q^k[x, y]$. It is obvious that

$$\mathscr{A}_q[x, y] = \mathscr{A}_q^0[x, y] \oplus \mathscr{A}_q^1[x, y] \oplus \cdots \oplus \mathscr{A}_q^{N-1}[x, y]. \qquad (18)$$

In particular a polynomial r of degree zero can be written as follows

$$r = \sum_{l=0}^{N-1} \beta_l x^l, \qquad \beta_l \in \mathbb{C}, \quad r \in \mathscr{A}_q^0[x, y]. \qquad (19)$$

Obviously the subspace of elements of degree zero $\mathscr{A}_q^0[x, y]$ is the subalgebra of $\mathscr{A}_q[x, y]$ generated by the variable x. Evidently the polynomial algebra $\mathscr{A}_q[x, y]$ of polynomials of a reduced quantum plane can be considered as a semi-commutative Galois extension of the subalgebra $\mathscr{A}_q^0[x, y]$ by means of the element y which satisfies the relation $y^N = \mathbb{1}$. The commutation relation $xy = q\,yx$ gives us a semi-commutativity of this extension.

Now we can endow the polynomial algebra $\mathscr{A}_q[x, y]$ with an N-differential d. Making use of Theorem 1 we define the N-differential by the following formula

$$dw = [y, w]_q = y\,w - q^{|w|}\,w\,y, \qquad (20)$$

where q is a primitive Nth root of unity and $w \in \mathscr{A}_q[x, y]$. Hence the algebra $\mathscr{A}_q[x, y]$ equipped with the N-differential d is a graded q-differential algebra.

In order to give a differential-geometric interpretation to the graded q-differential algebra structure of $\mathscr{A}_q[x, y]$ induced by the N-differential d_v we interpret the commutative subalgebra $\mathscr{A}_q^0[x, y]$ of the x-polynomials (19) of $\mathscr{A}_q[x, y]$ as an algebra of polynomial functions on a one dimensional space with coordinate x. Since $\mathscr{A}_q^k[x, y]$ for $k > 0$ is a $\mathscr{A}_q^0[x, y]$-bimodule we interpret this $\mathscr{A}_q^0[x, y]$-bimodule of the elements of degree k as a bimodule of differential forms of degree k and we shall call an element of this bimodule a differential k-form on a one dimensional space with coordinate x. The N-differential d can be interpreted as an exterior differential.

It is easy to show that in one dimensional case we have a simple situation when every bimodule $\mathscr{A}_q^k[x, y], k > 0$ of the differential k-forms is a free right module over the commutative algebra of functions $\mathscr{A}_q^0[x, y]$. Indeed if we write a differential k-form w as follows

$$w = y^k \sum_{l=0}^{N-1} \beta_l x^l = y^k r, \quad r = \sum_{l=0}^{N-1} \beta_l x^l \in \mathscr{A}_q^0[x, y], \qquad (21)$$

and take into account that the polynomial $r = (y^k)^{-1} w = y^{N-k} w$ is uniquely determined then we can conclude that $\mathscr{A}_q^k[x, y]$ is a free right module over $\mathscr{A}_q^0[x, y]$ generated by y^k.

As it was mentioned before a bimodule structure of a free right module over an algebra $\mathcal{B}$ generated freely by p generators is uniquely determined by the homomorphism from an algebra $\mathcal{B}$ to the algebra of $(p \times p)$-matrices over $\mathcal{B}$. In the case of a reduced quantum plane every right module $\mathscr{A}_q^k[x, y]$ is freely generated by one generator (for instant we can take y^k as a generator of this module). Thus its bimodule structure induces an endomorphism of the algebra of functions $\mathscr{A}_q^0[x, y]$ and denoting this endomorphism in the case of the generator y^k by $A_k : \mathscr{A}_q^0[x, y] \to \mathscr{A}_q^0[x, y]$ we get

$$r \, y^k = y^k A_k(r), \quad \text{(no summation over } k) \tag{22}$$

for any function $r \in \mathscr{A}_q^0[x, y]$. Making use of the commutation relations of variables x, y we easily find that $A_k(x) = q^k x$. Since the algebra of functions $\mathscr{A}_q^0[x, y]$ may be viewed as a bimodule over the same algebra we can consider the functions as degree zero differential forms, and the corresponding endomorphism is the identity mapping of $\mathscr{A}_q[x, y]$, i.e. $A_0 = I$, where $I : \mathscr{A}_q^0[x, y] \to \mathscr{A}_q^0[x, y]$ is the identity mapping. Thus the bimodule structures of the free right modules $\mathscr{A}_q^0[x, y], \mathscr{A}_q^1[x, y], \ldots, \mathscr{A}_q^{N-1}[x, y]$ of differential forms induce the associated endomorphisms $A_0, A_1, \ldots, A_{N-1}$ of the algebra $\mathscr{A}_q^0[x, y]$. It is easy to see that for any k it holds $A_k = A_1^k$.

Let us start with the first order differential calculus $(\mathscr{A}_q^0[x, y], \mathscr{A}_q^1[x, y], d)$ over the algebra of functions $\mathscr{A}_q^0[x, y]$ induced by the N-differential d, where $d : \mathscr{A}_q^0[x, y] \to \mathscr{A}_q^1[x, y]$ and $\mathscr{A}_q^1[x, y]$ is the bimodule over $\mathscr{A}_q^0[x, y]$. For any $w \in \mathscr{A}_q^0[x, y]$ we have

$$dw = yw - wy = yw - yA_1(w) = y(w - A_1(w)) = y \, \Delta_q(w), \tag{23}$$

where $\Delta_q = I - A_1 : \mathscr{A}_q^0[x, y] \to \mathscr{A}_q^0[x, y]$. It is easy to verify that for any two functions $w, w' \in \mathscr{A}_q^0[x, y]$ the mapping Δ_q has the following properties

$$\Delta_q(ww') = \Delta_q(w)w' + A_1(w)\Delta_q(w'), \tag{24}$$
$$\Delta_q(x^k) = (1 - q)[k]_q \, x^k. \tag{25}$$

Particularly $dx = y\Delta_q(x)$, and this formula shows that dx can be taken as a generator for the free right module $\mathscr{A}_q^1[x, y]$.

Since the bimodule $\mathscr{A}_q^1[x, y]$ of the first order differential calculus $(\mathscr{A}_q^0[x, y], \mathscr{A}_q^1[x, y], d)$ is a free right module we have a coordinate first order differential calculus over the algebra $\mathscr{A}_q^0[x, y]$, and in the case of a calculus of this kind the differential induces the derivative $\partial : \mathscr{A}_q^0[x, y] \to \mathscr{A}_q^0[x, y]$ which is defined by the formula $dw = dx \, \partial w, \forall w \in \mathscr{A}_q^0[x, y]$. Using this definition we find that for any function w it holds

$$\partial w = (1 - q)^{-1} x^{N-1} \Delta_q(w). \tag{26}$$

From this formula and (24), (25) it follows that this derivative satisfies the twisted Leibniz rule

$$\partial(ww') = \partial(w) \cdot w' + A_1(w) \cdot \partial(w'),\tag{27}$$

and

$$\partial x^k = [k]_q \, x^{k-1}.\tag{28}$$

Let us study the structure of the higher order exterior calculus on a reduced quantum plane or, by other words, the structure of the bimodule $\mathscr{A}_q^k[x, y]$ of differential k-forms, when $k > 1$. In this case we have a choice for the generator of the free right module. Indeed since the kth power of the exterior differential d is not equal to zero when $k < N$, i.e. $d^k \neq 0$ for $k < N$, a differential k-form w may be expressed either by means of $(dx)^k$ or by means of $d^k x$. Straightforward calculation shows that we have the following relation between these generators

$$d^k x = \frac{[k]_q}{q^{\frac{k(k-1)}{2}}} (dx)^k x^{1-k}.\tag{29}$$

We will use the generator $(dx)^k$ of the free right module $\mathscr{A}_q^k[x, y]$ as a basis in our calculations with differential k-forms. For any differential k-form $w \in \mathscr{A}_q^k[x, y]$ we have $dw \in \mathscr{A}_q^{k+1}[x, y]$. Let us express these two differential forms in terms of the generators of the modules $\mathscr{A}_q^k[x, y]$ and $\mathscr{A}_q^{k+1}[x, y]$. We have $w = (dx)^k r$, $dw = (dx)^{k+1} \tilde{r}$, where $r, \tilde{r} \in \mathscr{A}_q^0[x, y]$ are the functions. Making use of the definition of the exterior differential d we calculate the relation between the functions $r, \tilde{r}$ which is

$$\tilde{r} = (\Delta_q x)^{-1}(q^{-k} r - q^k A_1(r)),\tag{30}$$

where A_1 is the endomorphism of the algebra of functions $\mathscr{A}_q^0[x, y]$. This relation shows that the exterior differential d considered in the case of the differential k-forms induces the mapping $\Delta_q^{(k)} : \mathscr{A}_q^0[x, y] \to \mathscr{A}_q^0[x, y]$ of the algebra of the function which is defined by the formula

$$dw = (dx)^{k+1} \Delta_q^{(k)}(r),\tag{31}$$

where

$$w = (dx)^k r.\tag{32}$$

It is obvious that

$$\Delta_q^{(k)}(r) = (\Delta_q x)^{-1}(q^{-k} r - q^k A_1(r)).\tag{33}$$

It is obvious that for $k = 0$ the mapping $\Delta_q^{(0)}$ coincides with the derivative induced by the differential d in the first order calculus, i.e.

$$\Delta_q^{(0)}(r) = \partial r = (\Delta_q x)^{-1}(r - A_1(r)). \tag{34}$$

The higher order mappings $\Delta_q^{(k)}$, which we do not have in the case of a classical exterior calculus on a one dimensional space, have the derivation type property

$$\Delta_q^{(k)}(r\,r') = \Delta_q^{(k)}(r)\,r' + q^k A_1(r)\,\Delta_x^{(0)}(r'), \tag{35}$$

where $k = 0, 1, 2, \ldots, N - 1$. A higher order mapping $\Delta_q^{(k)}$ can be expressed in terms of the derivative ∂ as a differential operator on the algebra of functions as follows

$$\Delta_q^{(k)} = q^k\,\partial\ + \frac{q^{-k} - q^k}{1 - q}\,x^{-1}. \tag{36}$$

Thus we see that exterior calculus on a one dimensional space with coordinate x satisfying $x^N = 1$ generated by the exterior differential d satisfying $d^N = 0$ has the differential forms of higher order which are not presented in the case of a classical exterior calculus with $d^2 = 0$. The formula for the exterior differential of differential forms can be defined by means of contains not an a derivative which satisfies the twisted Leibniz rule (36).

Acknowledgement The authors is gratefully acknowledge the Estonian Science Foundation for financial support of this work under the Research Grant No. ETF9328. This research was also supported by institutional research funding IUT20-57 of the Estonian Ministry of Education and Research. The second author also acknowledges his gratitude to the Doctoral School in Mathematics and Statistics for financial support of his doctoral studies at the Institute of Mathematics, University of Tartu. The authors are also grateful for partial support from Linda Peeters Foundation for cooperation between Sweden and Estonia provided by Swedish Mathematical Society.

References

1. Abramov, V.: On a graded q-differential algebra. J. Nonlinear Math. Phys. **13**, 1–8 (2006)
2. Abramov, V.: Algebra forms with $d^N = 0$ on quantum plane. Generalized Clifford algebra approach. Adv. Appl. Clifford Algebr. **17**, 577–588 (2007)
3. Abramov, V., Kerner, R.: Exterior differentials of higher order and their covariant generalization. J. Math. Phys. **41**(8), 5598–5614 (2000)
4. Abramov, V., Liivapuu, O.: Connection on module over a graded q-differential algebra. J. Gen. Lie Theory Appl. **3**(2), 112–116 (2008)
5. Abramov, V., Liivapuu, O.: Generalization of connection on the concept of graded q-differential algebra. Proc. Estonian Acad. Sci. **59**(4), 256–264 (2010)
6. Abramov, V., Kerner, R., Le Roy, B.: Hypersymmetry: a $\mathbb{Z}_3$-graded generalization of supersymmetry. J. Math. Phys. **38**, 1650–1669 (1997)
7. Borowiec, A., Kharchenko, V.K.: Algebraic approach to calculus with partial derivatives. Sib. Adv. Math. **5**(2), 10–37 (1995)
8. Coquereaux, R., Garcia, A.O., Trinchero, R.: Differential calculus and connection on a quantum plane at a cubic root of unity. Rev. Math. Phys. **12**(02), 227–285 (2000)
9. Dubois-Violette, M., Kerner, R.: Universal q differential calculus and q analog of homological algebra. Acta Math. Univ. Comenian. **65**, 175 188 (1996)

10. Kapranov, M.: On the q-analog of homological algebra. Preprint Cornell University. arXiv:q-alg/9611005

11. Kerner, R., Abramov, V.: On certain realizations of q-deformed exterior differential calculus. Rep. Math. Phys. **43**(1–2), 179–194 (1999)

12. Kerner, R., Suzuki, O.: Internal symmetry groups of cubic algebras. Int. J. Geom. Methods Mod. Phys. **09**(6) (2012). doi:10.1142/S0219887812610075

13. Lawrynowicz, J., Nouno, K., Nagayama, D., Suzuki, O.: A method of noncommutative Galois theory for binary and ternary Clifford analysis. In: Sivasundaram, S. (ed.), 9th International Conference on Mathematical Problems in Engineering, Aerospace and Sciences: ICNPAA 2012, AIP Conference Proceedings **1493**, 1007–1014 (2012)

14. Lawrynowicz, J., Nôno, K., Nagayama, D., Suzuki, O.: A method of noncommutative Galois theory for construction of quark models (Kobayashi-Masukawa Model) I. Bulletin de la Société des Science et des Lettres de Łódź. **LXIII**, 95–112 (2013)

15. Trovon, A.: Noncommutative Galois extensions and ternary Clifford analysis. Advances in Applied Clifford Algebras. (to be published)

Valued Custom Skew Fields with Generalised PBW Property from Power Series Construction

Lars Hellström

Abstract This chapter describes a construction of associative algebras that, despite starting from a commutation relation that the user may customize quite extensively, still manages to produce algebras with a number of useful properties: they have a Poincaré–Birkhoff–Witt type basis, they are equipped with a norm (actually an ultranorm) that is trivial to compute for basis elements, they are topologically complete, and they satisfy their given commutation relation. In addition, parameters can be chosen so that the algebras will in fact turn out to be skew fields and the norms become valuations. The construction is basically that of a power series algebra with given commutation relation, stated to be effective enough that the other properties can be derived. What is worked out in detail here is the case of algebras with two generators, but only the analysis of the commutation relation is specific for that case.

Keywords Diamond Lemma · Commutation relation · Skew field construction · Ultranorm · Valuation · Irrational weighting of variables

1 Introduction

Power series is one of those concepts which can turn out to be very different things in different branches of mathematics. In algebra, power series is one of many constructions of new rings from old ones; depending on one's point of view, the results may be anywhere from exciting to rather trivial. A combinatorialist regards a power series mostly as a fancy way to present a sequence, which none the less is quite useful since it comes with a host of dirty tricks that boil down to bold applications of elementary algebra. Pre-modern calculus used power series all over the place, mixing spectacular successes with equally spectacular failures that eventually earned them a bad reputation. But in modern analysis, which was born out of the need to put

L. Hellström (✉)
Division of Applied Mathematics, School of Education, Culture and Communication,
Mälardalen University, Box 883, 721 23 Västerås, Sweden
e-mail: lars.hellstrom@mdh.se

© Springer International Publishing Switzerland 2016   33
S. Silvestrov and M. Rančić (eds.), *Engineering Mathematics II*,
Springer Proceedings in Mathematics & Statistics 179,
DOI 10.1007/978-3-319-42105-6_3

calculus on a rigorous foundation, the power series is just a special case of series: it *only* means something if it converges, and the ultimate judge of convergence is the (point-set) topology.

My need for the power series considered below arose in the context [3, pp. 100–101] of looking for commuting homogeneous elements in a q-deformed Heisenberg–Weyl algebra; concretely that algebra had two generators A and B satisfying the commutation relation $AB - qBA = 1$ for some nonzero scalar q, and the question was when two elements on the form

$$\sum_{i=0}^{\min\{k,l\}} r_i B^{k-i} A^{l-i}$$

(for different values of k, l, and scalars r_i) would commute with each other; the product of two such homogeneous elements is again a homogeneous element, and arbitrary algebra elements can be written as finite sums of homogeneous elements. It turns out that there is a simple necessary condition in terms of the exponents in the leading terms, and that when this condition is met and one homogeneous element is given, the problem of determining the scalars in the other element is a straightforward linear equation system with what is essentially a lower triangular matrix. The system is however overdetermined— after getting to the equation that determines the last scalar $r_{\min\{k,l\}}$, there remained a couple of equations that needed to be satisfied, which they sometimes were and at other times were not; there did not seem to be a simple condition that could determine beforehand in which case one would end up. But what if there were no last scalar? If there in each new equation is also a new r_i to absorb whatever remains after having substituted known values of all r_j with $j < i$, then the system will always have a solution and the known necessary condition becomes sufficient! This would however mean looking for a homogeneous element on the form $\sum_{i=0}^{\infty} r_i B^{k-i} A^{l-i}$, which is not something that can be found in the original algebra. Considering negative powers of the generators A and B may seem odd, but is actually not unheard of in the literature on this problem. Making the sum infinite is another matter: the proposed form is that of some kind of Laurent series— in two *noncommuting* variables! Does that even exist?

In Sweden, 20th century mathematics was very much dominated by analysis, and the shape that the following construction took is in a way a consequence of this: anything that looked like an infinite sum had to be rigorously justified, and the one true framework was that of analysis! Or so I believed, as a Ph.d. student; I have subsequently learnt of other ways, mathematically no less rigorous, in which that initial goal could have been achieved, but this very analytically flavoured approach to noncommutative power series turned out to have some unexpected advantages. In particular, several additional properties of the constructed object— some of which were called for in the motivating problem about commuting elements, whereas others were unexpected discoveries— follow with little extra effort once the foundation has been laid. The following result provides a nice sample of what can be had.

Theorem 1 *Consider the commutation relation*

$$AB - qBA = \sum_{i=1}^{n} r_i \prod_{j=1}^{m_i} B^{k_{ij}} A^{l_{ij}} \tag{1}$$

where n is a positive integer, $\{m_i\}_{i=1}^{n} \subset \mathbb{Z}_{>0}$, the coefficients $q \neq 0$ and $\{r_i\}_{i=1}^{n}$ are scalars taken from some field $\mathcal{R}$, and the exponents $\{k_{ij}, l_{ij}\}_{j=1; i=1}^{m_i; n} \subset \mathbb{Z}$ are arbitrary.

If there exists a straight line in $\mathbb{R}^2$ such that the point $(1, 1)$ is on one side of the line and all points $\left(\sum_{j=1}^{m_i} k_{ij}, \sum_{j=1}^{m_i} l_{ij} \right)$ for $i = 1, \ldots, n$ are on the other, then there exists an $\mathcal{R}$-algebra $\mathcal{A}$, a function $a \mapsto \|a\| : \mathcal{A} \longrightarrow \mathbb{R}$, two distinct elements $A, B \in \mathcal{A}$, and two constants $\alpha, \beta \in \mathbb{R}$ such that:

1. *The commutation relation (1) holds in $\mathcal{A}$.*
2. *The algebra $\mathcal{A}$ is a skew field, i.e., all nonzero elements in $\mathcal{A}$ are invertible.*
3. *$\|\cdot\|$ is an ultranorm on $\mathcal{A}$ and $\|a\| \, \|b\| = \|ab\|$ for all $a, b \in \mathcal{A}$.*
4. *$\mathcal{A}$ is complete in the topology induced by $\|\cdot\|$.*
5. *The set $\{B^k A^l\}_{k,l \in \mathbb{Z}}$ is an orthogonal Hilbert basis for $\mathcal{A}$ and $\|B^k A^l\| = 2^{l\alpha + k\beta}$.*
6. *Every nonzero $a \in \mathcal{A}$ has a unique leading term $r B^k A^l$, i.e., there exist unique $r \in \mathcal{R}$ and $k, l \in \mathbb{Z}$ such that $\|a - r B^k A^l\| < \|a\|$.*

It should be pointed out that this theorem does not exhaust the power of the construction, but rather provides a sample of the conclusions that can be drawn in the more advanced cases. Several variations are possible, such as relaxing the condition on the degrees in the right hand side at the price of instead adding conditions on the scalar coefficients of those terms. Not all conditions are needed for all of the conclusions, although the order in which the various conclusions are established comes with a couple of surprises.

As long as the intent is only to *construct* the algebra $\mathcal{A}$, it is even possible to proceed with only a few twists in addition to those anyway needed to produce some algebra with two elements A and B satisfying (1). Recall that the classical construction would be to:

1. Construct the free algebra $\mathcal{R}\langle X \rangle$ where $X = \{\mathsf{a}, \bar{\mathsf{a}}, \mathsf{b}, \bar{\mathsf{b}}\}$ is a set of four formal variables. Variables a and b will give rise to the named elements A and B, whereas $\bar{\mathsf{a}}$ and $\bar{\mathsf{b}}$ are used to ensure that these have multiplicative inverses.

   This free associative algebra $\mathcal{R}\langle X \rangle$— the algebra of "noncommutative poly-nomials" on X over $\mathcal{R}$— is in the literature also known as the tensor algebra $T\big(\mathrm{Span}_{\mathcal{R}}(X)\big)$, but that would be a far more awkward way of looking at it, con-sidering what lies ahead.

2. Quotient $\mathcal{R}\langle X \rangle$ by the two-sided ideal $\mathcal{J}$ generated by the five elements

$$\mathsf{ab} - q\mathsf{ba} - \sum_{i=1}^{n} r_i \prod_{j=1}^{m_i} \mathsf{b}^{k_{ij}} \mathsf{a}^{l_{ij}}, \quad \mathsf{a}\bar{\mathsf{a}} - 1, \quad \bar{\mathsf{a}}\mathsf{a} - 1, \quad \mathsf{b}\bar{\mathsf{b}} - 1, \quad \bar{\mathsf{b}}\mathsf{b} - 1.$$

Then $A = \mathsf{a} + \mathfrak{I} \in \mathcal{R}\langle X \rangle / \mathfrak{I}$ and $B = \mathsf{b} + \mathfrak{I} \in \mathcal{R}\langle X \rangle / \mathfrak{I}$ trivially satisfy (1). The quotient $\mathcal{R}\langle X \rangle / \mathfrak{I}$ is however typically nowhere near satisfying the other claims of Theorem 1.

The power series construction lengthens the above to:

1. Construct the free algebra $\mathcal{R}\langle X \rangle$ where $X = \{\mathsf{a}, \bar{\mathsf{a}}, \mathsf{b}, \bar{\mathsf{b}}\}$ is a set of four formal variables (as before).
2. Let $\alpha, \beta, \gamma \in \mathbb{R}$ be constants such that $\alpha/\beta \in \mathbb{R} \setminus \mathbb{Q}$, $\alpha + \beta > \gamma$, and $\beta \sum_{j=1}^{m_i} k_{ij} + \alpha \sum_{j=1}^{m_i} l_{ij} \leqslant \gamma$ for all $i = 1, \ldots, n$; this means $\beta k + \alpha l = \gamma$ is the equation of one such line in the kl-plane as was required. Let $\mathcal{R}$ be normed by the trivial norm (3). Define $v \colon X \longrightarrow \mathbb{R}$ by $v(\mathsf{a}) = -v(\bar{\mathsf{a}}) = \alpha$ and $v(\mathsf{b}) = -v(\bar{\mathsf{b}}) = \beta$, and let $\|\cdot\|$ be the v-degree norm on $\mathcal{R}\langle X \rangle$ (see Definition 3).
3. Construct the topological completion $\overline{\mathcal{R}\langle X \rangle}$ of $\mathcal{R}\langle X \rangle$ with respect to the norm $\|\cdot\|$. This completion is an $\mathcal{R}$-algebra containing $\mathcal{R}\langle X \rangle$, and in particular containing the elements a, $\bar{\mathsf{a}}$, b, and $\bar{\mathsf{b}}$.
4. Quotient $\overline{\mathcal{R}\langle X \rangle}$ by the topological closure $\mathfrak{I}$ of the two-sided ideal in $\overline{\mathcal{R}\langle X \rangle}$ that is generated by the five elements

$$\mathsf{ab} - q\mathsf{ba} - \sum_{i=1}^{n} r_i \prod_{j=1}^{m_i} \mathsf{b}^{k_{ij}} \mathsf{a}^{l_{ij}}, \quad \mathsf{a}\bar{\mathsf{a}} - 1, \quad \bar{\mathsf{a}}\mathsf{a} - 1, \quad \mathsf{b}\bar{\mathsf{b}} - 1, \quad \bar{\mathsf{b}}\mathsf{b} - 1. \quad (2)$$

Then $A - \mathsf{a} + \mathfrak{I} \in \overline{\mathcal{R}\langle X \rangle} / \mathfrak{I}$ and $B = \mathsf{b} + \mathfrak{I} \in \overline{\mathcal{R}\langle X \rangle} / \mathfrak{I}$ trivially satisfy (1), and less trivially also the other claims of Theorem 1.

In short, the extra steps are to construct a norm on $\mathcal{R}\langle X \rangle$, to form the completion $\overline{\mathcal{R}\langle X \rangle}$, and to remember to take the closure of the ideal before forming the quotient.

The rest of this chapter is essentially a long proof of Theorem 1, with numerous interspersed definitions of concepts that become relevant and (often informal) discussions of techniques that are employed. Section 2 introduces the analysis-inspired foundations for this power series algebra construction. Section 3 employs the Diamond Lemma for power series algebras to analyze the result, which in particular exhibits a basis of the quotient. The final Sect. 4 completes the proof, and goes on to sketch some generalizations of the argument.

2   Normed Algebras

The material in this section is essentially standard (even if it may be hard to find a Mathematics Subject Classification covering this body of knowledge). Therefore focus is primarily on giving full definitions for easy reference and secondarily on pointing out important features of the concepts defined. Proofs are mostly left as exercises to the reader, but the curious may find them in [4, Sect. 2.2–2.3].

Definition 1 Let $\mathcal{R}$ be a ring and let $\|\cdot\|$ be a function from $\mathcal{R}$ to $\mathbb{R}$. Then $\mathcal{R}$ is said to be a **ring with norm** $\|\cdot\|$ if the following conditions are satisfied:

1. $\|a\| \geqslant 0$ for all $a \in \mathcal{R}$, and $\|a\| = 0$ if and only if $a = 0$.
2. $\|a - b\| \leqslant \|a\| + \|b\|$ for all $a, b \in \mathcal{R}$.
3. $\|ab\| \leqslant \|a\| \, \|b\|$ for all $a, b \in \mathcal{R}$.

If $\mathcal{R}$ is a ring with norm $\|\cdot\|$, but the norm is known from the context, then one may simply say that $\mathcal{R}$ is a **normed ring**. If $\mathcal{R}$ is a ring with norm $\|\cdot\|$ then the function $\|\cdot\|$ is called the **ring norm** or simply the **norm**.

Condition 2 above is just a more compact combination of two more intuitive properties. One is that $\|-b\| = \|b\|$ for all $b \in \mathcal{R}$, since $\|-b\| = \|0 - b\| \leqslant \|0\| + \|b\| = \|b\|$. This property is needed for the corresponding metric $\sigma(a, b) = \|a - b\|$ to be symmetric. The other is the normal triangle inequality, which holds since $\|a + b\| = \|a - (-b)\| \leqslant \|a\| + \|-b\| = \|a\| + \|b\|$.

Functional analysis provides plenty of examples of normed rings, for example as Banach algebras, but those examples are *not* the ones which are of interest here. Instead, the following norm will be frequently used:

$$\|a\| = \begin{cases} 0 & \text{if } a = 0, \\ 1 & \text{otherwise.} \end{cases} \tag{3}$$

This norm, which is called the **trivial norm**, is a ring norm for all rings $\mathcal{R}$. The topology it introduces on the ring is not the trivial topology (where only $\varnothing$ and $\mathcal{R}$ itself are open sets) however, but the discrete topology (all subsets of $\mathcal{R}$ are open).

Definition 2 Let $\mathcal{R}$ be an associative and commutative normed ring with unit, and let $|\cdot|$ be the norm on $\mathcal{R}$. Let $\mathcal{A}$ be an associative $\mathcal{R}$-algebra. Then $\mathcal{A}$ is said to be a **normed $\mathcal{R}$-algebra** if there exists a function $\|\cdot\| \colon \mathcal{A} \longrightarrow \mathbb{R}$, called the **norm** or more precisely $\mathcal{R}$**-algebra norm**, such that $\mathcal{A}$ is a normed ring with ring norm $\|\cdot\|$ and

$$\|ra\| \leqslant |r| \, \|a\| \tag{4}$$

for all $r \in \mathcal{R}$ and $a \in \mathcal{A}$.

Analogously, an $\mathcal{R}$-module $\mathcal{M}$ is said to be a **normed $\mathcal{R}$-module** if there exists a function $\|\cdot\| \colon \mathcal{M} \longrightarrow \mathbb{R}$, called the **norm** or more precisely $\mathcal{R}$**-module norm**, such that the following conditions are satisfied:

1. $\|a\| \geqslant 0$ for all $a \in \mathcal{M}$, and $\|a\| = 0$ if and only if $a = 0$.
2. $\|a - b\| \leqslant \|a\| + \|b\|$ for all $a, b \in \mathcal{M}$.
3. $\|ra\| \leqslant |r| \, \|a\|$ for all $r \in \mathcal{R}$ and $a \in \mathcal{M}$.

It is easily checked that if $\mathcal{A}$ is any associative $\mathcal{R}$-algebra, and $\|\cdot\|$ and $|\cdot|$ are the trivial ring norms on $\mathcal{A}$ and $\mathcal{R}$ respectively, then $\mathcal{A}$ will be a normed $\mathcal{R}$-algebra with norm $\|\cdot\|$. The only normed modules that will be of interest here are normed algebras or submodules of normed algebras, but some of the concepts needed are more natural to define for normed modules in general.

The fact that (4) is an inequality, and not an equality, might seem strange at first. It is necessary in the case of a general ring $\mathcal{R}$ however, since if $r_1, r_2 \in \mathcal{R}$ and $a \in \mathcal{A}$ are nonzero and satisfy $r_1 r_2 = 0$ then

$$0 = \|0\| = \|0a\| = \|r_1 r_2 a\| \leqslant |r_1| \, \|r_2 a\| \leqslant |r_1| |r_2| \, \|a\| > 0.$$

The class of norms that will be most important in this paper are the v-*degree norms*, which are easy to define on the free algebra $\mathcal{R}\langle X \rangle$. By definition every element $a \in \mathcal{R}\langle X \rangle$ has a unique presentation as a sum $\sum_{\mu \in X^*} r_\mu \mu$, where $X^* \subset \mathcal{R}\langle X \rangle$ denotes the free monoid on X (i.e., the set of monomials in the noncommutative polynomial ring $\mathcal{R}\langle X \rangle$, or equivalently the set of all elements in $\mathcal{R}\langle X \rangle$ that are finite products of elements of X) and $\{r_\mu\}_{\mu \in X^*} \subseteq \mathcal{R}$ are the coefficients of those monomials; these sums are furthermore finite, in the sense that $r_\mu = 0$ for all but a finite set of monomials μ.

Definition 3 Let an associative and commutative ring with unit $\mathcal{R}$ be given, and let $\|\cdot\|$ be a norm on $\mathcal{R}$. Let X be a set and consider the free associative algebra $\mathcal{R}\langle X \rangle$. Any function $v \colon X \longrightarrow \mathbb{R}$ can be used as the *seed function* of a corresponding v-degree norm. Since X^* is the free monoid on X, the function v extends uniquely to a monoid homomorphism $(X^*, \cdot) \longrightarrow (\mathbb{R}, +)$. Then the v-**degree norm** on $\mathcal{R}\langle X \rangle$ is defined by

$$\left\| \sum_{\mu \in X^*} r_\mu \mu \right\| = \max_{\mu \in X^*} |r_\mu| 2^{v(\mu)} \tag{5}$$

where the maximum is surely attained since $|r_\mu|$ can be nonzero only for finitely many $\mu \in X^*$.

The trivial norm is recovered when $v(x) = 0$ for all $x \in X$. Variables x for which $v(x) < 0$ are power-series-like in that higher powers get smaller (in norm) whereas variables with $v(x) \geqslant 0$ behave more like the variables of an ordinary polynomial ring.

Definition 4 Let $\mathcal{R}$ be an associative unital ring with norm $|\cdot|$. Let $\mathcal{M}$ be an $\mathcal{R}$-module with norm $\|\cdot\|$.

A subset $U \subseteq \mathcal{M}$ is an **open subset** in the topology induced by the norm $\|\cdot\|$ if there for every $a \in U$ exists some real number $\varepsilon > 0$ such that any $b \in \mathcal{M}$ with $\|a - b\| < \varepsilon$ satisfies $b \in U$.

A sequence $\{a_n\}_{n=1}^\infty \subseteq \mathcal{M}$ is said to be a **convergent sequence** if there exists some $b \in \mathcal{M}$, called the **limit** of the sequence, with the property that there for every $\varepsilon > 0$ exists some integer N such that $\|a_n - b\| < \varepsilon$ for every $n \geqslant N$. A **limit point** of a set $U \subseteq \mathcal{M}$ is some $a \in \mathcal{M}$ with the property that there for every $\varepsilon > 0$ exists some $b \in U \setminus \{a\}$ such that $\|a - b\| < \varepsilon$. A set $U \subseteq \mathcal{M}$ is **closed set** in the topology induced by the norm $\|\cdot\|$ if every limit point of U is an element of U. The **topological closure** of $U \subseteq \mathcal{M}$, denoted $\overline{U}$, is the smallest closed subset of $\mathcal{M}$ that contains U.

The usual laws of general topology hold with the definitions above: a subset is closed if and only if its complement is open, an arbitrary union of open sets is again open, but only finite intersections of open sets will necessarily be open, and so on. Using more properties of a norm, one may show that the algebra operations — addition, subtraction, ring multiplication, scalar multiple, and even the norm itself — are all continuous with respect to the topology the norm induces; the algebraic and the topological structures play very nice together.

On the matter of continuity, it is also worth pointing out that the condition for this can be simplified considerably in the case of linear maps: an $\mathcal{R}$-linear map $f : \mathcal{M} \longrightarrow \mathcal{M}$ is continuous if there for every $\varepsilon > 0$ exists some $\delta > 0$ such that any $b \in \mathcal{M}$ with $\|b\| < \delta$ satisfies $\|f(b)\| < \varepsilon$; what happens is that continuity at 0 implies (uniform) continuity everywhere. The same thing happens for equicontinuity; normally a family F of maps $\mathcal{M} \longrightarrow \mathcal{M}$ is said to be **equicontinuous** if there for every $\varepsilon > 0$ exists some $\delta > 0$ such that it for all $f \in F$ and $a, b \in \mathcal{M}$ satisfying $\|b - a\| < \delta$ holds that $\|f(b) - f(a)\| < \varepsilon$, but when all the maps in F are linear (indeed, it suffices that they are homomorphisms of the additive group) it is sufficient to require this for $a = 0$.

Definition 5 Let $\mathcal{R}$ be an associative unital ring with norm $|\cdot|$. Let $\mathcal{M}$ be an $\mathcal{R}$-module with norm $\|\cdot\|$. A sequence $\{a_n\}_{n=1}^\infty \subseteq \mathcal{M}$ is said to be a **Cauchy sequence** if there for every $\varepsilon > 0$ exists some integer N such that $\|a_n - a_m\| < \varepsilon$ for all $m, n \geqslant N$. The set $\mathcal{M}$ is said to be **topologically complete** if every Cauchy sequence in it has a limit in $\mathcal{M}$. A subset $S \subseteq \mathcal{M}$ is said to be **dense** in $\mathcal{M}$ if there for every $a \in \mathcal{M}$ and every $\varepsilon > 0$ exists some $b \in S$ such that $\|b - a\| < \varepsilon$.

A key component in the power series algebra construction is the standard construction of the *completion* of a normed module (which can also be carried out in the greater generality of a metric space or alternatively that of a topological abelian group), which given any module $\mathcal{M}$ produces a topologically complete module $\overline{\mathcal{M}}$ containing $\mathcal{M}$ as a dense subspace; writing $\overline{\mathcal{M}}$ for the completion is borderline an abuse of notation, but as soon as one accepts that the completion exists as a topological space and contains $\mathcal{M}$, then it follows from $\mathcal{M}$ being dense in the completion that the completion is equal to the *closure* $\overline{\mathcal{M}}$. A nice feature of the completion is that any continuous map from the original set $\mathcal{M}$ to a topologically complete set extends uniquely to a continuous map defined on the whole completion. This can be used to extend the algebraic operations to the completion. Moreover, the continuity then implies that they still satisfy all the algebraic identities they had before extension, so the completion $\overline{\mathcal{R}}$ of a normed ring $\mathcal{R}$ is again a normed ring, the completion $\overline{\mathcal{M}}$ of a normed $\mathcal{R}$-module $\mathcal{M}$ is again a normed $\mathcal{R}$-module, and the completion $\overline{\mathcal{A}}$ of a normed $\mathcal{R}$-algebra $\mathcal{A}$ is again a normed $\mathcal{R}$-algebra.

Set-theoretically, the completion of $\mathcal{M}$ can be constructed as a set of equivalence classes of Cauchy sequences in $\mathcal{M}$, where two sequences $\{a_n\}_{n=1}^\infty$ and $\{b_n\}_{n=1}^\infty$ are equivalent if $\lim_{n \to \infty}(a_n - b_n) = 0$. The elements of $\mathcal{M}$ are then not strictly elements of the completion, but there is a canonical embedding of $\mathcal{M}$ into $\overline{\mathcal{M}}$ that maps $a \in \mathcal{M}$ to the equivalence class of the Cauchy sequence where all elements are a.

Definition 6 A module (or ring) norm $\|\cdot\|$ defined on some module $\mathcal{M}$ (or ring $\mathcal{R}$) is said to be a (module/ring) **ultranorm** if it satisfies the *strong triangle inequality*

$$\|a + b\| \leqslant \max\{\|a\|, \|b\|\} \tag{6}$$

for all $a, b \in \mathcal{M}$ (or $\mathcal{R}$).

The trivial norm is obviously an ultranorm. A v-degree norm will also be an ultranorm whenever the norm on the scalars is an ultranorm. Ultranorms are sometimes said to be *non-Archimedean*, since they have the property that any sequence $\left\{\sum_{i=1}^{n} a\right\}_{n=1}^{\infty}$ of integer multiples of an element a will be bounded. Classical algebraic examples of ultranorms are provided by the p-adic valuations on $\mathbb{Q}$ (and more generally on the field of p-adic numbers $\mathbb{Q}_p$).

A very striking property of ultrametric topology is the following "freshman's dream:"

Lemma 1 *If $\overline{\mathcal{M}}$ is the completion of the module $\mathcal{M}$ with respect to an ultranorm $\|\cdot\|$, then the extension of this norm to $\overline{\mathcal{M}}$ is also an ultranorm and a series $\sum_{n=1}^{\infty} a_n$ with terms in $\overline{\mathcal{M}}$ converges if and only if $\lim_{n \to \infty} \|a_n\| = 0$.*

In a sense, that is how one *wants* formal power series to behave: that there be no risk of divergence due to interactions between terms. The price one pays for this is however that the space becomes *totally disconnected*: every open ε-neighbourhood $\{b \in \mathcal{M} \mid \|b - a\| < \varepsilon\}$ of a point a is also topologically closed (since it is the complement of the union of all ε-neighbourhoods that do not contain a)! A good intuitive understanding of what an ultrametric space looks like can be had by imagining it as a Cantor set.

Topology aside, there is a concept of algebraic providence that is quite close to that of an ultranorm, namely that of a **valuation**, so it should be sorted out how the two compare. One advantage of norms is that the notation is more standardised.

codomain   Norms assume values in $\mathbb{R}$, whereas the definition of valuations typically permit an arbitrary totally ordered group, or even semigroup [6], as codomain of the valuation map.

This may seem like a significant generalisation, but in practice it is not. The reason is mainly that a total order on a semigroup gives rise to a canonical order-preserving homomorphism to the real numbers [4, Theorem 3.40], essentially via the Eudoxian theory of proportion. (Book V of *The Elements*, with Euclid's original proofs rather than modern arithmetic substitutes, becomes much more interesting if one allows oneself to consider magnitudes for which addition need *not* be commutative.)

direction of order   For norms, it is a well established standard that small elements have small norms (as measured by the standard order on $\mathbb{R}$). For a valuation V, it rather depends on the author; we find some writing $V(a) > V(b)$ to mean that b is smaller than a, whereas others take it to mean that a is smaller than b, and the strong triangle inequality might be written

$$V(a + b) \leqslant \max\{V(a), V(b)\} \quad \text{or} \quad V(a + b) \geqslant \min\{V(a), V(b)\}$$

with the latter probably being more common.

notation for group operation   For norms, it is a well established standard that multiplication in the ring corresponds to multiplication in $\mathbb{R}$, as in the inequality $\|ab\| \leqslant \|a\| \|b\|$. For valuations, there is a variation in that some authors denote the group operation as addition whereas others denote it as multiplication. Here, it is addition that probably is the more common convention.

treatment of zero   For norms, there is a clear convention that the norm of 0 is 0. For valuations, one may either leave the valuation V a partial function not defined for 0, or adjoin an extra element to the group to serve as $V(0)$. Under the small element has big value convention, it may be convenient to name that extra element ∞.

equality   For valuations (and assuming the additive convention), there is a strong preference that the multiplication axiom should be an equality: $V(ab) = V(a) + V(b)$; a consequence is that the existence of a valuation implies the absence of zero divisors. For norms, it is rather quite common that the multiplication axiom is an inequality, and a notable feature if something satisfies it with equality.

In the present construction, it is only at the very end that equality turns out to hold in the multiplication axiom for norms, so it makes sense to work with a concept that does not seem to imply it from the start. The main reason that one cannot assume equality is the step of forming the quotient.

The main advantage of using norms here and now is rather that they have real numbers as values, because elementary mathematics lets us do so much with real numbers; it is trivial to state $\|B^k A^l\| = 2^{l\alpha + k\beta}$ and require that α/β is irrational.

Definition 7 Let $\mathcal{M}$ be a normed $\mathcal{R}$-module with norm $\|\cdot\|$, and let $\mathcal{N}$ be a submodule of $\mathcal{M}$. Then the **quotient norm** $\|\cdot\|_{\mathcal{M}/\mathcal{N}}$ on $\mathcal{M}/\mathcal{N}$ is defined by

$$\|a + \mathcal{N}\|_{\mathcal{M}/\mathcal{N}} = \inf_{c \in \mathcal{N}} \|a + c\| \qquad \text{for all } a \in \mathcal{M}. \tag{7}$$

The quotient norm is an $\mathcal{R}$-module norm if and only if $\mathcal{N}$ is topologically closed. It will be an $\mathcal{R}$-algebra norm if $\mathcal{M}$ is a normed $\mathcal{R}$-algebra and $\mathcal{N}$ is a two-sided ideal.

The final piece of terminology in Theorem 1 that needs to be defined is that about the orthogonal basis. Here it is useful to first write down a definition of a *Hilbert basis*, since that is the basis concept that is predominant in this chapter. The concept of Hilbert basis should be contrasted to that of a Hamel basis, where one only considers finite linear combinations of basis elements.

Definition 8 Let $\mathcal{R}$ be an associative ring with unit and let $\mathcal{M}$ be a topological $\mathcal{R}$-module. Let $Z \subseteq \mathcal{M}$ be arbitrary. Recall that the notation $\mathrm{Span}(Z)$ denotes the set of all finite linear combinations of elements of Z. It is often convenient to have a simple notation for the topological closure of this set as well. Therefore define

$$\mathrm{Cspan}(Z) = \overline{\mathrm{Span}(Z)} \quad \text{for all sets } Z.$$

Linear independence also needs a topologized counterpart. Define the set Z to be **topologically linearly independent** if it is linearly independent and every countably infinite sequence $\{\mu_i\}_{i=1}^{\infty}$ of distinct elements from Z is such that: $0 \in \mathcal{M}$ is a limit point of the sequence $\left\{ \sum_{i=1}^{n} r_i \mu_i \right\}_{n \in \mathbb{Z}_{>0}}$, where $\{r_i\}_{i=1}^{\infty} \subseteq \mathcal{R}$, if and only if $r_i = 0$ for all $i \in \mathbb{Z}_{>0}$. The set Z is said to be a **Hilbert basis** for $\mathcal{M}$ if it is topologically linearly independent and $\mathcal{M} = \mathrm{Cspan}(Z)$.

In many cases, the most convenient way of showing that a set is a Hilbert basis is to show that it is an orthogonal basis. Contrary to popular opinion, the concept of orthogonality does not require an inner product; it can be defined in arbitrary normed spaces. The theory of orthogonality in normed spaces is however in many aspects different from the theory for inner product spaces. In particular the focus is shifted from elements to sets.

Definition 9 Let $\mathcal{M}$ be an $\mathcal{R}$-module with norm $\|\cdot\|$. A submodule $\mathcal{N}_1 \subseteq \mathcal{M}$ is said to be **orthogonal** to a submodule $\mathcal{N}_2 \subset \mathcal{M}$ if $\|a + b\| \geqslant \|a\|$ for all $a \in \mathcal{N}_1$ and $b \in \mathcal{N}_2$. A subset Y of $\mathcal{M}$ is said to be **orthogonal** if for every bipartition $Y_1 \cup Y_2$ of Y ($Y_1 \cap Y_2 = \varnothing$) the module $\mathrm{Span}(Y_1)$ is orthogonal to $\mathrm{Span}(Y_2)$.

An important example of an orthogonal set is the set X^* of monomials in the free algebra $\mathcal{R}\langle X \rangle$, when that is normed by a v-degree norm. A Hamel basis Y of a module $\mathcal{M}$ that is orthogonal with respect to the norm on $\mathcal{M}$ will be a Hilbert basis of the completion $\overline{\mathcal{M}}$.

One notable property of orthogonal bases that the present norm-derived concept shares with its counterpart in Hilbert spaces is the existence of an associated *dual basis*, or more informally of "Fourier coefficients" for every module element. Concretely, let $\mathcal{R}$ be a topologically complete normed associative ring with unit and $\mathcal{M}$ be a normed $\mathcal{R}$-module with orthogonal basis Y. Then there exists for every $\mu \in Y$ a continuous $\mathcal{R}$-module homomorphism $f_\mu \colon \overline{\mathcal{M}} \longrightarrow \mathcal{R}$ such that $f_\mu(\mu) = 1$ and $f_\mu(\rho) = 0$ for all $\rho \in Y \setminus \{\mu\}$; the dual basis consists of the family $\{f_\mu\}_{\mu \in Y}$ of these maps. The continuity of these maps is relied upon in (8) below, to prove that reductions are also continuous.

3   Rewriting and the Diamond Lemma

Rewriting is usually classified as a branch of computer science, but it touches upon fundamental logic enough to be relevant for all of mathematics, especially combinatorial algebra. What it contributes is in particular a framework for making certain operations *effective* and thus *decidable*, where the traditional constructions of abstract algebra would only produce an infinite set with no obvious algorithm for deciding membership. The application of rewriting that is of interest here is called *equational reasoning*, and addresses the quotient operation; we shall in particular deal with the quotient of an algebra by a two-sided ideal.

A key feature in rewriting is the use of *rewrite rules*, which abstractly is a relation $\to$ stating that the left hand side "may be changed into" the right hand side. In the case of equational reasoning, the external justification for having a particular rule is that both sides of the relation are equivalent, so applying a rule preserves everything of interest. On the other hand, there is also an expectation that the right hand side (in some, not necessarily obvious, way) is *simpler* than the left hand side, so that the application of a rule can be viewed as a step of algebraic simplification.

Rewriting comes in many flavours, distinguished by what the basic objects are that one rewrites. The one that corresponds to associative algebra (but also group theory) is called *word rewriting* since it operates on *words*, which in this case are defined to be finite sequences of symbols from some ground set X; those familiar with programming might find it more intuitive to read 'word' as 'string', since that is essentially what it is. For $X = \{a, b\}$, the first couple of words are 1 (the empty word, a sequence of length zero), a, b, aa $= a^2$, ab, ba, bb $= b^2$, a^3, and so on. The set of all words on X is in abstract algebra known as the free monoid on X, and conversely the operation of concatenating two words is denoted as multiplication because concatenation *is* the multiplication operation in the free monoid.

A standard trick for rewriting, when one aims to show something about an $\mathcal{R}$-algebra of some kind [1, 2], is to work not with the bare words, but with formal linear combinations of words; rewrite rules then end up transforming elements of $\mathcal{R}\langle X \rangle$ into other elements of $\mathcal{R}\langle X \rangle$. In the present setting, where the quotient to examine is not one of $\mathcal{R}\langle X \rangle$ but one of its completion $\overline{\mathcal{R}\langle X \rangle}$, it is necessary to take that one step further and add also a topological structure to the objects being rewritten. The end result is however not too bad, since the three structures (monoid, linear, and topological) combine quite nicely.

Getting more into the technicalities, it is convenient to consider a formalism where a **rewrite rule** is a pair (μ, a), where the left hand side μ is a word, but the corresponding right hand side a can be an arbitrary element of $\overline{\mathcal{R}\langle X \rangle}$. A rule is allowed to act upon any element of $\overline{\mathcal{R}\langle X \rangle}$ where μ occurs as a subexpression.

Definition 10 A **rewrite system** for $\overline{\mathcal{R}\langle X \rangle}$ is a set $S \subseteq X^* \times \overline{\mathcal{R}\langle X \rangle}$. The elements of S are called **rules**. Given any rule $s \in S$, the first component of s (also called the **left hand side**) will be written μ_s and the second component (also called the **right hand side**) will be written a_s; thus $s = (\mu_s, a_s)$.

For every rewrite system S is defined the corresponding ideal $\mathcal{I}(S)$, which is the least topologically closed two-sided ideal in $\overline{\mathcal{R}\langle X \rangle}$ that contains $\{\mu_s - a_s \mid s \in S\}$.

The rewrite system that will be of interest here has 8 rules, which for consistency with [4] will be named s_1 through s_8. We have almost seen five of those rules already, namely

$$s_1 = (\mathsf{ab}, q\mathsf{ba} + c) \quad \text{where} \quad c = \sum_{i=1}^{n} r_i \prod_{j=1}^{m_i} \mathsf{b}^{k_{ij}} \mathsf{a}^{l_{ij}},$$

$$s_2 = \left(\mathsf{b\bar{b}}, 1\right), \qquad\qquad\qquad s_5 = \left(\mathsf{a\bar{a}}, 1\right),$$

$$s_3 = \left(\mathsf{\bar{b}b}, 1\right), \qquad\qquad\qquad s_6 = \left(\mathsf{\bar{a}a}, 1\right);$$

the five elements (2) generating the ideal by which we wish to quotient are exactly $\mu_{s_i} - a_{s_i}$ for $i = 1, 2, 3, 5, 6$. The remaining rules are, with the same c as in s_1,

$$s_4 = \left(\mathsf{a\bar{b}}, q^{-1}\mathsf{\bar{b}a} - q^{-1}\mathsf{\bar{b}c\bar{b}}\right),$$

$$s_7 = \left(\mathsf{\bar{a}b}, q^{-1}\mathsf{b\bar{a}} - q^{-1}\mathsf{\bar{a}c\bar{a}}\right),$$

$$s_8 = \left(\mathsf{\bar{a}\bar{b}}, q\mathsf{\bar{b}\bar{a}} + \mathsf{\bar{a}\bar{b}c\bar{b}\bar{a}}\right).$$

These are needed for technical reasons that will be apparent later, but

$$\mathfrak{I}\big(\{s_1, s_2, s_3, s_4, s_5, s_6, s_7, s_8\}\big) = \mathfrak{I}\big(\{s_1, s_2, s_3, s_5, s_6\}\big)$$

so they do not change the constructed quotient algebra; concretely

$$\mu_{s_4} - a_{s_4} = q^{-1}\mathsf{\bar{b}}(\mu_{s_1} - a_{s_1})\mathsf{\bar{b}} - (\mu_{s_3} - u_{s_3})\mathsf{a\bar{b}} + q^{-1}\mathsf{\bar{b}a}(\mu_{s_2} - a_{s_2}),$$

$$\mu_{s_7} - a_{s_7} = -q^{-1}\mathsf{\bar{a}}(\mu_{s_1} - a_{s_1})\mathsf{\bar{a}} - \mathsf{\bar{a}b}(\mu_{s_5} - a_{s_5}) + q^{-1}(\mu_{s_6} - a_{s_6})\mathsf{b\bar{a}},$$

$$\mu_{s_8} - a_{s_8} = -q\mathsf{\bar{a}}(\mu_{s_4} - a_{s_4})\mathsf{\bar{a}} - \mathsf{\bar{a}b}(\mu_{s_5} - a_{s_5}) + q(\mu_{s_6} - a_{s_6})\mathsf{\bar{b}\bar{a}}.$$

Collectively, the purpose of rules s_1 through s_8 is to provide a rewrite simplification for every monomial that does not fit the PBW pattern— in this instance that pattern is $B^k A^l$, so "first everything B, then everything A"— by on the one hand moving any a or $\mathsf{\bar{a}}$ to the left of a b or $\mathsf{\bar{b}}$ to the right side of it (rules s_1, s_4, s_7, and s_8) and on the other hand making a b adjacent to $\mathsf{\bar{b}}$ or a a adjacent to $\mathsf{\bar{a}}$ cancel each other out (rules s_2, s_3, s_5, and s_6). Rule s_8 might seem like it partially fails to do this, on account of the $\mathsf{\bar{a}b}$ factor in the second term of its right hand side, but it will be all right in the limit. The more general pattern is that one needs two rules for every named generator of the algebra (s_5 and s_6 for A, s_2 and s_3 for B), and four rules for every commutation relation. Rewriting can be used to study also algebras whose defining relations are not simply commutation relations, but then the resulting basis will typically not be of PBW-type.

Of course, whereas the concept of some μ occurring as a subexpression might seem clear for something written down on paper, it is not obviously applicable for general elements of the completion $\overline{\mathcal{R}\langle X\rangle}$. Hence it is convenient to also have an alternative presentation in terms of a family of maps called *reductions*.

Definition 11 Let S be a rewriting system. Let $s = (\mu_s, a_s) \in S$ be an arbitrary rule and let $\lambda, \rho \in X^*$ be arbitrary monomials. Let $t_{\lambda s \rho}: \overline{\mathcal{R}\langle X\rangle} \longrightarrow \overline{\mathcal{R}\langle X\rangle}$ be defined by that

$$t_{\lambda s \rho}(b) = b + f_{\lambda \mu_s \rho}(b)\lambda(a_s - \mu_s)\rho \quad \text{for all } b \in \overline{\mathcal{R}\langle X \rangle} \tag{8}$$

(where $f_{\lambda \mu_s \rho}$ denotes one of the maps in the dual basis for $\overline{\mathcal{R}\langle X \rangle}$). This function $t_{\lambda s \rho}$ can alternatively be characterised as being the unique continuous $\mathcal{R}$-module homomorphism $\overline{\mathcal{R}\langle X \rangle} \longrightarrow \overline{\mathcal{R}\langle X \rangle}$ which satisfies

$$t_{\lambda s \rho}(\mu) = \begin{cases} \lambda a_s \rho & \text{if } \mu = \lambda \mu_s \rho \\ \mu & \text{otherwise} \end{cases}$$

for all $\mu \in X^*$.

Let $T_0(S) = \{\text{id}\}$, where $\text{id}: \overline{\mathcal{R}\langle X \rangle} \longrightarrow \overline{\mathcal{R}\langle X \rangle}$ is the identity map. Let

$$T_1(S) = \left\{ t_{\lambda s \rho} \,\middle|\, \lambda, \rho \in X^* \text{ and } s \in S \right\}.$$

Recursively define

$$T_{n+1}(S) = \left\{ t_1 \circ t_2 \,\middle|\, t_1 \in T_1(S) \text{ and } t_2 \in T_n(S) \right\}$$

for all $n \in \mathbb{Z}^+$. Set

$$T(S) = \bigcup_{n \in \mathbb{N}} T_n(S).$$

The elements of $T(S)$ are called **reductions** and the elements of $T_1(S)$ are called **simple reductions**. If $t(b) = b$ for some $b \in \overline{\mathcal{R}\langle X \rangle}$ then the reduction $t \in T(S)$ is said to **act trivially** on b.

In the reduction formalism, the counterpart of stating that b rewrites to b' is that there is some reduction t for which $t(b) = b'$. The fact that the simple reductions only act nontrivially on one monomial each and also only rewrite one occurrence of the rule's left hand side there implies that rewriting by reductions gives a very fine control of which rewrite steps are taken, something which will be useful later. There exist alternative rewriting formalisms which offer less control, for example requiring that all occurrences of a left hand side are recursively rewritten in each step (so-called generalised division) or requiring that it is always the leftmost occurrence of a left hand side in a monomial that is rewritten; such variations may remove some complications from the theory but introduce others (e.g. a leftmost occurrence rule may make it unclear whether the corresponding ideal automatically becomes two-sided).

It follows from the way reductions are defined that $a - t(a) \in \mathcal{I}(S)$ for every $a \in \overline{\mathcal{R}\langle X \rangle}$ and $t \in T(S)$. The ultimate point of the rewriting process is to exhibit a simplest expression for any element of the quotient algebra, or more technically, to find a unique representative in $\mathcal{R}\langle X \rangle$ for every equivalence class in the quotient $\mathcal{R}\langle X \rangle / \mathcal{I}(S)$. The way to recognise these representatives is that all reductions act trivially on them; if one cannot rewrite them to anything else, then apparently there is not anything simpler. Or can there be? In general, there are indeed ways that it can go wrong! The Diamond Lemma provides a checkable set of conditions which ensure that everything works out right.

Definition 12 Let a rewriting system S be given; the concepts defined below are all with respect to a particular rewriting system. Denote by $B_\varepsilon(b)$ the ε-neighbourhood $\left\{ a \in \overline{\mathcal{R}\langle X \rangle} \mid \|a - b\| < \varepsilon \right\}$ of some $b \in \overline{\mathcal{R}\langle X \rangle}$.

An $a \in \overline{\mathcal{R}\langle X \rangle}$ is said to be **irreducible** if $t(a) = a$ for all $t \in T(S)$. The set of all irreducible elements in $\overline{\mathcal{R}\langle X \rangle}$ is denoted $\mathrm{Irr}(S)$.

An $a \in \overline{\mathcal{R}\langle X \rangle}$ is said to be **stuck in** $F \subseteq \overline{\mathcal{R}\langle X \rangle}$ if $t(a) \in F$ for all $t \in T(S)$. An $a \in \overline{\mathcal{R}\langle X \rangle}$ is said to be **persistently reducible** if there for every $t_1 \in T(S)$ and $\varepsilon > 0$ exists some $t_2 \in T(S)$ and $b \in \mathrm{Irr}(S)$ such that $t_2(t_1(a))$ is stuck in $B_\varepsilon(b)$. The set of all elements in $\overline{\mathcal{R}\langle X \rangle}$ that are persistently reducible is denoted $\mathrm{Per}(S)$.

An $a \in \overline{\mathcal{R}\langle X \rangle}$ is said to be **uniquely reducible** if, for all $t_1, t_2 \in T(S)$, $b_1, b_2 \in \mathrm{Irr}(S)$, and $\varepsilon > 0$ such that $t_1(a)$ is stuck in $B_\varepsilon(b_1)$ and $t_2(a)$ is stuck in $B_\varepsilon(b_2)$, it holds that $\|b_1 - b_2\| < \varepsilon$. The set of all $a \in \overline{\mathcal{R}\langle X \rangle}$ which are both persistently and uniquely reducible is denoted $\mathrm{Red}(S)$.

A rewriting system S for which $\mathrm{Red}(S) = \overline{\mathcal{R}\langle X \rangle}$ is said to be **confluent**. The map $t^S \colon \mathrm{Red}(S) \longrightarrow \mathrm{Irr}(S)$ is defined by that, for any $a \in \mathrm{Red}(S)$ and $\varepsilon > 0$, there exists some $t \in T(S)$ such that $t(a)$ is stuck in $B_\varepsilon(t^S(a))$. The element $t^S(a)$ is called the **normal form** of a.

The map t^S constitutes a kind of limit of the set of all reductions $T(S)$; if $t^S(b) = b'$ then b' is the unique limit point in $\mathrm{Irr}(S)$ of $\{t(b)\}_{t \in T(S)}$. For a confluent rewriting system, t^S becomes a projection of $\overline{\mathcal{R}\langle X \rangle} = \mathrm{Red}(S)$ onto $\mathrm{Irr}(S)$ and $\ker t^S = \mathcal{J}(S)$. Hence $\mathrm{Irr}(S)$ is in that case isomorphic to the quotient $\overline{\mathcal{R}\langle X \rangle}/\mathcal{J}(S)$ as an $\mathcal{R}$-module, but much simpler to describe. In the case $S = \{s_1, s_2, s_3, s_4, s_5, s_6, s_7, s_8\}$,

$$\mathrm{Irr}(S) = \mathrm{Cspan}\left(\left\{ \mathsf{b}^i \mathsf{a}^j, \bar{\mathsf{b}}^i \mathsf{a}^j, \mathsf{b}^i \bar{\mathsf{a}}^j, \bar{\mathsf{b}}^i \bar{\mathsf{a}}^j \mid i, j \in \mathbb{Z}_{>0} \right\} \cup \left\{ \mathsf{b}^i, \bar{\mathsf{b}}^i, \mathsf{a}^i, \bar{\mathsf{a}}^i \mid i \in \mathbb{Z}_{>0} \right\} \cup \{1\} \right), \quad (9)$$

which is how it will follow that $\overline{\mathcal{R}\langle X \rangle}/\mathcal{J}(S)$ has a basis on the form $\{B^k A^l\}_{k,l \in \mathbb{Z}}$. But it still remains to prove that this rewriting system S is confluent.

The main obstacle is to prove unique reducibility, but there are some smaller ones that will need to be dealt with before that. First, it is useful to observe that $\|\mu_s\| = \|a_s\|$ for all $s \in S = \{s_1, s_2, s_3, s_4, s_5, s_6, s_7, s_8\}$; this is no accident, but the result of a deliberate design. For the formal inverse rules s_2, s_3, s_5, and s_6, it straightforwardly follows from $v(\bar{\mathsf{b}}) = -v(\mathsf{b})$ and $v(\bar{\mathsf{a}}) = -v(\mathsf{a})$. For the commutation relation s_1, it instead follows from $\|\mathsf{ab}\| = \|q\mathsf{ba}\| > \|c\|$, and the inequality here is a direct consequence of the condition in Theorem 1 about a straight line separating $(1, 1)$ from a bunch of other points. The significance of $(1, 1)$ here is that it is the (bi)degree of ab and ba, whereas the other points are the (bi)degrees of the terms making up c. The constants $\alpha = v(\mathsf{a})$ and $\beta = v(\mathsf{b})$ were chosen so that $\|c\| < \|q\mathsf{ba}\|$, which by the strong triangle inequality makes $\|a_{s_1}\| = \|q\mathsf{ba}\|$. (That α/β is irrational is not important for this step, but will be important later.) The corresponding calculations for s_4, s_7, and s_8 are slightly embellished forms of that for s_1.

A consequence of this is that $\|t(b)\| \leqslant \|b\|$ for all $b \in \overline{\mathcal{R}\langle X \rangle}$ and $t \in T(S)$ (first prove it for $t \in T_1(S)$ using (8); it is not possible to get equality since terms may cancel). In a confluent system, this implies that the infimum over $a \in \mathcal{I}(S)$ of $\|b + a\|$ is attained for $b + a = t^S(b)$, so the quotient norm on $\overline{\mathcal{R}\langle X \rangle}/\mathcal{I}(S)$ can be calculated from the norm of the normal form representative. This completes the claims in point 5 of Theorem 1.

A further consequence is that $T(S)$ is equicontinuous, which is a technical require-ment in the topologized Diamond Lemma. Among other things, it ensures that Per(S) and Red(S) are topologically closed, and that t^S is continuous. It follows already from their definitions that Irr(S), Per(S) and Red(S) are $\mathcal{R}$-modules, that Irr(S) is topologically closed, and that t^S is an $\mathcal{R}$-module homomorphism and projection.

The second obstacle is to prove that Per$(S) = \overline{\mathcal{R}\langle X \rangle}$, which morally constitutes the claim that there for every $b \in \overline{\mathcal{R}\langle X \rangle}$ and $\varepsilon > 0$ exists some sequence of rewrite steps which will remove all non-irreducible terms larger than ε from b; technically the definition of persistent reducibility has some extra twists to it, but those are there to make it more convenient in proofs. The way that this is proved is by induction over the monomials, since it given the above follows from $X^* \subseteq$ Per(S) that Per$(S) = \overline{\mathcal{R}\langle X \rangle}$.

More precisely, the induction is a form of well-founded induction, so it is carried out with respect to a partial order P on X^*. *This partial order is what determines what it means for one expression or element of $\overline{\mathcal{R}\langle X \rangle}$ to be 'simpler' than another.* A rewrite system S is said to be *compatible* with a partial order P if its right hand sides are smaller than its left hand sides.

Definition 13 Let P be a partial order on X^*. The **down-set module** of some $\mu \in X^*$ with respect to P is the set

$$\text{DSM}(\mu, P) = \text{Cspan}\left(\left\{\rho \in X^* \mid \rho < \mu \text{ in } P\right\}\right),$$

where '$\rho < \mu$ in P' means 'ρ is strictly less than μ according to the partial order P'. A rewrite rule $s = (\mu_s, a_s)$ is said to be **compatible** with P if $a_s \in \text{DSM}(\mu_s, P)$. A rewrite system is compatible with P if all rules in it are compatible with P.

What is needed for proving persistent reducibility is however that $t(\mu) \in \{\mu\} \cup$ DSM(μ, P) for all $\mu \in X^*$ and $t \in T(S)$. Transitivity of P makes this follow for general reductions once it has been established for simple reductions $t_{\lambda s \rho}$, but the step from rules to simple reductions require a bit more from the partial order P, namely that it is preserved under "padding" by arbitrary monomials λ and ρ.

Definition 14 A partial order P on a semigroup $\mathcal{S}$ is said to be a **semigroup partial order** if for all μ, ν, $\lambda \in \mathcal{S}$ it hold that $\mu < \nu$ in P implies $\lambda\mu < \lambda\nu$ in P and $\mu\lambda < \nu\lambda$ in P.

The construction of semigroup partial orders with which a given rewriting system can be compatible is something of an art in itself, but a useful technique is to layer different ordering criteria on top of each other, so that if the first ordering criterion does not distinguish two elements then the second is tried, and if that too considers them equal then a third is used, and so on. A convenient way to formalise this is to describe each layer as a *semigroup quasi-order* from a "toolbox" of simple generic constructions; the detail fitting to a particular rewriting system is achieved by on the one hand choosing parameters in the definitions of these quasi-orders, and on the other choosing how to combine the quasi-orders. See [4, Sect. 3.4] for a detailed treatment of this.

In the present case, there is one more concern that needs to be taken into account when designing P, namely the well-foundedness that is the basis for the induction. Technically, the condition that needs to be satisfied is the following.

Definition 15 If P is a partial order on some $M \subseteq \overline{\mathcal{R}\langle X \rangle}$ such that every strictly P-descending sequence $\{\rho_n\}_{n=1}^{\infty} \subseteq M$ — that is, $\rho_n > \rho_{n+1}$ in P for all $n \in \mathbb{Z}_{>0}$ — satisfies $\|\rho_n\| \to 0$ as $n \to \infty$, then P is said to **satisfy the descending chain condition in norm**, or to be **DCC in norm** for short.

This descending chain condition supports induction on the following form: if $L \subseteq M$ is a set such that

(basis)   any $\rho \in M$ with $\|\rho\| < \varepsilon$ satisfies $\rho \in L$, and
(step)   if $\rho \in M$ is such that any $\sigma < \rho$ in P satisfies $\sigma \in L$ (a kind of condition close to membership in DSM(ρ, P)), then $\rho \in L$

then $L = M$. The descending chain form is however often more intuitive in an algorithmic setting: rewriting ρ_1 might produce ρ_2, which in turn might rewrite to ρ_3, and so on; then the descending chain condition implies that in the limit the non-irreducible terms vanish, or more technically that a finite number of rewrite steps suffice for getting rid of all non-irreducible terms of norm larger than some arbitrary $\varepsilon > 0$. The linear structure adds some complications in that a rewrite step getting rid of one term may introduce several new terms, which will require tracing several descending chains, but this ends up not being a problem.

Thus having presented all the constraints on the partial order P, it is time to present its exact definition. First, one must choose some real number $\theta > 1$ such that $\|c\| \theta \leqslant \|ab\|$. Then $\mu < \rho$ in P for some $\mu, \rho \in X^*$ in the following three cases:

1. $\|\mu\| \theta \leqslant \|\rho\|$,
2. $\|\mu\| = \|\rho\|$ but the length of the word μ is strictly less than the length of the word ρ, or
3. $\|\mu\| = \|\rho\|$ and μ is the same length as ρ, but μ comes before ρ in the word lexicographic order which has $b < \bar{b} < a < \bar{a}$.

Compatibility-wise, case 1 takes care of the second terms of a_{s_1}, a_{s_4}, a_{s_7}, and a_{s_8}, i.e., it implies that $c \in \mathrm{DSM}(\mu_{s_1}, P)$, $-q^{-1}\bar{b}c\bar{b} \in \mathrm{DSM}(\mu_{s_4}, P)$, $-q^{-1}\bar{a}c\bar{a} \in \mathrm{DSM}(\mu_{s_7}, P)$, and $\bar{a}\bar{b}c\bar{b}\bar{a} \in \mathrm{DSM}(\mu_{s_8}, P)$. Case 2 takes care of a_{s_2}, a_{s_3}, a_{s_5}, and a_{s_6} since 1 has word length 0 whereas $\mu_{s_2} = bb$, $\mu_{s_3} = \bar{b}b$, $\mu_{s_5} = a\bar{a}$, and $\mu_{s_6} = \bar{a}a$ all have word length 2. Finally case 3 takes case of the first terms of a_{s_1}, a_{s_4}, a_{s_7}, and a_{s_8}; the referenced lexicographic order orders the length 2 monomials as

$$\mathsf{bb} < \mathsf{b\bar{b}} < \mathsf{ba} < \mathsf{b\bar{a}} < \mathsf{\bar{b}b} < \mathsf{\bar{b}\bar{b}} < \mathsf{\bar{b}a} < \mathsf{\bar{b}\bar{a}} < \mathsf{ab} < \mathsf{a\bar{b}} < \mathsf{aa} < \mathsf{a\bar{a}} < \mathsf{\bar{a}b} < \mathsf{\bar{a}\bar{b}} < \mathsf{\bar{a}a} < \mathsf{\bar{a}\bar{a}}$$

although for the composite order P it is likely that case 1 has preference for pairs of words that contain different sets of letters.

That the partial order P so defined is a semigroup partial order follows from its presentation as a lexicographic composition of semigroup quasi-orders, and will not be shown explicitly; see [4, Sect. 3.4] for the details of how it is done. More interesting is the way in which P gets to be DCC in norm. Basically, the combination of cases 2 and 3— the so-called *length-lexicographic order*— ends up being a well-order of X^*; hence any *infinite* P-descending chain must have infinitely many steps at which the strict descent is ruled according to case 1. Whenever that happens, the norm must decrease by a factor $\theta^{-1} < 1$, so in the limit the norm tends to 0. But why must this θ be explicit?

Had α/β been a rational number, then the set of possible norms would have been discrete, and the quotient between two distinct norm values would have been bounded away from 1; it would have been possible to state case 1 as $\|\mu\| < \|\rho\|$, since that would automatically have implied $\|\mu\| \theta \leqslant \|\rho\|$ for some fixed $\theta > 1$ depending on α and β. But when α/β is irrational the set of possible norms becomes dense in $\mathbb{R}_{>0}$; without an explicit minimal step θ, the order P would not have become DCC in norm. This has the consequence that P does not relate two elements of distinct but almost equal norm, so P is not a total order. This comes with a slight penalty, in that it precludes the use of some rewriting formalisms, in particular the standard bases formalism of Mora [6], for analysing the present power series algebra construction, but that seems unavoidable. Indeed, it turns out that several claims in Theorem 1 are true *precisely* in those cases which cannot be analysed using a monomial order that is both total and DCC in norm! Hence the use of *partial* rather than *total* orders really provides a practical advantage.

The above has cleared the second obstacle to confluence, so we can now go on to the main obstacle, which is the uniqueness of the normal forms. With respect to a random rewrite system, it is quite possible to find a, $b_1 = t_1(a)$ for some reduction t_1, and $b_2 = t_2(a)$ for some other reduction t_2 such that $b_1 \neq b_2$ are both irreducible; the rewrite system $\{s_1, s_2, s_3\}$ with $c = 1$ exhibits this for $a = \mathsf{abb}$. That there can in general be several different reductions which act nontrivially on an element facilitates the formation of such forks in the rewriting process.

The observation on which the Diamond Lemma is based is that such forks are not a problem unless they are final; as long as there is some common successor in both paths, we have not yet made an irrevocable decision on whether to go to b_1 or to b_2. Hence if no decision ever was final, then there could not be two distinct normal forms to choose between, and thus the normal forms would have to be unique! The classical 'diamond condition' (also known as local confluence) for which the Diamond Lemma is named states that for every element a (the apex of the diamond; 'diamond' here simply means the geometric figure of a quadrilateral standing on a corner) and pair of simple reductions t_1, t_2 acting nontrivially on a there exists general reductions t_3, t_4 such that $t_3\big(t_1(a)\big) = t_4\big(t_2(a)\big)$; the elements $a, t_1(a)$, and $t_2(a)$ are the top three corners of the diamond, whereas t_3 and t_4 contributes two additional sides that meet at the fourth corner $t_3(t_1(a)) = t_4(t_2(a))$, thereby "closing the diamond". Though strictly speaking, in this topologized setting one can only count on the two sides getting arbitrarily close, so the condition rather has to be that there for every $\varepsilon > 0$ exists general reductions t_3, t_4 such that $\left\| t_3(t_1(a)) - t_4(t_2(a)) \right\| < \varepsilon$. There are as always some technicalities involved, but by combining the diamond condition with induction over the monomials one can prove the unique reducibility of all elements of $\overline{\mathcal{R}\langle X \rangle}$.

The way in which one verifies the diamond condition is ultimately to do explicit calculations, but it would be impossible to do so for all $a \in \overline{\mathcal{R}\langle X \rangle}$. It is, in view of how induction was used to establish persistent reducibility, probably no surprise that it suffices to verify the diamond condition for monomials a, but that would still leave infinitely many cases to check. Each such case would however be of the form that a monomial μ is acted nontrivially on by two simple reductions t_1 and t_2 — an arrangement which is called an **ambiguity**. Since t_1 and t_2 are *simple* reductions, they can be expressed more explicitly as $t_1 = t_{\lambda_1 s_1 \rho_1}$ and $t_2 = t_{\lambda_2 s_2 \rho_2}$, and since they both act nontrivially on μ it must be the case that $\lambda_1 \mu_{s_1} \rho_1 = \mu = \lambda_2 \mu_{s_2} \rho_2$. Additional restrictions can be imposed on the monomial factors λ_1, μ_{s_1}, ρ_1, λ_2, μ_{s_2}, and ρ_2, because a lot of the separate cases that *can* be considered are in fact "padded" versions of a simpler case, where any common prefix of λ_1 and λ_2 (and similarly any common suffix of ρ_1 and ρ_2) has been shaved off; thanks to the fine control provided by the minimalistic definition of reductions, it is always possible to insert arbitrary padding into all four sides of a known diamond.

In the end, it turns out that the only cases one needs to check explicitly can be specified as a quintuplet $(s_1, s_2, v_1, v_2, v_3)$ where $s_1, s_2 \in S$ are rules and $v_1, v_2, v_3 \in X^*$ are monomials. The product $v_1 v_2 v_3$ is the apex corner of the diamond, v_1 is the part of this monomial which is acted upon by s_1 but not s_2, v_2 is the part of this monomial which is acted upon by both rules, and v_3 is again a part acted upon by only one of the rules (usually s_2 in practice, but s_1 is theoretically possible). Hence either $\mu_{s_1} = v_1 v_2$ and $\mu_{s_2} = v_2 v_3$, with the two simple reductions being $t_{1 s_1 v_3}$ and $t_{v_1 s_2 1}$, or $\mu_{s_1} = v_1 v_2 v_3$ and $\mu_{s_2} = v_2$, with the two simple reductions being $t_{1 s_1 1}$ and $t_{v_1 s_2 v_3}$. This means in particular that there for any given pair of rules (s_1, s_2) is only a finite number of ambiguities that need to be checked explicitly, so for any finite rewrite system S the number of cases to check is finite. For $S = \{s_1, s_2, s_3, s_4, s_5, s_6, s_7, s_8\}$, there turns out to be twelve of them.

What does it mean to perform such a check, though? It would suffice to produce reductions t_3 and t_4 to close the diamond, but in the case of this reduction system it turns out that this in some cases requires going to the limit, which is a bit awkward presentation-wise and also would require dealing with the explicit value of c. There is an alternative condition called *relative resolvability* which as far as the Diamond Lemma is concerned suffices just as well, but which requires introducing a few more concepts.

Definition 16 Let S be a rewriting system for $\overline{\mathcal{R}\langle X \rangle}$ and let P be a partial order on X^*. The **down-set ideal section** of $\rho \in X^*$ with respect to P and S is denoted $\mathrm{DIS}(\rho, P, S)$ and is characterised as being the least topologically closed

$\mathcal{R}$-submodule of $\overline{\mathcal{R}\langle X \rangle}$ that contains all $\lambda(\mu_s - a_s)\nu$ such that $\lambda\mu_s\nu < \rho$ in P for $\lambda, \nu \in X^*$ and $s \in S$. An ambiguity (t_1, μ, t_2) is said to be **resolvable relative to** P if $t_1(\mu) - t_2(\mu) \in \mathrm{DIS}(\mu, P, S)$.

Informally the difference between relative resolvability and the ordinary resolvability using a diamond is that instead of having a lower half with two sides meeting each other in one minimum, the two horizontal extrema are joined with a jagged line, that sometimes goes up and sometimes down; this will still be fine provided all intermediate peaks are strictly below the apex of the upper half (as measured by the partial order P).

Practically, a demonstration of relative resolvability can be presented in lines on the form

$$(s_1, s_2, \nu_1, \nu_2, \nu_3) \qquad\qquad a_{s_1}\nu_3 - \nu_1 a_{s_2} = \text{simplified} = \text{DIS-form}$$

where the quintuplet identifies the ambiguity $(t_{1s_1\nu_3}, \nu_1\nu_2\nu_3, t_{\nu_1 s_2 1})$ being resolved. Then comes an expression which is the difference $t_{1s_1\nu_3}(\nu_1\nu_2\nu_3) - t_{\nu_1 s_2 1}(\nu_1\nu_2\nu_3)$, and in the next step that is simplified. The final step gives a presentation of that same element of $\overline{\mathcal{R}\langle X \rangle}$ which makes it obvious that it belongs to $\mathrm{DIS}(\nu_1\nu_2\nu_3, P, S)$. More precisely, it is expressed as a linear combination of terms where each term contains a factor $e_i := \mu_{s_i} - a_{s_i}$, and also each term is labelled with a reason why this term is in $\mathrm{DSM}(\nu_1\nu_2\nu_3, P)$: 'Norm' refers to case 1 on p. 46 and 'Lex' refers to case 3; it turns out case 2 never comes into play in these comparisons.

$(s_1, s_2, \mathsf{a}, \mathsf{b}, \bar{\mathsf{b}})$ $\qquad a_{s_1}\bar{\mathsf{b}} - \mathsf{a}a_{s_2} = q\mathsf{bab}\bar{\mathsf{b}} + c\bar{\mathsf{b}} - \mathsf{a} = \underbrace{q\mathsf{b}e_4}_{\text{Lex}} + \underbrace{e_2\mathsf{a}}_{\text{Lex}} - \underbrace{e_2 c\bar{\mathsf{b}}}_{\text{Norm}}$

$(s_2, s_3, \mathsf{b}, \bar{\mathsf{b}}, \mathsf{b})$ $\qquad\qquad a_{s_2}\mathsf{b} - \mathsf{b}a_{s_3} = \mathsf{b} - \mathsf{b} = 0$

$(s_3, s_2, \bar{\mathsf{b}}, \mathsf{b}, \bar{\mathsf{b}})$ $\qquad\qquad a_{s_3}\bar{\mathsf{b}} - \bar{\mathsf{b}}a_{s_2} = \bar{\mathsf{b}} - \bar{\mathsf{b}} = 0$

$(s_4, s_3, \mathsf{a}, \bar{\mathsf{b}}, \mathsf{b})$ $\quad a_{s_4}\mathsf{b} - \mathsf{a}a_{s_3} = q^{-1}\bar{\mathsf{b}}\mathsf{ab} - q^{-1}\bar{\mathsf{b}}c\mathsf{bb} - \mathsf{a} = \underbrace{q^{-1}\bar{\mathsf{b}}e_1}_{\text{Lex}} + \underbrace{e_3\mathsf{a}}_{\text{Lex}} - \underbrace{q^{-1}\mathsf{b}ce_3}_{\text{Norm}}$

$(s_5, s_6, \mathsf{a}, \bar{\mathsf{a}}, \mathsf{a})$ $\qquad\qquad a_{s_5}\mathsf{a} - \mathsf{a}a_{s_6} = \mathsf{a} - \mathsf{a} = 0$

$(s_5, s_7, \mathsf{a}, \bar{\mathsf{a}}, \mathsf{b})$ $\quad a_{s_5}\mathsf{b} - \mathsf{a}a_{s_7} = \mathsf{b} - q^{-1}\mathsf{ab}\bar{\mathsf{a}} + q^{-1}\mathsf{a}\bar{\mathsf{a}}c\bar{\mathsf{a}} = \underbrace{-q^{-1}e_1\bar{\mathsf{a}}}_{\text{Lex}} - \underbrace{\mathsf{b}e_5}_{\text{Lex}} + \underbrace{q^{-1}e_5 c\bar{\mathsf{a}}}_{\text{Norm}}$

$(s_5, s_8, \mathsf{a}, \bar{\mathsf{a}}, \bar{\mathsf{b}})$ $\qquad a_{s_5}\bar{\mathsf{b}} - \mathsf{a}a_{s_8} = \bar{\mathsf{b}} - q\mathsf{ab}\bar{\mathsf{a}} - \mathsf{a}\bar{\mathsf{a}}\mathsf{b}c\bar{\mathsf{b}}\bar{\mathsf{a}} = \underbrace{-q e_4\bar{\mathsf{a}}}_{\text{Lex}} - \underbrace{\bar{\mathsf{b}}e_5}_{\text{Lex}} - \underbrace{e_5\bar{\mathsf{b}}c\mathsf{b}\bar{\mathsf{a}}}_{\text{Norm}}$

$(s_6, s_1, \bar{\mathsf{a}}, \mathsf{a}, \mathsf{b})$ $\qquad a_{s_6}\mathsf{b} - \bar{\mathsf{a}}a_{s_1} = \mathsf{b} - q\bar{\mathsf{a}}\mathsf{ba} - \bar{\mathsf{a}}c = \underbrace{-q e_7\mathsf{a}}_{\text{Lex}} - \underbrace{\mathsf{b}e_6}_{\text{Lex}} + \underbrace{\bar{\mathsf{a}}ce_6}_{\text{Norm}}$

$(s_6, s_4, \bar{\mathsf{a}}, \mathsf{a}, \bar{\mathsf{b}})$ $\qquad a_{s_6}\bar{\mathsf{b}} - \bar{\mathsf{a}}a_{s_4} = \bar{\mathsf{b}} - q^{-1}\bar{\mathsf{a}}\mathsf{ba} + q^{-1}\bar{\mathsf{a}}\mathsf{b}c\mathsf{b} =$

$$= \underbrace{-q^{-1}e_8\mathsf{a}}_{\text{Lex}} - \underbrace{\bar{\mathsf{b}}e_6}_{\text{Lex}} - \underbrace{q^{-1}\bar{\mathsf{a}}\mathsf{b}ce_6}_{\text{Norm}}$$

$(s_6, s_5, \bar{\mathsf{a}}, \mathsf{a}, \bar{\mathsf{a}})$ $\qquad\qquad a_{s_3}\bar{\mathsf{a}} - \bar{\mathsf{a}}a_{s_2} = \bar{\mathsf{a}} - \bar{\mathsf{a}} = 0$

$(s_7, s_2, \bar{\mathsf{a}}, \mathsf{b}, \bar{\mathsf{b}})$ $\quad a_{s_7}\bar{\mathsf{b}} - \bar{\mathsf{a}}a_{s_2} = q^{-1}\mathsf{b}\mathsf{a}\bar{\mathsf{b}} - q^{-1}\bar{\mathsf{a}}c\mathsf{a}\bar{\mathsf{b}} - \bar{\mathsf{a}} =$

$$= \underbrace{q^{-1}\mathsf{b}e_8}_{\text{Lex}} + \underbrace{e_2\bar{\mathsf{a}}}_{\text{Lex}} - \underbrace{e_7\mathsf{b}c\bar{\mathsf{b}}\bar{\mathsf{a}}}_{\text{Norm}} + \underbrace{\bar{\mathsf{a}}e_2 c\mathsf{b}\bar{\mathsf{a}}}_{\text{Norm}} - \underbrace{q^{-1}\bar{\mathsf{a}}ce_8}_{\text{Norm}}$$

$(s_8, s_3, \bar{a}, \bar{b}, b)$ $a_{s_8}b - \bar{a}a_{s_3} = q\bar{b}\bar{a}b + \bar{a}\bar{b}c\bar{b}\bar{a}b - \bar{a} =$

$$= \underbrace{q\bar{b}e_7}_{\text{Lex}} + \underbrace{e_3\bar{a}}_{\text{Lex}} + \underbrace{\bar{a}\bar{b}c\bar{b}e_7}_{\text{Norm}} + \underbrace{q^{-1}\bar{a}\bar{b}ce_3\bar{a}}_{\text{Norm}} + \underbrace{q^{-1}e_8c\bar{a}}_{\text{Norm}}$$

This fulfils the last condition in the Diamond Lemma, so it is hereby established that $\overline{\mathcal{R}\langle X\rangle} = \mathrm{Irr}(S) \oplus \mathcal{I}(S)$ and hence that the quotient $\overline{\mathcal{R}\langle X\rangle}/\mathcal{I}(S) \cong \mathrm{Irr}(S)$ as a vector space.

For clarity, it might however be worth giving a more exact reference to the particular Diamond Lemma that would be used. Theorem 3.30 of [4] will suffice, but it should be pointed out that the terminology used in this section is probably more in line with that of the later paper [5] where the two disagree. It is alternatively possible to use the more general Theorem 5.11 of [5] to establish the above conclusion; in that case one would make use of [5, Sect. 7] to set up the topological framework and [5, Ex. 6.10] to analyse the ambiguities. In both those statements of a Diamond Lemma, the first and second obstacles above address explicit conditions in the theorem statement, whereas the "main" obstacle appears as one of several claims equivalent to confluence and $\overline{\mathcal{R}\langle X\rangle} = \mathrm{Irr}(S) \oplus \mathcal{I}(S)$.

Another point that should addressed in this context is how the norm on the scalars affects the set-up of the Diamond Lemma machinery. As long as that scalar norm is trivial, it is sufficient to define P as a partial order on the set of monomials X^*, but if it is not then P rather has to be defined as a partial order on the set of all *terms $r\mu$* where $r \in \mathcal{R} \setminus \{0\}$ and $\mu \in X^*$; this is needed because a term with large $\|\mu\|$ might still be made small by a tiny $|r|$ (and vice versa). This setting of ordering the terms is explicitly that which was used in [4], whereas [5] takes a more abstract route, but for this chapter it seemed an additional complication that readers for the most part were better off without.

4   The Unreasonable Usefulness of Irrationality

Having pushed through the rather daunting mounds of technicalities in the previous section, this seems like a good place to pause and evaluate where we are with respect to proving the claims in Theorem 1. The algebra $\mathcal{A}$ is constructed as $\overline{\mathcal{R}\langle X\rangle}/\mathcal{I}(S)$, and it carries a quotient norm $\|\cdot\|$ inherited from the v-degree norm on $\overline{\mathcal{R}\langle X\rangle}$. The algebra $\mathcal{A}$ furthermore has two named elements $A = \mathsf{a} + \mathcal{I}(S)$ and $B = \mathsf{b} + \mathcal{I}(S)$. With that in mind, the claims were:

1. *The commutation relation* (1) *holds in $\mathcal{A}$.*
   That one has been obvious since Sect. 1.
2. *The algebra $\mathcal{A}$ is a skew field, i.e., all nonzero elements in $\mathcal{A}$ are invertible.*
   That one has conversely seen no progress at all.
3. $\|\cdot\|$ *is an ultranorm on $\mathcal{A}$ and* $\|a\| \, \|b\| = \|ab\|$ *for all* $a, b \in \mathcal{A}$.
   Here, there is partial progress. It was established in Sect. 3 that the quotient norm on $\mathcal{A}$ is in fact equal to the norm on $\mathrm{Irr}(S)$ (composed with the isomor-

phism between the two $\mathcal{R}$-modules) so it will be an ultranorm, but the claim that $\|a\| \|b\| = \|ab\|$ still needs to be proved.

4. *A is complete in the topology induced by $\|\cdot\|$.*

   This follows from the isomorphism with $\mathrm{Irr}(S)$.

5. *The set $\{B^k A^l\}_{k,l \in \mathbb{Z}}$ is an orthogonal Hilbert basis for A and $\|B^k A^l\| = 2^{l\alpha + k\beta}$.*

   This, too, follows from the isomorphism with $\mathrm{Irr}(S)$. It is in many ways the main conclusion from applying the Diamond Lemma machinery.

6. *Every nonzero $a \in A$ has a unique leading term $r B^k A^l$, i.e., there exist unique $r \in \mathcal{R}$ and $k, l \in \mathbb{Z}$ such that $\|a - r B^k A^l\| < \|a\|$.*

   This has not been proved, but we are now ready to do it!

And the key to getting further is the irrationality of α/β that up to this point has rather been a complication.

An immediate consequence of the ratio of α to β being irrational is that the map $\mathbb{Z} \times \mathbb{Z} \longrightarrow \mathbb{R} : (k, l) \mapsto l\alpha + k\beta$ is injective, and thus the map $\mathbb{Z} \times \mathbb{Z} \longrightarrow \mathbb{R} : (k, l) \mapsto \|B^k A^l\|$ is injective as well. Hence *every element of the basis for A has a distinct norm*, from which the unique leading term property immediately follows. Every $a \in A$ has, by the status of $\{B^k A^l\}_{k,l \in \mathbb{Z}}$ as a Hilbert basis, a unique presentation on the form

$$a = \sum_{k,l \in \mathbb{Z}} r_{k,l} B^k A^l \qquad \text{for some } \{r_{k,l}\}_{k,l \in \mathbb{Z}} \subseteq \mathcal{R}. \tag{10}$$

For any $\varepsilon > 0$, the set of (k, l) such that $\|r_{k,l} B^k A^l\| \geq \varepsilon$ is finite (because otherwise the series (10) would not converge) and for every $a \neq 0$ there is thus a unique term $r_{k,l} B^k A^l$ that is maximal in norm. Since all other terms have strictly smaller norm, it follows from the strong triangle inequality that $\|a\| = \|r_{k,l} B^k A^l\| > \|a - r_{k,l} B^k A^l\|$.

Unique leading terms, invertibility of basis elements, and invertibility of scalars is then all that is needed to prove invertibility of arbitrary nonzero elements, by using the old trick of viewing the formula for the sum of a geometric series as a formula for the inverse of things close to 1. Concretely, if $\|a\| = \|r B^k A^l\| > \|a - r B^k A^l\|$ then

$$a = r B^k A^l - (r B^k A^l - a) = \left(1 - (1 - r^{-1} a A^{-l} B^{-k})\right) \cdot r B^k A^l \tag{11}$$

where $\|1 - r^{-1} a A^{-l} B^{-k}\| \leq \|r B^k A^l - a\| \|A^{-l}\| \|r^{-1} B^{-k}\| < 1$ and thus

$$r^{-1} A^{-l} B^{-k} \sum_{n=0}^{\infty} (1 - r^{-1} a A^{-l} B^{-k})^n \tag{12}$$

converges. It follows from (11) that (12) is in fact the multiplicative inverse of a, and therefore any $a \in A \setminus \{0\}$ is invertible. Thus A is a skew field.

A second consequence of the expression (12) is that $\left\| a^{-1} \right\| \leqslant \left\| a \right\|^{-1}$, since

$$\left\| r^{-1} A^{-l} B^{-k} \sum_{n=0}^{\infty} (1 - r^{-1} a A^{-l} B^{-k})^n \right\| \leqslant$$

$$\leqslant |r^{-1}| \left\| A^{-l} \right\| \left\| B^{-k} \right\| \left\| \sum_{n=0}^{\infty} (1 - r^{-1} a A^{-l} B^{-k})^n \right\| =$$

$$= |r|^{-1} 2^{-\alpha l} 2^{-\beta k} \left\| 1 \right\| = \left\| r B^k A^l \right\|^{-1} = \left\| a \right\|^{-1}.$$

On the other hand,

$$1 = \left\| 1 \right\| = \left\| a a^{-1} \right\| \leqslant \left\| a \right\| \left\| a^{-1} \right\| \leqslant \left\| a \right\| \left\| a \right\|^{-1} = 1,$$

so it must in fact be the case that $\left\| a^{-1} \right\| = \left\| a \right\|^{-1}$. This makes it easy to prove that $\left\| a \right\| \left\| b \right\| = \left\| ab \right\|$ for all $a, b \in \mathcal{A}$. If a or b is 0 then that claim is trivial, whereas if a and b are both invertible then

$$\left\| a \right\| \left\| b \right\| = \left\| a^{-1} \right\|^{-1} \left\| b^{-1} \right\|^{-1} = \left(\left\| b^{-1} \right\| \left\| a^{-1} \right\| \right)^{-1} \leqslant$$

$$\leqslant \left\| b^{-1} a^{-1} \right\|^{-1} = \left\| (ab)^{-1} \right\|^{-1} = \left\| ab \right\|,$$

forcing equality also in this case; the algebra $\mathcal{A}$ is *valued* in the sense that the norm gives rise to a valuation, in the strict interpretation which requires equality in the multiplication axiom. This concludes the proof of claim 3, and thus also of Theorem 1 as a whole.

If instead of having the norm $|\cdot|$ on the field $\mathcal{R}$ be trivial one allows it to be an arbitrary valuation (ultranorm satisfying the multiplication axiom with equality), then the degree conditions on the terms in the remainder c can be relaxed. There is still the requirement that $\left\| c \right\| \theta \leqslant \left\| ab \right\|$ for some $\theta > 1$, but now that only boils down to $|r| \left\| \mu \right\| \theta \leqslant \left\| ab \right\|$ for every term $r\mu$ in c, and if $|r|$ is small then that may compensate for $\left\| \mu \right\|$ being large, but also vice versa. The coefficient q in the term $q\mathsf{ba}$ must satisfy $|q| = 1$, since rewriting requires both $\left\| \mathsf{ab} \right\| \geqslant \left\| q\mathsf{ba} \right\|$ and $\left\| \mathsf{a\bar{b}} \right\| \geqslant \left\| q^{-1}\mathsf{\bar{b}a} \right\|$.

That $\mathcal{R}$ is a field really only becomes important in the penultimate step of proving that $\mathcal{A}$ is a skew field; the uniqueness of the leading term follows even for a general coefficient ring $\mathcal{R}$. That q is invertible is on the other hand necessary already for setting up the rewriting system, since rules s_4 and s_7 in a sense have the roles of left hand side and leading term of the right hand side reversed from what they are in s_1.

From the point of view of non-power-series rewriting, it is somewhat surprising that the exact value of the "remainder" c turns out not to matter— as long as it is smaller (in norm) than the other terms ab and $q\mathsf{ba}$ of the commutation relation, it can be whatever one wants! Part of this is due to only having two generators A and B, since having several commutation relations can lead to them interacting with each other in nontrivial ways, but part of it is also due to the simplifying ability of power series; a finite sum $\sum_{k=0}^{n} r_k x^k$ is only a polynomial, but an infinite sum $\sum_{k=0}^{\infty} r_k x^k$ can reproduce any analytical function.

References

1. Bergman, G.M.: The diamond lemma for ring theory. Adv. Math. **29**, 178–218 (1978)
2. Bokut, L.A.: Embeddings into simple associative algebras (Russian). Algebra i Logika. **15**(2), 117–142, 245 (1976)
3. Hellström, L., Silvestrov, S.D.: Commuting Elements in q-deformed Heisenberg Algebras. World Scientific Publishing Co. Inc., River Edge (2000)
4. Hellström, L.: The diamond lemma for power series algebras (doctorate thesis), Umeå University, xviii+228 pp. http://www.risc.jku.at/Groebner-Bases-Bibliography/details.php?details_id=1354. ISBN 91-7305-327-9 (2002)
5. Hellström, L.: A generic framework for diamond lemmas. arXiv:0712.1142v1 [math.RA] (2007)
6. Mora, T.: Seven variations on standard bases, preprint 45, Dip. Mat. Genova, 81 pp. http://www.disi.unige.it/person/MoraF/publications.html (1988)

Computing Burchnall–Chaundy Polynomials with Determinants

Johan Richter and Sergei Silvestrov

Abstract In this expository paper we discuss a way of computing the Burchnall–Chaundy polynomial of two commuting differential operators using a determinant. We describe how the algorithm can be generalized to general Ore extensions, and which properties of the algorithm that are preserved.

Keywords Ore extensions · Burchnall-Chaundy theory · Determinants

1 Introduction

It is a classical result, going back to [2–4], that all pairs of commuting elements in the Weyl (Heisenberg) algebra are algebraically dependent over $\mathbb{C}$. This result was later rediscovered and applied to the study of non-linear partial differential equations [9, 10, 12].

In this paper we will describe a method, the Burchnall–Chaundy eliminant construction, for computing explicitly the algebraic relation satisfied by two commuting elements, and consider the generalisation to the class of rings known as Ore extensions. We will describe results showing that the eliminant construction partially generalises. We will also give counterexamples showing that these generalisations do not always retain all desired properties.

J. Richter (✉) · S. Silvestrov
Division of Applied Mathematics, School of Education,
Culture and Communication, Mälardalen University,
Box 883, 721 23 Västerås, Sweden
e-mail: johan.richter@mdh.se

S. Silvestrov
e-mail: sergei.silvestrov@mdh.se

© Springer International Publishing Switzerland 2016   57
S. Silvestrov and M. Rančić (eds.), *Engineering Mathematics II*,
Springer Proceedings in Mathematics & Statistics 179,
DOI 10.1007/978-3-319-42105-6_4

2 Definitions

We recall the following definition.

Definition 1 Let R be a ring, σ an endomorphism of R and δ an additive function, $R \to R$, satisfying

$$\delta(ab) = \sigma(a)\delta(b) + \delta(a)b$$

for all $a, b \in R$. (Such δ:s are known as σ-derivations.) The *Ore extension* $R[x; \sigma, \delta]$ is the polynomial ring $R[x]$ equipped with a new multiplication such that $xr = \sigma(r)x + \delta(r)$ for all $r \in R$. Every element of $R[x; \sigma, \delta]$ can be written uniquely as $\sum_i a_i x^i$ for some $a_i \in R$.

If $\sigma = \mathrm{id}$ then $R[x; \mathrm{id}_R, \delta]$ is called a *differential operator ring*. If $P = \sum_{i=0}^n a_i x^i$, with $a_n \neq 0$, we say that P has *degree* n. We say that the zero element has degree $-\infty$.

Ore extensions were defined by the Norwegian mathematician Ore [13] as a non-commutative analogue of polynomial rings.

Definition 2 The *Weyl* (or Heisenberg) algebra, can be defined as the Ore extension $\mathbb{C}[y][x; \mathrm{id}, \delta]$ where δ is the usual algebraic derivative on $\mathbb{C}[y]$.

The q-deformed Weyl algebra can be defined as the Ore extension $\mathbb{C}[y][x; \sigma, \delta]$ where $\sigma(\alpha) = \alpha$ and $\delta(\alpha) = 0$ for all $\alpha \in C$ and where $\sigma(y) = qy$ and $\delta(y) = 1$.

We will simply refer to a q-deformed Weyl algebra as a q-Weyl algebra. A q-Weyl algebra is thus an algebra over $\mathbb{C}$ with two generators, x and y, such that $xy = qyx + 1$. The Weyl algebra is the special case when $q = 1$.

If A is any algebra over a ring R and P, Q are two commuting elements of A then we say that P, Q are algebraically dependent if $f(P, Q) = 0$ for some non-zero polynomial $f(s, t) \in R[s, t]$ in two central indeterminates s and t over R. The polynomial f is called an annihilating polynomial.

3 Algebraic Dependence

In a series of papers in the 1920s and 30s [2–4], Burchnall and Chaundy studied the properties of commuting pairs of ordinary differential operators. The following theorem is essentially found in their papers. (Their paper is somewhat imprecise on formal details.)

Theorem 1 Let $P = \sum_{i=0}^n p_i D^i$ and $Q = \sum_{j=0}^m q_j D^j$ be two commuting elements of T with constant leading coefficients. Then there is a non-zero polynomial $f(s, t)$ in two commuting variables over $\mathbb{C}$ such that $f(P, Q) = 0$. Note that the fact that P and Q commute guarantees that $f(P, Q)$ is well-defined.

Burchnall's and Chaundy's work rely on analytical facts, such as the existence theorem for solutions of linear ordinary differential equations. However, it is possible to give algebraic proofs for the existence of the annihilating polynomial. This was done later by authors such as Amitsur [1] and Goodearl [5, 8]. Once one casts Burchnall's and Chaundy's results in an algebraic form one can also generalize them to a broader class of rings.

More specifically, one can prove Burchnall's and Chaundy's result for certain types of Ore extensions. We cite an important early result by Amitsur as an example.

Amitsur [1, Theorem 1] (following work of Flanders [7]) studied the case when R is a field of characteristic zero and δ is an arbitrary derivation on R. He obtained the following theorem.

Theorem 2 *Let K be a field of characteristic zero with a derivation δ. Let F denote the subfield of constants. Form the differential operator ring $S = K[x; \mathrm{id}, \delta]$, and let P be an element of S of degree n. Denote by by $F[P]$ the ring of polynomials in P with constant coefficients, $F[P] = \{\sum_{j=0}^{m} b_j P^j \mid b_j \in F\}$. Then $C_S(P)$ is a commutative subring of S and a free $F[P]$-module of rank at most n.*

The next corollary can be found in [1, Corollary 2].

Corollary 1 *Let P and Q be two commuting elements of $K[x; \mathrm{id}, \delta]$, where k is a field of characteristic zero. Then there is a nonzero polynomial $f(s, t)$, with coefficients in F, such that $f(P, Q) = 0$.*

Proof Let P have degree n. Since Q belongs to $C_S(P)$ we know that $1, Q, \ldots, Q^n$ are linearly dependent over $F[P]$ by Theorem 2. But this tells us that there are elements $\phi_0(P), \phi_1(P), \ldots \phi_n(P)$, in $F[P]$, of which not all are zero, such that

$$\phi_0(P) + \phi_1(P)Q + \cdots + \phi_n(P)Q^n = 0.$$

Setting $f(s, t) = \sum_{i=0}^{n} \phi_i(s)t^i$ the corollary is proved. □

4 The Determinant Construction

The cited result by Amitsur is an existence proof, but Burchnall and Chaundy also gave an algorithm for computing the annihilating polynomial in the case of differential operators. In this section we will describe this algorithm for the similar case of the q-Weyl algebra.

Let $P = \sum_{i=0}^{n} p_i(y)x^i$ and $Q = \sum_{j=0}^{m} q_j(y)x^j$ be commuting elements in a q-Weyl algebra. For $e = 0, 1, \ldots m - 1$ compute

$$x^e(P - s) = \sum_{i=0}^{n+m-1} p_{i,e}(y, s)x^i$$

and similarly, for $l = 0, 1, \ldots n - 1$ compute

$$x^l(Q - t) = \sum_{j=0}^{n+m-1} q_{j,l}(y, t)x^j.$$

Here the computation is done in the ring $C[y][x; \sigma, \delta][s, t]$, the polynomial ring in two central indeterminates over $C[y][x; \sigma, \delta]$. Form a square matrix of size $n + m$ with $p_{i,e}$ as the element in row $e + 1$ and column $n + m - i$. We let $q_{j,l}$ be the matrix element in row $l + m + 1$ and column $n + m - j$. The determinant of this matrix will be called the eliminant (of P and Q) and denoted $\Delta_{P,Q}$.

De Jeu, Svensson and Silvestrov [6] prove the following theorem.

Theorem 3 *Let K be a field, and q an element of K such that $\sum_{i=0}^{N} q^i \neq 0$ for all natural numbers N. (Note that such a q only exists if K is an infinite field.) Let $\Delta_{P,Q}$ denote the eliminant constructed above. (A polynomial in y, s and t.) Write $\Delta_{P,Q} = \sum f_i(s, t)y^i$. Then*

(i) *at least one of the f_i are non-zero;*
(ii) *$f_i(P, Q) = 0$ for all i.*

In the case when $K = \mathbb{R}$ and $q = 1$, this is the same method as Burchnall and Chaundy describe.

Example 4 That a condition on q is needed in the theorem can be seen as follows: if q is a primitive nth root of unity, where $n > 1$, then x^n and y^n both belong to the center of $\mathbb{C}[y][x; \sigma, \delta]$. But there is no non-zero polynomial over $\mathbb{C}$ that annihilates x^n and y^n.

Example 5 We describe an example of the eliminant when $q = 1$. Let $P = yx$ and $Q = y^2x^2$. Then

$$x^0(P - s) = yx - s,$$

$$x^1(P - s) = yx^2 + (1 - s)x,$$

and

$$x^0(Q - t) = y^2x^2 - t.$$

Thus

$$\Delta_{P,Q} = \begin{vmatrix} 0 & y & -s \\ y & (1 - s) & 0 \\ y^2 & 0 & -t \end{vmatrix} = (t + s(1 - s))y^2.$$

Since indeed $Q^2 + P(1 - P) = 0$, this is consistent with Theorem 3.

Example 6 We can describe a similar example in the q-Weyl algebra. Set $P = yx$ and $Q = y^2x^2$, again. One can check that

$$PQ = QP = q^2 y^3 x^3 + (q+1) y^2 x^2.$$

The eliminant becomes

$$\Delta_{P,Q} = \begin{vmatrix} 0 & y & -s \\ qy & (1-s) & 0 \\ y^2 & 0 & -t \end{vmatrix} = (qt + s(1-s)) y^2.$$

As expected we recover the result of the previous example by setting $q = 1$.

5 Generalisation to Ore Extensions

The eliminant construction can be generalised to any Ore extension in an obvious way. This was done by Larsson in [11] and described in more detail in [15].

To elaborate slightly, suppose that P and Q are commuting elements of some Ore extension $R[x; \sigma, \delta]$ with P having degree n and Q having degree m. For $e = 0, 1, \ldots m - 1$ compute

$$x^e(P - s) = \sum_{i=0}^{n+m-1} p_{i,e}(s)) x^i,$$

and similarly, for $l = 0, 1, \ldots n - 1$ compute

$$x^l(Q - t) = \sum_{j=0}^{n+m-1} q_{j,l}(t) x^j.$$

Then use the coefficients $p_{i,e}$ and $q_{j,l}$ to form the determinant like before.

The question whether the eliminant still computes an annihilating polynomial is answered in the following theorem, found in [15].

Theorem 7 *If P and Q are commuting elements of $R[x; \sigma, \delta]$ then*

$$f(s, t) = \Delta_{P,Q}(s, t)$$

is a polynomial in two commuting variables such that $f(P, Q) = 0$. If R is an integral domain and σ is an injective function then $\Delta_{P,Q}(s, t)$ is a non-zero polynomial.

We will illustrate the eliminant construction in a special class of Ore extensions. They will be of the form $K[y][x; \sigma, \delta]$ where K is a field, σ and δ are K-linear and $\deg_y(\sigma(y)) > 1$. This is closely similar to the construction of the q-Weyl algebra. Instead of the relation $xy = qyx + 1$, we now have a relation $xy = f(y)x + 1$, where $f(y)$ is some polynomial of degree larger than 1.

Example 8 Consider the case when $\sigma(y) = y^2$ and $\delta(y) = 1$. Then $P = yx$ and $Q = y^3x^2$ commute. The eliminant becomes

$$\Delta_{P,Q} = \begin{vmatrix} 0 & y & -s \\ y^2 & (1-s) & 0 \\ y^3 & 0 & -t \end{vmatrix} = (t + s(1-s))y^3.$$

It is true that $Q + P - P^2 = 0$ so the result is consistent with Theorem 7.

Example 9 Take $R = K[y]$ as before, and set $\sigma(y) = y^2 + 1$ and $\delta(y) = 0$. Then $P = y^2x - 1$ and $Q = (y^2x)^2$ commute. We find that $Q = y^2(y^2 + 1)^2x^2$ and that the eliminant is

$$\Delta_{P,Q} = \begin{vmatrix} 0 & y^2 & 1-s \\ (y^2+1)^2 & (1-s) & 0 \\ y^2(y^2+1)^2 & 0 & -t \end{vmatrix} = (t + (s-1)^2)y^2(y^2+1)^2.$$

We note that in the preceding examples we actually found an annihilating polynomial over K, not just $K[y]$. In [14] one can find it proven that such a polynomial always exist for commuting elements P and Q.

Theorem 10 *Let K be a field. Let σ be an endomorphism of $K[y]$ such that $\sigma(y) = p(y)$, where $\deg(p) > 1$, and let δ be a σ-derivation. Suppose that $\sigma(\alpha) = \alpha$ and $\delta(\alpha) = 0$ for all $\alpha \in k$. Let P, Q be two commuting elements of $K[y][x; \sigma, \delta]$. Then there is a nonzero polynomial $f(s, t) \in K[s, t]$ such that $f(P, Q) = 0$.*

One might hope that the eliminant construction might allow us to always compute an annihilating polynomial in the same way as in Theorem 3. We conjecture that this is true but have not been able to prove it.

We finish with an example where we do not have a theorem like Theorem 3.

Example 11 Consider the q-Weyl algebra with $q = -1$. Then $P = y^2x^2$ and $Q = x^4$ commute. The eliminant becomes

$$\Delta_{P,Q}(s, t) = -(s^2 - ty^4)^2.$$

This is still an annihilating polynomial over $K[y]$ but it does not give us an annihilating polynomial over K, which is expected since no such polynomial exists.

References

1. Amitsur, S.A.: Commutative linear differential operators. Pac. J. Math. **8**, 1–10 (1958)
2. Burchnall, J.L., Chaundy, T.W.: Commutative ordinary differential operators. Proc. Lond. Math. Soc. (Ser. 2) **21**, 420–440 (1922)

3. Burchnall, J.L., Chaundy, T.W.: Commutative ordinary differential operators. Proc. Roy. Soc. Lond. (Ser. A) **118**, 557–583 (1928)
4. Burchnall, J.L., Chaundy, T.W.: Commutative ordinary differential operators. II. The Identity $P^n = Q^m$. Proc. Roy. Soc. Lond. (Ser. A). **134**, 471–485 (1932)
5. Carlson, R.C., Goodearl, K.R.: Commutants of ordinary differential operators. J. Differ. Equs. **35**, 339–365 (1980)
6. De Jeu, M., Svensson, C., Silvestrov, S.: Algebraic curves for commuting elements in the q-deformed Heisenberg algebra. J. Algebr. **321**(4), 1239–1255 (2009)
7. Flanders, H.: Commutative linear differential operators. Technical Report 1, University of California, Berkely (1955)
8. Goodearl, K.R.: Centralizers in differential, pseudodifferential, and fractional differential operator rings. Rocky Mt J. Math. **13**, 573–618 (1983)
9. Krichever, I.M.: Integration of non-linear equations by the methods of algebraic geometry. Funktz. Anal. Priloz. **11**(1), 15–31 (1977)
10. Krichever, I.M.: Methods of algebraic geometry in the theory of nonlinear equations. Uspekhi Mat. Nauk. **32**(6), 183–208 (1977)
11. Larsson, D.: Burchnall–Chaundy theory, Ore extensions and σ-differential operators. Preprint U.U.D.M. Report vol. 45. Department of Mathematics, Uppsala University 13 pp (2008)
12. Mumford, D.: An algebro-geometric construction of commuting operators and of solutions to the Toda lattice equation, Korteweg–de Vries equation and related non-linear equations. In: Proceedings of International Symposium on Algebraic Geometry, Kyoto, Japan, pp. 115–153 (1978)
13. Ore, O.: Theory of non-commutative polynomials. Ann. Math. **34**(2, 3), 480–508 (1933)
14. Richter, J.: Burchnall–Chaundy theory for Ore extensions. In: Makhlouf, A., Paal, E., Silvestrov, S.D., Stolin, A. (eds.) Algebra, Geometry and Mathematical Physics, vol. 85, pp. 61–70. Springer Proceedings in Mathematics & Statistics, Mulhouse, France (2014)
15. Richter, J., Silvestrov, S.: Burchnall-Chaundy annihilating polynomials for commuting elements in Ore extension rings. J. Phys.: Conf. Ser. **346** (2012). doi:10.1088/1742-6596/346/1/012021

Centralizers and Pseudo-Degree Functions

Johan Richter

Abstract This paper generalizes a proof of certain results by Hellström and Silvestrov (J Algebr 314:17–41, 2007, [8]) on centralizers in graded algebras. We study centralizers in certain algebras with valuations. We prove that the centralizer of an element in these algebras is a free module over a certain ring. Under further assumptions we obtain that the centralizer is also commutative.

Keywords Ore extensions · Algebraic dependence · Commutative subrings

1 Introduction

The British mathematicians Burchnall and Chaundy studied, in a series of papers in the 1920s and 30s [3–5], the properties of commuting pairs of ordinary differential operators. The following theorem is essentially found in their papers.

Theorem 1 *Let* $P = \sum_{i=0}^{n} p_i D^i$ *and* $Q = \sum_{j=0}^{m} q_j D^j$ *be two commuting elements of* T *with constant leading coefficients. Then there is a non-zero polynomial* $f(s, t)$ *in two commuting variables over* $\mathbb{C}$ *such that* $f(P, Q) = 0$. *Note that the fact that* P *and* Q *commute guarantees that* $f(P, Q)$ *is well-defined.*

The result of Burchnall and Chaundy was rediscovered independently during the 70 s by researchers in the area of PDEs. It turns out that several important equations can be equivalently formulated as a condition that a pair of differential operators commute. These differential equations are completely integrable as a result, which roughly means that they possess an infinite number of conservation laws. In fact Theorem 1 was rediscovered by Kricherver [9] as part of his research into integrable systems.

To state some generalizations of Burchnall's and Chaundy's result we shall recall a definition.

J. Richter (✉)
Division of Applied Mathematics, School of Education,
Culture and Communication, Mälardalen University,
Box 883, 721 23 Västerås, Sweden
e-mail: johan.richter@mdh.se

© Springer International Publishing Switzerland 2016
S. Silvestrov and M. Rančić (eds.), *Engineering Mathematics II*,
Springer Proceedings in Mathematics & Statistics 179,
DOI 10.1007/978-3-319-42105-6_5

Definition 1 Let R be a ring, σ an endomorphism of R and δ an additive function, $R \to R$, satisfying

$$\delta(ab) = \sigma(a)\delta(b) + \delta(a)b$$

for all $a, b \in R$. (Such δ:s are known as σ-derivations.) The *Ore extension* $R[x; \sigma, \delta]$ is the polynomial ring $R[x]$ equipped with a new multiplication such that $xr = \sigma(r)x + \delta(r)$ for all $r \in R$. Every element of $R[x; \sigma, \delta]$ can be written uniquely as $\sum_i a_i x^i$ for some $a_i \in R$.

If $\sigma = \mathrm{id}$ then $R[x; \mathrm{id}_R, \delta]$ is called a *differential operator ring*. If $P = \sum_{i=0}^{n} a_i x^i$, with $a_n \neq 0$, we say that P has *degree* n. The degree of the the zero element is defined to be $-\infty$.

The ring of differential operators studied by Burchnall and Chaundy can be taken to be the Ore extension $T = C^\infty(\mathbb{R}, \mathbb{C})[D; \mathrm{id}, \delta]$, where δ is the ordinary derivation.

In a paper by Amitsur [1] one can find the following theorem.

Theorem 2 *Let K be a field of characteristic zero with a derivation δ. Let F denote the subfield of constants. (By a* constant *we mean an element that is mapped to zero by the derivation.) Form the differential operator ring $S = K[x; \mathrm{id}, \delta]$, and let P be an element of S of degree $n > 0$. Set $F[P] = \{\sum_{j=0}^{m} b_j P^j \mid b_j \in F\}$, the ring of polynomials in P with constant coefficients. Then the centralizer of P is a commutative subring of S and a free $F[P]$-module of rank at most n.*

Later authors have found other contexts where Amitsur's method of proof can be made to work. We mention an article by Goodearl and Carlson [6], and one by Goodearl alone [7], that generalize Amitsur's result to a wider class of rings. The proof has also been generalized by Bavula [2], Mazorchouk [10] and Tang [11], among other authors. As a corollary of these results, one can recover Theorem 1.

This paper is most directly inspired by a paper by Hellström and Silvestrov [8], however. Hellström and Silvestrov study graded algebras satisfying a condition they call l-BDHC (short for "Bounded-Dimension Homogeneous Centralizers").

Definition 2 Let K be a field, ℓ a positive integer and S a $\mathbb{Z}$-graded K-algebra. The homogeneous components of the gradation are denoted S_m, for $m \in \mathbb{Z}$. Let $\mathrm{Cen}(n, a)$, for $n \in \mathbb{Z}$ and $a \in S$, denote the elements in S_n that commute with a. We say that S has ℓ-BDHC if for all $n \in \mathbb{Z}$, nonzero $m \in \mathbb{Z}$ and nonzero $a \in S_m$, it holds that $\dim_K \mathrm{Cen}(n, a) \leq \ell$.

Hellström and Silvestrov apply the ideas of Amitsur's proof. They need to modify them however, especially to handle the case when $\ell > 1$.

To explain their results further, we introduce some more of their notation. Denote by π_n the projection, defined in the obvious way, from S to S_n. Hellström and Silvestrov define a function $\bar{\chi} : A \setminus \{0\} \to \mathbb{Z}$ by

$$\bar{\chi}(a) = \max\{n \in \mathbb{Z} \mid \pi_n(a) \neq 0\},$$

and set $\bar{\chi}(0) = -\infty$. Set further $\bar{\pi}(a) = \pi_{\bar{\chi}(a)}(a)$.

Now we have introduced enough notation to state the relevant results. The following result is the main part of Lemma 2.5 in their paper.

Theorem 3 *Assume S is a K-algebra with l-BDHC and that there are no zero divisors in S. If $a \in S \setminus S_0$ is such that $\bar{\chi}(a) = m > 0$ and $\bar{\pi}(a)$ is not invertible in S, then there exists a finite $K[a]$-module basis $\{b_1, \ldots, b_k\}$ for the centralizer of a. Furthermore $k \leq ml$.*

The reason they refer to it as a lemma is that their main interest is in the following corollary of this result, (which is proved the same way as Corollary 1 in this paper).

Theorem 4 *Let K be a field and assume the K-algebra S has l-BDHC and that there are no zero divisors in S. If $a \in S \setminus S_0$ and $b \in S$ are such that $ab = ba$, $\bar{\chi}(a) > 0$ and $\bar{\pi}(a)$ is not invertible in S, then there exists a nonzero polynomial P in two commuting variables with coefficients from K such that $P(a, b) = 0$.*

Theorem 4 is directly analogous to Theorem 1.

Hellström and Silvestrov also have a result asserting that certain centralizers are commutative. Their proof can be made to work in the case when A has 1-BDHC.

Theorem 5 *Assume the K-algebra S has 1-BDHC and that there are no zero divisors in S. If $a \in S \setminus S_0$ satisfies $\bar{\chi}(a) = m > 0$ and $\bar{\pi}(a)$ is not invertible in S, then there exists a finite $K[a]$-module basis $\{b_1, \ldots, b_k\}$ for the centralizer of a. The cardinality, k, of the basis divides m. Furthermore the centralizer of a is commutative.*

It shall be the goal of this paper to generalize the results we have cited from [8].

1.1 Notation and Conventions

$\mathbb{Z}$ will denote the integers.

If R is a ring then $R[x_1, x_2, \ldots x_n]$ denotes the ring of polynomials over R in central indeterminates $x_1, x_2, \ldots, x_n$.

All rings and algebras are assumed to be associative and unital.

Let R be a commutative ring and S an R-algebra. Two commuting elements, $p, q \in S$, are said to be *algebraically dependent* (over R) if there is a non-zero polynomial, $f(s, t) \in R[s, t]$, such that $f(p, q) = 0$, in which case f is called an annihilating polynomial.

If S is a ring and a is an element in S, the *centralizer* of a, denoted $C_S(a)$, is the set of all elements in S that commute with a.

By K we will always denote a field.

2 Centralizers in Algebras with Degree Functions

Upon reading the proofs in [8] closely it turns out that they are based upon certain properties of the function $\bar{\chi}$ they define. We shall axiomatize the properties that are needed to make their proof work.

Definition 3 Let K be a field and let S be a K-algebra. A function, χ, from S to $\mathbb{Z} \cup \{-\infty\}$ is called a pseudo-degree function if it satisfies the following conditions:

- $\chi(a) = -\infty$ iff $a = 0$,
- $\chi(ab) = \chi(a) + \chi(b)$ for all $a, b \in S$,
- $\chi(a + b) \leq \max(\chi(a), \chi(b))$,
- $\chi(a + b) = \chi(a)$ if $\chi(b) < \chi(a)$.

This is essentially a special case of the concept of a *valuation*.

We also need a condition that can replace l-BDHC. We formulate it next.

Definition 4 Let K be a field and S a K-algebra with a pseudo-degree function, χ, and let ℓ be a positive integer. A subalgebra, $B \subset A$, is said to satisfy condition $D(\ell)$ if $\chi(b) \geq 0$ for all non-zero $b \in B$ and if, whenever we have $\ell + 1$ elements $b_1, \ldots, b_{l+1} \in B$, all mapped to the same integer by χ, there exist $\alpha_1, \ldots, \alpha_{\ell+1} \in K$, not all zero, such that $\chi\left(\sum_{i=1}^{l+1} \alpha_i b_i\right) < \chi(b_1)$.

Remark 1 Note that the requirement that $\alpha_1, \ldots, \alpha_{l+1}$ are mapped to the same integer by χ excludes the possibility that they are equal to 0.

Remark 2 Suppose that S is a K-algebra and $a \in S$ is such that $C_S(a)$ satisfies condition $D(\ell)$ for some ℓ. If b is an invertible element then $\chi(b^{-1}) = -\chi(b)$. So all invertible elements of $C_S(a)$ must be mapped to zero by χ. In particular the non-zero scalars are all mapped to zero by χ.

Lemma 1 *Suppose that S is an K-algebra and χ is a pseudo-degree function on S that maps all the non-zero scalars to zero. Then if $a, b \in S$ are such that $\chi(b) < \chi(a)$, the identity*

$$\chi(a + b) = \chi(a) \tag{1}$$

holds.

Proof On the one hand we find $\chi(a + b) \leq \max(\chi(a), \chi(b)) = \chi(a)$. On the other hand $\chi(a) = \chi(a + b - b) \leq \max(\chi(a + b), \chi(b))$ Since $\chi(b) < \chi(a)$ we must have $\chi(a) \leq \chi(a + b)$. $\qquad \square$

We now proceed to prove an analogue of Theorem 3, using just the existence of some pseudo-degree function and the condition $D(\ell)$.

Theorem 6 *Let K be a field and let S be a K-algebra. Suppose S has a pseudo-degree function, χ.*

Let a be an element of S, with $m = \chi(a) > 0$, such that $C_S(a)$ satisfies condition $D(\ell)$ for some positive integer ℓ. Then $C_S(a)$ is a free $K[a]$-module of rank at most ℓm.

Proof Construct a sequence $b_1, b, \ldots$ by setting $b_1 = 1$ and choosing $b_{k+1} \in C_S(a)$ such that $\chi(b_{k+1})$ is minimal subject to the restriction that b_{k+1} does not lie in the $K[a]$-linear span of $\{b_1, \ldots, b_k\}$. We will show later in the proof that such a sequence has at most ℓm elements.

We first claim that

$$\chi\left(\sum_{i=1}^k \phi_i b_i\right) = \max_{i \leq k}(\chi(\phi_i) + \chi(b_i)), \qquad (2)$$

for any $\phi_1, \ldots \phi_k \in K[a]$. We show this by induction on $n = \max_{i \leq k}(\chi(\phi_i) + \chi(b_i))$. It is clear that the left-hand side of (2) is never greater than the right-hand side. When $n = -\infty$ Eq. (2) holds since in that case all $\phi_i = 0$. If $n = 0$, Equation (2) holds since $\chi(b) \geq 0$ for all non-zero $b \in C_S(a)$. That $\chi(b) \geq 0$ for all non-zero b in $C_S(a)$ also means that no value of n between $-\infty$ and 0 is possible.

For the induction step, assume (2) holds when the right-hand side is strictly less than n. To verify that it holds for n as well, we can assume without loss of generality that $\chi(\phi_k) + \chi(b_k) = n$, since if $\chi(\phi_j b_j) < n$ for some term $\phi_j b_j$ we can drop it without affecting either side of (2), by Lemma 1. If $\phi_k \in K$ then $\chi(\phi_k) = 0$, by Remark 2, and thus $\chi(b_k) = n$. By the choice of b_k it then follows that $\chi(\sum_{i=1}^k \phi_i b_i) \geq n$, as otherwise $\sum_{i=1}^k \phi_i b_i$ would have been picked instead of b_k. If $\phi_k \notin K$, then $\chi(b_k) < n$ and thus $\chi(b_i) < n$ for $i = 1, \ldots k$. Let $r_1, \ldots, r_k \in K$ and $\xi_1, \ldots, \xi_k \in K[a]$ be such that $\phi_i = a\xi_i + r_i$ for $i = 1, \ldots, k$. We have $\chi(\sum_{i=1}^k r_i b_i) < n$ and thus by Lemma 1 and the assumptions on χ we get

$$\chi\left(\sum_{i=1}^k \phi_i b_i\right) = \chi\left(\sum_{i=1}^k a\xi_i b_i + \sum_{i=1}^k r_i b_i\right) = \chi\left(a\sum_{i=1}^k \xi_i b_i\right) = m + \chi\left(\sum_{i=1}^k \xi_i b_i\right).$$

We also have that $\max_{i \leq k}(\chi(\phi_i) + \chi(b_i)) = m + \max_{i \leq k}(\chi(\xi_i) + \chi(b_i))$. By the induction hypothesis

$$\chi\left(\sum_{i=1}^k \xi_i b_i\right) = \max_{i \leq k}(\chi(\xi_i) + \chi(b_i)),$$

which completes the induction step.

We now show that if $\chi(b_i) = \chi(b_j)$ for some $i \leq j$ then $j - i < l$. Suppose $b_1, \ldots, b_{l+1}$ all are mapped to zero by χ. Then there exists $\alpha_1, \ldots, \alpha_{l+1}$, not all zero, such that

$$\chi \left(\sum_{i=1}^{l+1} \alpha_i b_i \right) < 0,$$

which is impossible since $\sum_{i=1}^{l+1} \alpha_i b_i \in C_S(a)$.

Suppose now instead that $b_j, \ldots, b_{j+l}$ are all mapped to the same positive integer, q, by χ. Then there exists $\alpha_j, \ldots, \alpha_{j+l} \in K$, not all zero, such that

$$\chi \left(\sum_{i=j}^{j+l} \alpha_i b_i \right) < q.$$

But this contradicts (2).

It remains only to show that the sequence (b_i) contains only lm elements. We will prove that every residue class $(\bmod\, m)$ can only contain at most l elements. Suppose to the contrary, that we had elements $c_1, \ldots, c_{l+1}$, belonging to the sequence (b_i) and all satisfying that $\chi(c_i) = n \pmod{m}$. Set $k = \max_{1 \le i \le l+1}(\chi(c_i))$ and define $\gamma_i = a^{\frac{k-\chi(c_i)}{m}}$. Then $\chi(\gamma_i c_i) = k$, for all $i \in \{1, \ldots, l+1\}$, which implies that there exists $\alpha_1, \ldots, \alpha_{l+1} \in K$, such that

$$\chi \left(\sum_{i=j}^{j+l} \alpha_i \gamma_i c_i \right) < k.$$

But this once again contradicts (2).   □

We can also prove a result on the algebraic dependence of pairs of commuting elements.

Corollary 1 *Let S be a K-algebra with a pseudo-degree function, χ. Let $a \in S$ be such that $C_S(a)$ satisfies Condition $D(l)$ for some $l > 0$. Let b be any element in $C_S(a)$. Then there exists a nonzero polynomial $P(s, t) \in K[s, t]$ such that $K(a, b) = 0$. (Note that $K(a, b)$ is well-defined when a, b commute.)*

Proof Since $C_S(a)$ has finite rank as a $K[a]$-module the elements $b, b^2, \ldots$ can not all be linearly independent over $K[a]$. Thus there exists $f_1(x), \ldots, f_k(x) \in K[x]$, not all zero, such that $\sum_{i=0}^{k} f_i(a)b^i = 0$. Then $P(s, t) = \sum_{i=0}^{k} f_i(s)t^i = 0$ is a polynomial with the desired property.   □

We can also prove a result asserting that certain centralizers are commutative, though for that we need to assume that $C_S(a)$ satisfies condition $D(1)$.

Theorem 7 *Let K be a field and suppose S is a K-algebra. Let S have a pseudo-degree function, χ. If $a \in S$ satisfies $\chi(a) = m > 0$ and $C_S(a)$ satisfies condition $D(1)$ then:*

1. $C_S(a)$ has a finite basis as a $K[a]$-module, the cardinality of which divides m.
2. $C_S(a)$ is a commutative algebra.

Proof By Theorem 6 it is clear that there is a subset H of $\{1, \ldots, m\}$ and elements $(b_i)_{i \in H}$ such that the b_i form a basis for $C_S(a)$. By the proof of Theorem 6 it is also clear that $\chi(b_i) \neq \chi(b_j)$ if $i \neq j$. Without loss of generality we can assume $\chi(b_i) = i$ for all $i \in H$. We can map H into $\mathbb{Z}_m$ in a natural way. Denote the image by G. We want to show G is a subgroup, for which it is enough to show that it is closed under addition.

Suppose $g, h \in G$. There exists $i, j \in H$, with $i \equiv g \pmod{m}$ and $j \equiv h \pmod{m}$. We can write $b_i b_j = \sum_{k \in H} \phi_k b_k$, for some $\{b_k\}$. It follows that

$$g + h \equiv i + j = \chi(b_i b_j) = \max(\chi(\phi_k) + \chi(b_k)) \equiv \chi(b_k) = k \pmod{m}$$

for some $k \in H$.

Since G is a subgroup of $\mathbb{Z}_m$ it is clear that the cardinality of G, which is also the cardinality of H, must divide m.

G is cyclic. Let g be a generator of G. Consider the algebra generated by b_i and a, where $i \equiv g \pmod{m}$. It is a commutative algebra and a sub-K-vector space of $C_S(a)$. Denote it by E. If c is any element of $C_S(a)$ we can write $c = e + f$, where $e \in E$ and $\chi(f) < mi$, since if $\chi(c) \geq mi$ then there exists $k \leq m$ and $j \in \mathbb{N}$ such that $\chi(a^j b_i^k) = \chi(c)$ and thus there exists $\alpha \in K$ such that $\chi(c - \alpha a^j b_i^k) < \chi(c)$.

Thus the quotient $C_S(a)/E$ is finite-dimensional. Each $f \in K[a]$ gives rise to an endomorphism on $C_S(a)/E$, by the action of multiplication by f. Since $K[a]$ is infinite-dimensional and the endomorphism ring of $C_S(a)/E$ is finite-dimensional, there is some nonzero $\phi \in K[a]$ that induces the zero endomorphism. But this means that $\phi c \in E$ for any $c \in C_S(a)$.

Now let c_1, c_2 be two arbitrary elements of $C_S(a)$. Since E is commutative, and everything in $C_S(a)$ commutes with ϕ, it follows that

$$\phi^2 c_1 c_2 = \phi c_1 \cdot \phi c_2 = \phi c_2 \cdot \phi c_1 = \phi^2 c_2 c_1.$$

Since $C_S(a)$ is a domain it follows that $c_1 c_2 = c_2 c_1$ and thus that $C_S(a)$ is commutative. $\qquad\square$

3 Examples

Theorems 3, 4 and 5 follow from our results combined with Lemmas 2.2 and 2.5 in [8]. But our results can also be applied in certain situations that are not covered by the results in [8].

Proposition 1 *Let K be a field. Set $R = K[y]$, let σ be an endomorphism of R such that $s = \deg_y(\sigma(y)) > 1$ and let δ be a σ-derivation. Form the Ore extension*

$S = R[x; \sigma, \delta]$. If $a \in S \setminus K$ then $C_S(a)$ is a free $K[a]$-module of finite rank and a commutative subalgebra of S.

Proof If $a \in K[y] \setminus K$ then $C_S(a) = K[y]$ and the claim is true. So suppose that $a \notin K[y]$. We shall apply Theorem 7. To do so we need a pseudo-degree function.

The notion of the degree of an element in S with respect to x was defined in the introduction of this article. Denote the degree of an element b by $\chi(b)$. It is easy to see that χ satisfies all the requirement to be a pseudo-degree function. We proceed to show that $C_S(a)$ satisfies condition $D(1)$. Certainly it is true that $\chi(b) \geq 0$ for all nonzero $b \in C_S(a)$.

Let b be a nonzero element of S that commutes with a, such that $\chi(b) = n$. Suppose $\chi(a) = m$. By equating the highest order coefficient of ab and ba we find that

$$a_m \sigma^m(b_n) = b_n \sigma^n(a_m), \qquad (3)$$

where a_m and b_n denote the highest order coefficients of a and b, respectively. (Recall that these are polynomials in y.) We equate the degree in y of both sides of (3) and find that

$$\deg_y(a_m) + s^m \deg_y(b_n) = \deg_y(b_n) + s^n \deg_y(a_m),$$

which determines the degree of b_m uniquely. It follows that the solutions of (3) form a K-sub space of $K[y]$ that is at most one-dimensional. This in turn implies that condition $D(1)$ is fulfilled.

We have now verified all the hypothesis necessary to apply Theorem 7. □

Acknowledgement This work was performed in part while the author was employed at Lund University and in part while the author was employed at Mälardalen University.

The author wishes to thank Lars Hellström and Johan Öinert for helpful discussions.

References

1. Amitsur, S.A.: Commutative linear differential operators. Pac. J. Math. **8**, 1–10 (1958)
2. Bavula, V.V.: Dixmier's problem 6 for the Weyl algebra (the generic type problem). Commun. Algebr. **34**(4), 1381–1406 (2006)
3. Burchnall, J.L., Chaundy, T.W.: Commutative ordinary differential operators. Proc. Lond. Math. Soc. (Ser. 2) **21**, 420–440 (1922)
4. Burchnall, J.L., Chaundy, T.W.: Commutative ordinary differential operators. Proc. Roy. Soc. Lond. (Ser. A) **118**, 557–583 (1928)
5. Burchnall, J.L., Chaundy, T.W.: Commutative ordinary differential operators. II. The Identity $P^n = Q^m$. Proc. Roy. Soc. Lond. (Ser. A). **134**, 471–485 (1932)
6. Carlson, R.C., Goodearl, K.R.: Commutants of ordinary differential operators. J. Differ. Equs. **35**, 339–365 (1980)
7. Goodearl, K.R.: Centralizers in differential, pseudodifferential, and fractional differential operator rings. Rocky Mt. J. Math. **13**, 573–618 (1983)
8. Hellström, L., Silvestrov, S.: Ergodipotent maps and commutativity of elements in non-commutative rings and algebras with twisted intertwining. J. Algebr. **314**, 17–41 (2007)

9. Krichever, I.M.: Methods of algebraic geometry in the theory of nonlinear equations. Uspekhi Mat. Nauk. **32**(6), 183–208 (1977)
10. Mazorchuk, V.: A note on centralizers in q-deformed Heisenberg algebras. AMA Algebra Montp. Announc. (electronic) 2, 6 (2001)
11. Tang, X.: Maximal commutative subalgebras of certain skew polynomial rings. http://faculty.uncfsu.edu/xtang/maxsubalgebras.pdf (2005)

Crossed Product Algebras for Piece-Wise Constant Functions

Johan Richter, Sergei Silvestrov, Vincent Ssembatya
and Alex Behakanira Tumwesigye

Abstract In this paper we consider algebras of functions that are constant on the sets of a partition. We describe the crossed product algebras of the mentioned algebras with $\mathbb{Z}$. We show that the function algebra is isomorphic to the algebra of all functions on some set. We also describe the commutant of the function algebra and finish by giving an example of piece-wise constant functions on a real line.

Keywords Piecewise constant · Crossed products · Maximal commutative subalgebra

1 Introduction

An important direction of investigation for any class of non-commutative algebras and rings, is the description of commutative subalgebras and commutative subrings. This is because such a description allows one to relate representation theory, non-commutative properties, graded structures, ideals and subalgebras, homological and other properties of non-commutative algebras to spectral theory, duality, algebraic geometry and topology naturally associated with commutative algebras. In representation theory, for example, semi-direct products or crossed products play a central

J. Richter · S. Silvestrov
Division of Applied Mathematics, School of Education,
Culture and Communication, Mälardalen University, Box 883, 721 23 Västerås, Sweden
e-mail: johan.richter@mdh.se

S. Silvestrov
e-mail: sergei.silvestrov@mdh.se

V. Ssembatya · A.B. Tumwesigye (✉)
Department of Mathematics, College of Natural Sciences, Makerere University,
Box 7062, Kampala, Uganda
e-mail: alexbt@cns.mak.ac.ug

V. Ssembatya
e-mail: vas@cns.mak.ac.ug

© Springer International Publishing Switzerland 2016
S. Silvestrov and M. Rančić (eds.), *Engineering Mathematics II*,
Springer Proceedings in Mathematics & Statistics 179,
DOI 10.1007/978-3-319-42105-6_6

role in the construction and classification of representations using the method of induced representations. When a non-commutative algebra is given, one looks for a subalgebra such that its representations can be studied and classified more easily and such that the whole algebra can be decomposed as a crossed product of this subalgebra by a suitable action.

When one has found a way to present a non-commutative algebra as a crossed product of a commutative subalgebra by some action on it, then it is important to know whether the subalgebra is maximal commutative, or if not, to find a maximal commutative subalgebra containing the given subalgebra. This maximality of a commutative subalgebra and related properties of the action are intimately related to the description and classification of representations of the non-commutative algebra.

Some work has been done in this direction [2, 4, 6] where the interplay between topological dynamics of the action on one had and the algebraic property of the commutative subalgebra in the C^*−crossed product algebra $C(X) \rtimes \mathbb{Z}$ being maximal commutative on the other hand are considered. In [4], an explicit description of the (unique) maximal commutative subalgebra containing a subalgebra $\mathcal{A}$ of $\mathbb{C}^X$ is given. In [3], properties of commutative subrings and ideals in non-commutative algebraic crossed products by arbitrary groups are investigated and a description of the commutant of the base coefficient subring in the crossed product ring is given. More results on commutants in crossed products and dynamical systems can be found in [1, 5] and the references therein.

In this article, we take a slightly different approach. We consider algebras of functions that are constant on the sets of a partition, describe the crossed product algebras of the mentioned algebras with $\mathbb{Z}$ and show that the function algebra is isomorphic to the algebra of all functions on some set. We also describe the commutant of the function algebra and finish by giving an example of piece-wise constant functions on a real line.

2 Definitions and a Preliminary Result

Let $\mathcal{A}$ be any commutative algebra. Using the notation in [4], we let $\psi : \mathcal{A} \to \mathcal{A}$ be any algebra automorphism on $\mathcal{A}$ and define

$$\mathcal{A} \rtimes_\psi \mathbb{Z} := \{f : \mathbb{Z} \to \mathcal{A} : f(n) = 0 \text{ except for a finite number of } n\}.$$

It can be proved that $\mathcal{A} \rtimes_\psi \mathbb{Z}$ is an associative $\mathbb{C}$−algebra with respect to point-wise addition, scalar multiplication and multiplication defined by *twisted convolution*, $*$ as follows;

$$(f * g)(n) = \sum_{k \in \mathbb{Z}} f(k).\psi^k(g(n-k)),$$

where ψ^k denotes the k−fold composition of ψ with itself for positive k and we use the obvious definition for $k \leq 0$.

Definition 1 $A \rtimes_\psi \mathbb{Z}$ as described above is called the crossed product algebra of A and $\mathbb{Z}$ under ψ.

A useful and convenient way of working with $A \rtimes_\psi \mathbb{Z}$, is to write elements $f, g \in A \rtimes_\psi \mathbb{Z}$ in the form $f = \sum_{n \in \mathbb{Z}} f_n \delta^n$ and $g = \sum_{n \in \mathbb{Z}} g_m \delta^m$ where $f_n = f(n)$, $g_m = g(m)$ and

$$\delta^n(k) = \begin{cases} 1, & \text{if } k = n, \\ 0, & \text{if } k \neq n. \end{cases}$$

Then addition and scalar multiplication are canonically defined and multiplication is determined by the relation

$$(f_n \delta^n) * (g_m \delta^m) = f_n \psi^n (g_m) \delta^{n+m}, \tag{1}$$

where $m, n \in \mathbb{Z}$ and $f_n, g_m \in A$.

Definition 2 By the commutant A' of A in $A \rtimes_\psi \mathbb{Z}$ we mean

$$A' := \{f \in A \rtimes_\psi \mathbb{Z} : fg = gf \text{ for every } g \in A\}.$$

It has been proven [4] that the commutant A' is commutative and thus, is the unique maximal commutative subalgebra containing A. For any $f, g \in A \rtimes_\psi \mathbb{Z}$, that is, $f = \sum_{n \in \mathbb{Z}} f_n \delta^n$ and $g = \sum_{m \in \mathbb{Z}} g_m \delta^m$, then $fg = gf$ if and only if

$$\forall r : \sum_{n \in \mathbb{Z}} f_n \phi^n (g_{r-m}) = \sum_{m \in \mathbb{Z}} g_m \phi^m (f_{r-m}).$$

Now let X be any set and A an algebra of complex valued functions on X. Let $\sigma : X \to X$ be any bijection such that A is invariant under σ and σ^{-1}, that is for every $h \in A$, $h \circ \sigma \in A$ and $h \circ \sigma^{-1} \in A$. Then (X, σ) is a discrete dynamical system and σ induces an automorphism $\tilde{\sigma} : A \to A$ defined by,

$$\tilde{\sigma}(f) = f \circ \sigma^{-1}.$$

Our goal is to describe the commutant of A in the crossed product algebra $A \rtimes_{\tilde{\sigma}} \mathbb{Z}$ for the case where A is the algebra of functions that are constant on the sets of a partition. First we have the following results.

Definition 3 For any nonzero $n \in \mathbb{Z}$, we set

$$Sep_A^n(X) = \{x \in X \mid \exists h \in A : h(x) \neq \tilde{\sigma}^n(h)(x)\}. \tag{2}$$

The following theorem has been proven in [4].

Theorem 1 *The unique maximal commutative subalgebra of $A \rtimes_{\tilde{\sigma}} \mathbb{Z}$ that contains A is precisely the set of elements*

$$A' = \left\{ \sum_{n \in \mathbb{Z}} f_n \delta^n \mid \textit{for all } n \in \mathbb{Z}: \ f_n|_{Sep^n_{\mathcal{A}}(X)} \equiv 0 \right\}. \tag{3}$$

We observe that since $\tilde{\sigma}(f) = f \circ \sigma^{-1}$, then

$$\tilde{\sigma}^2(f) = \tilde{\sigma}(f \circ \sigma^{-1}) = (f \circ \sigma^{-1}) \circ \sigma^{-1} = f \circ \sigma^{-2},$$

and hence for every $n \in \mathbb{Z}$, $\tilde{\sigma}^n(f) = f \circ \sigma^{-n}$. Therefore, by taking $X = \mathbb{R}$ and $\mathcal{A}$ as the algebra of constant functions on X we have: for every $x \in X$ and every $h \in \mathcal{A}$,

$$\tilde{\sigma}^n(h)(x) := h \circ \sigma^{-n}(x) = h(\sigma^{-n}(x)) = h(x),$$

since h is a constant function. It follows that in this case $Sep^n_{\mathcal{A}}(X) = \emptyset$. Therefore in this case, $\mathcal{A}' = \mathcal{A} \rtimes_{\tilde{\sigma}} \mathbb{Z}$.

3 Algebra of Piece-Wise Constant Functions

Let X be any set, J a countable set and $\mathbb{P} = \{X_j : j \in J\}$ be a partition of X; that is $X = \cup_{r \in J} X_r$ where $X_r \neq \emptyset$ and $X_r \cap X_{r'} = \emptyset$ if $r \neq r'$.

Let $\mathcal{A}$ be the algebra of piece-wise constant complex-valued functions on X. That is

$$\mathcal{A} = \{h \in \mathbb{C}^X : \text{ for every } j \in J : \ h(X_j) = \{c_j\}\}.$$

Let $\sigma : X \to X$ be a bijection on X. The lemma below gives the necessary and sufficient conditions for (X, σ) to be a dynamical system.

Lemma 1 *The following are equivalent.*

1. *The algebra $\mathcal{A}$ is invariant under σ and σ^{-1}.*
2. *For every $i \in J$ there exists $j \in J$ such that $\sigma(X_i) = X_j$.*

Proof We recall that the algebra $\mathcal{A}$ is invariant under σ if and only if for every $h \in \mathcal{A}$, $h \circ \sigma \in \mathcal{A}$.

Obviously, if for every $i \in J$ there exists a unique $j \in J$ such that $\sigma(X_i) = X_j$, then

$$(h \circ \sigma)(X_i) = h(\sigma(X_i)) = h(X_j) = \{c_j\}.$$

Thus $h \circ \sigma \in \mathcal{A}$.

Conversely, suppose $\mathcal{A}$ is invariant under σ but 2. does not hold. Let $x_1, x_2 \in X_j$ and $X_r, X_{r'} \in \mathbb{P}$ such that $\sigma(x_1) \in X_r$ and $\sigma(x_2) \in X_{r'}$. Let $h : X \to \mathbb{C}$ be the function defined by

$$h(x) = \begin{cases} 1 & \text{if } x \in X_r, \\ 0 & \text{otherwise.} \end{cases}$$

Then $h \in \mathcal{A}$. But $h \circ \sigma(x_1) = 1$ and $h \circ \sigma(x_2) = 0$. Thus $h \notin \mathcal{A}$, which contradicts the assumption. $\qquad\qquad\qquad\qquad\qquad\qquad\qquad\qquad\qquad\qquad\qquad\qquad\qquad\square$

The following lemma asserts that any bijection $\sigma_2 : X \to X$ that preserves the structure of a partition essentially produces the same algebra of functions.

Lemma 2 *Let* $\mathbb{P}_1 = \{X_j : j \in J\}$ *and* $\mathbb{P}_2 = \{Y_j : j \in J\}$ *be partitions of the sets* X *and* Y *respectively, and let*

$$\mathcal{A}_X = \{h \in \mathbb{C}^X : \text{for every } j \in J : h(X_j) = \{c_j\}\},$$

and

$$\mathcal{A}_Y = \{h \in \mathbb{C}^Y : \text{for every } j \in J : h(Y_j) = \{d_j\}\}.$$

Then $\mathcal{A}_X$ *is isomorphic to* $\mathcal{A}_Y$.

Proof Choose points $x_i \in X$ and $y_i \in Y$ such that $x_i \in X_i$ if and only if $y_i \in Y_i \; \forall \, i \in J$ and let $\mu : \mathcal{A}_X \to \mathcal{A}_Y$ be a function defined by

$$\mu(f)(y) = f(x_i) \text{ if } y \in Y_j, \; \forall \, j \in J. \tag{4}$$

It is enough to prove that μ is an algebra isomorphism.

- Let $f, g \in \mathcal{A}_X$ and let $\alpha, \beta \in \mathbb{C}$. Then if $y \in Y$, then $y \in Y_i$ for some $i \in J$, therefore,

$$\begin{aligned}
\mu(\alpha f + \beta g)(y) &= (\alpha f + \beta g)(x_i) \\
&= \alpha f(x_i) + \beta g(x_i) \\
&= \alpha \mu(f)(y) + \beta \mu(g)(y) \\
&= [\alpha \mu(f) + \beta \mu(g)](y).
\end{aligned}$$

Therefore μ is linear since y was arbitrary.
- For every $f, g \in \mathcal{A}_X$ and $y \in Y$ $(y \in Y_i)$,

$$\begin{aligned}
\mu(fg)(y) &= (fg)(x_i) \\
&= f(x_i)g(x_i) \\
&= \mu(f)(y)\mu(g)(y) \\
&= [\mu(f)\mu(g)](y).
\end{aligned}$$

Thus μ is a multiplicative homomorphism.
- Now, suppose $f, g \in \mathcal{A}_X$ such that $f \neq g$. Then there exists $i \in J$ such that $f(x_i) \neq g(x_i)$, $x_i \in X_i$. Therefore, if $y \in Y_i$,

$$\mu(f)(y) = f(x_i) \neq g(x_i) = \mu(g)y.$$

Therefore μ is injective.

- Finally, suppose $h \in \mathcal{A}_Y$ and let $f \in \mathcal{A}_X$ be defined by $f(x) = h(y_i)$. If $y \in Y$, then $y \in Y_i$ for some $i \in J$, and hence,

$$h(y) = h(y_i) = f(x) = f(x_i) = \mu(f)(y).$$

It follows that μ is onto and hence an algebra isomorphism.

$\square$

Theorem 2 *Let* $\mathbb{P}_1 = \{X_j : j \in J\}$ *and* $\mathbb{P}_2 = \{Y_j : j \in J\}$ *be partitions of two sets* X *and* Y *and* $\mathcal{A}_X$ *and* $\mathcal{A}_Y$ *be algebras of functions that are constant on the sets of the partitions* $\mathbb{P}_1$ *and* $\mathbb{P}_2$ *respectively. Let* $\sigma_1 : X \to X$ *and* $\sigma_2 : Y \to Y$ *be bijections such that* $\mathcal{A}_X$ *is invariant under* σ_1 *(and* σ_1^{-1}*) and* $\mathcal{A}_Y$ *is invariant under* σ_2 *(and* σ_2^{-1}*) and that* $\sigma_1(X_i) = X_j$ *whenever* $\sigma_2(Y_i) = Y_j$ *for all* $i, j \in J$. *Suppose* $\tilde{\sigma}_1 : \mathcal{A}_X \to \mathcal{A}_X$ *is the automorphism on* $\mathcal{A}_X$ *induced by* σ_1, *and* $\tilde{\sigma}_2 : \mathcal{A}_Y \to \mathcal{A}_Y$ *is the automorphism on* $\mathcal{A}_Y$ *induced by* σ_2. *Then*

$$\tilde{\sigma}_2 \circ \mu = \mu \circ \tilde{\sigma}_1. \tag{5}$$

where μ *is given by (4). Moreover, for every* $n \in \mathbb{Z}$,

$$\tilde{\sigma}_2^n \circ \mu = \mu \circ \tilde{\sigma}_1^n. \tag{6}$$

Proof Let $y \subset X$ such that $y \in Y_i$ for some $i \in J$. Then for every $f \in \mathcal{A}$,

$$\begin{aligned}
(\tilde{\sigma}_2 \circ \mu)(f)(y) &= (\mu f) \circ \sigma_2^{-1}(y) \\
&= (\mu f)(\sigma_2^{-1}(y)) \\
&= f(\sigma_1^{-1}(x_i)) \\
&= (f \circ \sigma_1^{-1})(x_i) \\
&= \mu(f \circ \sigma_1^{-1})(y) \\
&= \mu[\tilde{\sigma}_1(f)](y) \\
&= [\mu \circ \tilde{\sigma}_1](f)(y).
\end{aligned}$$

Since y is arbitrary, we have

$$(\tilde{\sigma}_2 \circ)(f)\mu = \mu \circ \tilde{\sigma}_1(f)$$

for every $f \in \mathcal{A}$. And since f is arbitrary,

$$\tilde{\sigma}_2 \circ \mu = \mu \circ \tilde{\sigma}_1.$$

Now from (5), we have

$$\tilde{\sigma}_2^2 \circ \mu = \tilde{\sigma}_2 \circ (\tilde{\sigma}_2 \circ \mu) = \tilde{\sigma}_2 \circ (\mu \circ \tilde{\sigma}_1) = (\tilde{\sigma}_2 \circ \mu) \circ \tilde{\sigma}_1 = (\mu \circ \tilde{\sigma}_1) \circ \tilde{\sigma}_1 = \mu \circ \tilde{\sigma}_1^2.$$

Therefore the relation (6) holds for $n = 2$.

Now suppose the relation (6) holds for k. Then:

$$\tilde{\sigma}_2^{k+1} \circ \mu = \tilde{\sigma}_2 \circ (\tilde{\sigma}_2^k \circ \mu) = \tilde{\sigma}_2 \circ (\mu \circ \tilde{\sigma}_1^k) = (\tilde{\sigma}_2 \circ \mu) \circ \tilde{\sigma}_1^k = (\mu \circ \tilde{\sigma}_1) \circ \tilde{\sigma}_1^k = \mu \circ \tilde{\sigma}_1^{k+1}.$$

Therefore, from the induction principle,

$$\tilde{\sigma}_2^n \circ \mu = \mu \circ \tilde{\sigma}_1^n. \qquad \Box$$

Remark 1 From Theorem 2 above, we get two nice results. The first is that if $\mathbb{P}_1 = \mathbb{P}_2$ are partitions of X and $\sigma_1, \sigma_2 : X \to X$ are bijections on X which preserve the structure of the partition, they will give rise to the same automorphism. That is, suppose $\mathbb{P}_1 = \{X_j : j \in J\}$ is a partition of X and $\sigma_1, \sigma_2 : X \to X$ are bijections on X such that, if $\sigma_1(X_i) = X_j$, then $\sigma_2(X_i) = X_j$, for all $i, j \in J$. Let $\tilde{\sigma} : \mathcal{A} \to \mathcal{A}$ be the automorphism on induced by σ, that is, for every $h \in \mathcal{A}$,

$$\tilde{\sigma}(h) = h \circ \sigma^{-1}.$$

Then for every $f \in \mathcal{A}$,

$$\tilde{\sigma}_1(f) = \tilde{\sigma}_2(f).$$

This is given by the fact that if $\mathbb{P}_1 = \mathbb{P}_2$, then in (5), we can take $\mu = id$.

The second is the following important theorem.

Theorem 3 *Let $\mathbb{P}_1 = \{X_j : j \in J\}$ and $\mathbb{P}_2 = \{Y_j : j \in J\}$ be partitions of two sets X and Y and $\mathcal{A}_X$ and $\mathcal{A}_Y$ be algebras of functions that are constant on the sets of the partitions $\mathbb{P}_1$ and $\mathbb{P}_2$ respectively. Let $\sigma_1 : X \to X$ and $\sigma_2 : Y \to Y$ be bijections such that $\mathcal{A}_X$ is invariant under σ_1 (and σ_1^{-1}) and $\mathcal{A}_Y$ is invariant under σ_2 (and σ_2^{-1}) and that $\sigma_1(X_i) = X_j$ whenever $\sigma_2(Y_i) = Y_j$ for all $i, j \in J$. Suppose $\tilde{\sigma}_1 : \mathcal{A}_X \to \mathcal{A}_X$ is the automorphism on $\mathcal{A}_X$ induced by σ_1, and $\tilde{\sigma}_2 : \mathcal{A}_Y \to \mathcal{A}_Y$ is the automorphism on $\mathcal{A}_Y$ induced by σ_2. Then the crossed product algebras $\mathcal{A} \rtimes_{\tilde{\sigma}_1} \mathbb{Z}$ and $\mathcal{A} \rtimes_{\tilde{\sigma}_2} \mathbb{Z}$ are isomorphic.*

Proof We need to construct the an isomorphism between the crossed product algebras $\mathcal{A}_X \rtimes_{\tilde{\sigma}_1} \mathbb{Z}$ and $\mathcal{A}_Y \rtimes_{\tilde{\sigma}_1} \mathbb{Z}$. Using the notation in [4], we let $f := \sum_{n \in \mathbb{Z}} f_n \delta^n$ be an element in $\mathcal{A}_X \rtimes_{\tilde{\sigma}_1} \mathbb{Z}$. Define a function $\mu : \mathcal{A}_X \rtimes_{\tilde{\sigma}_1} \mathbb{Z} \to \mathcal{A}_Y \rtimes_{\tilde{\sigma}_2} \mathbb{Z}$ be defined by

$$\tilde{\mu}\left(\sum_{n \in \mathbb{Z}} f_n \delta_1^n\right) = \sum_{n \in \mathbb{Z}} \mu(f_n) \delta_2^n, \qquad (7)$$

where μ is defined in (4). Then, since μ is an algebra isomorphism, it is enough to prove that $\tilde{\mu}$ is multiplicative. To this end, we let $f := \sum_{n \in \mathbb{Z}} f_n \delta_1^n$ and $g := \sum_{m \in \mathbb{Z}} g_m \delta_1^m$ be arbitrary elements in $\mathcal{A}_X \rtimes_{\tilde{\sigma}_1} \mathbb{Z}$, then we prove that $\tilde{\mu}$ is multiplicative on the generators $f_n \delta_1^n$ and $g_m \delta_1^m$ respectively. Using (1) we have

$$\tilde{\mu}((f_n\delta_1^n) * (g_m\delta_1^m)) = \tilde{\mu}(f_n\tilde{\sigma_1}^n(g_m)\delta_1^{n+m})$$
$$= \mu(f_n\tilde{\sigma_1}^n(g_m))\delta_2^{n+m}$$
$$= [\mu(f_n)\mu(\tilde{\sigma_1}^n(f_m))]\delta_2^{n+m}$$
$$= \mu(f_n)\tilde{\sigma_2}^n(\mu(f_m))\delta_2^{n+m} \quad \text{by (6)}$$
$$= \tilde{\mu}(f_n\delta_2^n) * \tilde{\mu}(f_m\delta_2^m).$$

Therefore $\tilde{\mu}$ is multiplicative on the generators $f_n\delta^n$ and since μ is linear, it is multiplicative on the elements $f = \sum_{n\in\mathbb{Z}} f_n\delta^n \in \mathcal{A}_X \rtimes_{\tilde{\sigma}} \mathbb{Z}$. □

Remark 2 In Lemma 1 we proved the necessary and sufficient condition on a bijection $\sigma : X \to X$ such that the algebra $\mathcal{A}_X$ is invariant under σ, that is, for every $i \in J$ there exists $j \in J$ such that $\sigma(X_i) = X_j$ where the X_i form a partition for X. From this, it can be shown that $\mathcal{A}$ is isomorphic to $\mathbb{C}^J$, where by $\mathbb{C}^J$ we denote the space of complex sequences indexed by J. This can be done by constructing an isomorphism between $\mathcal{A}_X$ and $\mathbb{C}^J$ via σ as follows.

Let $\tau : J \to J$ be a map such that $\tau(i) = j$ is equivalent to $\sigma(X_i) = X_j$ for all $i, j \in J$. Then τ is a bijection that plays the same role as σ_2 in Lemma 2. Therefore, using the same Lemma, we deduce that the algebra $\mathcal{A}$ is isomorphic to $\mathbb{C}^J$. In Theorem 3, we have shown a method of constructing an isomorphism between the crossed product algebras $\mathcal{A}_X \rtimes_{\tilde{\sigma}_1} \mathbb{Z}$ and $\mathcal{A}_Y \rtimes_{\tilde{\sigma}_2} \mathbb{Z}$, when $\mathcal{A}_X$ and $\mathcal{A}_Y$ are isomorphic. It follows that the crossed product algebra $\mathcal{A}_X \rtimes_{\tilde{\sigma}_1} \mathbb{Z}$ is isomorphic to $\mathbb{C}^J \rtimes_{\tilde{\tau}} \mathbb{Z}$, where $\tilde{\tau}$ follows the same definition as $\tilde{\sigma}$.

In the next section we describe the commutant of our algebra $\mathcal{A}_X$ in the crossed product algebra $\mathcal{A}_X \rtimes_{\tilde{\sigma}} \mathbb{Z}$.

3.1 Maximal Commutative Subalgebra

We take the same partition $\mathbb{P} = \cup_{j\in J} X_j$ and a bijection $\sigma : X \to X$ such that for all $i \in J$, there exists $j \in J$ such that $\sigma(X_i) = X_j$. For $k \in \mathbb{Z}_{>0}$, let

$$C_k := \{x \in X \mid k \text{ is the smallest positive integer such that } x, \sigma^k(x) \in X_j \quad (8)$$
$$\text{for some } j \in J\}.$$

According to Theorem 1, the unique maximal commutative subalgebra of $\mathcal{A} \rtimes_{\tilde{\sigma}} \mathbb{Z}$ that contains $\mathcal{A}$ is precisely the set of elements

$$A' = \left\{ \sum_{n\in\mathbb{Z}} f_n\delta^n \mid \text{ for all } n \in \mathbb{Z} : f_n|_{Sep_A^n(X)} \equiv 0 \right\}, \quad (9)$$

where $Sep_A^n(X)$ is given by (2). We have the following theorem which gives the description of $Sep_A^n(X)$ in this case and is crucial in the description if the maximal commutative subalgebra.

Theorem 4 *Let $\sigma : X \to X$ be a bijection on X as given above, $\tilde{\sigma} : A_X \to A_X$ be the automorphism on A_X induced by σ and C_k be given by (8). Then for every $n \in \mathbb{Z}$,*

$$Sep_A^n(X) = \left\{ \bigcup_{k \nmid n} C_k \cup C_\infty \right\}, \tag{10}$$

where

$$C_\infty = \{X_j \in \mathbb{P} \,:\, \sigma^k(X_j) \neq X_j \;\forall k \geq 1\}.$$

Proof 1. If $n \equiv 0 \pmod{k}$ and $x \in X_j \in C_k$, the we can write $n = mk$ for some $m \in \mathbb{Z}$. Then, since $\sigma^k(X_j) = X_j$ it follows that $\sigma^{-k}(X_j) = X_j$ and therefore for every $h \in A$,

$$\tilde{\sigma}^n(h)(x) = \tilde{\sigma}^{mk}(h)(x) = (h \circ \sigma^{-mk})(x) = h(\sigma^{-mk}(x)) = h(x),$$

since x and $\sigma^{-mk}(x) \in X_j$ for all $m \in \mathbb{Z}$.

2. If $n \not\equiv 0 \pmod{k}$, we can write $n = mk + j$ where $m, j \in \mathbb{Z}$ with $1 \leq j < k$. It follows that for every $x \in X_j \in C_k$,

$$\begin{aligned}
\tilde{\sigma}^n(h)(x) &= \tilde{\sigma}^{mk+j}(h)(x) \\
&= (h \circ \sigma^{-mk+j})(x) \\
&= h(\sigma^{-mk+j}(x)) \\
&= \tilde{\sigma}^j(h)(x).
\end{aligned}$$

But k is the smallest integer such that $\sigma^k(X_j) = X_j$. Therefore since $j < k$,

$$\tilde{\sigma}^j(h)(x) \neq h(x).$$

Hence

$$Sep_A^n(X) = \{x \in X \mid \exists\, h \in A : h(x) \neq \tilde{\sigma}^n(h)(x)\}$$

$$= \begin{cases} \{\cup_j : X_j \notin C_k X_j\} & \text{if } n \equiv 0 \pmod{k}, \\ \{\cup_j : X_j \in C_k X_j\} & \text{if } n \not\equiv 0 \pmod{k}, \end{cases}$$

and if $x \in C_\infty$, then obviously $x \in Sep_A^n$ for every $n \geq 1$, or simply

$$Sep_A^n(\mathbb{R}) = \bigcup_{k \nmid n} C_k \cup C_\infty.$$

From the above theorem, the description of the maximal commutative subalgebra in $A \rtimes_{\tilde{\sigma}} \mathbb{Z}$ can be done as follows.

Theorem 5 *Let A_X be the algebra of piece-wise constant functions $f : X \to \mathbb{C}$, $\sigma : X \to X$ any bijection on X, $\tilde{\sigma} : A_X \to A_X$ the automorphism on A_X induced by σ and C_k be as described above. Then the unique maximal commutative subalgebra of $A_X \rtimes_{\tilde{\sigma}} \mathbb{Z}$ that contains A_X is given by*

$$A' = \left\{ \sum_{n \in \mathbb{Z} \,:\, k|n} \left(\sum_{j_n \in J} a_{j_n} \chi_{X_{j_n}} \right) \delta^n \right\}.$$

Proof From (9) we have that the unique maximal commutative subalgebra of $A_X \rtimes_{\tilde{\sigma}} \mathbb{Z}$ that contains A_X is precisely the set of elements

$$A' = \left\{ \sum_{n \in \mathbb{Z}} f_n \delta^n \mid \text{for all } n \in \mathbb{Z} : \; f_n|_{Sep_A^n(X)} \equiv 0 \right\},$$

and from (2),

$$Sep_A^n(\mathbb{R}) = \bigcup_{k \nmid n} C_k. \tag{11}$$

Combining the two results and using the definition of $h_n \in A_X$ as

$$h_n = \sum_{j_n \in J} a_{j_n} \chi_{X_{j_n}},$$

we get

$$A' = \left\{ \sum_{n \in \mathbb{Z} \,:\, k|n} \left(\sum_{j_n \in J} a_{j_n} \chi_{X_{j_n}} \right) \delta^n \right\}.$$

$\square$

It can be observed from the results in Theorem 4 that it is possible to have $Sep_A^n(X) = X$ for all $n \in \mathbb{Z}$. For example, suppose J is infinite and let $\sigma : X \to X$ be a bijection such that $\sigma(X_j) = X_{j+1}$ for every $j \in J$. Then it is easily seen that in in this case $Sep_A^n(X) = X$. However, this is not possible if J is finite since in this case σ acts like a permutation on a finite group. In the following section, we treat one such a case. We let $X = \mathbb{R}$ and A_X be the algebra of piece-wise constant functions on $\mathbb{R}$ with N fixed jump points, where $N \geq 1$ is an integer. In order to work in the setting described before, we treat jump points as intervals of zero length. Then $\mathbb{R}$ is partitioned into $2N + 1$ sub-intervals.

4 Algebra of Piece-Wise Constant Functions on the Real Line with N Fixed Jump Points

Let A be the algebra of piece-wise constant functions $f : \mathbb{R} \to \mathbb{R}$ with N fixed jumps at points $t_1, t_2, \ldots, t_N$. Partition $\mathbb{R}$ into $N + 1$ intervals $I_0, I_1, \ldots, I_N$ where $I_\alpha =]t_\alpha, t_{\alpha+1}[$ with $t_0 = -\infty$ and $t_{N+1} = \infty$. By looking at jump points as intervals of zero length, we can write $\mathbb{R} = \cup I_\alpha$ where I_α is as described above for $\alpha = 0, 1, \ldots N$ and $I_\alpha = \{t_\alpha\}$ if $\alpha > N$. Then for every $h \in A$ we have

$$h(x) = \sum_{\alpha=0}^{2N} a_\alpha \chi_{I_\alpha}(x), \qquad (12)$$

where χ_{I_α} is the characteristic function of I_α. As in the preceding section, we let $\sigma : \mathbb{R} \to \mathbb{R}$ be any bijection on $\mathbb{R}$ and let $\tilde{\sigma} : A \to A$ be the automorphism on A induced by σ. Then we have the following lemma which gives the necessary and sufficient conditions for $(\mathbb{R}, \sigma)$ to be a discrete dynamical system.

Lemma 3 *The algebra A is invariant under both σ and σ^{-1} if and only if the following conditions hold.*

1. *σ (and σ^{-1}) maps the each jump point t_k, $k = 1, \ldots, N$ onto another jump point.*
2. *σ maps every interval I_α, $\alpha = 0, 1, \ldots N$ bijectively onto any of the other intervals $I_0, I_1 \ldots I_N$.*

Proof Obviously, if the two conditions hold, then A is invariant under σ. So we suppose that A is invariant under σ and prove that the two conditions must hold.

1. Suppose $\sigma(t_k) = t_0 \notin \{t_1, t_2, \ldots, t_N\}$ for some $k \in \{1, 2, \ldots, N\}$. Then, since σ is onto, there exists $x_0 \in \mathbb{R}$ such that $\sigma(x_0) = t_k$, that is, there exists a non jump point that is mapped onto a jump point. We show that this is not possible. Let

$$h(x) = \begin{cases} 1 & \text{if } x = t_k, \\ 0 & \text{otherwise.} \end{cases}$$

Then $h \in A$. But

$$h \circ \sigma(x) = \begin{cases} 1 & \text{if } \sigma(x) = t_k, \\ 0 & \text{otherwise,} \end{cases} = \begin{cases} 1 & \text{if } x = x_0, \\ 0 & \text{otherwise.} \end{cases}$$

Therefore $h \circ \sigma \notin A$ which is a contradiction, implying that σ does not map a non jump point onto a jump point, proving the first condition.

2. Consider the bijection $\sigma : \mathbb{R} \to \mathbb{R}$ defined by

$$\sigma(x) = \begin{cases} x & \text{if } x \neq t'_k \text{ or } t''_k, \\ t''_k & \text{if } x = t'_k, \\ t'_k & \text{if } x = t''_k, \end{cases} \tag{13}$$

where $t'_k \in I_k$ and $t''_k \in I_{k+1}$ for some $k \in \{1, 2, \ldots, N\}$. Then σ is a bijection that permutes the jump points. Let $h \in \mathcal{A}$. Then using (12) and for the σ in Eq. (13) above, we have:

$$\sigma(x) = \begin{cases} h(x) & \text{if } x \neq t'_k \text{ or } t''_k, \\ a_{k+1} & \text{if } x = t'_k, \\ a_k & \text{if } x = t''_k. \end{cases}$$

Therefore, $h \circ \sigma$ has jumps at points $t_1, \ldots, t_N, t'_k, t''_k$ implying that $h \circ \sigma \notin \mathcal{A}$.
□

The following theorem gives the description of $Sep^n_{\mathcal{A}}(\mathbb{R})$ for any $n \in \mathbb{Z}$.

Theorem 6 *Let $\mathcal{A}$ be an algebra of piece-wise constant functions with N fixed jumps at points $t_1, \ldots, t_N$, $\sigma : \mathbb{R} \to \mathbb{R}$ be any bijection on $\mathbb{R}$ such that $\mathcal{A}$ is invariant under σ and let $\tilde{\sigma} : \mathcal{A} \to \mathcal{A}$ be the automorphism on $\mathcal{A}$ induced by σ. Let*

$$C_k := \{x \in \mathbb{R} \mid k \text{ is the smallest positive integer such that } x, \sigma^k(x) \in I_\alpha \tag{14}$$
$$\text{for some } \alpha = 0, \ldots 2N\}.$$

Then for every $n \in \mathbb{Z}$,

$$Sep^n_{\mathcal{A}}(\mathbb{R}) = \bigcup_{k \nmid n} C_k. \tag{15}$$

Proof See Theorem 4 and observe that $C_\infty = \emptyset$ in this case.
□

Example 1 Let $\mathcal{A}$ be the algebra of piece-wise constant functions with 4–fixed jump points at t_1, t_2, t_3, t_4. Partition $\mathbb{R}$ into five subintervals $I_0, \ldots, I_4$ where $I_\alpha =]t_\alpha, t_{\alpha+1}[$ with $t_0 = -\infty$ and $t_5 = \infty$.

Let $\sigma : \mathbb{R} \to \mathbb{R}$ be a bijection such that $\sigma(I_0) = I_1$, $\sigma(I_1) = I_2$, $\sigma(I_2) = I_0$, $\sigma(I_3) = I_4$ and $\sigma(I_4) = I_3$. It follows that $\sigma^3(I_0) = I_0$, $\sigma^3(I_1) = I_1$ and $\sigma^3(I_2) = I_2$. But $\sigma^j(I_\alpha) \neq I_\alpha$ for $\alpha = 0, 1, 2$ and $1 \leq j < 3$.

Also $\sigma^2(I_3) = I_3$, $\sigma^2(I_4) = I_4$ but $\sigma^j(I_\alpha) \neq I_\alpha$ if $j \not\equiv 0 \pmod 2$ with $\alpha = 3, 4$. Therefore:

$$Sep^n_{\mathcal{A}}(\mathbb{R}) = \{x \in \mathbb{R} \mid \exists h \in \mathcal{A} : h(x) \neq \tilde{\sigma}^n(h)(x)\}$$
$$= \mathbb{R} \setminus \{\{I_3 \cup I_4\} \cup \{t_k : \sigma^2(t_k) = t_k, \ k = 1, 2, 3, 4\}\} \text{ if } n \equiv 0 \pmod 2$$
$$= \{I_0 \cup I_1 \cup I_2\} \cup \{t_k : \sigma^2(t_k) \neq t_k, \ k = 1, 2, 3, 4\} \text{ if } n \equiv 0 \pmod 2,$$

and

$$Sep_A^n(\mathbb{R}) = \{x \in \mathbb{R} \mid \exists\, h \in A : h(x) \neq \tilde{\sigma}^n(h)(x)\}$$
$$= \mathbb{R} \setminus \{I_0 \cup \{I_1 \cup I_2\} \cup \{t_k : \sigma^3(t_k) = t_k, \quad k = 1, 2, 3, 4\}\} \text{ if } n \equiv 0 \pmod 3$$
$$= \{I_3 \cup I_4\} \cup \{t_k : \sigma^3(t_k) \neq t_k, \quad k = 1, 2, 3, 4\} \text{ if } n \equiv 0 \pmod 3.$$

From these results we have the following theorem.

Theorem 7 *Let A be the algebra of piece-wise constant functions $f : \mathbb{R} \to \mathbb{R}$ with N fixed jumps at points $t_1, t_2, \ldots, t_N$. Partition $\mathbb{R}$ into $N + 1$ intervals $I_0, I_1, \ldots, I_N$ where $I_\alpha =]t_\alpha, t_{\alpha+1}[$ with $t_0 = -\infty$ and $t_{N+1} = \infty$ and $I_M = \{t_\alpha\}$ for $N + 1 \leqslant M \leqslant 2N$. Let $\sigma : \mathbb{R} \to \mathbb{R}$ be any bijection on $\mathbb{R}$ such that A is invariant under σ and let $\tilde{\sigma} : A \to A$ be the automorphism on A induced by σ. Let*

$$C_k := \{x \in \mathbb{R} \mid k \text{ is the smallest positive integer such that } x, \sigma^k(x) \in I_\alpha \quad (16)$$
$$\text{for some } \alpha = 0, \ldots 2N\}.$$

Then the unique maximal commutative subalgebra of $A \rtimes_{\tilde{\sigma}} \mathbb{Z}$ that contains A is given by

$$A' = \left\{ \sum_{n \in \mathbb{Z}\,:\,k|n} \left(\sum_{\alpha_n=0}^{2N} a_{\alpha_n} \chi_{I_{\alpha_n}} \right) \delta^n \right\}.$$

Proof From (9) we have that the unique maximal commutative subalgebra of $A \rtimes_{\tilde{\sigma}} \mathbb{Z}$ that contains A is precisely the set of elements

$$A' = \left\{ \sum_{n \in \mathbb{Z}} f_n \delta^n \mid \text{for all } n \in \mathbb{Z} : f_n|_{Sep_A^n(X)} \equiv 0 \right\},$$

and from (11),

$$Sep_A^n(\mathbb{R}) = \bigcup_{k \nmid n} C_k.$$

Combining the two results and using the definition of $h_n \in A$ as

$$h_n = \sum_{\alpha_n=0}^{2N} a_{\alpha_n} \chi_{I_{\alpha_n}},$$

we get

$$A' = \left\{ \sum_{n \in \mathbb{Z}\,:\,k|n} \left(\sum_{\alpha_n=0}^{2N} a_{\alpha_n} \chi_{I_{\alpha_n}} \right) \delta^n \right\}.$$

$\square$

5   Some Examples

In this section we give some examples of how our results hold for well known simple cases. We treat two cases of piece-wise constant functions on the real line; those with one fixed jump point and those with two fixed jump points.

5.1   Piece-Wise Constant Functions with One Jump Point

Let $\mathcal{A}$ be the collection of all piece-wise constant functions on the real line with one fixed jump point t_0. Following the methods in the previous section $\mathbb{R}$ is partitioned into three intervals $I_0 = (-\infty, t_0)$, $I_1 = (t_0, \infty)$ and $I_2 = \{t_0\}$. Then we can write $h \in \mathcal{A}$ as

$$h = \sum_{\alpha=0}^{2} a_\alpha \chi_{I_\alpha} = a_0 \chi_{I_0} + a_1 \chi_{I_1} + a_2 \chi_{I_2}. \tag{17}$$

Let $\sigma : \mathbb{R} \to \mathbb{R}$ be any bijection on $\mathbb{R}$ and let $\tilde{\sigma}$ be the automorphism on $\mathcal{A}$ induced by σ. Note that by the first part of Lemma 3, invariance of the algebra $\mathcal{A}$ implies that $\sigma(t_0) = t_0$. It follows therefore that $\sigma(I_0) = I_0$ or $\sigma(I_0) = I_1$. We treat these two cases below.

5.1.1 $\sigma(I_0) = I_0$

In this case (and by bijectivity of σ), we have that $\sigma(I_1) = I_1$ and since $\sigma(t_0) = t_0$, then for every $x \in \mathbb{R}$, $h \in \mathcal{A}$ and $n \in \mathbb{Z}$

$$\tilde{\sigma}^n h(x) := h \circ \sigma^{-n}(x) = h(x),$$

since x and $\sigma^{-n}(x)$ will lie in the same interval. Therefore, all intervals I_α, $\alpha = 0, 1, 2$ belong to C_1 and hence

$$Sep_{\mathcal{A}}^n(\mathbb{R}) = \bigcup_{k \nmid n} C_k = \emptyset.$$

Therefore, the maximal commutative subalgebra will be given by

$$\mathcal{A}' = \left\{ \sum_{n \in \mathbb{Z}} f_n \delta^n \mid \text{for all } n \in \mathbb{Z} : \ f_n|_{Sep_{\mathcal{A}}^n(X)} \equiv 0 \right\}$$

$$= \mathcal{A} \rtimes_{\tilde{\sigma}} \mathbb{Z}.$$

5.1.2 $\sigma(I_0) = I_1$

In this case (and by bijectivity of σ), we have that $\sigma(I_1) = I_0$ and since $\sigma(t_0) = t_0$, then for every $x \in \mathbb{R}$, $h \in \mathcal{A}$ and $n \in \mathbb{Z}$ such that $2 \mid n$ we have

$$\tilde{\sigma}^n h(x) := h \circ \sigma^{-n}(x) = h(x),$$

since x and $\sigma^{-n}(x)$ will lie in the same interval. And for odd n, $\tilde{\sigma}^n(h)(x) = h(x)$ if and only if $x = t_0$. Therefore, we have,

$$C_1 = \{I_\alpha \mid \sigma(I_\alpha) = I_\alpha\} = I_2,$$

and

$$C_2 = \{I_\alpha \mid \sigma^2(I_\alpha) = I_\alpha\} = I_0 \cup I_1.$$

Therefore,

$$Sep_{\mathcal{A}}^n(\mathbb{R}) = \bigcup_{k \nmid n} C_k = \begin{cases} C_2 & \text{if } k = 1, \\ \emptyset & \text{if } k = 2. \end{cases}$$

Therefore, the maximal commutative subalgebra will be given by

$$\mathcal{A}' = \left\{ \sum_{n \in \mathbb{Z}\,:\,k \mid n} \left(\sum_{\alpha_n=0}^2 a_{\alpha_n} \chi_{I_{\alpha_n}} \right) \delta^n \right\}$$

$$= \left\{ \sum_{n \in \mathbb{Z}\,:\,2 \mid n} \left(\sum_{\alpha=0}^2 a_{\alpha_n} \chi_{I_\alpha} \right) \delta^n \right\}$$

$$= \left\{ \sum_{m \in \mathbb{Z}} (a_{0,m} \chi_{I_0} + a_{1,m} \chi_{I_1} + a_{2,m} \chi_{I_2}) \delta^{2m} + \sum_{m \in \mathbb{Z}} (a_{2,m} \chi_{I_2}) \delta^{2m+1} \right\}.$$

5.2 Piece-Wise Constant Functions with Two Jump Points

Let $\mathcal{A}$ be the collection of all piece-wise constant functions on the real line with two fixed jump points at t_0 and t_1. Following the methods in the previous section $\mathbb{R}$ is partitioned into intervals $I_0 =]-\infty, t_0[$, $I_1 =]t_0, t_1[$, $I_2 =]t_1, \infty[$, $I_3 = \{t_0\}$ and $I_4 = \{t_1\}$. Then we can write $h \in \mathcal{A}$ as

$$h = \sum_{\alpha=0}^4 a_\alpha \chi_{I_\alpha}. \tag{18}$$

Let $\sigma : \mathbb{R} \to \mathbb{R}$ be any bijection on $\mathbb{R}$ and let $\tilde{\sigma}$ be the automorphism on $\mathcal{A}$ induced by σ. Note that by the first part of Lemma 3, invariance of the algebra $\mathcal{A}$ implies that $\sigma(t_0) = t_0$ (and $\sigma(t_1) = t_1$) or $\sigma(t_0) = t_1$ (in which case $\sigma(t_1) = t_0$). Below we give a description for the maximal commutative subalgebra of $\mathcal{A} \rtimes_{\tilde{\sigma}} \mathbb{Z}$ for different types of σ.

5.2.1 $\sigma(I_\alpha) = I_\alpha$ for all $\alpha = 0, \ldots, 4$

This case is similar to the one in Sect. 5.1.1 in the sense that, for every $x \in \mathbb{R}$, $h \in \mathcal{A}$ and $n \in \mathbb{Z}$

$$\tilde{\sigma}^n h(x) := h \circ \sigma^{-n}(x) = h(x),$$

since x and $\sigma^{-n}(x)$ will lie in the same interval. Therefore, all intervals I_α, $\alpha = 0, \ldots, 4$ belong to C_1 and hence

$$Sep_{\mathcal{A}}^n(\mathbb{R}) = \bigcup_{k \nmid n} C_k = \emptyset.$$

Therefore, the maximal commutative subalgebra will be given by

$$\mathcal{A}' - \left\{ \sum_{n \in \mathbb{Z}} f_n \delta^n \mid \text{for all } n \in \mathbb{Z}, \cdot \ f_n|_{Sep_{\mathcal{A}}^n(X)} \equiv 0 \right\}$$
$$= \mathcal{A} \rtimes_{\tilde{\sigma}} \mathbb{Z}.$$

5.2.2 $\sigma(I_0) = I_1$, $\sigma(I_1) = I_0$ and $\sigma(I_\alpha) = I_\alpha$, $\alpha = 2, 3, 4$

In this case (and by bijectivity of σ), we have that $\sigma(I_1) = I_0$ and therefore for every $x \in \mathbb{R}$, $h \in \mathcal{A}$ and $n \in \mathbb{Z}$ such that $2 \mid n$ we have

$$\tilde{\sigma}^n h(x) := h \circ \sigma^{-n}(x) = h(x),$$

since x and $\sigma^{-n}(x)$ will lie in the same interval. And for odd n, $\tilde{\sigma}^n(h)(x) = h(x)$ if and only if $x \in I_2 \cup I_3 \cup I_4$. Therefore, we have,

$$C_1 = \{I_\alpha \mid \sigma(I_\alpha) = I_\alpha\} = I_2 \cup I_3 \cup I_4,$$

and

$$C_2 = \{I_\alpha \mid \sigma^2(I_\alpha) = I_\alpha\} = I_0 \cup I_1.$$

Therefore,

$$Sep_{\mathcal{A}}^n(\mathbb{R}) = \bigcup_{k \nmid n} C_k = \begin{cases} C_2 & \text{if } k = 1, \\ \emptyset & \text{if } k = 2. \end{cases}$$

It follows that for $n \in \mathbb{Z}$ such that $2 \mid n$, the maximal commutative subalgebra will be given by

$$\mathcal{A}_1' = \left\{ \sum_{n \in \mathbb{Z} \,:\, k \mid n} \left(\sum_{\alpha_n=0}^{2N} a_{\alpha_n} \chi_{I_{\alpha_n}} \right) \delta^n \right\}$$

$$= \left\{ \sum_{n \in \mathbb{Z} \,:\, 2 \mid n} \left(\sum_{\alpha_n=0}^{2N} a_{\alpha_n} \chi_{I_{\alpha_n}} \right) \delta^n \right\}$$

$$= \sum_{m \in \mathbb{Z}} \left(\sum_{\alpha=0}^{4} a_{\alpha,m} \chi_{I_\alpha} \right) \delta^{2m}.$$

And for odd n, we have

$$\mathcal{A}_2' = \sum_n (a_{2,m} \chi_{I_2} + a_{3,m} \chi_{I_3} + a_{4,m} \chi_{I_4}) \delta^n.$$

Therefore, the commutant $\mathcal{A}'$ is given by:

$$\mathcal{A} = \left\{ \sum_{m \in \mathbb{Z}} \left(\sum_{\alpha=0}^{4} a_{\alpha,m} \chi_{I_\alpha} \right) \delta^{2m} + \sum_{m \in \mathbb{Z}} (a_{2,m} \chi_{I_2} + a_{3,m} \chi_{I_3} + a_{4,m} \chi_{I_4}) \delta^{2m+1} \right\}.$$

Similar results can be obtained for the following cases

1. $\sigma(I_0) = I_1$, $\sigma(I_1) = I_0$, $\sigma(I_3) = I_4$, $\sigma(I_4) = I_3$ and $\sigma(I_2) = I_2$.
2. $\sigma(I_0) = I_2$, $\sigma(I_2) = I_0$ and $\sigma(I_\alpha) = I_\alpha$, $\alpha = 1, 3, 4$.
3. $\sigma(I_0) = I_2$, $\sigma(I_2) = I_0$, $\sigma(I_3) = I_4$ $\sigma(I_4) = I_3$ and $\sigma(I_1) = I_1$.
4. $\sigma(I_1) = I_2$, $\sigma(I_2) = I_1$ and $\sigma(I_\alpha) = I_\alpha$ $\alpha = 0, 3, 4$.
5. $\sigma(I_1) = I_2$, $\sigma(I_2) = I_1$, $\sigma(I_3) = I_4$, $\sigma(I_4) = I_3$ and $\sigma(I_0) = I_0$.

Since in all these cases, $\sigma^2(I_\alpha) = I_\alpha$, $\alpha = 0, \ldots, 4$.

5.2.3 $\sigma(I_0) = I_1$, $\sigma(I_1) = I_2$, $\sigma(I_2) = I_0$ and $\sigma(I_\alpha) = I_\alpha$, $\alpha = 3, 4$

In this case, using similar methods we have,

$$C_1 = \{I_\alpha \mid \sigma(I_\alpha) = I_\alpha\} = \cup I_3 \cup I_4, \quad C_2 = \emptyset,$$

and

$$C_3 = \{I_\alpha \mid \sigma^3(I_\alpha) = I_\alpha\} = I_0 \cup I_1 \cup I_2.$$

Therefore,

$$Sep_A^n(\mathbb{R}) = \bigcup_{k \nmid n} C_k = \begin{cases} C_3 & \text{if } k \neq 3, \\ \varnothing & \text{if } k = 3. \end{cases}$$

It follows that for $n \in \mathbb{Z}$ such that $3 \mid n$, the maximal commutative subalgebra will be given by

$$
\begin{aligned}
\mathcal{A}_1' &= \left\{ \sum_{n \in \mathbb{Z}\,:\,k|n} \left(\sum_{\alpha_n=0}^{2N} a_{\alpha_n} \chi_{I_{\alpha_n}} \right) \delta^n \right\} \\
&= \left\{ \sum_{n \in \mathbb{Z}\,:\,3|n} \left(\sum_{\alpha=0}^{4} a_{\alpha_n} \chi_{I_\alpha} \right) \delta^n \right\} \\
&= \sum_{m \in \mathbb{Z}} \left(\sum_{\alpha=0}^{4} a_{\alpha,m} \chi_{I_\alpha} \right) \delta^{3m}.
\end{aligned}
$$

If $3 \nmid n$, then

$$\mathcal{A}_2' = \sum_n (a_{3,n} \chi_{I_3} + a_{4,n} \chi_{I_4}) \delta^n.$$

Therefore:

$$\mathcal{A}' = \left\{ \sum_{m \in \mathbb{Z}} \left(\sum_{\alpha=0}^{4} a_{\alpha,m} \chi_{I_\alpha} \right) \delta^{3m} + \sum_n (a_{3,n} \chi_{I_3} + a_{4,n} \chi_{I_4}) \delta^n \right\}.$$

5.2.4 $\sigma(I_0) = I_1,\ \sigma(I_1) = I_2,\ \sigma(I_2) = I_0$ and $\sigma(I_3) = I_4,\ \sigma(I_4) = I_3$

In this case, using similar methods we have,

$$C_1 = \varnothing, \quad C_2 = I_3 \cup I_4,$$

and

$$C_3 = \{I_\alpha \mid \sigma^3(I_\alpha) = I_\alpha\} = I_0 \cup I_1 \cup I_2.$$

Therefore,

$$Sep_A^n(\mathbb{R}) = \bigcup_{k \nmid n} C_k = \begin{cases} \mathbb{R} \setminus C_3 & \text{if } k = 3, \\ \mathbb{R} \setminus C_2 & \text{if } k = 2, \\ \mathbb{R} & \text{if } k = 1. \end{cases}$$

It follows that for $n \in \mathbb{Z}$ such that $3 \mid n$, the maximal commutative subalgebra will be given by

$$
\begin{aligned}
\mathcal{A}_1' &= \left\{ \sum_{n \in \mathbb{Z} \,:\, k|n} \left(\sum_{\alpha_n=0}^{2N} a_{\alpha_n} \chi_{I_{\alpha_n}} \right) \delta^n \right\} \\
&= \left\{ \sum_{n \in \mathbb{Z} \,:\, 3|n} \left(\sum_{\alpha_n=0}^{2N} a_{\alpha_n} \chi_{I_{\alpha_n}} \right) \delta^n \right\} \\
&= \sum_{m \in \mathbb{Z}} \left(a_{0,m} \chi_{I_0} + a_{1,m} \chi_{I_1} + a_{2,m} \chi_{I_2} \right) \delta^{3m}.
\end{aligned}
$$

If $2 \mid n$, then

$$
\mathcal{A}_2' = \sum_{m \in \mathbb{Z}} (a_{3,m} \chi_{I_3} + a_{4,m} \chi_{I_4}) \delta^{2m},
$$

and for all other values of n, $\mathcal{A}' = \mathcal{A}$. Hence:

$$
\mathcal{A}' = \left\{ \sum_{m \in \mathbb{Z}} \left(a_{0,m} \chi_{I_0} + a_{1,m} \chi_{I_1} + a_{2,m} \chi_{I_2} \right) \delta^{3m} + \sum_{m \in \mathbb{Z}} (a_{3,m} \chi_{I_3} + a_{4,m} \chi_{I_4}) \delta^{2m} \right\}.
$$

Acknowledgements This work was partially supported by the Swedish Sida Foundation - International Science Program. Alex Behakanira Tumwesigye thanks the Research environment MAM in Mathematics and Applied Mathematics, Division of Applied Mathematics, School of Education, Culture and Communication, Mälardalens University for providing an excellent environment for research and education.

References

1. Carlsen, T.M., Silvestrov, S.D.: On the excel crossed product of topological covering maps. Acta Appl. Math. **108**(3), 573–583 (2009)
2. Li, B.-R.: Introduction to Operator Algebras. World Scientific, Singapore - New Jersey - Hong Kong - London (1992)
3. Öinert, J., Silvestrov, S.D.: Commutativity and ideals in algebraic crossed products. J. Gen. Lie Theory Appl. **2**(4), 287–302 (2008)
4. Svensson, C., Silvestrov, S.D., de Jeu, M.: Dynamical systems and commutants in crossed products. Int. J. Math. **18**, 455–471 (2007)
5. Svensson, C., Silvestrov, S.D., de Jeu, M.: Dynamical systems associated with crossed products. Acta Appl. Math. **108**(3), 547–559 (2009)
6. Tomiyama, J.: Invitation to C^*–Algebras and Topological Dynamics. World Scientific, Singapore - New Jersey - Hong Kong - London (1987)

Commutants in Crossed Product Algebras for Piece-Wise Constant Functions

Johan Richter, Sergei Silvestrov and Alex Behakanira Tumwesigye

Abstract In this paper we consider crossed product algebras of algebras of piece-wise constant functions on the real line with $\mathbb{Z}$. For an increasing sequence of algebras (in which case the commutants form a decreasing sequence), we describe the set difference between the corresponding commutants.

Keywords Piecewise constant · Crossed products · Commutant

1 Introduction

An important direction of investigation for any class of non-commutative algebras and rings, is the description of commutative subalgebras and commutative subrings. This is because such a description allows one to relate representation theory, non-commutative properties, graded structures, ideals and subalgebras, homological and other properties of non-commutative algebras to spectral theory, duality, algebraic geometry and topology naturally associated with commutative algebras. In representation theory, for example, semi-direct products or crossed products play a central role in the construction and classification of representations using the method of induced representations. When a non-commutative algebra is given, one looks for a subalgebra such that its representations can be studied and classified more easily

J. Richter · S. Silvestrov
Division of Applied Mathematics, School of Education,
Culture and Communication, Mälardalen University, Box 883, 721 23 Västerås, Sweden
e-mail: johan.richter@mdh.se

S. Silvestrov
e-mail: sergei.silvestrov@mdh.se

A.B. Tumwesigye (✉)
Department of Mathematics, College of Natural Sciences,
Makerere University, Box 7062, Kampala, Uganda
e-mail: alexbt@cns.mak.ac.ug

© Springer International Publishing Switzerland 2016
S. Silvestrov and M. Rančić (eds.), *Engineering Mathematics II*,
Springer Proceedings in Mathematics & Statistics 179,
DOI 10.1007/978-3-319-42105-6_7

and such that the whole algebra can be decomposed as a crossed product of this subalgebra by a suitable action.

When one has found a way to present a non-commutative algebra as a crossed product of a commutative subalgebra by some action on it, then it is important to know whether the subalgebra is maximal commutative, or if not, to find a maximal commutative subalgebra containing the given subalgebra. This maximality of a commutative subalgebra and related properties of the action are intimately related to the description and classification of representations of the non-commutative algebra.

Some work has been done in this direction [2, 4, 6] where the interplay between topological dynamics of the action on one hand and the algebraic property of the commutative subalgebra in the C^*-crossed product algebra $C(X) \rtimes \mathbb{Z}$ being maximal commutative on the other hand are considered. In [4], an explicit description of the (unique) maximal commutative subalgebra containing a subalgebra A of $\mathbb{C}^X$ is given. In [3], properties of commutative subrings and ideals in non-commutative algebraic crossed products by arbitrary groups are investigated and a description of the commutant of the base coefficient subring in the crossed product ring is given. More results on commutants in crossed products and dynamical systems can be found in [1, 5] and the references therein.

In this article, we consider algebras of piece-wise constant functions on the real line. In [7], a description of the maximal commutative subalgebra of the crossed product algebra of the said algebra with $\mathbb{Z}$ was given for the case where we have N fixed jumps. Given the algebras $\mathcal{A}_{t_i}$ of piece-wise constant functions with a fixed jump at t_i we take a sum of M such algebras. This yields an algebra of piece-wise constant functions with at most M jumps at points $t_1, \ldots, t_M$. Since $\mathcal{A}_{t_1, \ldots, t_k}$, $k = 1, \ldots, M$ is an increasing sequence of algebras, the commutants $\mathcal{A}'_{t_1, \ldots, t_k}$, $k = 1, \ldots, M$ form a decreasing sequence of algebras so we compute the difference between the said commutants.

2   Definitions and a Preliminary Result

Let $\mathcal{A}$ be any commutative algebra. Using the notation in [4], we let $\phi : \mathcal{A} \to \mathcal{A}$ be any algebra automorphism on $\mathcal{A}$ and define

$$\mathcal{A} \rtimes_\phi \mathbb{Z} := \{f : \mathbb{Z} \to \mathcal{A} : f(n) = 0 \text{ except for a finite number of } n\}.$$

It can be shown that $\mathcal{A} \rtimes_\phi \mathbb{Z}$ is an associative $\mathbb{C}-$ algebra with respect to point-wise addition, scalar multiplication and multiplication defined by *twisted convolution*, $*$ as follows:

$$(f * g)(n) = \sum_{k \in \mathbb{Z}} f(k)\phi^k(g(n - k)),$$

where ϕ^k denotes the $k-$fold composition of ϕ with itself for positive k and we use the obvious definition for $k \leq 0$.

Definition 1 $A \rtimes_\phi \mathbb{Z}$ as described above is called the crossed product algebra of A and $\mathbb{Z}$ under ϕ.

A useful and convenient way of working with $A \rtimes_\phi \mathbb{Z}$, is to write elements $f, g \in A \rtimes_\phi \mathbb{Z}$ in the form $f = \sum_{n \in \mathbb{Z}} f_n \delta^n$ and $g = \sum_{n \in \mathbb{Z}} g_m \delta^m$ where $f_n = f(n)$, $g_m = g(m)$ and

$$\delta^n(k) = \begin{cases} 1, & \text{if } k = n, \\ 0, & \text{if } k \neq n. \end{cases}$$

In the sum $\sum_{n \in \mathbb{Z}} f_n \delta^n$, we implicitly assume that $f_n = 0$ except for a finite number of n. Addition and scalar multiplication are canonically defined by the usual pointwise operations and multiplication is determined by the relation

$$(f_n \delta^n) * (g_m \delta^m) = f_n \phi^n(g_m) \delta^{n+m}, \tag{1}$$

where $m, n \in \mathbb{Z}$ and $f_n, g_m \in A$.

Definition 2 By the commutant A' of A in $A \rtimes_\phi \mathbb{Z}$ we mean

$$A' := \{f \in A \rtimes_\phi \mathbb{Z} : fg = gf \text{ for every } g \in A\}.$$

It has been proven [4] that the commutant A' is commutative and thus, is the unique maximal commutative subalgebra containing A. For any $f, g \in A \rtimes_\phi \mathbb{Z}$, that is, $f = \sum_{n \in \mathbb{Z}} f_n \delta^n$ and $g = \sum_{m \in \mathbb{Z}} g_m \delta^m$, $fg = gf$ if and only if

$$\forall r : \sum_{n \in \mathbb{Z}} f_n \phi^n(g_{r-m}) = \sum_{m \in \mathbb{Z}} g_m \phi^m(f_{r-m}).$$

Now let X be any set and A an algebra of complex valued functions on X. Let $\sigma : X \to X$ be any bijection such that A is invariant under σ and σ^{-1}, that is for every $h \in A$, $h \circ \sigma \in A$ and $h \circ \sigma^{-1} \in A$. Then (X, σ) is a discrete dynamical system and σ induces an automorphism $\tilde{\sigma} : A \to A$ defined by, $\tilde{\sigma}(f) = f \circ \sigma^{-1}$. In [7], a description of the commutant of A' in the crossed product algebra $A \rtimes_{\tilde{\sigma}} \mathbb{Z}$ for the case where A is the algebra of functions that are constant on the sets of a partition was given. Below are some definitions and results that will be important in our study. The proofs of the theorems can be found in [7] and the references in there.

Definition 3 For any nonzero $n \in \mathbb{Z}$, we set

$$Sep_A^n(X) = \{x \in X \mid \exists h \in A : h(x) \neq \tilde{\sigma}^n(h)(x)\}, \tag{2}$$

The following theorem has been proven in [4].

Theorem 1 *The unique maximal commutative subalgebra of $A \rtimes_{\tilde{\sigma}} \mathbb{Z}$ that contains A is precisely the set of elements*

$$\mathcal{A}' = \left\{ \sum_{n \in \mathbb{Z}} f_n \delta^n \mid \text{for all } n \in \mathbb{Z}: \ f_n | Sep^n_{\mathcal{A}}(X) \equiv 0 \right\}. \tag{3}$$

We observe that since $\tilde{\sigma} = f \circ \sigma^{-1}$, then

$$\tilde{\sigma}^2(f) = \tilde{\sigma}(f \circ \sigma^{-1}) = (f \circ \sigma^{-1}) \circ \sigma^{-1} = f \circ \sigma^{-2},$$

and hence $\tilde{\sigma}^n(f) = f \circ \sigma^{-n}$.

3 Algebra of Piece-Wise Constant Functions on the Real Line with N Fixed Jump Points

Let $\mathcal{A}$ be the algebra of piece-wise constant functions $f : \mathbb{R} \to \mathbb{R}$ with N fixed jumps at points $t_1, t_2, \ldots, t_N$. Partition $\mathbb{R}$ into $N + 1$ intervals $I_0, I_1, \ldots, I_N$ where $I_\alpha = (t_\alpha, t_{\alpha+1})$ with $t_0 = -\infty$ and $t_{N+1} = \infty$. By looking at jump points as intervals of zero length, we can write $\mathbb{R} = \cup I_\alpha$ where I_α is as described above for $\alpha = 0, 1, \ldots, N$ and $I_M = \{t_\alpha\}$ for $N + 1 \leqslant M \leqslant 2N$. Then for every $h \in \mathcal{A}$ we have

$$h(x) = \sum_{\alpha=0}^{2N} a_\alpha \chi_{I_\alpha}(x), \tag{4}$$

where χ_{I_α} is the characteristic function of I_α and a_α are some constants. As in the preceding section, we let $\sigma : \mathbb{R} \to \mathbb{R}$ be any bijection on $\mathbb{R}$ such that $\mathcal{A}$ is invariant under σ and let $\tilde{\sigma} : \mathcal{A} \to \mathcal{A}$ be the automorphism on $\mathcal{A}$ induced by σ. Then we have the following lemma which gives the necessary and sufficient conditions for $(\mathbb{R}, \sigma)$ to be a discrete dynamical system.

Lemma 1 *The algebra $\mathcal{A}$ is invariant under both σ and σ^{-1} if and only if the following conditions hold.*

1. *σ (and σ^{-1}) maps each jump point t_k, $k = 1, \ldots, N$ onto another jump point.*
2. *σ maps every interval I_α, $\alpha = 0, 1, \ldots, N$ bijectively onto any of the other intervals $I_0, I_1, \ldots, I_N$.*

The following theorem gives the description of $Sep^n_{\mathcal{A}}(\mathbb{R})$ for any $n \in \mathbb{Z}$.

Theorem 2 *Let $\sigma : \mathbb{R} \to \mathbb{R}$ be any bijection on $\mathbb{R}$ and let $\tilde{\sigma} : \mathcal{A} \to \mathcal{A}$ be the automorphism on $\mathcal{A}$ induced by σ. Let*

$$C_k := \left\{ x \in \mathbb{R} \mid k \text{ is the smallest positive integer such that } x, \sigma^k(x) \in I_\alpha \right. \tag{5}$$
$$\left. \text{for some } \alpha = 0, 1, \ldots, 2N \right\}.$$

Then for every $n \in \mathbb{Z}$,

$$Sep_{\mathcal{A}}^n(\mathbb{R}) = \bigcup_{k \nmid n} C_k. \tag{6}$$

Theorem 3 *Let $\mathcal{A}$ be the algebra of piece-wise constant functions $f : \mathbb{R} \to \mathbb{R}$ with N fixed jumps at points $t_1, t_2, \ldots, t_N$. We partition $\mathbb{R}$ into $N + 1$ intervals $I_0, I_1, \ldots, I_N$ where $I_\alpha = (t_\alpha, t_{\alpha+1})$ with $t_0 = -\infty$ and $t_{N+1} = \infty$. Let $\sigma : \mathbb{R} \to \mathbb{R}$ be any bijection on $\mathbb{R}$ such that $\mathcal{A}$ is invariant under both σ and σ^{-1} and let $\tilde{\sigma} : \mathcal{A} \to \mathcal{A}$ be the automorphism on $\mathcal{A}$ induced by σ. Then the unique maximal commutative subalgebra of $\mathcal{A} \rtimes_{\tilde{\sigma}} \mathbb{Z}$ that contains $\mathcal{A}$ is given by,*

$$\mathcal{A}' = \left\{ \sum_{n \in \mathbb{Z}} f_n \delta^n \mid f_n \equiv 0 \text{ on } C_k \text{ if } k \nmid n \right\},$$

where C_k is as defined in (5).

Proof From Theorem 1, we have that the unique maximal commutative subalgebra of $\mathcal{A} \rtimes_{\tilde{\sigma}} \mathbb{Z}$ that contains $\mathcal{A}$ is precisely the set of elements

$$\mathcal{A}' = \left\{ \sum_{n \in \mathbb{Z}} f_n \delta^n \mid \text{for all } n \in \mathbb{Z} : f_n|_{Sep_{\mathcal{A}}^n(X)} \equiv 0 \right\},$$

and from (6),

$$Sep_{\mathcal{A}}^n(\mathbb{R}) = \bigcup_{k \nmid n} C_k.$$

Combining the two results, we get

$$\mathcal{A}' = \left\{ \sum_{n \in \mathbb{Z}} f_n \delta^n \mid f_n \equiv 0 \text{ on } C_k \text{ if } k \nmid n \right\}.$$

$\square$

4 Comparison of Commutants

In this section, we give an explicit description of the set difference between commutants of an increasing finite sequence of algebras of piece-wise constant functions. Starting with an algebra of piece-wise constant functions with N fixed jumps at points $t_1, \ldots, t_N, \mathcal{A}_{\{t_1,\ldots,t_N\}}$, we add a finite number of jump points into one of the intervals (without loss of generality, the last one) and then take the sum of $\mathcal{A}_{\{t_1,\ldots,t_N\}}$ and the algebras of piece-wise constant functions at these points. In this way, we obtain a finite increasing sequence of algebras whose commutants (under some conditions), form a decreasing sequence. We give the description as follows.

Let $T = \{t_1, \ldots, t_N, t_{N+1}, \ldots, t_{N+M}\}$ for some $N, M \in \mathbb{N}$ such that $t_i < t_j$ if $i < j$. For $t_\alpha \in T$, let $\mathcal{A}_{t_\alpha}$ be the algebra of piece-wise constant functions with a fixed jump at t_α and for $K \in \{1, \ldots, N + M\}$, let

$$\mathcal{A}_{\{t_1,\ldots,t_K\}} := \left\{ \sum_{\alpha=1}^{K} f_\alpha ; \text{ where } f_\alpha \in \mathcal{A}_{t_\alpha} \right\}.$$

That is,

$$\mathcal{A}_{\{t_1,\ldots,t_K\}} = \sum_{\alpha=1}^{K} \mathcal{A}_{t_\alpha}.$$

Then $\mathcal{A}_{\{t_1,\ldots,t_K\}}$ consists of piece-wise constant functions with at most K jump points at points $t_1, \ldots, t_K$. It follows immediately that $\mathcal{A}_{\{t_1,\ldots,t_J\}} \subseteq \mathcal{A}_{\{t_1,\ldots,t_K\}}$ and therefore the commutants satisfy the relation $\mathcal{A}'_{\{t_1,\ldots,t_K\}} \subseteq \mathcal{A}'_{\{t_1,\ldots,t_J\}}$ for every $J, K \in \{1, 2, \ldots, N + M\}$ such that $J \leqslant K$. Observe that if $\mathcal{A}$ is a subalgebra of an algebra $\mathcal{B}$ of functions and σ is a bijection such that both $\mathcal{A}$ and $\mathcal{B}$ are invariant under σ, then $\mathcal{A} \rtimes_{\tilde{\sigma}} \mathbb{Z}$ is a subalgebra of $\mathcal{B} \rtimes_{\tilde{\sigma}} \mathbb{Z}$. In our case, we take $\mathcal{A} = \mathcal{A}_{\{t_1,\ldots,t_N\}}$ the algebra of piece-wise constant functions with jumps at $t_1, \ldots, t_N$ and $\mathcal{B} = \mathcal{A}_{\{t_1,\ldots,t_{N+M}\}}$, the algebra of piece-wise constant functions with jumps at $t_1, \ldots, t_{N+M}$. In Lemma 2 we give a sufficient condition on σ such that the algebras $\mathcal{A}_{\{t_1,\ldots,t_N\}}$ and $\mathcal{A}_{\{t_1,\ldots,t_{N+M}\}}$ are both invariant under σ. In this case $\mathcal{A}_{\{t_1,\ldots,t_N\}} \rtimes_{\tilde{\sigma}} \mathbb{Z} \subseteq \mathcal{A}_{\{t_1,\ldots,t_{N+M}\}} \rtimes_{\tilde{\sigma}} \mathbb{Z}$ and therefore we can compare the commutants $\mathcal{A}'_{\{t_1,\ldots,t_N\}}$ and $\mathcal{A}'_{\{t_1,\ldots,t_{N+M}\}}$ respectively.

For $\alpha \in \{0, 1, \ldots, N\}$, let $I_\alpha = (t_\alpha, t_{\alpha+1})$ with $t_0 = -\infty$ and $t_{N+1} = +\infty$ such that $I_N = (t_N, \infty)$ and for $i = 0, 1, \ldots, M$, let $I_N^i = (t_{N+i}, t_{N+i+1})$ with $t_{N+M+1} = \infty$. In order to be in the setting in [7], we define, for $\alpha = 1, \ldots, N$, $I_{N+\alpha} = \{t_\alpha\}$ and for $i = 1, \ldots, M$, $I_N^{M+i} = \{t_{N+i}\}$. Then we have the following.

Lemma 2 *Let $\sigma : \mathbb{R} \to \mathbb{R}$ be a bijection such that $\mathcal{A}_{\{t_1,\ldots,t_N\}}$ and $\mathcal{A}_{\{t_1,\ldots,t_{N+M}\}}$ are both invariant under σ. Then $\sigma(I_N) = I_N$.*

Proof Suppose $\sigma(I_N) \neq I_N$. Then by Lemma 1, $\sigma(I_N) = I_\alpha$ for some $\alpha \in \{0, 1, \ldots, N - 1\}$ and since $t_{N+i} \in I_N$ for each $i = 1, \ldots, M$, and since $\mathcal{A}_{\{t_1,\ldots,t_{N+M}\}}$ is invariant under σ (and $\sigma(I_N) \neq I_N$), then $\sigma(t_{N+i}) \in \{t_1, \ldots, t_N\}$ for every $i \in \{1, \ldots, M\}$. But also $\sigma(t_\alpha) \in \{t_1, \ldots, t_N\}$ for all $\alpha = 1, \ldots, N$.

Therefore $\sigma(\{t_1, \ldots, t_N, t_{N+1}, \ldots, t_{N+M}\}) = \{t_1, \ldots, t_N\}$ which contradicts bijectivity of σ. $\qquad\square$

Below we give a comparison of the commutants $\mathcal{A}'_{\{t_1,\ldots,t_N\}}$ and $\mathcal{A}'_{\{t_1,\ldots,t_{N+M}\}}$. Let $\tilde{\sigma}_1 : \mathcal{A}_{\{t_1,\ldots,t_N\}} \to \mathcal{A}_{\{t_1,\ldots,t_N\}}$ be the automorphism on $\mathcal{A}_{\{t_1,\ldots,t_N\}}$ induced by σ, that is for every $f \in \mathcal{A}_{\{t_1,\ldots,t_N\}}$, $\tilde{\sigma}_1(f) = f \circ \sigma^{-1}$. Similarly we define the automorphism $\tilde{\sigma}_2$ on $\mathcal{A}_{\{t_1,\ldots,t_{N+M}\}}$ induced by σ. Consider the following sets

$$Sep^n_{\mathcal{A}_{\{t_1,\ldots,t_N\}}}(\mathbb{R}) := \{x \in \mathbb{R} \mid \exists\, h \in \mathcal{A}_{\{t_1,\ldots,t_N\}} \mid h(x) \neq \tilde{\sigma}_1^n(h)(x)\},$$

and

$$Sep^n_{\mathcal{A}_{\{t_1,\ldots,t_{N+M}\}}}(\mathbb{R}) := \{x \in \mathbb{R} \mid \exists\, h \in \mathcal{A}_{\{t_1,\ldots,t_{N+M}\}} \mid h(x) \neq \tilde{\sigma}^n_2(h)(x)\}.$$

Then it can easily be seen that

$$Sep^n_{\mathcal{A}_{\{t_1,\ldots,t_N\}}}(\mathbb{R}) \subset Sep^n_{\mathcal{A}_{\{t_1,\ldots,t_{N+M}\}}}(\mathbb{R}).$$

Using the notation in [7], we let

$$C_k := \{x \in \mathbb{R} \mid k \text{ is the smallest positive integer such that } x, \sigma^k(x) \in I_\alpha$$
$$\text{for some } \alpha = 0, 1, \ldots, 2N\}.$$

Let also

$$\tilde{C}_k := \{x \in I_N \mid k \text{ is the smallest positive integer such that } x, \sigma^k(x) \in I^i_N$$
$$\text{for some } i = 1, \ldots, M\}.$$

Then we have the following Theorem 4.

Theorem 4

1. For every $n \in \mathbb{Z}$

$$Sep^n_{\mathcal{A}_{\{t_1,\ldots,t_{N+M}\}}}(\mathbb{R}) = Sep^n_{\mathcal{A}_{\{t_1,\ldots,t_N\}}}(\mathbb{R}) \bigcup \left(\cup_{k \nmid n} \tilde{C}_k \right),$$

2. and therefore the commutants satisfy:

$$\mathcal{A}'_{\{t_1,\ldots,t_{N+M}\}} = \mathcal{A}'_{\{t_1,\ldots,t_N\}} \setminus \left\{ \sum_{n \in \mathbb{Z}} f_n \delta^n \mid \text{ for some } n \in \mathbb{Z}, \ f_n \neq 0 \right.$$
$$\left. \text{on some } \tilde{C}_k \text{ with } k \nmid n \right\}.$$

Proof

1. By Lemma 2, $\sigma(I_N) = I_N$ and hence $I_N \not\subset Sep^n_{\mathcal{A}_{\{t_1,\ldots,t_N\}}}(\mathbb{R})$ for any $n \in \mathbb{Z}$. Also, for every $n \in \mathbb{Z}$,

$$Sep^n_{\mathcal{A}_{\{t_1,\ldots,t_{N+M}\}}}(\mathbb{R}) = \left(\cup_{k \nmid n} C_k \right) \bigcup \left(\cup_{k \nmid n} \tilde{C}_k \right)$$
$$= Sep^n_{\mathcal{A}_{\{t_1,\ldots,t_N\}}}(\mathbb{R}) \bigcup \left(\cup_{k \nmid n} \tilde{C}_k \right).$$

2. Now let us consider the commutants. By Theorem 1, the commutant $\mathcal{A}_{\{t_1,\ldots,t_N\}}$ is given by

$$\mathcal{A}'_{\{t_1,\dots,t_N\}} = \left\{ \sum_{n\in\mathbb{Z}} f_n \delta^n \mid \text{for all } n \in \mathbb{Z}: \ f_n\big|_{Sep^n_{\mathcal{A}_{t_1,\dots,t_N}}(\mathbb{R})} \equiv 0 \right\}.$$

Since $I_N \not\subset Sep^n_{\mathcal{A}_{\{t_1,\dots,t_N\}}}(\mathbb{R})$ for any $n \in \mathbb{Z}$, therefore,

$$\left\{ \sum_{n\in\mathbb{Z}} f_n \delta^n \mid f_n(x) \neq 0 \text{ if } x \in I_N \right\} \subset \mathcal{A}'_{\{t_1,\dots,t_N\}}.$$

Now

$$\mathcal{A}'_{\{t_1,\dots,t_{N+M}\}} = \left\{ \sum_{n\in\mathbb{Z}} f_n \delta^n \mid \text{for all } n \in \mathbb{Z}: \ f_n\big|_{Sep^n_{\mathcal{A}_{\{t_1,\dots,t_{N+M}\}}}(\mathbb{R})} \equiv 0 \right\}$$

$$= \mathcal{A}'_{\{t_1,\dots,t_N\}} \setminus \left\{ \sum_{n\in\mathbb{Z}} f_n \delta^n \mid \text{for some } n \in \mathbb{Z}, \ f_n \neq 0 \right.$$

$$\left. \text{on some } \tilde{C}_k \text{ with } k \nmid n \right\}.$$

□

4.1 An Example

As application of our results, we consider the case when only one jump point is introduced and give an explicit description. Suppose $\mathcal{A}_{\{t_1,t_2,\dots,t_N\}}$ and $\mathcal{A}_{t_1,t_2,\dots,t_{N+1}}$ are algebras defined respectively as follows:

$$\mathcal{A}_{\{t_1,t_2,\dots,t_N\}} = \left\{ \sum_{i=1}^{N} f_i \mid f_i \in \mathcal{A}_{t_i}, \ i = 1, \dots, N \right\},$$

and

$$\mathcal{A}_{\{t_1,t_2,\dots,t_{N+1}\}} = \left\{ \sum_{i=1}^{N} f_i \mid f_i \in \mathcal{A}_{f_i}, \ i = 1, \dots, N+1 \right\} = \mathcal{A}_{\{t_1,t_2,\dots,t_N\}} + \{\mathcal{A}_{t_{N+1}}\}.$$

Corollary 1 *Let $\sigma : \mathbb{R} \to \mathbb{R}$ be a bijection such that $\mathcal{A}_{\{t_1,t_2,\dots,t_N\}}$ and $\mathcal{A}_{\{t_1,t_2,\dots,t_{N+1}\}}$ are both invariant under σ. Let $I'_N = (t_N, t_{N+1})$ and $I''_N = (t_{N+1}, \infty)$. Then, we have the following.*

1. $Sep^n_{\mathcal{A}_{\{t_1,t_2,\ldots,t_N\}}}(\mathbb{R}) \subseteq Sep^n_{\mathcal{A}_{\{t_1,t_2,\ldots,t_{N+1}\}}}(\mathbb{R})$ *for every* $n \in \mathbb{Z}$ *and moreover*

$$Sep^n_{\mathcal{A}_{\{t_1,t_2,\ldots,t_{N+1}\}}}(\mathbb{R}) \setminus Sep^n_{\mathcal{A}_{\{t_1,t_2,\ldots,t_N\}}}(\mathbb{R}) = \begin{cases} \emptyset, & \text{if } \sigma(I'_N) = I'_N, \\ I'_N \cup I''_N, & \text{if } \sigma(I'_N) = I''_N \\ & \text{and } n \text{ is odd}, \\ \emptyset, & \text{if } \sigma(I'_N) = I''_N \\ & \text{and } n \text{ is even}. \end{cases}$$

2. *If* $\sigma(I'_N) = I''_N$, *then*

$$\mathcal{A}'_{\{t_1,t_2,\ldots,t_{N+1}\}} = \mathcal{A}'_{\{t_1,t_2,\ldots,t_N\}} \setminus \left\{ \sum_{m \in \mathbb{Z}} f_m \delta^m \mid f_{2m+1} \neq 0 \text{ on } I'_N \cup I''_N \right\}.$$

Proof

1. Since $\mathcal{A}_{\{t_1,t_2,\ldots,t_N\}}$ and $\mathcal{A}_{\{t_1,t_2,\ldots,t_{N+1}\}}$ are both invariant under σ, then by Lemma 2 we have that $\sigma(I_N) = I_N$ and hence $\sigma(t_{N+1}) = t_{N+1}$ where $I_N = (t_N, \infty)$. It follows therefore that $I_N \not\subset Sep^n_{\mathcal{A}_{\{t_n,\ldots,t_N\}}}(\mathbb{R})$ for every $n \in \mathbb{Z}$. Observe that if $I_k \subset Sep^n_{\mathcal{A}_{\{t_n,\ldots,t_N\}}}(\mathbb{R})$ for some $n \in \mathbb{Z}$, $k = 0, 1, \ldots, N - 1$, then $I_k \subset Sep^n_{\mathcal{A}_{\{t_n,\ldots,t_{N+1}\}}}(\mathbb{R})$. Now consider the action of σ on I_N and let $I'_N = (t_N, t_{N+1})$ and $I''_N = (t_{N+1}, \infty)$. Then we have the following:

   a. If $\sigma(I'_N) = I'_N$, then since $\sigma(t_{N+1}) = t_{N+1}$ (and σ is a bijection), then $I_N \not\subset Sep^n_{\mathcal{A}_{\{t_n,\ldots,t_{N+1}\}}}(\mathbb{R})$ for every $n \in \mathbb{Z}$ and hence,

   $$Sep^n_{\mathcal{A}_{\{t_n,\ldots,t_N\}}}(\mathbb{R}) = Sep^n_{\mathcal{A}_{\{t_n,\ldots,t_{N+1}\}}}(\mathbb{R})$$

   for every $n \in \mathbb{Z}$.
   b. If $\sigma(I'_N) = I''_N$, then $I'_N, I''_N \subset Sep^1_{\mathcal{A}_{\{t_n,\ldots,t_{N+1}\}}}(\mathbb{R})$ and hence

   $$I'_N, I''_N \subset Sep^n_{\mathcal{A}_{\{t_n,\ldots,t_N\}}}(\mathbb{R})$$

   for each odd $n \in \mathbb{Z}$. It follows therefore that

   $$Sep^n_{\mathcal{A}_{\{t_1,t_2,\ldots,t_N\}}}(\mathbb{R}) \subseteq Sep^n_{\mathcal{A}_{\{t_1,t_2,\ldots,t_{N+1}\}}}(\mathbb{R})$$

   for every $n \in \mathbb{Z}$.

2. Using Theorem 1, the commutant $\mathcal{A}'_{\{t_1,t_2,\ldots,t_{N+1}\}}$ is given by

$$\mathcal{A}'_{\{t_1,t_2,\ldots,t_{N+1}\}} = \left\{ \sum_{n \in \mathbb{Z}} f_n \delta^n \mid f_n |_{Sep^n_{\mathcal{A}_{\{t_1,t_2,\ldots,t_{N+1}\}}}(\mathbb{R})} \equiv 0 \right\}.$$

We have the following cases:

a. If $\sigma(I'_N) = I'_N$, then

$$Sep^n_{\mathcal{A}_{\{t_n,\ldots,t_N\}}}(\mathbb{R}) = Sep^n_{\mathcal{A}_{\{t_n,\ldots,t_{N+1}\}}}(\mathbb{R})$$

for every $n \in \mathbb{Z}$, and hence

$$\mathcal{A}'_{\{t_1,t_2,\ldots,t_{N+1}\}} = \mathcal{A}'_{\{t_1,t_2,\ldots,t_N\}}.$$

b. If $\sigma(I'_N) = I''_N$, then

$$Sep^{2m+1}_{\mathcal{A}_{\{t_n,\ldots,t_{N+1}\}}}(\mathbb{R}) = Sep^{2m+1}_{\mathcal{A}_{\{t_n,\ldots,t_N\}}}(\mathbb{R}) \cup \left(I'_N \cup I''_N\right),$$

and hence

$$\mathcal{A}'_{\{t_1,t_2,\ldots,t_{N+1}\}} = \mathcal{A}'_{\{t_1,t_2,\ldots,t_N\}} \setminus \left\{ \sum_{m \in \mathbb{Z}} f_m \delta^m \mid \text{for some } m \in \mathbb{Z}, \right.$$

$$\left. f_{2m+1} \neq 0 \text{ on } I'_N \cup I''_N \right\}.$$

$\square$

5   Description of the Center

Below we give the description of the center of our crossed product algebra $\mathcal{A} \rtimes_{\tilde{\sigma}} \mathbb{Z}$. This center will be the commutant of some subset of the crossed product as can be seen from Remark 1 below. The lemma below will be important in our considerations.

Lemma 3 *Let $B \subseteq A$ be a subset of an associative $\mathbb{C}$−algebra A and let $\mathcal{B}$ be the algebra generated by B. Then $B' = \mathcal{B}'$ where B' and $\mathcal{B}'$ denote the commutants of B and $\mathcal{B}$ respectively.*

The following theorem whose proof can be found in [4], gives the description of the center of a crossed product algebra $\mathcal{A} \rtimes_{\tilde{\sigma}} \mathbb{Z}$ where $\mathcal{A} = \mathbb{C}^X$.

Theorem 5 *Let $A \subseteq \mathbb{C}^X$ be an algebra of functions that is invariant under a bijection $\sigma : X \to X$. An element $g = \sum_{m \in \mathbb{Z}} g_m \delta^m$ is in $Z(\mathcal{A} \rtimes_{\tilde{\sigma}} \mathbb{Z})$ if and only if both of the following conditions are satisfied:*

1. *for all $m \in \mathbb{Z}$, g_m is $\mathbb{Z}$−invariant, and*
2. *for all $m \in \mathbb{Z}$, $g_m|_{Sep^m_{\mathcal{A}}(X)} \equiv 0$.*

In the theorem below we give the description of the center $Z(\mathcal{A} \rtimes_{\tilde{\sigma}} \mathbb{Z})$ for the case when $\mathcal{A}$ is the algebra of piece-wise constant functions with N jumps. First we make a few observations.

Recall the definition of C_k as given by (5). By Lemma 1 and bijectivity of σ, if $I_\alpha \subset C_k$, then $\sigma^k(I_\alpha) = I_\alpha$. Therefore each C_k consists of cycles of intervals which we denote by O_k^i, and each O_k^i can be written as

$$O_k^i = \{I_{\alpha_1}^i, \ldots, I_{\alpha_k}^i\}$$

such that $\sigma(I_{\alpha_j}^i) = I_{\alpha_{j+1}}^i$ for $j = 1, \ldots, k-1$ and $\sigma(I_{\alpha_k}^i) = I_{\alpha_1}^i$. Using these cycles and Theorem 5 we give a description of the center below.

Theorem 6 *Let A be the algebra of piece-wise constant functions $f : \mathbb{R} \to \mathbb{R}$ with N fixed jumps at points $t_1, t_2, \ldots, t_N$ as described in section 3. Let $\sigma : \mathbb{R} \to \mathbb{R}$ be any bijection on $\mathbb{R}$ such that A is invariant under σ and σ^{-1} and let $\tilde{\sigma} : A \to A$ be the automorphism on A induced by σ. Then*

$$Z(A \rtimes_{\tilde{\sigma}} \mathbb{Z}) = \left\{ \sum_{n \in \mathbb{Z}} f_n \delta^n \mid f_n \text{ is constant on every cycle } O_k^i \text{ in } C_k, \right.$$

$$\left. \text{for all } k \text{ such that } k \mid n \right\}.$$

Proof By the second part of Theorem 5, an element $f = \sum_{n \in \mathbb{Z}} f_n \delta^n \in Z(A \rtimes_{\tilde{\sigma}} \mathbb{Z})$ only if for all $m \in \mathbb{Z}$, $g_m|_{Sep_A^m(X)} \equiv 0$. From (6), we have that $Sep_A^n(\mathbb{R}) = \bigcup_{k \nmid n} C_k$. Therefore, $f \in Z(A \rtimes_{\tilde{\sigma}} \mathbb{Z})$ only if $f_n \equiv 0$ on $I_\alpha : \sigma^n(I_\alpha) \neq I_\alpha$. Or equivalently, $f \in Z(A \rtimes_{\tilde{\sigma}} \mathbb{Z})$ only if $f_n \equiv 0$ on C_k for all $k \nmid n$.

Also by the first part of Theorem 5, assuming the condition above holds, then $f \in Z(A \rtimes_{\tilde{\sigma}} \mathbb{Z})$ if and only if for all $n \in \mathbb{Z}$, f_n is $\mathbb{Z}$−invariant, that is, for all $n \in \mathbb{Z}$ and all $x \in \mathbb{R}$, $f_n(\sigma(x)) = f_n(x)$. From above, $f_n \equiv 0$ on each C_k such that $k \nmid n$. Now consider C_k such that $k \mid n$. As observed above, such C_k consists of cycles of intervals of length k denoted by O_k^i, where each O_k^i can be written as

$$O_k^i = \{I_{\alpha_1}^i, \ldots, I_{\alpha_k}^i\}$$

such that $\sigma(I_{\alpha_j}^i) = I_{\alpha_{j+1}}^i$ for $j = 1, \ldots, k-1$ and $\sigma(I_{\alpha_k}^i) = I_{\alpha_1}^i$. Since for each n, f_n is a piece-wise constant function and $f_n \equiv 0$ on each C_k, for which $k \nmid n$, then f_n being $\mathbb{Z}$−invariant is equivalent to saying f_n takes a constant value on each of the cycles O_k^i in C_k. □

Remark 1 One question of interest would be to compare the set difference of the commutant $A'_{\{t_1,\ldots,t_N\}} \setminus A'_{\{t_1,\ldots,t_{N+M}\}}$ with the commutant $\left(A_{\{t_1,\ldots,t_{N+M}\}} \setminus A_{\{t_1,\ldots,t_N\}}\right)'$ in the crossed product algebra $A_{\{t_1,\ldots,t_{N+M}\}} \rtimes_{\tilde{\sigma}} \mathbb{Z}$.

Observe that $\left(A_{\{t_1,\ldots,t_{N+M}\}} \setminus A_{\{t_1,\ldots,t_N\}}\right)$ is not a subalgebra of $A_{t_1,t_2,\ldots,t_{N+M}}$ but still it's commutant would be a subalgebra of the crossed product algebra $A_{t_1,\ldots,t_{N+M}} \rtimes_{\tilde{\sigma}} \mathbb{Z}$. By Lemma 3, if B is any subset of an associative $\mathbb{C}$−algebra A and $\mathcal{B}$ is the algebra generated by B then $B' = \mathcal{B}'$. Therefore if we let $A = A_{t_1,\ldots,t_{N+M}} \rtimes_{\tilde{\sigma}} \mathbb{Z}$ and $B = A_{t_1,\ldots,t_{N+M}} \rtimes_{\tilde{\sigma}} \mathbb{Z} \setminus A_{t_1,\ldots,t_N} \rtimes_{\tilde{\sigma}} \mathbb{Z}$, then the commutant $B' = \mathcal{B}'$ where $\mathcal{B}$ is the algebra generated by B. Now, it is easily seen that if $C = A_{t_1,\ldots,t_{N+M}} \setminus A_{t_1,\ldots,t_N}$, then the algebra $\mathcal{C}$ generated by C is $\mathcal{A}_{t_1,\dots,t_{N+M}}$. It follows therefore that the algebra generated by B is the whole crossed product algebra $\mathcal{A}_{t_1,\dots,t_{N+M}} \rtimes_{\tilde{\sigma}} \mathbb{Z}$. Therefore to find the commutant B' of B is the equivalent to finding the center $Z(\mathcal{A}_{t_1,\dots,t_{N+M}} \rtimes_{\tilde{\sigma}} \mathbb{Z})$.

6   Jump Points Added into Different Intervals

Let $\{t_1, \dots, t_N\}$ be a set of points in $\mathbb{R}$ such that $t_1 < t_2 < \dots < t_N$ and let $\mathcal{A}$ be an algebra of piece-wise constant functions with N fixed jumps at points $t_1, \dots, t_N$. Let $S = \{s_1, \dots, s_m\}$ be a set in $\mathbb{R}$, $m \leqslant N$ such that $t_{j-1} < s_j < t_j$, $j = 1, \dots, m$. Let $\mathcal{A}_{s_j}$, $j = 1, \dots, m$ be the algebra of piece-wise constant with a fixed jump at s_j and define $\mathcal{A}_S$ by

$$\mathcal{A}_S = \mathcal{A} + \sum_{j=1}^{m} \mathcal{A}_{s_j}.$$

Then $\mathcal{A}_S$ is the algebra of piece-wise constant functions with at most $N + m$ jumps at points $t_1, \dots, t_N, s_1, \dots, s_m$. It can be seen obviously that $\mathcal{A} \subset \mathcal{A}_S$ and therefore $\mathcal{A}'_S \subset \mathcal{A}'$, (under some conditions), where $\mathcal{A}'$ and $\mathcal{A}'_S$ denote the commutant for $\mathcal{A}$ and $\mathcal{A}_S$ respectively. In this section we describe the set of separation points for $\mathcal{A}_S$ and then compare the commutants $\mathcal{A}'$ and $\mathcal{A}'_S$.

Using the methods in [7], let $I_\alpha = (t_\alpha, t_{\alpha+1})$ for $\alpha = 0, 1, \dots, N$ with $t_0 = -\infty$ and $t_{N+1} = \infty$ and $I_{N+\alpha+1} = \{t_\alpha\}$, $\alpha = 1, \dots, N$. Now, for functions in $\mathcal{A}_S$, a jump point is introduced in each of the intervals I_α, $\alpha = 0, \dots, m$, therefore each of these intervals is divided into three subintervals $I'_\alpha = (t_\alpha, s_{\alpha+1})$, $I''_\alpha = (s_{\alpha+1}, t_{\alpha+1})$, and $I'''_\alpha = \{s_{\alpha+1}\}$. We have the following.

Lemma 4 *Let $\sigma : \mathbb{R} \to \mathbb{R}$ be a bijection such that the algebras $\mathcal{A}$ and $\mathcal{A}_S$ are both invariant under σ. Then*

$$\sigma \left(\cup_{\alpha=0}^{m} I_\alpha \right) = \cup_{\alpha=0}^{m} I_\alpha.$$

Proof Suppose for some $\alpha \in \{0, 1, \dots, m\}$, $\sigma(I_\alpha) = I_\beta$ for some $\beta > m$. Since $s_{\alpha+1} \in I_\alpha$, then $\sigma(s_{\alpha+1}) \in I_\beta$. By invariance of $\mathcal{A}_S$ under σ, $\sigma(s_{\alpha+1})$ must be jump point for some function $f \in \mathcal{A}_S$, which is a contradiction.   □

6.1   *Description of $Sep^n_{\mathcal{A}_S}(\mathbb{R})$ and the Commutant $\mathcal{A}'_S$*

Below we give a description of $Sep^n_{\mathcal{A}_S}(\mathbb{R})$ in terms of $Sep^n_{\mathcal{A}}(\mathbb{R})$ for $n \in \mathbb{Z}$. As before, we let C_k be as defined in (5), that is

$$C_k := \{ x \in \mathbb{R} \mid k \text{ is the smallest positive integer such that } x, \sigma^k(x) \in I_\alpha$$
$$\text{for some } \alpha = 0, 1, \dots, 2N \}.$$

Let also

$$\tilde{C}_k := \{x \in C_k \mid k \text{ is the smallest positive integer such that } x, \sigma^k(x) \in I_\alpha$$
$$\text{for some } \alpha = 0, 1, \ldots, 2N\},$$

and

$$\bar{C}_k := \{x \in C_k \mid 2k \text{ is the smallest positive integer such that } x, \sigma^{2k}(x) \in I_\alpha$$
$$\text{for some } \alpha = 0, 1, \ldots, 2N\}.$$

We give the descriptions in the following theorem.

Theorem 7 *For every* $n \in \mathbb{Z}$

$$Sep_{\mathcal{A}_S}^n(\mathbb{R}) = \bigcup_{k \nmid n} \left(\tilde{C}_k \cup \bar{C}_{k/2} \right), \tag{7}$$

and the commutant is given by

$$\mathcal{A}'_S = \mathcal{A}' \setminus \left\{ \sum_{n \in \mathbb{Z}} f_n \delta^n \mid \text{for some } n, k \text{ such that } k \text{ is even, } k \nmid n \text{ and } k \mid 2n \right.$$
$$\left. f_n \neq 0 \text{ on } \bar{C}_{k/2} \right\}. \tag{8}$$

Proof Using (6), we have that

$$Sep_{\mathcal{A}}^n(\mathbb{R}) = \bigcup_{k \nmid n} C_k.$$

From the definitions of $\tilde{C}_k$ and $\bar{C}_k$ it follows immediately that

$$C_k = \tilde{C}_k \cup \bar{C}_{k/2},$$

and this proves (7). Now consider the commutant $\mathcal{A}'_S$.

Again, from Theorem 1, we have

$$\mathcal{A}' = \left\{ \sum_{n \in \mathbb{Z}} f_n \delta^n \mid \text{for all } n \in \mathbb{Z} : f_n | Sep_{\mathcal{A}}^n(X) \equiv 0 \right\}. \tag{9}$$

Looking at (7), we observe that if k is odd, then $\bar{C}_{k/2} = \emptyset$ and nothing changes on the commutant. Therefore taking even k and combining (7) and (9), we get

$$\mathcal{A}'_S = \mathcal{A}' \setminus \left\{ \sum_{n \in \mathbb{Z}} f_n \delta^n \mid \text{for some } n, k \text{ such that } k \text{ is even}, k \nmid n \text{ and } k \mid 2n, \right.$$

$$\left. f_n \neq 0 \text{ on } \bar{C}_{k/2} \right\}.$$

$\square$

Acknowledgements This work was partially supported by the Swedish SIDA Foundation - International Science Program. Alex Behakanira Tumwesigye thanks the Research environment MAM in Mathematics and Applied Mathematics, Division of Applied Mathematics, School of Education, Culture and Communication, Mälardalens University for providing an excellent environment for research and education.

References

1. Carlsen, T.M., Silvestrov, S.D.: On the Excel crossed product of topological covering maps. Acta Appl. Math. **108**(3), 573–583 (2009)
2. Li, B.-R.: Introduction to Operator Algebras. World Scientific, Singapore (1992)
3. Öinert, J., Silvestrov, S.D.: Commutativity and ideals in algebraic crossed products. J. Gen. Lie Theory Appl. **2**(4), 287–302 (2008)
4. Svensson, C., Silvestrov, S.D., de Jeu, M.: Dynamical systems and commutants in crossed products. Int. J. Math. **18**, 455–471 (2007)
5. Svensson, C., Silvestrov, S.D., de Jeu, M.: Dynamical systems associated with crossed products. Acta Appl. Math. **108**(3), 547–559 (2009)
6. Tomiyama, J.: Invitation to C^*−Algebras and Topological Dynamics. World Scientific, Singapore (1987)
7. Tumwesigye, A.B., Richter, J., Silvestrov, S.D., Ssembatya, V.A.: Crossed product algebras for piece-wise constant functions. In: Silvestrov, S., Rančić, M. (eds.) Engineering Mathematics and Algebraic, Analysis and Stochastic Structures for Networks, Data Classification and Optimization. Springer Proceedings in Mathematics and Statistics. Springer, Heidelberg (2016)

Asymptotic Expansions for Moment Functionals of Perturbed Discrete Time Semi-Markov Processes

Mikael Petersson

Abstract In this paper we study moment functionals of mixed power-exponential type for non-linearly perturbed semi-Markov processes in discrete time. Conditions under which the moment functionals of interest can be expanded in asymptotic power series with respect to the perturbation parameter are given. We show how the coefficients in these expansions can be computed from explicit recursive formulas. In particular, the results of the present paper have applications for studies of quasi-stationary distributions.

Keywords Semi-Markov process · Perturbation · Asymptotic expansion · Renewal equation · Solidarity property · First hitting time

1 Introduction

The aim of this paper is to present asymptotic power series expansions for some important moment functionals of non-linearly perturbed semi-Markov processes in discrete time and to show how the coefficients in these expansions can be calculated from explicit recursive formulas. These asymptotic expansions play a fundamental role for the main result in [6], which is a sequel of the present paper.

For each $\varepsilon \geq 0$, we let $\xi^{(\varepsilon)}(n)$, $n = 0, 1, \ldots$, be a discrete time semi-Markov process on the state space $X = \{0, 1, \ldots, N\}$. It is assumed that the process $\xi^{(\varepsilon)}(n)$ depends on ε in the sense that its transition probabilities $Q_{ij}^{(\varepsilon)}(n)$ are continuous at $\varepsilon = 0$ when considered as a function of ε. Thus, we can, for $\varepsilon > 0$, interpret the process $\xi^{(\varepsilon)}(n)$ as a perturbation of $\xi^{(0)}(n)$.

Throughout the paper, we consider the case where the states $\{1, \ldots, N\}$ is a communicating class of states for ε small enough. Transitions to state 0 may, or may not, be possible both for the perturbed process and the limiting process. It will also

M. Petersson (✉)
Department of Mathematics, Stockholm University, 106 91 Stockholm, Sweden
e-mail: mikpe@math.su.se

© Springer International Publishing Switzerland 2016
S. Silvestrov and M. Rančić (eds.), *Engineering Mathematics II*,
Springer Proceedings in Mathematics & Statistics 179,
DOI 10.1007/978-3-319-42105-6_8

be natural to consider state 0 as an absorbing state but the results hold even if this is not the case.

Our main objects of study are the following mixed power-exponential moment functionals,

$$\phi_{ij}^{(\varepsilon)}(\rho, r) = \sum_{n=0}^{\infty} n^r e^{\rho n} g_{ij}^{(\varepsilon)}(n), \quad \omega_{ijs}^{(\varepsilon)}(\rho, r) = \sum_{n=0}^{\infty} n^r e^{\rho n} h_{ijs}^{(\varepsilon)}(n), \qquad (1)$$

where $\rho \in \mathbb{R}, r = 0, 1, \ldots, i, j, s \in X,$

$$g_{ij}^{(\varepsilon)}(n) = \mathsf{P}_i \{ \mu_j^{(\varepsilon)} = n, \ \mu_0^{(\varepsilon)} > \mu_j^{(\varepsilon)} \},$$

$$h_{ijs}^{(\varepsilon)}(n) = \mathsf{P}_i \{ \xi^{(\varepsilon)}(n) = s, \ \mu_0^{(\varepsilon)} \wedge \mu_j^{(\varepsilon)} > n \},$$

and $\mu_j^{(\varepsilon)}$ is the first hitting time of state j.

As is well known, power moments, exponential moments, and, as in (1), a mixture of power and exponential moments, often play important roles in various applications. One reason that the moments defined by Eq. (1) is of interest is that the probabilities $P_{ij}^{(\varepsilon)}(n) = \mathsf{P}_i \{ \xi^{(\varepsilon)}(n) = j, \ \mu_0^{(\varepsilon)} > 0 \}$ satisfy the following discrete time renewal equation,

$$P_{ij}^{(\varepsilon)}(n) = h_{iij}^{(\varepsilon)}(n) + \sum_{k=0}^{n}{}' P_{ij}^{(\varepsilon)}(n-k) g_{ii}^{(\varepsilon)}(n), \ n = 0, 1, \ldots.$$

This can, for example, be used in studies of quasi-stationary distributions as is illustrated in [6].

Under the assumption that mixed power-exponential moments for transition probabilities can be expanded in asymptotic power series with respect to the perturbation parameter, we obtain corresponding asymptotic expansions for the moment functionals in Eq. (1). These expansions together with explicit formulas for calculating the coefficients in the expansions are the main results of this paper.

In order to achieve this, we use methods from [2] where corresponding moment functionals for continuous time semi-Markov processes are studied. These methods are based on first deriving recursive systems of linear equations connecting the moments of interest with moments of transition probabilities and then successively build expansions for solutions of such systems.

Analysis of perturbed Markov chains and semi-Markov processes constitutes a large branch of research in applied probability, see, for example, the books [2–4, 7], and [1]. More detailed comments on this and additional references are given in [6].

Let us now briefly outline the structure of the present paper. In Sect. 2 we define perturbed discrete time semi-Markov processes and formulate our basic conditions. Then, systems of linear equations for exponential moment functionals are derived in Sect. 3 and in Sect. 4 we show convergence for the solutions of these systems.

Finally, in Sect. 5, we present the main results which give asymptotic expansions for mixed power-exponential moment functionals.

2 Perturbed Semi-Markov Processes

In this section we define perturbed discrete time semi-Markov processes and formulate some basic conditions.

For every $\varepsilon \geq 0$, let $(\eta_n^{(\varepsilon)}, \kappa_n^{(\varepsilon)})$, $n = 0, 1, \ldots$, be a discrete time Markov renewal process, i.e., a homogeneous Markov chain with state space $X \times \mathbb{N}$, where $X = \{0, 1, \ldots, N\}$ and $\mathbb{N} = \{1, 2, \ldots\}$, an initial distribution $Q_i^{(\varepsilon)} = \mathsf{P}\{\eta_0^{(\varepsilon)} = i\}$, $i \in X$, and transition probabilities which do not depend on the current value of the second component, given by

$$Q_{ij}^{(\varepsilon)}(k) = \mathsf{P}\left\{\eta_{n+1}^{(\varepsilon)} = j, \ \kappa_{n+1}^{(\varepsilon)} = k \mid \eta_n^{(\varepsilon)} = i, \ \kappa_n^{(\varepsilon)} = l\right\}, \ k, l \in \mathbb{N}, \ i, j \in X.$$

In this case, it is known that $\eta_n^{(\varepsilon)}$, $n = 0, 1, \ldots$, is also a Markov chain with state space X and transition probabilities,

$$p_{ij}^{(\varepsilon)} = \mathsf{P}\left\{\eta_{n+1}^{(\varepsilon)} = j \mid \eta_n^{(\varepsilon)} = i\right\} = \sum_{k=1}^{\infty} Q_{ij}^{(\varepsilon)}(k), \ i, j \in X.$$

Let us define $\tau^{(\varepsilon)}(0) = 0$ and $\tau^{(\varepsilon)}(n) = \kappa_1^{(\varepsilon)} + \cdots + \kappa_n^{(\varepsilon)}$, for $n \in \mathbb{N}$. Furthermore, for $n = 0, 1, \ldots$, we define $\nu^{(\varepsilon)}(n) = \max\{k : \tau^{(\varepsilon)}(k) \leq n\}$. The discrete time semi-Markov process associated with the Markov renewal process $(\eta_n^{(\varepsilon)}, \kappa_n^{(\varepsilon)})$ is defined by the following relation,

$$\xi^{(\varepsilon)}(n) = \eta_{\nu^{(\varepsilon)}(n)}^{(\varepsilon)}, \ n = 0, 1, \ldots,$$

and we will refer to $Q_{ij}^{(\varepsilon)}(k)$ as the transition probabilities of this process.

In the semi-Markov process defined above, we have that (i) $\kappa_n^{(\varepsilon)}$ are the times between successive moments of jumps, (ii) $\tau^{(\varepsilon)}(n)$ are the moments of the jumps, (iii) $\nu^{(\varepsilon)}(n)$ are the number of jumps in the interval $[0, n]$, and (iv) $\eta_n^{(\varepsilon)}$ is the embedded Markov chain.

It is sometimes convenient to write the transition probabilities of the semi-Markov process as $Q_{ij}^{(\varepsilon)}(k) = p_{ij}^{(\varepsilon)} f_{ij}^{(\varepsilon)}(k)$, where

$$f_{ij}^{(\varepsilon)}(k) = \mathsf{P}\left\{\kappa_{n+1}^{(\varepsilon)} = k \mid \eta_n^{(\varepsilon)} = i, \ \eta_{n+1}^{(\varepsilon)} = j\right\}, \ k \in \mathbb{N}, \ i, j \in X,$$

are the conditional distributions of transition times.

We now define random variables for first hitting times. For each $j \in X$, let $v_j^{(\varepsilon)} = \min\{n \geq 1 : \eta_n^{(\varepsilon)} = j\}$ and $\mu_j^{(\varepsilon)} = \tau(v_j^{(\varepsilon)})$. Then, $v_j^{(\varepsilon)}$ and $\mu_j^{(\varepsilon)}$ are the first hitting times of state j for the embedded Markov chain and the semi-Markov process, respectively. Note that the random variables $v_j^{(\varepsilon)}$ and $\mu_j^{(\varepsilon)}$, which may be improper, take values in the set $\{1, 2, \ldots, \infty\}$.

Let us define

$$g_{ij}^{(\varepsilon)}(n) = \mathsf{P}_i \left\{ \mu_j^{(\varepsilon)} = n, \; v_0^{(\varepsilon)} > v_j^{(\varepsilon)} \right\}, \; n = 0, 1, \ldots, \; i, j \in X,$$

and

$$g_{ij}^{(\varepsilon)} = \mathsf{P}_i \left\{ v_0^{(\varepsilon)} > v_j^{(\varepsilon)} \right\}, \; i, j \in X.$$

Here, and in what follows, we write $\mathsf{P}_i(A^{(\varepsilon)}) = \mathsf{P}\{A^{(\varepsilon)} \mid \eta_0^{(\varepsilon)} = i\}$ for any event $A^{(\varepsilon)}$. Corresponding notation for conditional expectation will also be used.

Moment generating functions for distributions of first hitting times are defined by

$$\phi_{ij}^{(\varepsilon)}(\rho) = \sum_{n=0}^{\infty} e^{\rho n} g_{ij}^{(\varepsilon)}(n) = \mathsf{E}_i e^{\rho \mu_j^{(\varepsilon)}} \chi \left(v_0^{(\varepsilon)} > v_j^{(\varepsilon)} \right), \; \rho \in \mathbb{R}, \; i, j \in X. \quad (2)$$

Furthermore, let us define the following exponential moment functionals for transition probabilities,

$$p_{ij}^{(\varepsilon)}(\rho) = \sum_{n=0}^{\infty} e^{\rho n} Q_{ij}^{(\varepsilon)}(n), \; \rho \in \mathbb{R}, \; i, j \in X,$$

where we define $Q_{ij}^{(\varepsilon)}(0) = 0$.

Let us now introduce the following conditions, which we will refer to frequently throughout the paper:

A: (a) $p_{ij}^{(\varepsilon)} \to p_{ij}^{(0)}$, as $\varepsilon \to 0$, $i \neq 0$, $j \in X$.

   (b) $f_{ij}^{(\varepsilon)}(n) \to f_{ij}^{(0)}(n)$, as $\varepsilon \to 0$, $n \in \mathbb{N}$, $i \neq 0$, $j \in X$.

B: $g_{ij}^{(0)} > 0$, $i, j \neq 0$.

C: There exists $\beta > 0$ such that:

   (a) $\limsup_{0 \leq \varepsilon \to 0} p_{ij}^{(\varepsilon)}(\beta) < \infty$, for all $i \neq 0$, $j \in X$.

   (b) $\phi_{ii}^{(0)}(\beta_i) \in (1, \infty)$, for some $i \neq 0$ and $\beta_i \leq \beta$.

It follows from conditions **A** and **B** that $\{1, \ldots, N\}$ is a communicating class of states for sufficiently small ε. Let us also remark that if $p_{i0}^{(0)} = 0$ for all $i \neq 0$, it can be shown that part **(b)** of condition **C** always holds under conditions **A**, **B**, and **C(a)**.

3 Systems of Linear Equations

In this section we derive systems of linear equations for exponential moment functionals.

We first consider the moment generating functions $\phi_{ij}^{(\varepsilon)}(\rho)$, defined by Eq. (2). By conditioning on $(\eta_1^{(\varepsilon)}, \kappa_1^{(\varepsilon)})$, we get for each $i, j \neq 0$,

$$\phi_{ij}^{(\varepsilon)}(\rho) = \sum_{l \in X} \sum_{k=1}^{\infty} \mathsf{E}_i \left(e^{\rho \mu_j^{(\varepsilon)}} \chi \left(v_0^{(\varepsilon)} > v_j^{(\varepsilon)} \right) | \eta_1^{(\varepsilon)} = l, \; \kappa_1^{(\varepsilon)} = k \right) Q_{il}^{(\varepsilon)}(k) \qquad (3)$$

$$= \sum_{k=1}^{\infty} e^{\rho k} Q_{ij}^{(\varepsilon)}(k) + \sum_{l \neq 0, j} \sum_{k=1}^{\infty} \mathsf{E}_l e^{\rho(k + \mu_j^{(\varepsilon)})} \chi \left(v_0^{(\varepsilon)} > v_j^{(\varepsilon)} \right) Q_{il}^{(\varepsilon)}(k).$$

Relation (3) gives us the following system of linear equations,

$$\phi_{ij}^{(\varepsilon)}(\rho) = p_{ij}^{(\varepsilon)}(\rho) + \sum_{l \neq 0, j} p_{il}^{(\varepsilon)}(\rho) \phi_{lj}^{(\varepsilon)}(\rho), \; i, j \neq 0. \qquad (4)$$

In what follows it will often be convenient to use matrix notation. Let us introduce the following column vectors,

$$\Phi_j^{(\varepsilon)}(\rho) = \left[\phi_{1j}^{(\varepsilon)}(\rho) \cdots \phi_{Nj}^{(\varepsilon)}(\rho) \right]^T, \; j \neq 0, \qquad (5)$$

$$\mathbf{p}_j^{(\varepsilon)}(\rho) = \left[p_{1j}^{(\varepsilon)}(\rho) \cdots p_{Nj}^{(\varepsilon)}(\rho) \right]^T, \; j \in X. \qquad (6)$$

For each $j \neq 0$, we also define $N \times N$-matrices $_j\mathbf{P}^{(\varepsilon)}(\rho) = \| _j p_{ik}^{(\varepsilon)}(\rho) \|$ where the elements are given by

$$_j p_{ik}^{(\varepsilon)}(\rho) = \begin{cases} p_{ik}^{(\varepsilon)}(\rho) & i = 1, \ldots, N, \; k \neq j, \\ 0 & i = 1, \ldots, N, \; k = j. \end{cases} \qquad (7)$$

Using (5)–(7), we can write the system (4) in the following matrix form,

$$\Phi_j^{(\varepsilon)}(\rho) = \mathbf{p}_j^{(\varepsilon)}(\rho) + _j\mathbf{P}^{(\varepsilon)}(\rho) \Phi_j^{(\varepsilon)}(\rho), \; j \neq 0. \qquad (8)$$

Note that the relations given above hold for all $\rho \in \mathbb{R}$ even in the case where some of the quantities involved take the value infinity. In this case we use the convention $0 \cdot \infty = 0$ and the equalities may take the form $\infty = \infty$.

Let us now derive a similar type of system for the following exponential moment functionals,

$$\omega_{ijs}^{(\varepsilon)}(\rho) = \sum_{n=0}^{\infty} e^{\rho n} \mathsf{P}_i \left\{ \xi^{(\varepsilon)}(n) = s, \ \mu_0^{(\varepsilon)} \wedge \mu_j^{(\varepsilon)} > n \right\}, \ \rho \in \mathbb{R}, \ i, j, s \in X.$$

First, note that

$$\omega_{ijs}^{(\varepsilon)}(\rho) = \mathsf{E}_i \sum_{n=0}^{\infty} e^{\rho n} \chi \left(\xi^{(\varepsilon)}(n) = s, \ \mu_0^{(\varepsilon)} \wedge \mu_j^{(\varepsilon)} > n \right)$$

$$= \mathsf{E}_i \sum_{n=0}^{\mu_0^{(\varepsilon)} \wedge \mu_j^{(\varepsilon)} - 1} e^{\rho n} \chi(\xi^{(\varepsilon)}(n) = s).$$

We now decompose $\omega_{ijs}^{(\varepsilon)}(\rho)$ into two parts,

$$\omega_{ijs}^{(\varepsilon)}(\rho) = \mathsf{E}_i \sum_{n=0}^{\kappa_1^{(\varepsilon)} - 1} e^{\rho n} \chi(\xi^{(\varepsilon)}(n) = s) + \mathsf{E}_i \sum_{n=\kappa_1^{(\varepsilon)}}^{\mu_0^{(\varepsilon)} \wedge \mu_j^{(\varepsilon)} - 1} e^{\rho n} \chi(\xi^{(\varepsilon)}(n) = s). \quad (9)$$

Let us first rewrite the first term on the right hand side of Eq. (9). By conditioning on $\kappa_1^{(\varepsilon)}$ we get, for $i, s \neq 0$,

$$\mathsf{E}_i \sum_{n=0}^{\kappa_1^{(\varepsilon)} - 1} e^{\rho n} \chi(\xi^{(\varepsilon)}(n) = s)$$

$$= \sum_{k=1}^{\infty} \mathsf{E}_i \left(\sum_{n=0}^{\kappa_1^{(\varepsilon)} - 1} e^{\rho n} \chi(\xi^{(\varepsilon)}(n) = s) \ \Big| \ \kappa_1^{(\varepsilon)} = k \right) \mathsf{P}_i \{\kappa_1^{(\varepsilon)} = k\}$$

$$= \sum_{k=1}^{\infty} \delta(i, s) \left(\sum_{n=0}^{k-1} e^{\rho n} \right) \mathsf{P}_i \{\kappa_1^{(\varepsilon)} = k\}.$$

It follows that

$$\mathsf{E}_i \sum_{n=0}^{\kappa_1^{(\varepsilon)} - 1} e^{\rho n} \chi(\xi^{(\varepsilon)}(n) = s) = \delta(i, s) \varphi_i^{(\varepsilon)}(\rho), \ i, s \neq 0, \quad (10)$$

where

$$\varphi_i^{(\varepsilon)}(\rho) = \begin{cases} \mathsf{E}_i \kappa_1^{(\varepsilon)} & \rho = 0, \\ (\mathsf{E}_i e^{\rho \kappa_1^{(\varepsilon)}} - 1)/(e^\rho - 1) & \rho \neq 0. \end{cases} \quad (11)$$

Let us now consider the second term on the right hand side of Eq. (9). By conditioning on $(\eta_1^{(\varepsilon)}, \kappa_1^{(\varepsilon)})$ we get, for $i, j, s \neq 0$,

$$
\mathsf{E}_i \sum_{n=\kappa_1^{(\varepsilon)}}^{\mu_0^{(\varepsilon)} \wedge \mu_j^{(\varepsilon)} -1} e^{\rho n} \chi(\xi^{(\varepsilon)}(n) = s)
$$

$$
= \sum_{l \neq 0, j} \sum_{k=1}^{\infty} \mathsf{E}_i \left(\sum_{n=\kappa_1^{(\varepsilon)}}^{\mu_0^{(\varepsilon)} \wedge \mu_j^{(\varepsilon)} -1} e^{\rho n} \chi(\xi^{(\varepsilon)}(n) = s) \,\Big|\, \eta_1^{(\varepsilon)} = l, \; \kappa_1^{(\varepsilon)} = k \right) Q_{il}^{(\varepsilon)}(k)
$$

$$
= \sum_{l \neq 0, j} \sum_{k=1}^{\infty} \mathsf{E}_l \left(\sum_{n=0}^{\mu_0^{(\varepsilon)} \wedge \mu_j^{(\varepsilon)} -1} e^{\rho(k+n)} \chi(\xi^{(\varepsilon)}(n) = s) \right) Q_{il}^{(\varepsilon)}(k).
$$

It follows that

$$
\mathsf{E}_i \sum_{n=\kappa_1^{(\varepsilon)}}^{\mu_0^{(\varepsilon)} \wedge \mu_j^{(\varepsilon)} -1} e^{\rho n} \chi(\xi^{(\varepsilon)}(n) = s) = \sum_{l \neq 0, j} p_{il}^{(\varepsilon)}(\rho) \omega_{ljs}^{(\varepsilon)}(\rho), \; i, j, s \neq 0. \tag{12}
$$

From (9), (10), and (12) we now get the following system of linear equations,

$$
\omega_{ijs}^{(\varepsilon)}(\rho) = \delta(i, s)\varphi_i^{(\varepsilon)}(\rho) + \sum_{l \neq 0, j} p_{il}^{(\varepsilon)}(\rho) \omega_{ljs}^{(\varepsilon)}(\rho), \; i, j, s \neq 0. \tag{13}
$$

In order to write this system in matrix form, let us define the following column vectors,

$$
\widehat{\boldsymbol{\varphi}}_s^{(\varepsilon)}(\rho) = \left[\delta(1, s)\varphi_1^{(\varepsilon)}(\rho) \; \cdots \; \delta(N, s)\varphi_N^{(\varepsilon)}(\rho) \right]^T, \; s \neq 0, \tag{14}
$$

$$
\boldsymbol{\Omega}_{js}^{(\varepsilon)}(\rho) = \left[\omega_{1js}^{(\varepsilon)}(\rho) \; \cdots \; \omega_{Njs}^{(\varepsilon)}(\rho) \right]^T, \; j, s \neq 0. \tag{15}
$$

Using (7), (14), and (15), the system (13) can be written in the following matrix form,

$$
\boldsymbol{\Omega}_{js}^{(\varepsilon)}(\rho) = \widehat{\boldsymbol{\varphi}}_s^{(\varepsilon)}(\rho) + {}_j\mathbf{P}^{(\varepsilon)}(\rho)\boldsymbol{\Omega}_{js}^{(\varepsilon)}(\rho), \; j, s \neq 0. \tag{16}
$$

We close this section with a lemma which will be important in what follows.

Lemma 1 *Assume that we for some $\varepsilon \geq 0$ and $\rho \in \mathbb{R}$ have that $g_{ik}^{(\varepsilon)} > 0$, $i, k \neq 0$ and $p_{ik}^{(\varepsilon)}(\rho) < \infty$, $i \neq 0$, $k \in X$. Then, for any $j \neq 0$, the following statements are equivalent:*

(a) $\Phi_j^{(\varepsilon)}(\rho) < \infty$.

(b) $\Omega_{js}^{(\varepsilon)}(\rho) < \infty$, $s \neq 0$.

(c) *The inverse matrix $(\mathbf{I} - {}_j\mathbf{P}^{(\varepsilon)}(\rho))^{-1}$ exists.*

Proof For each $j \neq 0$, let us define a matrix valued function $_j\mathbf{A}^{(\varepsilon)}(\rho) = \|_j a_{ik}^{(\varepsilon)}(\rho)\|$ by the relation

$$_j\mathbf{A}^{(\varepsilon)}(\rho) = \mathbf{I} + {}_j\mathbf{P}^{(\varepsilon)}(\rho) + ({}_j\mathbf{P}^{(\varepsilon)}(\rho))^2 + \cdots, \quad \rho \in \mathbb{R}. \tag{17}$$

Since each term on the right hand side of (17) is non-negative, it follows that the elements $_j a_{ik}^{(\varepsilon)}(\rho)$ are well defined and take values in the set $[0, \infty]$. Furthermore, the elements can be written in the following form which gives a probabilistic interpretation,

$$_j a_{ik}^{(\varepsilon)}(\rho) = \mathsf{E}_i \sum_{n=0}^{\infty} e^{\rho \tau^{(\varepsilon)}(n)} \chi (v_0^{(\varepsilon)} \wedge v_j^{(\varepsilon)} > n, \ \eta_n^{(\varepsilon)} = k), \ i, k \neq 0. \tag{18}$$

Let us now show that

$$\Phi_j^{(\varepsilon)}(\rho) = {}_j\mathbf{A}^{(\varepsilon)}(\rho)\mathbf{p}_j^{(\varepsilon)}(\rho), \ \rho \in \mathbb{R}, \ j \neq 0. \tag{19}$$

In order to do this, first note that, for $j \neq 0$,

$$\chi (v_0^{(\varepsilon)} > v_j^{(\varepsilon)}) = \sum_{n=0}^{\infty} \sum_{k \neq 0} \chi \left(v_0^{(\varepsilon)} \wedge v_j^{(\varepsilon)} > n, \ \eta_n^{(\varepsilon)} = k, \ \eta_{n+1}^{(\varepsilon)} = j \right). \tag{20}$$

Using (20) and the regenerative property of the semi-Markov process, the following is obtained, for $i, j \neq 0$,

$$\phi_{ij}^{(\varepsilon)}(\rho) = \sum_{n=0}^{\infty} \sum_{k \neq 0} \mathsf{E}_i e^{\rho \mu_j^{(\varepsilon)}} \chi \left(v_0^{(\varepsilon)} \wedge v_j^{(\varepsilon)} > n, \ \eta_n^{(\varepsilon)} = k, \ \eta_{n+1}^{(\varepsilon)} = j \right) \tag{21}$$

$$= \sum_{n=0}^{\infty} \sum_{k \neq 0} \mathsf{E}_i e^{\rho \tau^{(\varepsilon)}(n)} \chi \left(v_0^{(\varepsilon)} \wedge v_j^{(\varepsilon)} > n, \ \eta_n^{(\varepsilon)} = k \right) p_{kj}^{(\varepsilon)}(\rho).$$

From (18) and (21) we get

$$\phi_{ij}^{(\varepsilon)}(\rho) = \sum_{k \neq 0} {}_j a_{ik}^{(\varepsilon)}(\rho) p_{kj}^{(\varepsilon)}(\rho), \ i, j \neq 0, \tag{22}$$

and this proves (19).

Let us now define

$$\omega_{ij}^{(\varepsilon)}(\rho) = \sum_{s \neq 0} \omega_{ijs}^{(\varepsilon)}(\rho) = \sum_{n=0}^{\infty} e^{\rho n} \mathsf{P}_i \left\{ \mu_0^{(\varepsilon)} \wedge \mu_j^{(\varepsilon)} > n \right\}, \ \rho \in \mathbb{R}, \ i, j \neq 0. \tag{23}$$

Then, we have

$$\omega_{ij}^{(\varepsilon)}(\rho) = \begin{cases} \mathsf{E}_i(\mu_0^{(\varepsilon)} \wedge \mu_j^{(\varepsilon)}) & \rho = 0, \\ (\mathsf{E}_i e^{\rho(\mu_0^{(\varepsilon)} \wedge \mu_j^{(\varepsilon)})} - 1)/(e^\rho - 1) & \rho \neq 0. \end{cases} \qquad (24)$$

Also notice that

$$\mathsf{E}_i e^{\rho(\mu_0^{(\varepsilon)} \wedge \mu_j^{(\varepsilon)})} = \mathsf{E}_i e^{\rho\mu_j^{(\varepsilon)}} \chi(v_0^{(\varepsilon)} > v_j^{(\varepsilon)}) + \mathsf{E}_i e^{\rho\mu_0^{(\varepsilon)}} \chi(v_0^{(\varepsilon)} < v_j^{(\varepsilon)}), \ i, j \neq 0. \quad (25)$$

Using similar calculations as above, it can be shown that

$$\mathsf{E}_i e^{\rho\mu_0^{(\varepsilon)}} \chi(v_0^{(\varepsilon)} < v_j^{(\varepsilon)}) = \sum_{k \neq 0} {}_j a_{ik}^{(\varepsilon)}(\rho) p_{k0}^{(\varepsilon)}(\rho), \ i, j \neq 0. \qquad (26)$$

It follows from (22), (25), and (26) that

$$\mathsf{E}_i e^{\rho(\mu_0^{(\varepsilon)} \wedge \mu_j^{(\varepsilon)})} = \sum_{k \neq 0} {}_j a_{ik}^{(\varepsilon)}(\rho) \left(p_{kj}^{(\varepsilon)}(\rho) + p_{k0}^{(\varepsilon)}(\rho) \right), \ i, j \neq 0. \qquad (27)$$

Let us now show that (**a**) implies (**b**).
By iterating relation (8) we obtain,

$$\Phi_j^{(\varepsilon)}(\rho) = \left(\mathbf{I} + {}_j\mathbf{P}^{(\varepsilon)}(\rho) + \cdots + \left({}_j\mathbf{P}^{(\varepsilon)}(\rho) \right)^n \right) \mathbf{p}_j^{(\varepsilon)}(\rho) \qquad (28)$$

$$+ \left({}_j\mathbf{P}^{(\varepsilon)}(\rho) \right)^{n+1} \Phi_j^{(\varepsilon)}(\rho), \ n = 1, 2, \ldots$$

Since $\Phi_j^{(\varepsilon)}(\rho) < \infty$, it follows from (28) that

$$\left({}_j\mathbf{P}^{(\varepsilon)}(\rho) \right)^{n+1} \Phi_j^{(\varepsilon)}(\rho) \to \mathbf{0}, \ \text{as } n \to \infty. \qquad (29)$$

The assumptions of the lemma guarantee that $\Phi_j^{(\varepsilon)}(\rho) > 0$. From this and relation (29) we can conclude that $({}_j\mathbf{P}^{(\varepsilon)}(\rho))^{n+1} \to \mathbf{0}$, as $n \to \infty$. It is known that this holds if and only if the matrix series (17) converges in norms, that is, ${}_j\mathbf{A}^{(\varepsilon)}(\rho)$ is finite. From this and relations (23), (24), and (27) it follows that (**b**) holds.

Next we show that (**b**) implies (**c**).
By summing over all $s \neq 0$ in relation (16) it follows that

$$\Omega_j^{(\varepsilon)}(\rho) = \varphi^{(\varepsilon)}(\rho) + {}_j\mathbf{P}^{(\varepsilon)}(\rho)\Omega_j^{(\varepsilon)}(\rho), \ \rho \in \mathbb{R}, \qquad (30)$$

where

$$\Omega_j^{(\varepsilon)}(\rho) = \left[\omega_{1j}^{(\varepsilon)}(\rho) \cdots \omega_{Nj}^{(\varepsilon)}(\rho) \right]^T, \ j \neq 0,$$

and

$$\boldsymbol{\varphi}^{(\varepsilon)}(\rho) = \left[\varphi_1^{(\varepsilon)}(\rho) \cdots \varphi_N^{(\varepsilon)}(\rho)\right]^T.$$

By iterating relation (30) we get

$$\Omega_j^{(\varepsilon)}(\rho) = (\mathbf{I} + {}_j\mathbf{P}^{(\varepsilon)}(\rho) + \cdots + ({}_j\mathbf{P}^{(\varepsilon)}(\rho))^n)\boldsymbol{\varphi}^{(\varepsilon)}(\rho) \qquad (31)$$
$$+ ({}_j\mathbf{P}^{(\varepsilon)}(\rho))^{n+1}\Omega_j^{(\varepsilon)}(\rho), \ n = 1, 2, \ldots$$

It follows from (**b**) and the definition of $\omega_{ij}^{(\varepsilon)}(\rho)$ that $0 < \Omega_j^{(\varepsilon)}(\rho) < \infty$. So, letting $n \to \infty$ in (31) and using similar arguments as above, it follows that the matrix series (17) converges in norms. It is then known that the inverse matrix $(\mathbf{I} - {}_j\mathbf{P}^{(\varepsilon)}(\rho))^{-1}$ exists, that is, (**c**) holds.

Let us finally argue that (**c**) implies (**a**).

If $(\mathbf{I} - {}_j\mathbf{P}^{(\varepsilon)}(\rho))^{-1}$ exists, then the following relation holds,

$$(\mathbf{I} - {}_j\mathbf{P}^{(\varepsilon)}(\rho))^{-1} = \mathbf{I} + {}_j\mathbf{P}^{(\varepsilon)}(\rho)(\mathbf{I} - {}_j\mathbf{P}^{(\varepsilon)}(\rho))^{-1}. \qquad (32)$$

Iteration of (32) gives

$$(\mathbf{I} - {}_j\mathbf{P}^{(\varepsilon)}(\rho))^{-1} = \mathbf{I} + {}_j\mathbf{P}^{(\varepsilon)}(\rho) + ({}_j\mathbf{P}^{(\varepsilon)}(\rho))^2 + \cdots + ({}_j\mathbf{P}^{(\varepsilon)}(\rho))^n \qquad (33)$$
$$+ ({}_j\mathbf{P}^{(\varepsilon)}(\rho))^{n+1}(\mathbf{I} - {}_j\mathbf{P}^{(\varepsilon)}(\rho))^{-1}, \ n = 1, 2, \ldots.$$

Letting $n \to \infty$ in (33) it follows that ${}_j\mathbf{A}^{(\varepsilon)}(\rho) = (\mathbf{I} - {}_j\mathbf{P}^{(\varepsilon)}(\rho))^{-1} < \infty$. From (19) we now see that (**a**) holds. $\qquad\qquad\square$

4   Convergence of Moment Functionals

In this section it is shown that the solutions of the systems derived in Sect. 3 converge as the perturbation parameter tends to zero. In addition, we prove some properties for the solution of a characteristic equation.

Let us define

$$_k\phi_{ij}^{(\varepsilon)}(\rho) = \mathsf{E}_i e^{\rho\mu_j^{(\varepsilon)}} \chi(v_0^{(\varepsilon)} \wedge v_k^{(\varepsilon)} > v_j^{(\varepsilon)}), \ \rho \in \mathbb{R}, \ i, j, k \in X.$$

If the states $\{1, \ldots, N\}$ is a communicating class and $\phi_{ii}^{(\varepsilon)}(\rho) \leq 1$ for some $i \neq 0$, then it can be shown (see, for example, [5]) that the following relation holds for all $j \neq 0$,

$$\left(1 - \phi_{ii}^{(\varepsilon)}(\rho)\right)\left(1 - {}_i\phi_{jj}^{(\varepsilon)}(\rho)\right) = \left(1 - \phi_{jj}^{(\varepsilon)}(\rho)\right)\left(1 - {}_j\phi_{ii}^{(\varepsilon)}(\rho)\right). \qquad (34)$$

Relation (34) is useful in order to prove various solidarity properties for semi-Markov processes. In particular, if $\phi_{ii}^{(\varepsilon)}(\rho) = 1$, relation (34) reduces to

$$\left(1 - \phi_{jj}^{(\varepsilon)}(\rho)\right)\left(1 - {}_j\phi_{ii}^{(\varepsilon)}(\rho)\right) = 0. \tag{35}$$

From the regenerative property of the semi-Markov process it follows that

$$\phi_{ii}^{(\varepsilon)}(\rho) = {}_j\phi_{ii}^{(\varepsilon)}(\rho) + {}_i\phi_{ij}^{(\varepsilon)}(\rho)\phi_{ji}^{(\varepsilon)}(\rho), \quad j \neq 0, i. \tag{36}$$

Since $\{1, \ldots, N\}$ is a communicating class, we have ${}_i\phi_{ij}^{(\varepsilon)}(\rho) > 0$ and $\phi_{ji}^{(\varepsilon)}(\rho) > 0$. So, if $\phi_{ii}^{(\varepsilon)}(\rho) = 1$ it follows from (36) that ${}_j\phi_{ii}^{(\varepsilon)}(\rho) < 1$. From this and (35) we can conclude that $\phi_{jj}^{(\varepsilon)}(\rho) = 1$ for all $j \neq 0$. Thus, we have the following lemma:

Lemma 2 *Assume that we for some $\varepsilon \geq 0$ have that $g_{kj}^{(\varepsilon)} > 0$ for all $k, j \neq 0$. Then, if we for some $i \neq 0$ and $\rho \in \mathbb{R}$, have that $\phi_{ii}^{(\varepsilon)}(\rho) = 1$, it follows that $\phi_{jj}^{(\varepsilon)}(\rho) = 1$ for all $j \neq 0$.*

Let us now define the following characteristic equation,

$$\phi_{ii}^{(\varepsilon)}(\rho) = 1, \tag{37}$$

where $i \neq 0$ is arbitrary. The root of Eq. (37) plays an important role for the asymptotic behaviour of the corresponding semi-Markov process, see, for example, [6].

The following lemma gives limits of moment functionals and properties for the root of the characteristic equation.

Lemma 3 *If conditions A–C hold, then there exists $\delta \in (0, \beta]$ such that the following holds:*

(i) $\phi_{kj}^{(\varepsilon)}(\rho) \to \phi_{kj}^{(0)}(\rho) < \infty$, *as* $\varepsilon \to 0$, $\rho \leq \delta$, $k, j \neq 0$.

(ii) $\omega_{kjs}^{(\varepsilon)}(\rho) \to \omega_{kjs}^{(0)}(\rho) < \infty$, *as* $\varepsilon \to 0$, $\rho \leq \delta$, $k, j, s \neq 0$.

(iii) $\phi_{jj}^{(0)}(\delta) \in (1, \infty)$, $j \neq 0$.

(iv) *For sufficiently small ε, there exists a unique non-negative root $\rho^{(\varepsilon)}$ of the characteristic equation (37) which does not depend on i.*

(v) $\rho^{(\varepsilon)} \to \rho^{(0)} < \delta$ *as* $\varepsilon \to 0$.

Proof Let $i \neq 0$ and $\beta_i \leq \beta$ be the values given in condition **C**. It follows from conditions **B** and **C** that $\phi_{ii}^{(0)}(\rho)$ is a continuous and strictly increasing function for $\rho \leq \beta_i$. Since $\phi_{ii}^{(0)}(0) = g_{ii}^{(0)} \leq 1$ and $\phi_{ii}^{(0)}(\beta_i) > 1$, there exists a unique $\rho' \in [0, \beta_i)$ such that $\phi_{ii}^{(0)}(\rho') = 1$. Moreover, by Lemma 2,

$$\phi_{jj}^{(0)}(\rho') = 1, \quad j \neq 0. \tag{38}$$

For all $j \neq 0$, we have

$$\phi_{jj}^{(0)}(\rho') = {}_k\phi_{jj}^{(0)}(\rho') + {}_j\phi_{jk}^{(0)}(\rho')\phi_{kj}^{(0)}(\rho'), \ k \neq 0, j. \tag{39}$$

It follows from (38) and (39), and condition **B**, that

$$\phi_{kj}^{(0)}(\rho') < \infty, \ k, j \neq 0. \tag{40}$$

From (40) and Lemma 1 we get that $\det(\mathbf{I} - {}_j\mathbf{P}^{(0)}(\rho')) \neq 0$, for $j \neq 0$. Under condition **C**, the elements of $\mathbf{I} - {}_j\mathbf{P}^{(0)}(\rho)$ are continuous functions for $\rho \leq \beta$. This implies that we for each $j \neq 0$ can find $\beta_j \in (\rho', \beta_i]$ such that $\det(\mathbf{I} - {}_j\mathbf{P}^{(0)}(\beta_j)) \neq 0$. By condition **C** we also have that $p_{kj}^{(0)}(\beta_j) < \infty$ for $k \neq 0, j \in X$. It now follows from Lemma 1 that $\phi_{kj}^{(0)}(\beta_j) < \infty, k, j \neq 0$. If we define $\delta = \min\{\beta_1, \ldots, \beta_N\}$, it follows that

$$\phi_{kj}^{(0)}(\rho) < \infty, \ \rho \leq \delta, \ k, j \neq 0. \tag{41}$$

Now, let $\rho \leq \delta$ be fixed. Relation (41) and Lemma 1 imply that

$$\det(\mathbf{I} - {}_j\mathbf{P}^{(0)}(\rho)) \neq 0, \ j \neq 0. \tag{42}$$

Note that we have

$$p_{kj}^{(\varepsilon)}(\rho) = p_{kj}^{(\varepsilon)} \sum_{n=0}^{\infty} e^{\rho n} f_{kj}^{(\varepsilon)}(n), \ k, j \in X. \tag{43}$$

Since $f_{kj}^{(\varepsilon)}(n)$ are proper probability distributions, it follows from (43) and conditions **A** and **C** that

$$p_{kj}^{(\varepsilon)}(\rho) \to p_{kj}^{(0)}(\rho) < \infty, \text{ as } \varepsilon \to 0, \ k \neq 0, \ j \in X. \tag{44}$$

It follows from (42) and (44) that there exists $\varepsilon_1 > 0$ such that we for all $\varepsilon \leq \varepsilon_1$ have that $\det(\mathbf{I} - {}_j\mathbf{P}^{(\varepsilon)}(\rho)) \neq 0$ and $p_{kj}^{(\varepsilon)}(\rho) < \infty$, for all $k, j \neq 0$. Using Lemma 1 once again, it now follows that $\phi_{kj}^{(\varepsilon)}(\rho) < \infty, k, j \neq 0$, for all $\varepsilon \leq \varepsilon_1$. Moreover, in this case, the system of linear equations (8) has a unique solution for $\varepsilon \leq \varepsilon_1$ given by

$$\Phi_j^{(\varepsilon)}(\rho) = (\mathbf{I} - {}_j\mathbf{P}^{(\varepsilon)}(\rho))^{-1}\mathbf{p}_j^{(\varepsilon)}(\rho), \ j \neq 0. \tag{45}$$

From (44) and (45) it follows that

$$\phi_{kj}^{(\varepsilon)}(\rho) \to \phi_{kj}^{(0)}(\rho) < \infty, \text{ as } \varepsilon \to 0, \ k, j \neq 0.$$

This completes the proof of part (**i**).

For the proof of part (ii) we first note that, since $\phi_{kj}^{(\varepsilon)}(\rho) < \infty$ for $\varepsilon \le \varepsilon_1, k, j \ne 0$, it follows from Lemma 1 that $\omega_{kjs}^{(\varepsilon)}(\rho) < \infty$ for $\varepsilon \le \varepsilon_1, k, j, s \ne 0$. From this, and arguments given above, we see that the system of linear equations given by relation (16) has a unique solution for $\varepsilon \le \varepsilon_1$ given by

$$\Omega_{js}^{(\varepsilon)}(\rho) = (\mathbf{I} - {}_j\mathbf{P}^{(\varepsilon)}(\rho))^{-1}\widehat{\boldsymbol{\varphi}}_s^{(\varepsilon)}(\rho), \quad j, s \ne 0. \tag{46}$$

Now, since $\mathsf{E}_i e^{\rho\kappa_1^{(\varepsilon)}} = \sum_{j \in X} p_{ij}^{(\varepsilon)}(\rho)$, it follows from (11) and (44) that $\varphi_i^{(\varepsilon)}(\rho) \to \varphi_i^{(0)}(\rho) < \infty$ as $\varepsilon \to 0, i \ne 0$. Using this and relations (44) and (46) we can conclude that part (ii) holds.

By part (i) we have, in particular, $\phi_{jj}^{(\varepsilon)}(\delta) \to \phi_{jj}^{(0)}(\delta) < \infty$ as $\varepsilon \to 0$, for all $j \ne 0$. Furthermore, since $\rho' < \delta$ and $\phi_{jj}^{(0)}(\rho)$ is strictly increasing for $\rho \le \delta$, it follows from (38) that $\phi_{jj}^{(0)}(\delta) > 1, j \ne 0$. This proves part (iii).

Let us now prove part (iv).

It follows from (i) and (iii) that we can find $\varepsilon_2 > 0$ such that $\phi_{jj}^{(\varepsilon)}(\delta) \in (1, \infty)$, $j \ne 0$, for all $\varepsilon \le \varepsilon_2$. By conditions **A** and **B** there exists $\varepsilon_3 > 0$ such that, for each $i \ne 0$ and $\varepsilon \le \varepsilon_3$, the function $g_{ii}^{(\varepsilon)}(n)$ is not concentrated at zero. Thus, for every $i \ne 0$ and $\varepsilon \le \min\{\varepsilon_2, \varepsilon_3\}$, we have that $\phi_{ii}^{(\varepsilon)}(\rho)$ is a continuous and strictly increasing function for $\rho \in [0, \delta]$. Since $\phi_{ii}^{(\varepsilon)}(0) = g_{ii}^{(\varepsilon)} \le 1$ and $\phi_{ii}^{(\varepsilon)}(\delta) > 1$, there exists a unique $\rho_i^{(\varepsilon)} \in [0, \delta)$ such that $\phi_{ii}^{(\varepsilon)}(\rho_i^{(\varepsilon)}) = 1$. By Lemma 2, the root of the characteristic equation does not depend on i so we can write $\rho^{(\varepsilon)}$ instead of $\rho_i^{(\varepsilon)}$. This proves part (iv).

Finally, we show that $\rho^{(\varepsilon)} \to \rho^{(0)}$ as $\varepsilon \to 0$.

Let $\gamma > 0$ such that $\rho^{(0)} + \gamma \le \delta$ be arbitrary. Then $\phi_{ii}^{(0)}(\rho^{(0)} - \gamma) < 1$ and $\phi_{ii}^{(0)}(\rho^{(0)} + \gamma) > 1$. From this and part (i) we get that there exists $\varepsilon_4 > 0$ such that $\phi_{ii}^{(\varepsilon)}(\rho^{(0)} - \gamma) < 1$ and $\phi_{ii}^{(\varepsilon)}(\rho^{(0)} + \gamma) > 1$, for all $\varepsilon \le \varepsilon_4$. So, it follows that $|\rho^{(\varepsilon)} - \rho^{(0)}| < \gamma$ for $\varepsilon \le \min\{\varepsilon_2, \varepsilon_3, \varepsilon_4\}$. This completes the proof of Lemma 3. $\square$

5 Expansions of Moment Functionals

In this section, asymptotic expansions for mixed power-exponential moment functionals are constructed. The main results are given by Theorems 1 and 2.

Let us define the following mixed power-exponential moment functionals for distributions of first hitting times,

$$\phi_{ij}^{(\varepsilon)}(\rho, r) = \sum_{n=0}^{\infty} n^r e^{\rho n} g_{ij}^{(\varepsilon)}(n), \quad \rho \in \mathbb{R}, \ r = 0, 1, \dots, \ i, j \in X.$$

By definition, $\phi_{ij}^{(\varepsilon)}(\rho, 0) = \phi_{ij}^{(\varepsilon)}(\rho)$.

We also define the following mixed power-exponential moment functionals for transition probabilities,

$$p_{ij}^{(\varepsilon)}(\rho, r) = \sum_{n=0}^{\infty} n^r e^{\rho n} Q_{ij}^{(\varepsilon)}(n), \ \rho \in \mathbb{R}, \ r = 0, 1, \ldots, \ i, j \in X.$$

By definition, $p_{ij}^{(\varepsilon)}(\rho, 0) = p_{ij}^{(\varepsilon)}(\rho)$.

It follows from conditions **A–C** and Lemma 3 that, for $\rho < \delta$ and sufficiently small ε, the functions $\phi_{ij}^{(\varepsilon)}(\rho)$ and $p_{ij}^{(\varepsilon)}(\rho)$ are arbitrarily many times differentiable with respect to ρ, and the derivatives of order r are given by $\phi_{ij}^{(\varepsilon)}(\rho, r)$ and $p_{ij}^{(\varepsilon)}(\rho, r)$, respectively.

Recall from Sect. 3 that the following system of linear equations holds,

$$\phi_{ij}^{(\varepsilon)}(\rho) = p_{ij}^{(\varepsilon)}(\rho) + \sum_{l \neq 0, j} p_{il}^{(\varepsilon)}(\rho) \phi_{lj}^{(\varepsilon)}(\rho), \ i, j \neq 0. \tag{47}$$

Differentiating relation (47) gives

$$\phi_{ij}^{(\varepsilon)}(\rho, r) = \lambda_{ij}^{(\varepsilon)}(\rho, r) + \sum_{l \neq 0, j} p_{il}^{(\varepsilon)}(\rho) \phi_{lj}^{(\varepsilon)}(\rho, r), \ r = 1, 2, \ldots, \ i, j \neq 0, \tag{48}$$

where

$$\lambda_{ij}^{(\varepsilon)}(\rho, r) = p_{ij}^{(\varepsilon)}(\rho, r) + \sum_{m=1}^{r} \binom{r}{m} \sum_{l \neq 0, j} p_{il}^{(\varepsilon)}(\rho, m) \phi_{lj}^{(\varepsilon)}(\rho, r - m). \tag{49}$$

In order to write relations (47)–(49) in matrix form, let us define the following column vectors,

$$\Phi_j^{(\varepsilon)}(\rho, r) = \left[\phi_{1j}^{(\varepsilon)}(\rho, r) \cdots \phi_{Nj}^{(\varepsilon)}(\rho, r) \right]^T, \ j \neq 0, \tag{50}$$

$$\mathbf{p}_j^{(\varepsilon)}(\rho, r) = \left[p_{1j}^{(\varepsilon)}(\rho, r) \cdots p_{Nj}^{(\varepsilon)}(\rho, r) \right]^T, \ j \neq 0, \tag{51}$$

$$\Lambda_j^{(\varepsilon)}(\rho, r) = \left[\lambda_{1j}^{(\varepsilon)}(\rho, r) \cdots \lambda_{Nj}^{(\varepsilon)}(\rho, r) \right]^T, \ j \neq 0. \tag{52}$$

Let us also, for $j \neq 0$, define $N \times N$-matrices $_j\mathbf{P}^{(\varepsilon)}(\rho, r) = \| _j p_{ik}^{(\varepsilon)}(\rho, r) \|$ where the elements are given by

$$_j p_{ik}^{(\varepsilon)}(\rho, r) = \begin{cases} p_{ik}^{(\varepsilon)}(\rho, r) \ i = 1, \ldots, N, \ k \neq j, \\ 0 \qquad\quad i = 1, \ldots, N, \ k = j. \end{cases} \tag{53}$$

Using (47)–(53) we can for any $j \neq 0$ write the following recursive systems of linear equations,

$$\Phi_j^{(\varepsilon)}(\rho) = \mathbf{p}_j^{(\varepsilon)}(\rho) + {}_j\mathbf{P}^{(\varepsilon)}(\rho)\Phi_j^{(\varepsilon)}(\rho), \tag{54}$$

and, for $r = 1, 2, \ldots,$

$$\Phi_j^{(\varepsilon)}(\rho, r) = \Lambda_j^{(\varepsilon)}(\rho, r) + {}_j\mathbf{P}^{(\varepsilon)}(\rho)\Phi_j^{(\varepsilon)}(\rho, r), \tag{55}$$

where

$$\Lambda_j^{(\varepsilon)}(\rho, r) = \mathbf{p}_j^{(\varepsilon)}(\rho, r) + \sum_{m=1}^{r} \binom{r}{m} {}_j\mathbf{P}^{(\varepsilon)}(\rho, m)\Phi_j^{(\varepsilon)}(\rho, r - m). \tag{56}$$

Let us now introduce the following perturbation condition, which is assumed to hold for some $\rho < \delta$, where δ is the parameter in Lemma 3:

$\mathbf{P_k^*}$: $p_{ij}^{(\varepsilon)}(\rho, r) = p_{ij}^{(0)}(\rho, r) + p_{ij}[\rho, r, 1]\varepsilon + \cdots + p_{ij}[\rho, r, k - r]\varepsilon^{k-r} + o(\varepsilon^{k-r})$, for $r = 0, \ldots, k$, $i \neq 0$, $j \in X$, where $|p_{ij}[\rho, r, n]| < \infty$, for $r = 0, \ldots, k$, $n = 1, \ldots, k - r$, $i \neq 0$, $j \in X$.

For convenience, we denote $p_{ij}^{(0)}(\rho, r) = p_{ij}[\rho, r, 0]$, for $r = 0, \ldots, k$.

Note that if condition $\mathbf{P_k^*}$ holds, then, for $r = 0, \ldots, k$, we have the following asymptotic matrix expansions,

$$_j\mathbf{P}^{(\varepsilon)}(\rho, r) = {}_j\mathbf{P}[\rho, r, 0] + {}_j\mathbf{P}[\rho, r, 1]\varepsilon + \cdots + {}_j\mathbf{P}[\rho, r, k - r]\varepsilon^{k-r} + \mathbf{o}(\varepsilon^{k-r}), \tag{57}$$

$$\mathbf{p}_j^{(\varepsilon)}(\rho, r) = \mathbf{p}_j[\rho, r, 0] + \mathbf{p}_j[\rho, r, 1]\varepsilon + \cdots + \mathbf{p}_j[\rho, r, k - r]\varepsilon^{k-r} + \mathbf{o}(\varepsilon^{k-r}). \tag{58}$$

Here, and in what follows, $\mathbf{o}(\varepsilon^p)$ denotes a matrix-valued function of ε where all elements are of order $o(\varepsilon^p)$. The coefficients in (57) are $N \times N$-matrices $_j\mathbf{P}[\rho, r, n] = \|_j p_{ik}[\rho, r, n]\|$ with elements given by

$$_j p_{ik}[\rho, r, n] = \begin{cases} p_{ik}[\rho, r, n] & i = 1, \ldots, N, \ k \neq j, \\ 0 & i = 1, \ldots, N, \ k = j, \end{cases}$$

and the coefficients in (58) are column vectors defined by

$$\mathbf{p}_j[\rho, r, n] = \begin{bmatrix} p_{1j}[\rho, r, n] & \cdots & p_{Nj}[\rho, r, n] \end{bmatrix}^T.$$

Let us now define the following matrix, which will play an important role in what follows,

$$_j\mathbf{U}^{(\varepsilon)}(\rho) = (\mathbf{I} - {}_j\mathbf{P}^{(\varepsilon)}(\rho))^{-1}.$$

Under conditions **A–C**, it follows from Lemmas 1 and 3 that $_j\mathbf{U}^{(\varepsilon)}(\rho)$ is well defined for $\rho \leq \delta$ and sufficiently small ε.

The following lemma gives an asymptotic expansion for $_j\mathbf{U}^{(\varepsilon)}(\rho)$.

Lemma 4 *Assume that conditions* **A–C** *and* $\mathbf{P_k^*}$ *hold. Then we have the following asymptotic expansion,*

$$_j\mathbf{U}^{(\varepsilon)}(\rho) = {}_j\mathbf{U}[\rho, 0] + {}_j\mathbf{U}[\rho, 1]\varepsilon + \cdots + {}_j\mathbf{U}[\rho, k]\varepsilon^k + \mathbf{o}(\varepsilon^k), \qquad (59)$$

where

$$_j\mathbf{U}[\rho, n] = \begin{cases} (\mathbf{I} - {}_j\mathbf{P}^{(0)}(\rho))^{-1} & n = 0, \\ {}_j\mathbf{U}[\rho, 0]\sum_{q=1}^{n} {}_j\mathbf{P}[\rho, 0, q]_j\mathbf{U}[\rho, n-q] & n = 1, \dots, k. \end{cases} \qquad (60)$$

Proof As already mentioned above, conditions **A–C** ensure us that the inverse $_j\mathbf{U}^{(\varepsilon)}(\rho)$ exists for sufficiently small ε. In this case, it is known that the expansion (59) exists under condition $\mathbf{P_k^*}$. To see that the coefficients are given by (60), first note that

$$\mathbf{I} = (\mathbf{I} - {}_j\mathbf{P}^{(\varepsilon)}(\rho))_j\mathbf{U}^{(\varepsilon)}(\rho) \qquad (61)$$
$$= (\mathbf{I} - {}_j\mathbf{P}^{(0)}(\rho) - {}_j\mathbf{P}[\rho, 0, 1]\varepsilon - \cdots - {}_j\mathbf{P}[\rho, 0, k]\varepsilon^k + \mathbf{o}(\varepsilon^k))$$
$$\times (_j\mathbf{U}[\rho, 0] + {}_j\mathbf{U}[\rho, 1]\varepsilon + \cdots + {}_j\mathbf{U}[\rho, k]\varepsilon^k + \mathbf{o}(\varepsilon^k)).$$

By first expanding both sides of Eq. (61) and then, for $n = 0, 1, \dots, k$, equating coefficients of ε^n in the left and right hand sides, we get formula (60). $\qquad \square$

We are now ready to construct asymptotic expansions for $\Phi_j^{(\varepsilon)}(\rho, r)$.

Theorem 1 *Assume that conditions* **A–C** *and* $\mathbf{P_k^*}$ *hold. Then:*

(i) *We have the following asymptotic expansion,*

$$\Phi_j^{(\varepsilon)}(\rho) = \Phi_j[\rho, 0, 0] + \Phi_j[\rho, 0, 1]\varepsilon + \cdots + \Phi_j[\rho, 0, k]\varepsilon^k + \mathbf{o}(\varepsilon^k),$$

where

$$\Phi_j[\rho, 0, n] = \begin{cases} \Phi_j^{(0)}(\rho) & n = 0, \\ \sum_{q=0}^{n} {}_j\mathbf{U}[\rho, q]\mathbf{p}_j[\rho, 0, n-q] & n = 1, \dots, k. \end{cases}$$

(ii) *For* $r = 1, \dots, k$, *we have the following asymptotic expansions,*

$$\Phi_j^{(\varepsilon)}(\rho, r) = \Phi_j[\rho, r, 0] + \Phi_j[\rho, r, 1]\varepsilon + \cdots + \Phi_j[\rho, r, k-r]\varepsilon^{k-r} + \mathbf{o}(\varepsilon^{k-r}),$$

where

$$\Phi_j[\rho, r, n] = \begin{cases} \Phi_j^{(0)}(\rho, r) & n = 0, \\ \sum_{q=0}^{n} {}_j\mathbf{U}[\rho, q]\Lambda_j[\rho, r, n-q] & n = 1, \dots, k-r, \end{cases}$$

and, for $t = 0, \ldots, k - r$,

$$\Lambda_j[\rho, r, t] = \mathbf{p}_j[\rho, r, t] + \sum_{m=1}^{r} \binom{r}{m} \sum_{q=0}^{t} {}_j P[\rho, m, q] \Phi_j[\rho, r - m, t - q].$$

Before proceeding with the proof of Theorem 1 we would like to comment on the reason that the theorem is stated in such a way that $\Phi_j^{(\varepsilon)}(\rho, r)$, for $r = 1, \ldots, k$, has an expansion of order $k - r$. The reason is that this is exactly what we need for the main result in [6], which, we remind, is a sequel of the present paper. However, it is possible to construct asymptotic expansions of different orders than the ones stated in the theorem. In that case, appropriate changes in the perturbation condition should be made. The same remark applies to Lemma 5 and Theorem 2.

Proof Under conditions **A–C**, we have, for sufficiently small ε, that the recursive systems of linear equations given by relations (54)–(56), all have finite components. Moreover, the inverse matrix ${}_j U^{(\varepsilon)}(\rho) = (I - {}_j P^{(\varepsilon)}(\rho))^{-1}$ exists, so these systems have unique solutions.

It follows from (54), Lemma 4, and condition $\mathbf{P}_k^*$ that

$$\Phi_j^{(\varepsilon)}(\rho) = {}_j U^{(\varepsilon)}(\rho) \mathbf{p}_j^{(\varepsilon)}(\rho) \tag{62}$$

$$= ({}_j U[\rho, 0] + {}_j U[\rho, 1]\varepsilon + \cdots + {}_j U[\rho, k]\varepsilon^k + \mathbf{o}(\varepsilon^k))$$
$$\times (\mathbf{p}_j[\rho, 0, 0] + \mathbf{p}_j[\rho, 0, 1]\varepsilon + \cdots + \mathbf{p}_j[\rho, 0, k]\varepsilon^k + \mathbf{o}(\varepsilon^k)).$$

By expanding the right hand side of Eq. (62), we see that part (**i**) of Theorem 1 holds.

With $r = 1$, relation (56) takes the form

$$\Lambda_j^{(\varepsilon)}(\rho, 1) = \mathbf{p}_j^{(\varepsilon)}(\rho, 1) + {}_j P^{(\varepsilon)}(\rho, 1) \Phi_j^{(\varepsilon)}(\rho). \tag{63}$$

From (63), condition $\mathbf{P}_k^*$, and part (**i**), we get

$$\Lambda_j^{(\varepsilon)}(\rho, 1) = \mathbf{p}_j[\rho, 1, 0] + \cdots + \mathbf{p}_j[\rho, 1, k-1]\varepsilon^{k-1} + \mathbf{o}(\varepsilon^{k-1}) \tag{64}$$
$$+ ({}_j P[\rho, 1, 0] + \cdots + {}_j P[\rho, 1, k-1]\varepsilon^{k-1} + \mathbf{o}(\varepsilon^{k-1}))$$
$$\times (\Phi_j[\rho, 0, 0] + \cdots + \Phi_j[\rho, 0, k-1]\varepsilon^{k-1} + \mathbf{o}(\varepsilon^{k-1})).$$

Expanding the right hand side of (64) gives

$$\Lambda_j^{(\varepsilon)}(\rho, 1) = \Lambda_j[\rho, 1, 0] + \Lambda_j[\rho, 1, 1]\varepsilon + \cdots + \Lambda_j[\rho, 1, k-1]\varepsilon^{k-1} + \mathbf{o}(\varepsilon^{k-1}), \tag{65}$$

where

$$\Lambda_j[\rho, 1, t] = \mathbf{p}_j[\rho, 1, t] + \sum_{q=0}^{t} {}_j P[\rho, 1, q] \Phi_j[\rho, 0, t - q], \quad t = 0, \ldots, k - 1.$$

It now follows from (55), (65), and Lemma 4 that

$$\Phi_j^{(\varepsilon)}(\rho, 1) = {}_j \mathbf{U}^{(\varepsilon)}(\rho) \Lambda_j^{(\varepsilon)}(\rho, 1) \tag{66}$$
$$= ({}_j \mathbf{U}[\rho, 0] + \cdots + {}_j \mathbf{U}[\rho, k-1]\varepsilon^{k-1} + \mathbf{o}(\varepsilon^{k-1}))$$
$$\times (\Lambda_j[\rho, 1, 0] + \cdots + \Lambda_j[\rho, 1, k-1]\varepsilon^{k-1} + \mathbf{o}(\varepsilon^{k-1})).$$

By expanding the right hand side of Eq. (66) we get the expansion in part (ii) for $r = 1$. If $k = 1$, this concludes the proof. If $k \geq 2$, we can repeat the steps above, successively, for $r = 2, \ldots, k$. This gives the expansions and formulas given in part (ii). $\qquad\qquad\qquad\qquad\qquad\qquad\qquad\qquad\qquad\qquad\qquad\qquad\qquad\qquad\square$

Let us now define the following mixed power exponential moment functionals, for $i, j, s \in X$,

$$\omega_{ijs}^{(\varepsilon)}(\rho, r) = \sum_{n=0}^{\infty} n^r e^{\rho n} \mathsf{P}_i \{\xi^{(\varepsilon)}(n) = s, \ \mu_0^{(\varepsilon)} \wedge \mu_j^{(\varepsilon)} > n\}, \ \rho \in \mathbb{R}, \ r = 0, 1, \ldots.$$

Notice that $\omega_{ijs}^{(\varepsilon)}(\rho, 0) = \omega_{ijs}^{(\varepsilon)}(\rho)$.

It follows from conditions **A–C** and Lemma 3 that for $\rho < \delta$ and sufficiently small ε, the functions $\omega_{ijs}^{(\varepsilon)}(\rho)$ and $p_{ij}^{(\varepsilon)}(\rho)$ are arbitrarily many times differentiable with respect to ρ, and the derivatives of order r are given by $\omega_{ijs}^{(\varepsilon)}(\rho, r)$ and $p_{ij}^{(\varepsilon)}(\rho, r)$, respectively. Under these conditions we also have that the functions $\varphi_i^{(\varepsilon)}(\rho)$, defined by Eq. (11), are differentiable. Let us denote the corresponding derivatives by $\varphi_i^{(\varepsilon)}(\rho, r)$.

Recall from Sect. 3 that the functions $\omega_{ijs}^{(\varepsilon)}(\rho)$ satisfy the following system of linear equations,

$$\omega_{ijs}^{(\varepsilon)}(\rho) = \delta(i, s)\varphi_i^{(\varepsilon)}(\rho) + \sum_{l \neq 0, j} p_{il}^{(\varepsilon)}(\rho)\omega_{ljs}^{(\varepsilon)}(\rho), \ i, j, s \neq 0. \tag{67}$$

Differentiating relation (67) gives

$$\omega_{ijs}^{(\varepsilon)}(\rho, r) = \theta_{ijs}^{(\varepsilon)}(\rho, r) + \sum_{l \neq 0, j} p_{il}^{(\varepsilon)}(\rho)\omega_{ljs}^{(\varepsilon)}(\rho, r), \ r = 1, 2, \ldots, \ i, j, s \neq 0, \tag{68}$$

where

$$\theta_{ijs}^{(\varepsilon)}(\rho, r) = \delta(i, s)\varphi_i^{(\varepsilon)}(\rho, r) + \sum_{m=1}^{r} \binom{r}{m} \sum_{l \neq 0, j} p_{il}^{(\varepsilon)}(\rho, m)\omega_{ljs}^{(\varepsilon)}(\rho, r - m). \tag{69}$$

In order to rewrite these systems in matrix form, we define the following column vectors,

$$\Omega_{js}^{(\varepsilon)}(\rho, r) = \left[\omega_{1js}^{(\varepsilon)}(\rho, r) \cdots \omega_{Njs}^{(\varepsilon)}(\rho, r)\right]^{T}, \ j, s \neq 0, \tag{70}$$

$$\Theta_{js}^{(\varepsilon)}(\rho, r) = \left[\theta_{1js}^{(\varepsilon)}(\rho, r) \cdots \theta_{Njs}^{(\varepsilon)}(\rho, r)\right]^{T}, \ j, s \neq 0, \tag{71}$$

$$\widehat{\boldsymbol{\varphi}}_{s}^{(\varepsilon)}(\rho, r) = \left[\delta(1, s)\varphi_{1}^{(\varepsilon)}(\rho, r) \cdots \delta(N, s)\varphi_{N}^{(\varepsilon)}(\rho, r)\right]^{T}, \ s \neq 0. \tag{72}$$

Using (53) and (67)–(72), we can for cach $j, s \neq 0$ write the following recursive systems of linear equations,

$$\Omega_{js}^{(\varepsilon)}(\rho) = \widehat{\boldsymbol{\varphi}}_{s}^{(\varepsilon)}(\rho) + {}_{j}\mathbf{P}^{(\varepsilon)}(\rho)\Omega_{js}^{(\varepsilon)}(\rho), \tag{73}$$

and, for $r = 1, 2, \ldots,$

$$\Omega_{js}^{(\varepsilon)}(\rho, r) = \Theta_{js}^{(\varepsilon)}(\rho, r) + {}_{j}\mathbf{P}^{(\varepsilon)}(\rho)\Omega_{js}^{(\varepsilon)}(\rho, r), \tag{74}$$

where

$$\Theta_{js}^{(\varepsilon)}(\rho, r) = \widehat{\boldsymbol{\varphi}}_{s}^{(\varepsilon)}(\rho, r) + \sum_{m=1}^{r} \binom{r}{m} {}_{j}\mathbf{P}^{(\varepsilon)}(\rho, m)\Omega_{js}^{(\varepsilon)}(\rho, r - m). \tag{75}$$

In order to construct asymptotic expansions for the vectors $\Omega_{js}^{(\varepsilon)}(\rho, r)$, we can use the same technique as in Theorem 1. However, a preliminary step needed in this case is to construct asymptotic expansions for the functions $\varphi_{i}^{(\varepsilon)}(\rho, r)$. In order to do this, we first derive an expression for these functions.

Let us define

$$\psi_{i}^{(\varepsilon)}(\rho, r) = \sum_{n=0}^{\infty} n^{r} e^{\rho n} \mathbf{P}_{i}\{\kappa_{1}^{(\varepsilon)} = n\}, \ \rho \in \mathbb{R}, \ r = 0, 1, \ldots, \ i \in X. \tag{76}$$

Note that

$$\psi_{i}^{(\varepsilon)}(\rho, r) = \sum_{j \in X} p_{ij}^{(\varepsilon)}(\rho, r), \ \rho \in \mathbb{R}, \ r = 0, 1, \ldots, \ i \in X. \tag{77}$$

Thus, the functions $\psi_{i}^{(\varepsilon)}(\rho, 0)$ are arbitrarily many times differentiable with respect to ρ and the corresponding derivatives are given by $\psi_{i}^{(\varepsilon)}(\rho, r)$.

The function $\varphi_{i}^{(\varepsilon)}(\rho)$, defined by Eq. (11), can be written as

$$\varphi_{i}^{(\varepsilon)}(\rho) = \begin{cases} \psi_{i}^{(\varepsilon)}(0, 1) & \rho = 0, \\ (\psi_{i}^{(\varepsilon)}(\rho, 0) - 1)/(e^{\rho} - 1) & \rho \neq 0. \end{cases} \tag{78}$$

From (76) and (78) it follows that

$$\psi_i^{(\varepsilon)}(\rho, 0) = (e^\rho - 1)\varphi_i^{(\varepsilon)}(\rho) + 1, \ \rho \in \mathbb{R}. \tag{79}$$

Differentiating both sides of (79) gives

$$\psi_i^{(\varepsilon)}(\rho, r) = (e^\rho - 1)\varphi_i^{(\varepsilon)}(\rho, r) + e^\rho \sum_{m=0}^{r-1} \binom{r}{m} \varphi_i^{(\varepsilon)}(\rho, m), \ r = 1, 2, \ldots . \tag{80}$$

If $\rho = 0$, Eq. (80) implies

$$\psi_i^{(\varepsilon)}(0, r) = r\varphi_i^{(\varepsilon)}(0, r - 1) + \sum_{m=0}^{r-2} \binom{r}{m} \varphi_i^{(\varepsilon)}(0, m), \ r = 2, 3, \ldots .$$

From this it follows that, for $r = 1, 2, \ldots$,

$$\varphi_i^{(\varepsilon)}(0, r) = \frac{1}{r+1}\left(\psi_i^{(\varepsilon)}(0, r + 1) - \sum_{m=0}^{r-1} \binom{r+1}{m} \varphi_i^{(\varepsilon)}(0, m) \right). \tag{81}$$

If $\rho \neq 0$, Eq. (80) gives, for $r = 1, 2, \ldots$,

$$\varphi_i^{(\varepsilon)}(\rho, r) = \frac{1}{e^\rho - 1}\left(\psi_i^{(\varepsilon)}(\rho, r) - e^\rho \sum_{m=0}^{r-1} \binom{r}{m} \varphi_i^{(\varepsilon)}(\rho, m) \right). \tag{82}$$

Using relations (77), (81), and (82), we can recursively calculate the derivatives of $\varphi_i^{(\varepsilon)}(\rho)$. Furthermore, it follows directly from these formulas that we can construct asymptotic expansions for these derivatives. The formulas are given in the following lemma.

Lemma 5 *Assume that conditions **A–C** hold.*

(i) *If, in addition, condition $\mathbf{P_k^*}$ holds, then for each $i \neq 0$ and $r = 0, \ldots, k$ we have the following asymptotic expansion,*

$$\psi_i^{(\varepsilon)}(\rho, r) = \psi_i[\rho, r, 0] + \psi_i[\rho, r, 1]\varepsilon + \cdots + \psi_i[\rho, r, k - r]\varepsilon^{k-r} + o(\varepsilon^{k-r}),$$

where
$$\psi_i[\rho, r, n] = \sum_{j \in X} p_{ij}[\rho, r, n], \ n = 0, \ldots, k - r.$$

(ii) *If, in addition, $\rho = 0$ and condition $\mathbf{P_{k+1}^*}$ holds, then for each $i \neq 0$ and $r = 0, \ldots, k$ we have the following asymptotic expansion,*

$$\varphi_i^{(\varepsilon)}(0, r) = \varphi_i[0, r, 0] + \varphi_i[0, r, 1]\varepsilon + \cdots + \varphi_i[0, r, k - r]\varepsilon^{k-r} + o(\varepsilon^{k-r}),$$

where, for $n = 0, \ldots, k - r$,

$$\varphi_i[0, r, n] = \frac{1}{r+1} \left(\psi_i[0, r+1, n] - \sum_{m=0}^{r-1} \binom{r+1}{m} \varphi_i[0, m, n] \right).$$

(iii) *If, in addition, $\rho \neq 0$ and condition $\mathbf{P_k^*}$ holds, then for each $i \neq 0$ and $r = 0, \dots, k$ we have the following asymptotic expansion,*

$$\varphi_i^{(\varepsilon)}(\rho, r) = \varphi_i[\rho, r, 0] + \varphi_i[\rho, r, 1]\varepsilon + \cdots + \varphi_i[\rho, r, k-r]\varepsilon^{k-r} + o(\varepsilon^{k-r}),$$

where, for $n = 0, \dots, k - r$,

$$\varphi_i[\rho, r, n] = \frac{1}{e^\rho - 1} \left(\psi_i[\rho, r, n] - e^\rho \sum_{m=0}^{r-1} \binom{r}{m} \varphi_i[\rho, m, n] \right).$$

Using (72) and Lemma 5 we can now construct the following asymptotic expansions, for $r = 0, \dots, k$, and $s \neq 0$,

$$\widehat{\varphi}_s^{(\varepsilon)}(\rho, r) = \widehat{\varphi}_s[\rho, r, 0] + \widehat{\varphi}_s[\rho, r, 1]\varepsilon + \cdots + \widehat{\varphi}_s[\rho, r, k-r]\varepsilon^{k-r} + o(\varepsilon^{k-r}). \tag{83}$$

The next lemma gives asymptotic expansions for $\Omega_{js}^{(\varepsilon)}(\rho, r)$.

Theorem 2 *Assume that conditions **A–C** hold. If $\rho = 0$, we also assume that condition $\mathbf{P_{k+1}^*}$ holds. If $\rho \neq 0$, we also assume that condition $\mathbf{P_k^*}$ holds. Then:*

(i) *We have the following asymptotic expansion,*

$$\Omega_{js}^{(\varepsilon)}(\rho) = \Omega_{js}[\rho, 0, 0] + \Omega_{js}[\rho, 0, 1]\varepsilon + \cdots + \Omega_{js}[\rho, 0, k]\varepsilon^k + o(\varepsilon^k),$$

where

$$\Omega_{js}[\rho, 0, n] = \begin{cases} \Omega_{js}^{(0)}(\rho) & n = 0, \\ \sum_{q=0}^{n} {}_j U[\rho, q] \widehat{\varphi}_s[\rho, 0, n-q] & n = 1, \dots, k. \end{cases}$$

(ii) *For $r = 1, \dots, k$, we have the following asymptotic expansions,*

$$\Omega_{js}^{(\varepsilon)}(\rho, r) = \Omega_{js}[\rho, r, 0] + \Omega_{js}[\rho, r, 1]\varepsilon + \cdots + \Omega_{js}[\rho, r, k-r]\varepsilon^{k-r} + o(\varepsilon^{k-r}),$$

where

$$\Omega_{js}[\rho, r, n] = \begin{cases} \Omega_{js}^{(0)}(\rho, r) & n = 0, \\ \sum_{q=0}^{n} {}_j U[\rho, q] \Theta_{js}[\rho, r, n-q] & n = 1, \dots, k-r, \end{cases}$$

and, for $t = 0, \dots, k - r$,

$$\Theta_{js}[\rho, r, t] = \widehat{\varphi}_s[\rho, r, t] + \sum_{m=1}^{r} \binom{r}{m} \sum_{q=0}^{t} {}_j\mathbf{P}[\rho, m, q]\Omega_{js}[\rho, r - m, t - q].$$

Proof Under conditions **A–C**, we have, for sufficiently small ε, that the recursive systems of linear equations given by relations (73)–(75), all have finite components. Moreover, the inverse matrix ${}_j\mathbf{U}^{(\varepsilon)}(\rho) = (\mathbf{I} - {}_j\mathbf{P}^{(\varepsilon)}(\rho))^{-1}$ exists, so these systems have unique solutions. Since we, by Lemma 5, have the expansions given in Eq. (83), the proof is from this point analogous to the proof of Theorem 1. $\square$

References

1. Avrachenkov, K.E., Filar, J.A., Howlett, P.G.: Analytic Perturbation Theory and Its Applications. SIAM, Philadelphia (2013)
2. Gyllenberg, M., Silvestrov, D.S.: Quasi-Stationary Phenomena in Nonlinearly Perturbed Stochastic Systems. De Gruyter Expositions in Mathematics, vol. 44. Walter de Gruyter, Berlin (2008)
3. Kartashov, M.V.: Strong Stable Markov Chains. VSP, Utrecht and TBiMC, Kiev (1996)
4. Koroliuk, V.S., Limnios, N.: Stochastic Systems in Merging Phase Space. World Scientific, Singapore (2005)
5. Petersson, M.: Quasi-stationary asymptotics for perturbed semi-Markov processes in discrete time. Research Report 2015:2, Department of Mathematics, Stockholm University, 36 pp. (2015)
6. Petersson, M.: Asymptotics for quasi-stationary distributions of perturbed discrete time semi-Markov processes. In: Silvestrov, S., Rančić, M. (eds.) Engineering Mathematics II. Algebraic, Stochastic and Analysis Structures for Networks, Data Classification and Optimization. Springer, Berlin (2016)
7. Yin, G., Zhang, Q.: Continuous-Time Markov chains and applications. A Singular Perturbation Approach. Applications of Mathematics, vol. 37. Springer, New York (1998)

Asymptotics for Quasi-stationary Distributions of Perturbed Discrete Time Semi-Markov Processes

Mikael Petersson

Abstract In this paper we study quasi-stationary distributions of non-linearly perturbed semi-Markov processes in discrete time. This type of distributions are of interest for analysis of stochastic systems which have finite lifetimes but are expected to persist for a long time. We obtain asymptotic power series expansions for quasi-stationary distributions and it is shown how the coefficients in these expansions can be computed from a recursive algorithm. As an illustration of this algorithm, we present a numerical example for a discrete time Markov chain.

Keywords Semi-Markov process · Perturbation · Quasi-stationary distribution · Asymptotic expansion · Renewal equation · Markov chain

1 Introduction

This paper is a sequel of [22] where recursive algorithms for computing asymptotic expansions of moment functionals for non-linearly perturbed semi-Markov processes in discrete time are presented. Here, these expansions play a fundamental role for constructing asymptotic expansions of quasi-stationary distributions for such processes. Let us remark that all notation, conditions, and key results which we need here are repeated. However, some extensive formulas needed for computation of coefficients in certain asymptotic expansions are not repeated. Thus, the present paper is essentially self-contained.

Quasi-stationary distributions are useful for studies of stochastic systems with random lifetimes. Usually, for such systems, the evolution of some quantity of interest is described by some stochastic process and the lifetime of the system is the first time this process hits some absorbing subset of the state space. For such processes, the stationary distribution will be concentrated on this absorbing subset. However, if we expect that the system will persist for a long time, the stationary distribution may

M. Petersson (✉)
Department of Mathematics, Stockholm University, SE-106 91, Stockholm, Sweden
e-mail: mikpe274@gmail.com

© Springer International Publishing Switzerland 2016
S. Silvestrov and M. Rančić (eds.), *Engineering Mathematics II*,
Springer Proceedings in Mathematics & Statistics 179,
DOI 10.1007/978-3-319-42105-6_9

not be an appropriate measure for describing the long time behaviour of the process. Instead, it might be more relevant to consider so-called quasi-stationary distributions. This type of distributions is obtained by taking limits of transition probabilities which are conditioned on the event that the process has not yet been absorbed.

Models of the type described above arise in many areas of applications such as epidemics, genetics, population dynamics, queuing theory, reliability, and risk theory. For example, in population dynamics models the number of individuals may be modelled by some stochastic process and we can consider the extinction time of the population as the lifetime. In epidemic models, the process may describe the evolution of the number of infected individuals and we can regard the end of the epidemic as the lifetime.

We consider, for every $\varepsilon \geq 0$, a discrete time semi-Markov process $\xi^{(\varepsilon)}(n)$, $n = 0, 1, \ldots$, on a finite state space $X = \{0, 1, \ldots, N\}$. It is assumed that the process $\xi^{(\varepsilon)}(n)$ depends on ε in such a way that its transition probabilities are functions of ε which converge pointwise to the transition probabilities for the limiting process $\xi^{(0)}(n)$. Thus, we can interpret $\xi^{(\varepsilon)}(n)$, for $\varepsilon > 0$, as a perturbation of $\xi^{(0)}(n)$. Furthermore, it is assumed that the states $\{1, \ldots, N\}$ is a communicating class for ε small enough.

Under conditions mentioned above, some additional assumptions of finite exponential moments of distributions of transition times, and a condition which guarantees that the limiting semi-Markov process is non-periodic, a unique quasi-stationary distribution, independent of the initial state, can be defined for each sufficiently small ε by the following relation,

$$\pi_j^{(\varepsilon)} = \lim_{n \to \infty} P_i \left\{ \xi^{(\varepsilon)}(n) = j \mid \mu_0^{(\varepsilon)} > n \right\}, \quad i, j \neq 0,$$

where $\mu_0^{(\varepsilon)}$ is the first hitting time of state 0.

In the present paper, we are interested in the asymptotic behaviour of the quasi-stationary distribution as the perturbation parameter ε tends to zero. Specifically, an asymptotic power series expansion of the quasi-stationary distribution is constructed.

We allow for nonlinear perturbations, i.e., the transition probabilities may be nonlinear functions of ε. We do, however, restrict our consideration to smooth perturbations by assuming that certain mixed power-exponential moment functionals for transition probabilities, up to some order k, can be expanded in asymptotic power series with respect to ε.

In this case, we show that the quasi-stationary distribution has the following asymptotic expansion,

$$\pi_j^{(\varepsilon)} = \pi_j^{(0)} + \pi_j[1]\varepsilon + \cdots + \pi_j[k]\varepsilon^k + o(\varepsilon^k), \quad j \neq 0, \tag{1}$$

where the coefficients $\pi_j[1], \ldots, \pi_j[k]$, can be calculated from explicit recursive formulas. These formulas are functions of the coefficients in the expansions of the moment functionals mentioned above. The existence of the expansion (1) and the algorithm for computing the coefficients in this expansion is the main result of this paper.

It is worth mentioning that the asymptotic relation given by Eq. (1) simultaneously cover three different cases. In the simplest case, there exists $\varepsilon_0 > 0$ such that transitions to state 0 are not possible for any $\varepsilon \in [0, \varepsilon_0]$. In this case, relation (1) gives asymptotic expansions for stationary distributions. Then, we have an intermediate case where transitions to state 0 are possible for all $\varepsilon \in (0, \varepsilon_0]$ but not possible for $\varepsilon = 0$. In this case we have that $\mu_0^{(\varepsilon)} \to \infty$ in probability as $\varepsilon \to 0$. In the mathematically most difficult case, we have that transitions to state 0 are possible for all $\varepsilon \in [0, \varepsilon_0]$. In this case, the random variables $\mu_0^{(\varepsilon)}$ are stochastically bounded as $\varepsilon \to 0$.

The expansion (1) is given for continuous time semi-Markov processes in [13, 14]. However, the discrete time case is interesting in its own right and deserves a special treatment. In particular, a discrete time model is often a natural choice in applications where measures of some quantity of interest are only available at given time points, for example days or months. The proof of the result for the continuous time case, as well as the proofs in the present paper, is based on the theory of non-linearly perturbed renewal equations. For results related to continuous time in this line of research, we refer to the comprehensive book [14], which also contains an extensive bibliography of work in related areas. The corresponding theory for discrete time renewal equations has been developed in [9, 12, 19–21, 25].

Quasi-stationary distributions have been studied extensively since the 1960s. For some of the early works on Markov chains and semi-Markov processes, see, for example, [4, 5, 7, 10, 16, 24, 30]. A survey of quasi-stationary distributions for models with discrete state spaces and more references can be found in [29].

Studies of asymptotic properties for first hitting times, stationary distributions, and other characteristics for Markov chains with linear, polynomial, and analytic perturbations have attracted a lot of attention, see, for example, [1–3, 6, 8, 11, 15, 17, 18, 23, 27, 28, 31, 32]. Recently, some of the results of these papers have been extended to non-linearly perturbed semi-Markov processes. Using a method of sequential phase space reduction, asymptotic expansions for expected first hitting times and stationary distributions are given in [26]. This paper also contains an extensive bibliography.

Let us now briefly comment on the structure of the present paper. In Sect. 2, most of the notation we need are introduced and the main result is formulated. We apply the discrete time renewal theorem in order to get a formula for the quasi-stationary distribution in Sect. 3 and then the proof of the main result is presented in Sect. 4. Finally, in Sect. 5, we illustrate the results in the special case of discrete time Markov chains.

2 Main Result

In this section we first introduce most of the notation that will be used in the present paper and then we formulate the main result.

For each $\varepsilon > 0$, let $\xi^{(\varepsilon)}(n), n = 0, 1, \ldots,$ be a discrete time semi-Markov process on the state space $X = \{0, 1, \ldots, N\}$, generated by the discrete time Markov renewal process $(\eta_n^{(\varepsilon)}, \kappa_n^{(\varepsilon)}), n = 0, 1, \ldots,$ having state space $X \times \{1, 2, \ldots\}$ and transition probabilities

$$Q_{ij}^{(\varepsilon)}(k) = \mathsf{P}\left\{\eta_{n+1}^{(\varepsilon)} = j, \ \kappa_{n+1}^{(\varepsilon)} = k \mid \eta_n^{(\varepsilon)} = i, \ \kappa_n^{(\varepsilon)} = l\right\}, \ k, l = 1, 2, \ldots, \ i, j \in X.$$

We can write the transition probabilities as $Q_{ij}^{(\varepsilon)}(k) = p_{ij}^{(\varepsilon)} f_{ij}^{(\varepsilon)}(k)$, where $p_{ij}^{(\varepsilon)}$ are transition probabilities for the embedded Markov chain $\eta_n^{(\varepsilon)}$ and

$$f_{ij}^{(\varepsilon)}(k) = \mathsf{P}\left\{\kappa_{n+1}^{(\varepsilon)} = k \mid \eta_n^{(\varepsilon)} = i, \ \eta_{n+1}^{(\varepsilon)} = j\right\}, \ k = 1, 2, \ldots, \ i, j \in X,$$

are conditional distributions of transition times.

Let us here remark that definitions of discrete time semi-Markov processes and Markov renewal processes can be found in, for example, [22].

For each $j \in X$, let $v_j^{(\varepsilon)} = \min\{n \geq 1 : \eta_n^{(\varepsilon)} = j\}$ and $\mu_j^{(\varepsilon)} = \kappa_1^{(\varepsilon)} + \cdots + \kappa_{v_j^{(\varepsilon)}}^{(\varepsilon)}$. By definition, $v_j^{(\varepsilon)}$ and $\mu_j^{(\varepsilon)}$ are the first hitting times of state j for the embedded Markov chain and the semi-Markov process, respectively.

In what follows, we use P_i and E_i to denote probabilities and expectations conditioned on the event $\{\eta_0^{(\varepsilon)} = i\}$.

Let us define

$$g_{ij}^{(\varepsilon)}(n) = \mathsf{P}_i\left\{\mu_j^{(\varepsilon)} = n, \ v_0^{(\varepsilon)} > v_j^{(\varepsilon)}\right\}, \ n = 0, 1, \ldots, \ i, j \in X,$$

and

$$g_{ij}^{(\varepsilon)} = \mathsf{P}_i\left\{v_0^{(\varepsilon)} > v_j^{(\varepsilon)}\right\}, \ i, j \in X.$$

The functions $g_{ij}^{(\varepsilon)}(n)$ define discrete probability distributions which may be improper, i.e., $\sum_{n=0}^{\infty} g_{ij}^{(\varepsilon)}(n) = g_{ij}^{(\varepsilon)} \leq 1$.

Let us also define the following mixed power-exponential moment functionals,

$$p_{ij}^{(\varepsilon)}(\rho, r) = \sum_{n=0}^{\infty} n^r e^{\rho n} Q_{ij}^{(\varepsilon)}(n), \ \rho \in \mathbb{R}, \ r = 0, 1, \ldots, \ i, j \in X,$$

$$\phi_{ij}^{(\varepsilon)}(\rho, r) = \sum_{n=0}^{\infty} n^r e^{\rho n} g_{ij}^{(\varepsilon)}(n), \ \rho \in \mathbb{R}, \ r = 0, 1, \ldots, \ i, j \in X,$$

$$\omega_{ijs}^{(\varepsilon)}(\rho, r) = \sum_{n=0}^{\infty} n^r e^{\rho n} h_{ijs}^{(\varepsilon)}(n), \ \rho \in \mathbb{R}, \ r = 0, 1, \ldots, \ i, j, s \in X,$$

where $h_{ijs}^{(\varepsilon)}(n) = P_i\{\xi^{(\varepsilon)}(n) = s, \; \mu_0^{(\varepsilon)} \wedge \mu_j^{(\varepsilon)} > n\}$. For convenience, we denote

$$p_{ij}^{(\varepsilon)}(\rho) = p_{ij}^{(\varepsilon)}(\rho, 0), \quad \phi_{ij}^{(\varepsilon)}(\rho) = \phi_{ij}^{(\varepsilon)}(\rho, 0), \quad \omega_{ijs}^{(\varepsilon)}(\rho) = \omega_{ijs}^{(\varepsilon)}(\rho, 0).$$

We now introduce the following conditions:

A: (a) $p_{ij}^{(\varepsilon)} \to p_{ij}^{(0)}$, as $\varepsilon \to 0, i \neq 0, j \in X$.

    (b) $f_{ij}^{(\varepsilon)}(n) \to f_{ij}^{(0)}(n)$, as $\varepsilon \to 0, n = 1, 2, \ldots, i \neq 0, j \in X$.

B: $g_{ij}^{(0)} > 0, i, j \neq 0$.

C: There exists $\beta > 0$ such that:

    (a) $\lim\sup_{0 \leq \varepsilon \to 0} p_{ij}^{(\varepsilon)}(\beta) < \infty$, for all $i \neq 0, j \in X$.

    (b) $\phi_{ii}^{(0)}(\beta_i) \in (1, \infty)$, for some $i \neq 0$ and $\beta_i \leq \beta$.

D: $g_{ii}^{(0)}(n)$ is a non-periodic distribution for some $i \neq 0$.

Under the conditions stated above, there exists, for sufficiently small ε, so-called quasi-stationary distributions, which are independent of the initial state $i \neq 0$, and given by the relation

$$\pi_j^{(\varepsilon)} = \lim_{n \to \infty} P_i \left\{ \xi^{(\varepsilon)}(n) = j \mid \mu_0^{(\varepsilon)} > n \right\}, \; j \neq 0. \tag{2}$$

An important role for the quasi-stationary distribution is played by the following characteristic equation,

$$\phi_{ii}^{(\varepsilon)}(\rho) = 1, \tag{3}$$

where $i \neq 0$ is arbitrary.

The following lemma summarizes some important properties for the root of Eq. (3). A proof is given in [22].

Lemma 9.1 *Under conditions A–C there exists, for sufficiently small ε, a unique non-negative solution $\rho^{(\varepsilon)}$ of the characteristic equation (3) which is independent of i. Moreover, $\rho^{(\varepsilon)} \to \rho^{(0)}$, as $\varepsilon \to 0$.*

In order to construct an asymptotic expansion for the quasi-stationary distribution, we need a perturbation condition for the transition probabilities $Q_{ij}^{(\varepsilon)}(k)$ which is stronger than **A**. This condition is formulated in terms of the moment functionals $p_{ij}^{(\varepsilon)}(\rho^{(0)}, r)$.

P$_k$: $p_{ij}^{(\varepsilon)}(\rho^{(0)}, r) = p_{ij}^{(0)}(\rho^{(0)}, r) + p_{ij}[\rho^{(0)}, r, 1]\varepsilon + \cdots + p_{ij}[\rho^{(0)}, r, k - r]\varepsilon^{k-r} + o(\varepsilon^{k-r})$, for $r = 0, \ldots, k, i \neq 0, j \in X$, where $|p_{ij}[\rho^{(0)}, r, n]| < \infty$, for $r = 0, \ldots, k, n = 1, \ldots, k - r, i \neq 0, j \in X$.

The following theorem is the main result of this paper. The proof is given in Sect. 4.

Theorem 9.1 *If conditions* **A–D** *and* $\mathbf{P_{k+1}}$ *hold, then we have the following asymptotic expansion,*

$$\pi_j^{(\varepsilon)} = \pi_j^{(0)} + \pi_j[1]\varepsilon + \cdots + \pi_j[k]\varepsilon^k + o(\varepsilon^k), \ j \neq 0,$$

where $\pi_j[n]$, $n = 1, \ldots, k$, $j \neq 0$, *can be calculated from a recursive algorithm which is described in Sect. 4.*

3   Quasi-stationary Distributions

In this section we use renewal theory in order to get a formula for the quasi-stationary distribution.

The probabilities $P_{ij}^{(\varepsilon)}(n) = \mathsf{P}_i\{\xi^{(\varepsilon)}(n) = j, \ \mu_0^{(\varepsilon)} > n\}$, $i, j \neq 0$, satisfy the following discrete time renewal equation,

$$P_{ij}^{(\varepsilon)}(n) = h_{ij}^{(\varepsilon)}(n) + \sum_{k=0}^{n} P_{ij}^{(\varepsilon)}(n - k) g_{ii}^{(\varepsilon)}(k), \ n = 0, 1, \ldots, \tag{4}$$

where

$$h_{ij}^{(\varepsilon)}(n) = \mathsf{P}_i\{\xi^{(\varepsilon)}(n) = j, \ \mu_0^{(\varepsilon)} \wedge \mu_i^{(\varepsilon)} > n\}.$$

Since $\sum_{n=0}^{\infty} g_{ii}^{(\varepsilon)}(n) = g_{ii}^{(\varepsilon)} \leq 1$, relation (4) defines a possibly improper renewal equation.

Let us now, for each $n = 0, 1, \ldots$, multiply both sides of (4) by $e^{\rho^{(\varepsilon)} n}$, where $\rho^{(\varepsilon)}$ is the root of the characteristic equation $\phi_{ii}^{(\varepsilon)}(\rho) = 1$. Then, we get

$$\widetilde{P}_{ij}^{(\varepsilon)}(n) = \widetilde{h}_{ij}^{(\varepsilon)}(n) + \sum_{k=0}^{n} \widetilde{P}_{ij}^{(\varepsilon)}(n - k) \widetilde{g}_{ii}^{(\varepsilon)}(k), \ n = 0, 1, \ldots, \tag{5}$$

where

$$\widetilde{P}_{ij}^{(\varepsilon)}(n) = e^{\rho^{(\varepsilon)} n} P_{ij}^{(\varepsilon)}(n), \ \widetilde{h}_{ij}^{(\varepsilon)}(n) = e^{\rho^{(\varepsilon)} n} h_{ij}^{(\varepsilon)}(n), \ \widetilde{g}_{ii}^{(\varepsilon)}(n) = e^{\rho^{(\varepsilon)} n} g_{ii}^{(\varepsilon)}(n).$$

By the definition of the root of the characteristic equation, relation (5) defines a proper renewal equation.

In order to prove our next result, we first formulate an auxiliary lemma. A proof can be found in [22].

Lemma 9.2 *Assume that conditions* **A–C** *hold. Then there exists* $\delta > \rho^{(0)}$ *such that:*

(i) $\phi_{kj}^{(\varepsilon)}(\rho) \to \phi_{kj}^{(0)}(\rho) < \infty$, *as* $\varepsilon \to 0$, $\rho \leq \delta$, $k, j \neq 0$.

(ii) $\omega_{kjs}^{(\varepsilon)}(\rho) \to \omega_{kjs}^{(0)}(\rho) < \infty$, *as* $\varepsilon \to 0$, $\rho \leq \delta$, $k, j, s \neq 0$.

We can now use the classical discrete time renewal theorem in order to get a formula for the quasi-stationary distribution.

Lemma 9.3 *Assume that conditions **A**–**D** hold. Then:*

(i) *For sufficiently small ε, the quasi stationary distribution $\pi_j^{(\varepsilon)}$, given by relation (2), have the following representation,*

$$\pi_j^{(\varepsilon)} = \frac{\omega_{iij}^{(\varepsilon)}(\rho^{(\varepsilon)})}{\omega_{ii1}^{(\varepsilon)}(\rho^{(\varepsilon)}) + \cdots + \omega_{iiN}^{(\varepsilon)}(\rho^{(\varepsilon)})}, \ i, j \neq 0. \tag{6}$$

(ii) *For $j = 1, \ldots, N$, we have*

$$\pi_j^{(\varepsilon)} \to \pi_j^{(0)}, \ as \ \varepsilon \to 0.$$

Proof Under condition **D**, the functions $g_{ii}^{(0)}(n)$ are non-periodic for all $i \neq 0$. By Lemma 9.2 we have that $\phi_{ii}^{(\varepsilon)}(\rho) \to \phi_{ii}^{(0)}(\rho)$ as $\varepsilon \to 0$, for $\rho \leq \delta$, $i \neq 0$. From this it follows that $g_{ii}^{(\varepsilon)}(n) \to g_{ii}^{(0)}(n)$ as $\varepsilon \to 0$, for $n \geq 0$, $i \neq 0$. Thus, we can conclude that there exists $\varepsilon_1 > 0$ such that the functions $\widetilde{g}_{ii}^{(\varepsilon)}(n)$, $i \neq 0$, are non-periodic for all $\varepsilon \leq \varepsilon_1$.

Now choose γ such that $\rho^{(0)} < \gamma < \delta$. Using Lemmas 9.1 and 9.2, we get the following for all $i \neq 0$,

$$\limsup_{0 \leq \varepsilon \to 0} \sum_{n=0}^{\infty} n\widetilde{g}_{ii}^{(\varepsilon)}(n) \leq \limsup_{0 \leq \varepsilon \to 0} \sum_{n=0}^{\infty} ne^{\gamma n} g_{ii}^{(\varepsilon)}(n)$$

$$\leq \left(\sup_{n \geq 0} ne^{-(\delta-\gamma)n} \right) \phi_{ii}^{(0)}(\delta) < \infty.$$

Thus, there exists $\varepsilon_2 > 0$ such that the distributions $\widetilde{g}_{ii}^{(\varepsilon)}(n)$, $i \neq 0$, have finite mean for all $\varepsilon \leq \varepsilon_2$.

Furthermore, it follows from Lemmas 9.1 and 9.2 that, for all $i, j \neq 0$,

$$\limsup_{0 \leq \varepsilon \to 0} \sum_{n=0}^{\infty} \widetilde{h}_{ij}^{(\varepsilon)}(n) \leq \limsup_{0 \leq \varepsilon \to 0} \sum_{n=0}^{\infty} e^{\delta n} h_{ij}^{(\varepsilon)}(n) = \omega_{iij}^{(0)}(\delta) < \infty,$$

so there exists $\varepsilon_3 > 0$ such that $\sum_{n=0}^{\infty} \widetilde{h}_{ij}^{(\varepsilon)}(n) < \infty$, $i, j \neq 0$, for all $\varepsilon \leq \varepsilon_3$.

Now, let $\varepsilon_0 = \min\{\varepsilon_1, \varepsilon_2, \varepsilon_3\}$. For all $\varepsilon \leq \varepsilon_0$, the assumptions of the discrete time renewal theorem are satisfied for the renewal equation defined by (5). This yields

$$\widetilde{P}_{ij}^{(\varepsilon)}(n) \to \frac{\sum_{k=0}^{\infty} \widetilde{h}_{ij}^{(\varepsilon)}(k)}{\sum_{k=0}^{\infty} k\widetilde{g}_{ii}^{(\varepsilon)}(k)}, \ as \ n \to \infty, \ i, j \neq 0, \ \varepsilon \leq \varepsilon_0. \tag{7}$$

Note that we have

$$\mathsf{P}_i\{\xi^{(\varepsilon)}(n) = j \mid \mu_0^{(\varepsilon)} > n\} = \frac{\widetilde{P}_{ij}^{(\varepsilon)}(n)}{\sum_{k=1}^N \widetilde{P}_{ik}^{(\varepsilon)}(n)}, \quad n = 0, 1, \ldots, \; i, j \neq 0. \qquad (8)$$

It follows from (7) and (8) that, for $\varepsilon \leq \varepsilon_0$,

$$\mathsf{P}_i\{\xi^{(\varepsilon)}(n) = j \mid \mu_0^{(\varepsilon)} > n\} \to \frac{\omega_{iij}^{(\varepsilon)}(\rho^{(\varepsilon)})}{\sum_{k=1}^N \omega_{iik}^{(\varepsilon)}(\rho^{(\varepsilon)})}, \quad \text{as } n \to \infty, \; i, j \neq 0.$$

This proves part (**i**).

For the proof of part (**ii**), first note that,

$$0 \leq \limsup_{0 \leq \varepsilon \to 0} \sum_{n=N}^{\infty} e^{\rho^{(\varepsilon)} n} h_{ij}^{(\varepsilon)}(n) \qquad (9)$$

$$\leq \limsup_{0 \leq \varepsilon \to 0} \sum_{n=N}^{\infty} e^{\gamma n} h_{ij}^{(\varepsilon)}(n)$$

$$\leq e^{-(\delta - \gamma)N} \omega_{iij}^{(0)}(\delta) < \infty, \quad N = 1, 2, \ldots, \; i, j \neq 0.$$

Relation (9) implies that

$$\lim_{N \to \infty} \limsup_{0 \leq \varepsilon \to 0} \sum_{n=N}^{\infty} e^{\rho^{(\varepsilon)} n} h_{ij}^{(\varepsilon)}(n) = 0, \; i, j \neq 0. \qquad (10)$$

It follows from Lemma 9.1 that

$$\rho^{(\varepsilon)} \to \rho^{(0)}, \quad \text{as } \varepsilon \to 0. \qquad (11)$$

Since $h_{ij}^{(\varepsilon)}(n)$, for each $n = 0, 1, \ldots$, can be written as a finite sum where each term in the sum is a continuous function of the quantities given in condition **A**, we have

$$h_{ij}^{(\varepsilon)}(n) \to h_{ij}^{(0)}(n), \quad \text{as } \varepsilon \to 0, \; i, j \neq 0. \qquad (12)$$

It now follows from (10)–(12) that

$$\omega_{iij}^{(\varepsilon)}(\rho^{(\varepsilon)}) \to \omega_{iij}^{(0)}(\rho^{(0)}), \quad \text{as } \varepsilon \to 0, \; i, j \neq 0. \qquad (13)$$

Relations (6) and (13) show that part (**ii**) of Lemma 9.3 holds. $\qquad\qquad\square$

4 Proof of the Main Result

In this section we prove Theorem 9.1.

Throughout this section, it is assumed that conditions **A–D** and $\mathbf{P_{k+1}}$ hold.

The proof is given in a sequence of lemmas. For the proof of the first lemma, we refer to [22].

Lemma 9.4 *For $r = 0, \ldots, k$ and $i, j \neq 0$ we have the following asymptotic expansions,*

$$\omega_{iij}^{(\varepsilon)}(\rho^{(0)}, r) = a_{ij}[r, 0] + a_{ij}[r, 1]\varepsilon + \cdots + a_{ij}[r, k - r]\varepsilon^{k-r} + o(\varepsilon^{k-r}), \quad (14)$$

$$\phi_{ii}^{(\varepsilon)}(\rho^{(0)}, r) = b_i[r, 0] + b_i[r, 1]\varepsilon + \cdots + b_i[r, k - r]\varepsilon^{k-r} + o(\varepsilon^{k-r}), \quad (15)$$

where the coefficients in these expansions can be calculated from lemmas and theorems given in [22].

Let us now recall from Sect. 3 that the quasi-stationary distribution, for sufficiently small ε, has the following representation,

$$\pi_j^{(\varepsilon)} = \frac{\omega_{iij}^{(\varepsilon)}(\rho^{(\varepsilon)})}{\omega_{ii1}^{(\varepsilon)}(\rho^{(\varepsilon)}) + \cdots + \omega_{iiN}^{(\varepsilon)}(\rho^{(\varepsilon)})}, \quad j = 1, \ldots, N. \quad (16)$$

The construction of the asymptotic expansion for the quasi-stationary distribution will be realized in three steps. First, we use the coefficients in the expansions given by (15) to build an asymptotic expansion for $\rho^{(\varepsilon)}$, the root of the characteristic equation. Then, the coefficients in this expansion and the coefficients in the expansions given by (14) are used to construct asymptotic expansions for $\omega_{iij}^{(\varepsilon)}(\rho^{(\varepsilon)})$. Finally, relation (16) is used to complete the proof.

We formulate these steps in the following three lemmas. Let us here remark that the proof of Lemma 9.5 is given in [25] in the context of general discrete time renewal equations and the proofs of Lemmas 9.6 and 9.7 are given in [20] in the context of quasi-stationary distributions for discrete time regenerative processes. In order to make the paper more self-contained, we also give the proofs here, in slightly reduced forms.

Lemma 9.5 *The root of the characteristic equation has the following asymptotic expansion,*

$$\rho^{(\varepsilon)} = \rho^{(0)} + c_1\varepsilon + \cdots + c_k\varepsilon^k + o(\varepsilon^k),$$

where $c_1 = -b_i[0, 1]/b_i[1, 0]$ and, for $n = 2, \ldots, k$,

$$c_n = -\frac{1}{b_i[1,0]}\left(b_i[0,n] + \sum_{q=1}^{n-1} b_i[1,n-q]c_q\right.$$

$$\left.+\sum_{m=2}^{n}\sum_{q=m}^{n} b_i[m,n-q]\cdot\sum_{n_1,\ldots,n_{q-1}\in D_{m,q}}\prod_{p=1}^{q-1}\frac{c_p^{n_p}}{n_p!}\right),$$

where $D_{m,q}$ is the set of all non-negative integer solutions of the system

$$n_1 + \cdots + n_{q-1} = m, \quad n_1 + 2n_2 + \cdots + (q-1)n_{q-1} = q.$$

Proof Let $\Delta^{(\varepsilon)} = \rho^{(\varepsilon)} - \rho^{(0)}$. It follows from the Taylor expansion of the exponential function that, for $n = 0, 1, \ldots,$

$$e^{\rho^{(\varepsilon)}n} = e^{\rho^{(0)}n}\left(\sum_{r=0}^{k}\frac{(\Delta^{(\varepsilon)})^r n^r}{r!} + \frac{(\Delta^{(\varepsilon)})^{k+1}n^{k+1}}{(k+1)!}e^{|\Delta^{(\varepsilon)}|n}\zeta_{k+1}^{(\varepsilon)}(n)\right), \qquad (17)$$

where $0 \le \zeta_{k+1}^{(\varepsilon)}(n) \le 1$.

If we multiply both sides of (17) by $g_{ii}^{(\varepsilon)}(n)$, sum over all n, and use that $\rho^{(\varepsilon)}$ is the root of the characteristic equation, we get

$$1 = \sum_{r=0}^{k}\frac{(\Delta^{(\varepsilon)})^r}{r!}\phi_{ii}^{(\varepsilon)}(\rho^{(0)},r) + (\Delta^{(\varepsilon)})^{k+1}M_{k+1}^{(\varepsilon)}, \qquad (18)$$

where

$$M_{k+1}^{(\varepsilon)} = \frac{1}{(k+1)!}\sum_{n=0}^{\infty}n^{k+1}e^{(\rho^{(0)}+|\Delta^{(\varepsilon)}|)n}\zeta_{k+1}^{(\varepsilon)}(n)g_{ii}^{(\varepsilon)}(n). \qquad (19)$$

Let $\delta > \rho^{(0)}$ be the value from Lemma 9.2. It follows from Lemma 9.1 that $|\Delta^{(\varepsilon)}| \to 0$ as $\varepsilon \to 0$, so there exist $\beta > 0$ and $\varepsilon_1(\beta) > 0$ such that

$$\rho^{(0)} + |\Delta^{(\varepsilon)}| \le \beta < \delta, \quad \varepsilon \le \varepsilon_1(\beta). \qquad (20)$$

Since $\beta < \delta$, Lemma 9.2 implies that there exists $\varepsilon_2(\beta) > 0$ such that

$$\phi_{ii}^{(\varepsilon)}(\beta,r) < \infty, \quad r = 0, 1, \ldots, \quad \varepsilon \le \varepsilon_2(\beta). \qquad (21)$$

Let $\varepsilon_0 = \varepsilon_0(\beta) = \min\{\varepsilon_1(\beta), \varepsilon_2(\beta)\}$. Then, relations (19)–(21) imply that

$$M_{k+1}^{(\varepsilon)} \le \frac{1}{(k+1)!}\phi_{ii}^{(\varepsilon)}(\beta,k+1) < \infty, \quad \varepsilon \le \varepsilon_0. \qquad (22)$$

It follows from (22) that we can rewrite (18) as

$$1 = \sum_{r=0}^{k} \frac{(\Delta^{(\varepsilon)})^r}{r!} \phi_{ii}^{(\varepsilon)}(\rho^{(0)}, r) + (\Delta^{(\varepsilon)})^{k+1} M_{k+1} \zeta_{k+1}^{(\varepsilon)}, \qquad (23)$$

where $M_{k+1} = \sup_{\varepsilon \leq \varepsilon_0} M_{k+1}^{(\varepsilon)} < \infty$ and $0 \leq \zeta_{k+1}^{(\varepsilon)} \leq 1$.

From relation (23) we can successively construct the asymptotic expansion for the root of the characteristic equation.

Let us first assume that $k = 1$. In this case (23) implies that

$$1 = \phi_{ii}^{(\varepsilon)}(\rho^{(0)}, 0) + \Delta^{(\varepsilon)} \phi_{ii}^{(\varepsilon)}(\rho^{(0)}, 1) + (\Delta^{(\varepsilon)})^2 O(1). \qquad (24)$$

Using (15), (24), and that $\Delta^{(\varepsilon)} \to 0$ as $\varepsilon \to 0$, it follows that

$$- b_i[0, 1]\varepsilon = \Delta^{(\varepsilon)}(b_i[1, 0] + o(1)) + o(\varepsilon). \qquad (25)$$

Dividing both sides of Eq. (25) by ε and letting ε tend to zero, we can conclude that $\Delta^{(\varepsilon)}/\varepsilon \to -b_i[0, 1]/b_i[1, 0]$ as $\varepsilon \to 0$. From this it follows that we have the representation

$$\Delta^{(\varepsilon)} = c_1 \varepsilon + \Delta_1^{(\varepsilon)}, \qquad (26)$$

where $c_1 = -b_i]0, 1]/b_i[1, 0]$ and $\Delta_1^{(\varepsilon)}/\varepsilon \to 0$ as $\varepsilon \to 0$.

This proves Lemma 9.5 for the case $k = 1$.

Let us now assume that $k = 2$. In this case relation (23) implies that

$$1 = \phi_{ii}^{(\varepsilon)}(\rho^{(0)}, 0) + \Delta^{(\varepsilon)} \phi_{ii}^{(\varepsilon)}(\rho^{(0)}, 1) + \frac{(\Delta^{(\varepsilon)})^2}{2} \phi_{ii}^{(\varepsilon)}(\rho^{(0)}, 2) + (\Delta^{(\varepsilon)})^3 O(1). \quad (27)$$

Using (15) and (26) in relation (27) and rearranging gives

$$- \left(b_i[0, 2] + b_i[1, 1]c_1 + \frac{b_i[2, 0]c_1^2}{2} \right) \varepsilon^2 = \Delta_1^{(\varepsilon)}(b_i[1, 0] + o(1)) + o(\varepsilon^2). \quad (28)$$

Dividing both sides of Eq. (28) by ε^2 and letting ε tend to zero, we can conclude that $\Delta_1^{(\varepsilon)}/\varepsilon^2 \to c_2$ as $\varepsilon \to 0$, where

$$c_2 = -\frac{1}{b_i[1, 0]} \left(b_i[0, 2] + b_i[1, 1]c_1 + \frac{b_i[2, 0]c_1^2}{2} \right).$$

From this and (26) it follows that we have the representation

$$\Delta^{(\varepsilon)} = c_1 \varepsilon + c_2 \varepsilon^2 + \Delta_2^{(\varepsilon)},$$

where $\Delta_2^{(\varepsilon)}/\varepsilon^2 \to 0$ as $\varepsilon \to 0$.

This proves Lemma 9.5 for the case $k = 2$.

Continuing in this way we can prove the lemma for any positive integer k. However, once it is known that the expansion exists, the coefficients can be obtained in a simpler way. From (15) and (23) we get the following formal equation,

$$- (b_i[0, 1]\varepsilon + b_i[0, 2]\varepsilon^2 + \cdots) \tag{29}$$
$$= (c_1\varepsilon + c_2\varepsilon^2 + \cdots)(b_i[1, 0] + b_i[1, 1]\varepsilon + \cdots)$$
$$+ (1/2!)(c_1\varepsilon + c_2\varepsilon^2 + \cdots)^2(b_i[2, 0] + b_i[2, 1]\varepsilon + \cdots) + \cdots .$$

By expanding the right hand side of (29) and then equating coefficients of equal powers of ε in the left and right hand sides, we obtain the formulas given in Lemma 9.5. □

Lemma 9.6 *For any i, $j \neq 0$, we have the following asymptotic expansion,*

$$\omega_{iij}^{(\varepsilon)}(\rho^{(\varepsilon)}) = \omega_{iij}^{(0)}(\rho^{(0)}) + d_{ij}[1]\varepsilon + \cdots + d_{ij}[k]\varepsilon^k + o(\varepsilon^k),$$

where $d_{ij}[1] = a_{ij}[0, 1] + a_{ij}[1, 0]c_1$, and, for $n = 2, \ldots, k$,

$$d_{ij}[n] = a_{ij}[0, n] + \sum_{q=1}^{n} a_{ij}[1, n - q]c_q + \sum_{m=2}^{n} \sum_{q=m}^{n} a_{ij}[m, n - q] \cdot \sum_{n_1, \ldots, n_{q-1} \in D_{m,q}} \prod_{p=1}^{q-1} \frac{c_p^{n_p}}{n_p!},$$

where $D_{m,q}$ is the set of all non-negative integer solutions of the system

$$n_1 + \cdots + n_{q-1} = m, \quad n_1 + 2n_2 + \cdots + (q - 1)n_{q-1} = q.$$

Proof Let us again use relation (17) given in the proof of Lemma 9.5. Multiplying both sides of (17) by $h_{ij}^{(\varepsilon)}(n)$ and summing over all n we get

$$\omega_{iij}^{(\varepsilon)}(\rho^{(\varepsilon)}) = \sum_{r=0}^{k} \frac{(\Delta^{(\varepsilon)})^r}{r!} \omega_{iij}^{(\varepsilon)}(\rho^{(0)}, r) + (\Delta^{(\varepsilon)})^{k+1} \widetilde{M}_{k+1}^{(\varepsilon)}, \tag{30}$$

where

$$\widetilde{M}_{k+1}^{(\varepsilon)} = \frac{1}{(k + 1)!} \sum_{n=0}^{\infty} n^{k+1} e^{(\rho^{(0)} + |\Delta^{(\varepsilon)}|)n} \zeta_{k+1}^{(\varepsilon)}(n) h_{ij}^{(\varepsilon)}(n).$$

Using similar arguments as in the proof of Lemma 9.5 we can rewrite (30) as

$$\omega_{iij}^{(\varepsilon)}(\rho^{(\varepsilon)}) = \sum_{r=0}^{k} \frac{(\Delta^{(\varepsilon)})^r}{r!} \omega_{iij}^{(\varepsilon)}(\rho^{(0)}, r) + (\Delta^{(\varepsilon)})^{k+1} \widetilde{M}_{k+1} \zeta_{k+1}^{(\varepsilon)}, \tag{31}$$

where $\widetilde{M}_{k+1} = \sup_{\varepsilon \leq \varepsilon_0} \widetilde{M}_{k+1}^{(\varepsilon)} < \infty$, for some $\varepsilon_0 > 0$, and $0 \leq \zeta_{k+1}^{(\varepsilon)} \leq 1$.

From Lemma 9.5 we have the following asymptotic expansion,

$$\Delta^{(\varepsilon)} = c_1\varepsilon + \cdots + c_k\varepsilon^k + o(\varepsilon^k). \tag{32}$$

Substituting the expansions (14) and (32) into relation (31) yields

$$\omega_{iij}^{(\varepsilon)}(\rho^{(\varepsilon)}) = \omega_{iij}^{(0)}(\rho^{(0)}) + a_{ij}[0, 1]\varepsilon + \cdots + a_{ij}[0, k]\varepsilon^k + o(\varepsilon^k) \tag{33}$$
$$+ (c_1\varepsilon + \cdots + c_k\varepsilon^k + o(\varepsilon^k))$$
$$\times (a_{ij}[1, 0] + a_{ij}[1, 1]\varepsilon + \cdots + a_{ij}[1, k-1]\varepsilon^{k-1} + o(\varepsilon^{k-1}))$$
$$+ \cdots +$$
$$+ (1/k!)(c_1\varepsilon + \cdots + c_k\varepsilon^k + o(\varepsilon^k))^k (a_{ij}[k, 0] + o(1)).$$

By expanding the right hand side of (33) and grouping coefficients of equal powers of ε we get the expansions and formulas given in Lemma 9.6. □

Lemma 9.7 *For any $j \neq 0$, we have the following asymptotic expansion,*

$$\pi_j^{(\varepsilon)} = \pi_j^{(0)} + \pi_j[1]\varepsilon + \cdots + \pi_j[k]\varepsilon^k + o(\varepsilon^k). \tag{34}$$

The coefficients $\pi_j[n]$, $n = 1, \ldots, k$, $j \neq 0$, are for any $i \neq 0$ given by the following recursive formulas,

$$\pi_j[n] = \frac{1}{e_i[0]}\left(d_{ij}[n] - \sum_{q=0}^{n-1} e_i[n-q]\pi_j[q] \right), \quad n = 1, \ldots, k,$$

where $\pi_j[0] = \pi_j^{(0)}$, $d_{ij}[0] = \omega_{iij}^{(0)}(\rho^{(0)})$, and $e_i[n] = \sum_{j \neq 0} d_{ij}[n]$, $n = 0, \ldots, k$.

Proof It follows from formula (16) and Lemma 9.6 that we for all $i, j \neq 0$ have

$$\pi_j^{(\varepsilon)} = \frac{d_{ij}[0] + d_{ij}[1]\varepsilon + \cdots + d_{ij}[k]\varepsilon^k + o(\varepsilon^k)}{e_i[0] + e_i[1]\varepsilon + \cdots + e_i[k]\varepsilon^k + o(\varepsilon^k)}. \tag{35}$$

Since $e_i[0] > 0$, it follows from (35) that the expansion (34) exists. From this and (35) we get the following equation,

$$(e_i[0] + e_i[1]\varepsilon + \cdots + e_i[k]\varepsilon^k + o(\varepsilon^k)) \tag{36}$$
$$\times (\pi_j[0] + \pi_j[1]\varepsilon + \cdots + \pi_j[k]\varepsilon^k + o(\varepsilon^k))$$
$$= d_{ij}[0] + d_{ij}[1]\varepsilon + \cdots + d_{ij}[k]\varepsilon^k + o(\varepsilon^k).$$

By expanding the left hand side of (36) and then equating coefficients of equal powers of ε in the left and right hand sides, we obtain the coefficients given in Lemma 9.7. □

5   Perturbed Markov Chains

In this section it is shown how the results of the present paper can be applied in the special case of perturbed discrete time Markov chains. As an illustration, we present a simple numerical example.

For every $\varepsilon \geq 0$, let $\eta_n^{(\varepsilon)}$, $n = 0, 1, \ldots$, be a homogeneous discrete time Markov chain with state space $X = \{0, 1, \ldots, N\}$, an initial distribution $p_i^{(\varepsilon)} = \mathsf{P}\{\eta_0^{(\varepsilon)} = i\}$, $i \in X$, and transition probabilities

$$p_{ij}^{(\varepsilon)} = \mathsf{P}\left\{\eta_{n+1}^{(\varepsilon)} = j \mid \eta_n^{(\varepsilon)} = i\right\}, \ i, j \in X.$$

This model is a particular case of a semi-Markov process. In this case, the transition probabilities are given by

$$Q_{ij}^{(\varepsilon)}(n) = p_{ij}^{(\varepsilon)} \chi(n = 1), \ n = 1, 2, \ldots, \ i, j \in X.$$

Furthermore, mixed power-exponential moment functionals for transition probabilities take the following form,

$$p_{ij}^{(\varepsilon)}(\rho, r) = \sum_{n=0}^{\infty} n^r e^{\rho n} Q_{ij}^{(\varepsilon)}(n) = e^{\rho} p_{ij}^{(\varepsilon)}, \ \rho \in \mathbb{R}, \ r = 0, 1, \ldots, \ i, j \in X. \quad (37)$$

Conditions **A–D** and $\mathbf{P_k}$ imposed in Sect. 2 now hold if the following conditions are satisfied:

A′: $g_{ij}^{(0)} > 0, i, j \neq 0$.
B′: $g_{ii}^{(0)}(n)$ is non-periodic for some $i \neq 0$.
P′$_\mathbf{k}$: $p_{ij}^{(\varepsilon)} = p_{ij}^{(0)} + p_{ij}[1]\varepsilon + \cdots + p_{ij}[k]\varepsilon^k + o(\varepsilon^k)$, $i, j \neq 0$, where $|p_{ij}[n]| < \infty$, $n = 1, \ldots, k, i, j \neq 0$.

Let us here remark that in order to construct an asymptotic expansion of order k for the quasi-stationary distribution of a Markov chain, it is sufficient to assume that the perturbation condition holds for the parameter k, and not for $k + 1$ as needed for semi-Markov processes. The stronger perturbation condition with parameter $k + 1$ is needed in order to construct the asymptotic expansions given in Eq. (14). However, for Markov chains these expansions can be constructed under the weaker perturbation condition. This follows from results given in [22].

It follows from (37) and **P′$_\mathbf{k}$** that the coefficients in the perturbation condition $\mathbf{P_k}$ are given by

$$p_{ij}[\rho^{(0)}, r, n] = e^{\rho^{(0)}} p_{ij}[n], \ r = 0, \ldots, k, \ n = 0, \ldots, k - r, \ i, j \neq 0. \quad (38)$$

Let us illustrate the remarks made above by means of a simple numerical example where we compute the asymptotic expansion of second order for the quasi-stationary distribution of a Markov chain with four states. We consider the simplest case where transitions to state 0 is not possible for the limiting Markov chain. In this case, exact computations can be made and we can focus on the algorithm itself and need not need to consider possible numerical issues.

We consider a perturbed Markov chain $\eta_n^{(\varepsilon)}$, $n = 0, 1, \ldots$, on the state space $X = \{0, 1, 2, 3\}$ with a matrix of transition probabilities given by

$$
\| p_{ij}^{(\varepsilon)} \| = \begin{bmatrix} 1 & 0 & 0 & 0 \\ 1 - e^{-\varepsilon} & 0 & e^{-\varepsilon} & 0 \\ 1 - e^{-\varepsilon} & 0 & 0 & e^{-\varepsilon} \\ 1 - e^{-2\varepsilon} & \frac{1}{2}e^{-2\varepsilon} & \frac{1}{2}e^{-2\varepsilon} & 0 \end{bmatrix}, \quad \varepsilon \geq 0. \tag{39}
$$

First, using the well known asymptotic expansion for the exponential function, we obtain the coefficients in condition $\mathbf{P'_k}$. The non-zero coefficients in this condition take the following numerical values,

$$
\begin{array}{llll}
p_{12}[0] = 1, & p_{23}[0] = 1, & p_{31}[0] = 1/2, & p_{32}[0] = 1/2, \\
p_{12}[1] = -1, & p_{23}[1] = -1, & p_{31}[1] = -1, & p_{32}[1] = -1, \\
p_{12}[2] = 1/2, & p_{23}[2] = 1/2, & p_{31}[2] = 1, & p_{32}[2] = 1.
\end{array} \tag{40}
$$

Then, the root of the characteristic equation for the limiting Markov chain needs to be found. Since $\phi_{ii}^{(0)}(0) = \mathsf{P}_i\{v_0^{(0)} > v_i^{(0)}\} = 1$, we have $\rho^{(0)} = 0$. In the case where transitions to state 0 is possible also for the limiting Markov chain, the root $\rho^{(0)}$ needs to be computed numerically.

Now, using that $\rho^{(0)} = 0$ and relations (38) and (40), we obtain the coefficients in condition $\mathbf{P_k}$.

Next step is to determine the coefficients in the expansions given in Eqs. (14) and (15) for the case where $k = 2$ and i is some fixed state which we can choose arbitrarily. Let us choose $i = 1$. In order to compute these coefficients we apply the results given in [22]. According to these results, we can, based on the coefficients in condition $\mathbf{P_k}$, compute the following asymptotic vector expansions,

$$
\Phi_1^{(\varepsilon)}(0, 0) = \Phi_1[0, 0, 0] + \Phi_1[0, 0, 1]\varepsilon + \Phi_1[0, 0, 2]\varepsilon^2 + \mathbf{o}(\varepsilon^2), \tag{41}
$$
$$
\Phi_1^{(\varepsilon)}(0, 1) = \Phi_1[0, 1, 0] + \Phi_1[0, 1, 1]\varepsilon + \mathbf{o}(\varepsilon),
$$
$$
\Phi_1^{(\varepsilon)}(0, 2) = \Phi_1[0, 2, 0] + \mathbf{o}(1),
$$

and, for $j = 1, 2, 3$,

$$
\Omega_{1j}^{(\varepsilon)}(0, 0) = \Omega_{1j}[0, 0, 0] + \Omega_{1j}[0, 0, 1]\varepsilon + \Omega_{1j}[0, 0, 2]\varepsilon^2 + \mathbf{o}(\varepsilon^2), \tag{42}
$$
$$
\Omega_{1j}^{(\varepsilon)}(0, 1) = \Omega_{1j}[0, 1, 0] + \Omega_{1j}[0, 1, 1]\varepsilon + \mathbf{o}(\varepsilon),
$$
$$
\Omega_{1j}^{(\varepsilon)}(0, 2) = \Omega_{1j}[0, 2, 0] + \mathbf{o}(1),
$$

where

$$\Phi_1^{(\varepsilon)}(0, r) = \left[\phi_{11}^{(\varepsilon)}(0, r) \ \phi_{21}^{(\varepsilon)}(0, r) \ \phi_{31}^{(\varepsilon)}(0, r)\right]^T, \ r = 0, 1, 2,$$

and

$$\Omega_{1j}^{(\varepsilon)}(0, r) = \left[\omega_{11j}^{(\varepsilon)}(0, r) \ \omega_{21j}^{(\varepsilon)}(0, r) \ \omega_{31j}^{(\varepsilon)}(0, r)\right]^T, \ r = 0, 1, 2, \ j = 1, 2, 3.$$

For example, the coefficients in (41) take the following numerical values,

$$\Phi_1[0, 0, 0] = \begin{bmatrix} 1 \\ 1 \\ 1 \end{bmatrix}, \ \Phi_1[0, 0, 1] = \begin{bmatrix} -7 \\ -6 \\ -5 \end{bmatrix}, \ \Phi_1[0, 0, 2] = \begin{bmatrix} 67/2 \\ 27 \\ 43/2 \end{bmatrix},$$

$$\Phi_1[0, 1, 0] = \begin{bmatrix} 5 \\ 4 \\ 3 \end{bmatrix}, \ \Phi_1[0, 1, 1] = \begin{bmatrix} -47 \\ -36 \\ -27 \end{bmatrix}, \ \Phi_1[0, 2, 0] = \begin{bmatrix} 33 \\ 24 \\ 17 \end{bmatrix}. \tag{43}$$

In particular, from (41) and (42) we can extract the following asymptotic expansions,

$$\phi_{11}^{(\varepsilon)}(0, 0) = b_1[0, 0] + b_1[0, 1]\varepsilon + b_1[0, 2]\varepsilon^2 + o(\varepsilon^2),$$
$$\phi_{11}^{(\varepsilon)}(0, 1) = b_1[1, 0] + b_1[1, 1]\varepsilon + o(\varepsilon),$$
$$\phi_{11}^{(\varepsilon)}(0, 2) = b_1[2, 0] + o(1),$$

and, for $j = 1, 2, 3$,

$$\omega_{11j}^{(\varepsilon)}(0, 0) = a_{1j}[0, 0] + a_{1j}[0, 1]\varepsilon + a_{1j}[0, 2]\varepsilon^2 + o(\varepsilon^2),$$
$$\omega_{11j}^{(\varepsilon)}(0, 1) = a_{1j}[1, 0] + a_{1j}[1, 1]\varepsilon + o(\varepsilon),$$
$$\omega_{11j}^{(\varepsilon)}(0, 2) = a_{1j}[2, 0] + o(1).$$

From (41) and (43) it follows that

$$b_1[0, 0] = 1, \ b_1[0, 1] = -7, \quad b_1[0, 2] = 67/2,$$
$$b_1[1, 0] = 5, \ b_1[1, 1] = -47, \ b_1[2, 0] = 33. \tag{44}$$

By first calculating the coefficients in (42), we then get the following numerical values,

$$\begin{aligned}
&a_{11}[0, 0] = 1, \ a_{12}[0, 0] = 2, \quad a_{13}[0, 0] = 2, \\
&a_{11}[0, 1] = 0, \ a_{12}[0, 1] = -8, \ a_{13}[0, 1] = -10, \\
&a_{11}[0, 2] = 0, \ a_{12}[0, 2] = 34, \quad a_{13}[0, 2] = 43, \\
&a_{11}[1, 0] = 0, \ a_{12}[1, 0] = 6, \quad a_{13}[1, 0] = 8, \\
&a_{11}[1, 1] = 0, \ a_{12}[1, 1] = -48, \ a_{13}[1, 1] = -64, \\
&a_{11}[2, 0] = 0, \ a_{12}[2, 0] = 34, \quad a_{13}[2, 0] = 48.
\end{aligned} \tag{45}$$

The asymptotic expansion for the quasi-stationary distribution can now be computed from the coefficients in Eqs. (44) and (45) by applying the lemmas in Sect. 4.

From Lemma 9.5 we get that the asymptotic expansion for the root of the characteristic equation is given by

$$\rho^{(\varepsilon)} = c_1\varepsilon + c_2\varepsilon^2 + o(\varepsilon^2),$$

where

$$c_1 = -\frac{b_1[0, 1]}{b_1[1, 0]} = \frac{7}{5}, \quad c_2 = -\frac{b_1[0, 2] + b_1[1, 1]c_1 + b_1[2, 0]c_1^2/2}{b_1[1, 0]} = -\frac{1}{125}.$$
(46)

Then, Lemma 9.6 gives us the following asymptotic expansions,

$$\omega_{11j}^{(\varepsilon)}(\rho^{(\varepsilon)}) = d_{1j}[0] + d_{1j}[1]\varepsilon + d_{1j}[2]\varepsilon^2 + o(\varepsilon^2), \quad j = 1, 2, 3,$$

where

$$d_{1j}[0] = a_{1j}[0, 0], \qquad (47)$$
$$d_{1j}[1] = a_{1j}[0, 1] + a_{1j}[1, 0]c_1,$$
$$d_{1j}[2] = a_{1j}[0, 2] + a_{1j}[1, 1]c_1 + a_{1j}[1, 0]c_2 + a_{1j}[2, 0]c_1^2/2.$$

From (45)–(47), we calculate

$$\begin{aligned}
d_{11}[0] &= 1, \ d_{12}[0] = 2, \quad d_{13}[0] = 2, \\
d_{11}[1] &= 0, \ d_{12}[1] = 2/5, \quad d_{13}[1] = 6/5, \\
d_{11}[2] &= 0, \ d_{12}[2] = 9/125, \ d_{13}[2] = 47/125.
\end{aligned}$$
(48)

Finally, let us use Lemma 9.7. First, using (48), we get

$$e_1[0] = d_{11}[0] + d_{12}[0] + d_{13}[0] = 5, \qquad (49)$$
$$e_1[1] = d_{11}[1] + d_{12}[1] + d_{13}[1] = 8/5,$$
$$e_1[2] = d_{11}[2] + d_{12}[2] + d_{13}[2] = 56/125.$$

Then, we can construct the asymptotic expansion for the quasi-stationary distribution,

$$\pi_j^{(\varepsilon)} = \pi_j[0] + \pi_j[1]\varepsilon + \pi_j[2]\varepsilon^2 + o(\varepsilon^2), \quad j = 1, 2, 3,$$

where

$$\pi_j[0] = d_{1j}[0]/e_1[0], \qquad (50)$$
$$\pi_j[1] = (d_{1j}[1] - e_1[1]\pi_j[0])/e_1[0],$$
$$\pi_j[2] = (d_{1j}[2] - e_1[2]\pi_j[0] - e_1[1]\pi_j[1])/e_1[0].$$

Using (48)–(50), the following numerical values are obtained,

$$\pi_1[0] = 1/5, \qquad \pi_2[0] = 2/5, \qquad \pi_3[0] = 2/5,$$
$$\pi_1[1] = -8/125, \quad \pi_2[1] = -6/125, \quad \pi_3[1] = 14/125,$$
$$\pi_1[2] = 8/3125, \quad \pi_2[2] = -19/3125, \quad \pi_3[2] = 11/3125.$$

Note here that $(\pi_1[0], \pi_2[0], \pi_3[0])$ is the stationary distribution of the limiting Markov chain. It is also worth noticing that $\pi_1[n] + \pi_2[n] + \pi_3[n] = 0$ for $n = 1, 2$, as expected.

References

1. Altman, E., Avrachenkov, K.E., Núñez-Queija, R.: Perturbation analysis for denumerable Markov chains with application to queueing models. Adv. Appl. Prob. **36**, 839–853 (2004)
2. Avrachenkov, K.E., Haviv, M.: The first Laurent series coefficients for singularly perturbed stochastic matrices. Linear Algebra Appl. **386**, 243–259 (2004)
3. Avrachenkov, K.E., Filar, J.A., Howlett, P.G.: Analytic Perturbation Theory and Its Applications. SIAM, Philadelphia (2013)
4. Cheong, C.K.: Ergodic and ratio limit theorems for α-recurrent semi-Markov processes. Z. Wahrscheinlichkeitstheorie verw. Geb. **9**, 270–286 (1968)
5. Cheong, C.K.: Quasi-stationary distributions in semi-Markov processes. J. Appl. Prob. **7**, 388–399 (1970). (Correction in J. Appl. Prob. **7**, 788)
6. Courtois, P.J., Louchard, G.: Approximation of eigencharacteristics in nearly-completely decomposable stochastic systems. Stoch. Process. Appl. **4**, 283–296 (1976)
7. Darroch, J.N., Seneta, E.: On quasi-stationary distributions in absorbing discrete-time finite Markov chains. J. Appl. Prob. **2**, 88–100 (1965)
8. Delebecque, F.: A reduction process for perturbed Markov chains. SIAM J. Appl. Math. **43**, 325–350 (1983)
9. Englund, E., Silvestrov, D.S.: Mixed large deviation and ergodic theorems for regenerative processes with discrete time. In: Jagers, P., Kulldorff, G., Portenko, N., Silvestrov, D. (eds.) Proceedings of the Second Scandinavian-Ukrainian Conference in Mathematical Statistics, vol. I, Umeå (1997) (Also in: Theory Stoch. Process. **3(19)**, no. 1–2, 164–176 (1997))
10. Flaspohler, D.C., Holmes, P.T.: Additional quasi-stationary distributions for semi-Markov processes. J. Appl. Prob. **9**, 671–676 (1972)
11. Gaïtsgori, V.G., Pervozvanskiĭ, A.A.: Aggregation of states in a Markov chain with weak interaction. Cybernetics **11**, 441–450 (1975)
12. Gyllenberg, M., Silvestrov, D.S.: Quasi-stationary distributions of stochastic metapopulation model. J. Math. Biol. **33**, 35–70 (1994)
13. Gyllenberg, M., Silvestrov, D.S.: Quasi-stationary phenomena for semi-Markov processes. In: Janssen, J., Limnios, N. (eds.) Semi-Markov Models and Applications, pp. 33–60. Kluwer, Dordrecht (1999)
14. Gyllenberg, M., Silvestrov, D.S.: Quasi-Stationary Phenomena in Nonlinearly Perturbed Stochastic Systems. De Gruyter Expositions in Mathematics, vol. 44. Walter de Gruyter, Berlin (2008)
15. Hassin, R., Haviv, M.: Mean passage times and nearly uncoupled Markov chains. SIAM J. Discrete Math. **5**(3), 386–397 (1992)
16. Kingman, J.F.C.: The exponential decay of Markov transition probabilities. Proc. Lond. Math. Soc. **13**, 337–358 (1963)

17. Latouche, G.: First passage times in nearly decomposable Markov chains. In: Stewart, W.J. (ed.) Numerical Solution of Markov Chains. Probability: Pure and Applied, vol. 8, pp. 401–411. Marcel Dekker, New York (1991)
18. Latouche, G., Louchard, G.: Return times in nearly-completely decomposable stochastic processes. J. Appl. Prob. **15**, 251–267 (1978)
19. Petersson, M.: Asymptotics of ruin probabilities for perturbed discrete time risk processes. In: Silvestrov, D., Martin-Löf, A. (eds.) Modern Problems in Insurance Mathematics. EAA Series, pp. 95–112. Springer, Cham (2014)
20. Petersson, M.: Quasi-stationary distributions for perturbed discrete time regenerative processes. Theory Probab. Math. Statist. **89**, 153–168 (2014)
21. Petersson, M.: Quasi-stationary asymptotics for perturbed semi-Markov processes in discrete time. Research Report 2015:2, Department of Mathematics, Stockholm University, 36 pp. (2015)
22. Petersson, M.: Asymptotic expansions for moment functionals of perturbed discrete time semi-Markov processes. In: Silvestrov, S., Rančić, M. (eds.) Engineering Mathematics II. Algebraic, Stochastic and Analysis Structures for Networks, Data Classification and Optimization. Springer, Berlin (2016)
23. Schweitzer, P.J.: Perturbation theory and finite Markov chains. J. Appl. Prob. **5**, 401–413 (1968)
24. Seneta, E., Vere-Jones, D.: On quasi-stationary distributions in discrete-time Markov chains with a denumerable infinity of states. J. Appl. Prob. **3**, 403–434 (1966)
25. Silvestrov, D.S., Petersson, M.: Exponential expansions for perturbed discrete time renewal equations. In: Frenkel, I., Karagrigoriou, A., Lisnianski, A., Kleyner, A. (eds.) Applied Reliability Engineering and Risk Analysis: Probabilistic Models and Statistical Inference, pp. 349–362. Wiley, Chichester (2013)
26. Silvestrov, D., Silvestrov S.: Asymptotic expansions for stationary distributions of perturbed semi-Markov processes. Research Report 2015:9, Department of Mathematics, Stockholm University, 75 pp. (2015)
27. Simon, H.A., Ando, A.: Aggregation of variables in dynamic systems. Econometrica **29**, 111–138 (1961)
28. Stewart, G.W.: On the sensitivity of nearly uncoupled Markov chains. In: Stewart, W.J. (ed.) Numerical Solution of Markov Chains. Probability: Pure and Applied, vol. 8, pp. 105–119. Marcel Dekker, New York (1991)
29. van Doorn, E.A., Pollett, P.K.: Quasi-stationary distributions for discrete-state models. Eur. J. Oper. Res. **230**, 1–14 (2013)
30. Vere-Jones, D.: Geometric ergodicity in denumerable Markov chains. Q. J. Math. **13**, 7–28 (1962)
31. Yin, G., Zhang, Q.: Continuous-Time Markov Chains and Applications. A Singular Perturbation Approach. Applications of Mathematics, vol. 37. Springer, New York (1998)
32. Yin, G., Zhang, Q.: Discrete-time singularly perturbed Markov chains. In: Yao, D.D., Zhang, H., Zhou, X.Y. (eds.) Stochastic Modelling and Optimization, pp. 1–42. Springer, New York (2003)

Asymptotic Expansions for Stationary Distributions of Perturbed Semi-Markov Processes

Dmitrii Silvestrov and Sergei Silvestrov

Abstract New algorithms for computing asymptotic expansions for stationary distributions of nonlinearly perturbed semi-Markov processes are presented. The algorithms are based on special techniques of sequential phase space reduction, which can be applied to processes with asymptotically coupled and uncoupled finite phase spaces.

Keywords Semi-Markov process · Birth-death-type process · Stationary distribution · Hitting time · Nonlinear perturbation · Laurent asymptotic expansion

1 Introduction

In this paper, we present new algorithms for construction asymptotic expansions for stationary distributions of nonlinearly perturbed semi-Markov processes with a finite phase space.

We consider models, where the phase space of embedded Markov chains for pre-limiting perturbed semi-Markov processes is one class of communicative states, while the phase space for the limiting embedded Markov chain can consist of one or several closed classes of communicative states and, possibly, a class of transient states.

The initial perturbation conditions are formulated in the forms of Taylor asymptotic expansions for transition probabilities of the corresponding embedded Markov chains and Laurent asymptotic expansions for expectations of sojourn times for

D. Silvestrov (✉)
Department of Mathematics, Stockholm University, 106 81 Stockholm, Sweden
e-mail: silvestrov@math.su.se

S. Silvestrov
Division of Applied Mathematics, School of Education, Culture and Communication,
Mälardalen University, Box 883, 721 23 Västerås, Sweden
e-mail: sergei.silvestrov@mdh.se

© Springer International Publishing Switzerland 2016
S. Silvestrov and M. Rančić (eds.), *Engineering Mathematics II*,
Springer Proceedings in Mathematics & Statistics 179,
DOI 10.1007/978-3-319-42105-6_10

perturbed semi-Markov processes. Two forms of these expansions are considered, with remainders given without or with explicit upper bounds.

The algorithms are based on special time-space screening procedures for sequential phase space reduction and algorithms for re-calculation of asymptotic expansions and upper bounds for remainders, which constitute perturbation conditions for the semi-Markov processes with reduced phase spaces.

The final asymptotic expansions for stationary distributions of nonlinearly perturbed semi-Markov processes are given in the form of Taylor asymptotic expansions with remainders given without or with explicit upper bounds.

The model of perturbed Markov chains and semi-Markov processes, in particular, in the most difficult case of so-called singularly perturbed Markov chains and semi-Markov processes with absorption and asymptotically uncoupled phase spaces, attracted attention of researchers in the mid of the 20th century.

The first works related to asymptotical problems for the above models are Meshalkin [221], Simon and Ando [323], Hanen [106–109], Kingman [169], Darroch and Seneta [65, 66], Keilson [160, 161], Seneta [273–275], Schweitzer [265] and Korolyuk [177].

Here and henceforth, references in groups are given in the chronological order.

The methods used for construction of asymptotic expansions for stationary distributions and related functionals such as moments of hitting times can be split in three groups.

The first and the most widely used methods are based on analysis of generalized matrix and operator inverses of resolvent type for transition matrices and operators for singularly perturbed Markov chains and semi-Markov processes. Mainly models with linear, polynomial and analytic perturbations have been objects of studies. We refer here to works by Schweitzer [265], Turbin [345], Poliščuk and Turbin [256], Koroljuk, Brodi and Turbin [179], Pervozvanskiĭand Smirnov [247], Courtois and Louchard [59], Korolyuk and Turbin [195, 196], Courtois [57], Latouche and Louchard [209], Kokotović, Phillips and Javid [170], Korolyuk, Penev and Turbin [190], Phillips and Kokotović [253], Delebecque [67], Abadov [1], Silvestrov and Abadov [310–312], Kartashov [151, 158], Haviv [112], Korolyuk [178], Stewart and Sun [339], Haviv, Ritov and Rothblum [121], Haviv and Ritov [119], Schweitzer and Stewart [272], Stewart [335, 336], Yin and Zhang [354–357], Avrachenkov [26, 27], Avrachenkov and Lasserre [34], Korolyuk, V.S. and Korolyuk, V.V. [180], Yin, G., Zhang, Yang and Yin, K. [359], Avrachenkov and Haviv [31, 32], Craven [64], Bini, Latouche and Meini [46], Korolyuk and Limnios [187] and Avrachenkov, Filar and Howlett [30].

Aggregation/disaggregation methods based on various modification of Gauss elimination method and space screening procedures for perturbed Markov chains have been employed for approximation of stationary distributions for Markov chains in works by Coderch, Willsky, Sastry and Castañon [53], Delebecque [67], Gaĭtsgori and Pervozvanskiĭ [89], Chatelin and Miranker [52], Courtois and Semal [61], Seneta [277], Cao and Stewart [51], Vantilborgh [347], Feinberg and Chiu [82], Haviv [113, 115, 116], Sumita and Reiders [342], Meyer [224], Schweitzer [269], Stewart and

Zhang [340], Stewart [333], Kim and Smith [168], Marek and Pultarová [218], Marek, Mayer and Pultarová [217] and Avrachenkov, Filar and Howlett [30].

Alternatively, the methods based on regenerative properties of Markov chains and semi-Markov processes, in particular, relations which link stationary probabilities and expectations of return times have been used for getting approximations for expectations of hitting times and stationary distributions in works by Grassman, Taksar and Heyman [94], Hassin and Haviv [111] and Hunter [140]. Also, the above mentioned relations and methods based on asymptotic expansions for nonlinearly perturbed regenerative processes developed in works by Silvestrov [301, 304, 305], Englund and Silvestrov [77], Gyllenberg and Silvestrov [99, 100, 102, 104], Englund [75, 76], Ni, Silvestrov and Malyarenko [243], Ni [238–242], Petersson [248, 252] and Silvestrov and Petersson [318] have been used for getting asymptotic expansions for stationary and quasi-stationary distributions for nonlinearly perturbed Markov chains and semi-Markov processes with absorption.

We would like to mention that the present paper contains also a more extended bibliography of works in the area supplemented by short bibliographical remarks given in the last section of the paper.

In the present paper, we combine methods based on stochastic aggregation/disaggregation approach with methods based on asymptotic expansions for perturbed regenerative processes applied to perturbed semi-Markov processes.

In the above mentioned works based on stochastic aggregation/disaggregation approach, space screening procedures for discrete time Markov chains are used. A Markov chain with a reduced phase space is constructed from the initial one as the sequence of its states at sequential moment of hitting into the reduced phase space. Times between sequential hitting of a reduced phase space are ignored. Such screening procedure preserves ratios of hitting frequencies for states from the reduced phase space and, thus, the ratios of stationary probabilities are the same for the initial and the reduced Markov chains. This implies that the stationary probabilities for the reduced Markov chain coincide with the corresponding stationary probabilities for the initial Markov chain up to the change of the corresponding normalizing factors.

We use another more complex type of time-space screening procedures for semi-Markov processes. In this case, a semi-Markov process with a reduced phase space is constructed from the initial one as the sequence of its states at sequential moment of hitting into the reduced phase space and times between sequential jumps of the reduced semi-Markov process are times between sequential hitting of the reduced space by the initial semi-Markov process. Such screening procedure preserves transition times between states from the reduced phase space, i.e., these times and, thus, their expectations are the same for the initial and the reduced semi-Markov processes.

We also formulate perturbation conditions in terms of asymptotic expansions for transition characteristics of perturbed semi-Markov processes. The remainders in these expansions and, thus, the transition characteristics of perturbed semi-Markov processes can be non-analytical functions of perturbation parameters that makes difference with the results for models with linear, polynomial and analytical perturbations.

We employ the methods of asymptotic analysis for nonlinearly perturbed regenerative processes developed in works by Silvestrov [301, 304, 305] and Gyllenberg and Silvestrov [99, 100, 102, 104] and applied to nonlinearly perturbed semi-Markov processes. However, we use techniques of more general Laurent asymptotic expansions instead of Taylor asymptotic expansions used in the above mentioned works and combine these methods with the aggregation/disaggregation approach instead of using the approach based on generalized matrix inverses. This permits us consider perturbed semi-Markov processes with an arbitrary communication structure of the phase space for the limiting semi-Markov process, including the general case, where this phase space may consist from one or several closed classes of communicative states and, possibly, a class of transient states.

Another new element is that we consider asymptotic expansions with remainders given not only in the form $o(\cdot)$, but, also, with explicit upper bounds.

It should be mentioned that the semi-Markov setting is an adequate and necessary element of the method proposed in the paper. Even in the case, where the initial process is a discrete or continuous time Markov chain, the time-space screening procedure of phase space reduction results in a semi-Markov process, since times between sequential hitting of the reduced space by the initial process have distributions which can differ of geometrical or exponential ones.

Also, the use of Laurent asymptotic expansions for expectations of sojourn times of perturbed semi-Markov processes is also a necessary element of the method. Indeed, even in the case, where expectations of sojourn times for all states of the initial semi-Markov process are asymptotically bounded and represented by Taylor asymptotic expansions, the exclusion of an asymptotically absorbing state from the initial phase space can cause appearance of states with asymptotically unbounded expectations of sojourn times represented by Laurent asymptotic expansions, for the reduced semi-Markov processes.

The method proposed in the paper can be considered as a stochastic analogue of the Gauss elimination method. It is based on the procedure of sequential exclusion of states from the phase space of a perturbed semi-Markov process accompanied by re-calculation of asymptotic expansions penetrating perturbation conditions for semi-Markov processes with reduced phase spaces. The corresponding algorithms are based on some kind of "operational calculus" for Laurent asymptotic expansions with remainders given in two forms, without or with explicit upper bounds.

The corresponding computational algorithms have an universal character. As was mentioned above, they can be applied to perturbed semi-Markov processes with an arbitrary asymptotic communicative structure and are computationally effective due to recurrent character of computational procedures.

In conclusion, we would like to point out that, by our opinion, the results presented in the paper have a good potential for continuation of studies (asymptotic expansions for high order power and exponential moments for hitting times, aggregated time-space screening procedures, asymptotic expansions for quasi-stationary distributions, etc.). We comment some prospective directions for future studies in the end of the paper.

The paper includes seven sections. In Sect. 2, we present so-called operational rules for Laurent asymptotic expansions. In Sect. 3, we formulate basic perturbation conditions for Markov chains and semi-Markov processes and give basic formulas for stationary distributions for semi-Markov processes, in particular, formulas connecting stationary distributions with expectations of return times. In Sect. 4, we present an one-step procedure of phase space reduction for semi-Markov processes and algorithms for re-calculation of asymptotic expansions for transition characteristics of perturbed semi-Markov processes with a reduced phase space. In Sect. 5, we present algorithms of sequential reduction of phase space for semi-Markov processes. In Sect. 6, we present algorithms for construction of asymptotic expansions for stationary distributions for nonlinearly perturbed semi-Markov processes and main results of this paper formulated in Theorems 10.8 and 10.9. In Sect. 7, we present some directions for future studies and short bibliographical remarks concerned works in the area.

We would like to conclude the introduction with the remark that the present paper is a slightly improved version of the research report Silvestrov, D. and Silvestrov S. [320].

2 Laurent Asymptotic Expansions

In this section, we present so-called operational rules for Laurent asymptotic expansions. We consider the corresponding results as possibly known, except, some of explicit formulas for remainders, in particular, those related to product, reciprocal and quotient rules.

2.1 Definition of Laurent Asymptotic Expansions

Let $A(\varepsilon)$ be a real-valued function defined on an interval $(0, \varepsilon_0]$, for some $0 < \varepsilon_0 \leq 1$, and given on this interval by a Laurent asymptotic expansion,

$$A(\varepsilon) = a_{h_A}\varepsilon^{h_A} + \cdots + a_{k_A}\varepsilon^{k_A} + o_A(\varepsilon^{k_A}), \tag{1}$$

where (a) $-\infty < h_A \leq k_A < \infty$ are integers, (b) the coefficients $a_{h_A}, \ldots, a_{k_A}$ are real numbers, (c) function $o_A(\varepsilon^{k_A})/\varepsilon^{k_A} \to 0$ as $\varepsilon \to 0$.

We refer to such Laurent asymptotic expansion as a (h_A, k_A)-expansion.

We say that (h_A, k_A)-expansion $A(\varepsilon)$ is pivotal if it is known that $a_{h_A} \neq 0$.

A Laurent asymptotic expansion $A(\varepsilon)$ can also be referred as a Taylor asymptotic expansion if $h_A \geq 0$.

We also say that (h_A, k_A)-expansion $A(\varepsilon)$ is a $(h_A, k_A, \delta_A, G_A, \varepsilon_A)$-expansion if its remainder $o_A(\varepsilon^{k_A})$ satisfies the following inequalities (d) $|o_A(\varepsilon^{k_A})| \leq G_A\varepsilon^{k_A+\delta_A}$, for $0 < \varepsilon \leq \varepsilon_A$, where (e) $0 < \delta_A \leq 1, 0 < G_A < \infty$ and $0 < \varepsilon_A \leq \varepsilon_0$.

In what follows, $[a]$ is the integer part of a real number a.

Also, the indicator of relation $A = B$ is denoted as $I(A = B)$. It equals to 1, if $A = B$, or 0, if $A \neq B$.

It is useful to note that there is no sense to consider, it seems, a more general case of upper bounds for the remainder $o_A(\varepsilon^{k_A})$, with parameter $\delta_A > 1$. Indeed, let us define $k_A' = k_A + [\delta_A] - I(\delta_A = [\delta_A])$ and $\delta_A' = \delta_A - [\delta_A] + I(\delta_A = [\delta_A]) \in (0, 1]$.

The $(h_A, k_A, \delta_A, G_A, \varepsilon_A)$-expansion (1) can be re-written in the equivalent form of the $(h_A, k_A', \delta_A', G_A, \varepsilon_A)$-expansion,

$$A(\varepsilon) = a_{h_A}\varepsilon^{h_A} + \cdots + a_{k_A}\varepsilon^{k_A} + 0\varepsilon^{k_A+1} + \cdots + 0\varepsilon^{k_A'} + o_A'(\varepsilon^{k_A'}), \qquad (2)$$

with the remainder term $o_A'(\varepsilon^{k_A'}) = o_A(\varepsilon^{k_A})$, which satisfies inequalities $|o_A'(\varepsilon^{k_A'})| = |o_A(\varepsilon^{k_A})| \leq G_A\varepsilon^{k_A+\delta_A} = G_A\varepsilon^{k_A'+\delta_A'}$, for $0 < \varepsilon \leq \varepsilon_A$.

Relation (2) implies that the asymptotic expansion $A(\varepsilon)$ can be represented in different forms. In such cases, we consider a more informative form with larger parameters h_A and k_A. As far as parameters δ_A, G_A and ε_A are concerned, we consider as a more informative form, first, with larger value of parameter δ_A, second, with smaller values of parameter G_A and, third, with the larger values of parameter ε_A.

In what follows, $a \vee b = \max(a, b), a \wedge b = \min(a, b)$, for real numbers a and b.

It is useful to note that formula (1) uniquely defines coefficients $a_{h_A}, \ldots, a_{k_A}$.

Lemma 10.1 *If function* $A(\varepsilon) = a_{h_A'}'\varepsilon^{h_A'} + \cdots + a_{k_A'}'\varepsilon^{k_A'} + o_A'(\varepsilon^{k_A'}) = a_{h_A''}''\varepsilon^{h_A''} + \cdots + a_{k_A''}''\varepsilon^{k_A''} + o_A''(\varepsilon^{k_A''}), \varepsilon \in (0, \varepsilon_0]$ *can be represented as, respectively, (h_A', k_A') and (h_A'', k_A'')-expansion, then the asymptotic expansion for function $A(\varepsilon)$ can be represented in the following the most informative form $A(\varepsilon) = a_{h_A}\varepsilon^{h_A} + \cdots + a_{k_A}\varepsilon^{k_A} + o_A(\varepsilon^{k_A}), \varepsilon \in (0, \varepsilon_0]$ of (h_A, k_A)-expansion, with parameters $h_A = h_A' \vee h_A'', k_A = k_A' \vee k_A''$, and coefficients $a_{h_A}, \ldots, a_{k_A}$ and remainder $o_A(\varepsilon^{k_A})$ given by the following relations:*

(i) $a_l', a_l'' = 0$, *for* $l < h_A$.

(ii) $a_l = a_l' = a_l''$, *for* $h_A \leq l \leq \tilde{k}_A = k_A' \wedge k_A''$.

(iii) $a_l = a_l''$, *for* $\tilde{k}_A = k_A' < l \leq k_A$ *if* $k_A' < k_A''$.

(iv) $a_l = a_l'$, *for* $\tilde{k}_A = k_A'' < l \leq k_A$ *if* $k_A'' < k_A'$.

(v)   *The remainder term* $o_A(\varepsilon^{k_A})$ *is given by the following relation,*

$$o_A(\varepsilon^{k_A}) = \begin{cases} o_A''(\varepsilon^{k_A''}) & \text{if } k_A' < k_A'', \\ o_A'(\varepsilon^{k_A'}) = o_A''(\varepsilon^{k_A''}) & \text{if } k_A' = k_A'', \\ o_A'(\varepsilon^{k_A'}) & \text{if } k_A' > k_A''. \end{cases} \qquad (3)$$

The latter asymptotical expansion is pivotal if and only if $a_{h_A} = a_{h_A}' = a_{h_A}'' \neq 0$.

It is useful to make some additional remarks.

The case $\tilde{k}_A < h_A$ is possible. In this case, the set of integers l such that $h_A \leq l \leq \tilde{k}_A$ is empty. This can happen if $k_A' < h_A''$ or $k_A'' < h_A'$. In the first case, all coefficients $a_l' = 0, l = h_A', \ldots, k_A'$ while $h_A = h_A'', k = k_A'', a_l'' = a_l, l = h_A, \ldots, k_A$.

In the second case, all coefficients $a_l'' = 0, l = h_A'', \ldots, k_A''$ while $h_A = h_A', k_A = k_A'$, $a_l = a_l', l = h_A, \ldots, k_A$.

If $k_A' = k_A''$ then $h_A \le \tilde{k}_A = k_A$ and the set of integers l such that $\tilde{k}_A < l \le k_A$ is empty. In this case, all coefficients $a_l = a_l' = a_l'', l = h_A, \ldots, k_A$.

If $a_{h_A}' \ne 0$ then $h_A = h_A'$ and $a_{h_A} = a_{h_A'}' \ne 0$. If $a_{h_A''}'' \ne 0$ then $h_A = h_A''$ and $a_{h_A} = a_{h_A''}'' \ne 0$. If $a_{h_A'}', a_{h_A''}'' \ne 0$ then $h_A = h_A' = h_A''$ and $a_{h_A} = a_{h_A'}' = a_{h_A''}'' \ne 0$.

The following proposition supplements Lemma 10.1.

Lemma 10.2 *If* $\quad A(\varepsilon) = a_{h_A'}' \varepsilon^{h_A'} + \cdots + a_{k_A'}' \varepsilon^{k_A'} + o_A'(\varepsilon^{k_A'}) = a_{h_A''}'' \varepsilon^{h_A''} + \cdots + a_{k_A''}''$
$\varepsilon^{k_A''} + o_A''(\varepsilon^{k_A''}), \varepsilon \in (0, \varepsilon_0]$ *can be represented as, respectively,* $(h_A', k_A', \delta_A', G_A', \varepsilon_A')$- *and* $(h_A'', k_A'', \delta_A'', G_A'', \varepsilon_A'')$-*expansion, then:*

(i) *The asymptotic expansion* $A(\varepsilon) = a_{h_A} \varepsilon^{h_A} + \cdots + a_{k_A} \varepsilon^k + o_A(\varepsilon^{k_A}), \varepsilon \in (0, \varepsilon_0]$
given in Lemma 10.1 is an $(h_A, k_A, \delta_A, G_A, \varepsilon_A)$-*expansion with parameters* G_A, δ_A *and* ε_A *which can be chosen in the following way consistent with the priority order described above:*

$$
(\delta_A, G_A, \varepsilon_A) = \begin{cases}
(\delta_A'', G_A'', \varepsilon_A'') & \text{if } k_A' < k_A'', \\
(\delta_A'', G_A'', \varepsilon_A'') & \text{if } k_A' = k_A'', \delta_A' < \delta_A'', \\
(\delta_A' = \delta_A'', G_A' \wedge G_A'', \varepsilon_A' \wedge \varepsilon_A'') & \text{if } k_A' = k_A'', \delta_A' = \delta_A'', \\
(\delta_A', G_A', \varepsilon_A') & \text{if } k_A' = k_A'', \delta_A' > \delta_A'', \\
(\delta_A', G_A', \varepsilon_A') & \text{if } k_A' > k_A''.
\end{cases} \quad (4)
$$

(ii) *The asymptotic expansion* $A(\varepsilon)$ *can also be represented in the form* $A(\varepsilon) = a_{\tilde{h}_A} \varepsilon^{\tilde{h}_A} + \cdots + a_{\tilde{k}_A} \varepsilon^{\tilde{k}_A} + \tilde{o}_A(\varepsilon^{\tilde{k}_A})$ *of an* $(\tilde{h}_A, \tilde{k}_A, \tilde{\delta}_A, \tilde{G}_A, \tilde{\varepsilon}_A)$-*expansion, with parameters* $\tilde{h}_A = h_A, \tilde{k}_A = k_A' \wedge k_A''$ *and parameters* $\tilde{\delta}_A, \tilde{G}_A, \tilde{\varepsilon}_A$ *given by the following formulas,*

$$
\tilde{\delta}_A = \begin{cases}
\delta_A' & \text{if } k_A' < k_A'', \\
\delta_A' \wedge \delta_A'' & \text{if } k_A' = k_A'', \\
\delta_A'' & \text{if } k_A' > k_A'',
\end{cases}
$$

$$
\tilde{G}_A = \begin{cases}
G_A' \wedge \left(\sum_{k_A' < l \le k_A''} |a_l''| \tilde{\varepsilon}_A^{l - k_A' - \delta_A'} + G_A'' \tilde{\varepsilon}_A^{k_A'' + \delta_A'' - k_A' - \delta_A'} \right) & \text{if } k_A' < k_A'', \\
G_A' \tilde{\varepsilon}_A^{\delta_A' - \tilde{\delta}_A} \wedge G_A'' \tilde{\varepsilon}_A^{\delta_A'' - \tilde{\delta}_A} & \text{if } k_A' = k_A'', \\
G_A'' \wedge (\sum_{k_A'' < l \le k_A'} |a_l'| \tilde{\varepsilon}_A^{l - k_A'' - \delta_A''} + G_A' \tilde{\varepsilon}_A^{k_A' + \delta_A' - k_A'' - \delta_A''}) & \text{if } k_A' > k_A''.
\end{cases}
$$

$$
\tilde{\varepsilon}_A = \varepsilon_A' \wedge \varepsilon_A''. \quad (5)
$$

(iii) *The remainders* $o_A'(\varepsilon^{k_A'}), o_A''(\varepsilon^{k_A''}), o_A(\varepsilon^{k_A})$ *and* $\tilde{o}_A(\varepsilon^{k_A})$ *are connected by the following relations:*

$$
\tilde{o}_A(\varepsilon^{\tilde{k}_A}) = o_A(\varepsilon^{k_A}) + \sum_{\tilde{k}_A < l \le k_A} a_l \varepsilon^l
$$

$$= \begin{cases} o'_A(\varepsilon^{k'_A}) & \text{if } k'_A < k''_A, \\ o'_A(\varepsilon^{k'_A}) = o''_A(\varepsilon^{k''_A}) & \text{if } k'_A = k''_A, \\ o''_A(\varepsilon^{k''_A}) & \text{if } k'_A > k''_A. \end{cases} \tag{6}$$

2.2 Operational Rules for Laurent Asymptotic Expansions

Let us consider four Laurent asymptotic expansions, $A(\varepsilon) = a_{h_A}\varepsilon^{h_A} + \cdots + a_{k_A}\varepsilon^{k_A} + o_A(\varepsilon^{k_A})$, $B(\varepsilon) = b_{h_B}\varepsilon^{h_B} + \cdots + b_{k_B}\varepsilon^{k_B} + o_B(\varepsilon^{k_B})$, $C(\varepsilon) = c_{h_C}\varepsilon^{h_C} + \cdots + c_{k_C}\varepsilon^{k_C} + o_C(\varepsilon^{k_C})$, and $D(\varepsilon) = d_{h_D}\varepsilon^{h_D} + \cdots + d_{k_D}\varepsilon^{k_D} + o_D(\varepsilon^{k_D})$ defined for $0 < \varepsilon \leq \varepsilon_0$, for some $0 < \varepsilon_0 \leq 1$.

The following lemma presents "operational" rules for Laurent asymptotic expansions.

Lemma 10.3 *The above asymptotic expansions have the following operational rules for computing coefficients*:

(i) *If $A(\varepsilon)$, $\varepsilon \in (0, \varepsilon_0]$ is a (h_A, k_A)-expansion and c is a constant, then $C(\varepsilon) = cA(\varepsilon)$, $\varepsilon \in (0, \varepsilon_0]$ is a (h_C, k_C)-expansion with parameters $h_C = h_A, k_C = k_A$ and coefficients,*

$$c_{h_C+r} = ca_{h_C+r}, r = 0, \ldots, k_C - h_C. \tag{7}$$

This expansion is pivotal if and only if $c_{h_C} = ca_{h_A} \neq 0$.

(ii) *If $A(\varepsilon)$, $\varepsilon \in (0, \varepsilon_0]$ is a (h_A, k_A)-expansion and $B(\varepsilon)$, $\varepsilon \in (0, \varepsilon_0]$ is a (h_B, k_B)-expansion, then $C(\varepsilon) = A(\varepsilon) + B(\varepsilon)$, $\varepsilon \in (0, \varepsilon_0]$ is a (h_C, k_C)-expansion with parameters $h_C = h_A \wedge h_B, k_C = k_A \wedge k_B$, and coefficients,*

$$c_{h_C+r} = a_{h_C+r} + b_{h_C+r}, r = 0, \ldots, k_C - h_C, \tag{8}$$

where $a_{h_C+r} = 0$ for $0 \leq r < h_A - h_C$ and $b_{h_C+r} = 0$ for $0 \leq r < h_B - h_C$. This expansion is pivotal if and only if $c_{h_C} = a_{h_C} + b_{h_C} \neq 0$.

(iii) *If $A(\varepsilon)$, $\varepsilon \in (0, \varepsilon_0]$ is a (h_A, k_A)-expansion and $B(\varepsilon)$, $\varepsilon \in (0, \varepsilon_0]$ is a (h_B, k_B)-expansion, then $C(\varepsilon) = A(\varepsilon) \cdot B(\varepsilon)$, $\varepsilon \in (0, \varepsilon_0]$ is a (h_C, k_C)-expansion with parameters $h_C = h_A + h_B, k_C = (h_A + k_B) \wedge (h_B + k_A)$, and coefficients,*

$$c_{h_C+r} = \sum_{0 \leq i \leq r} a_{h_A+i} b_{h_B+r-i}, r = 0, \ldots, k_C - h_C. \tag{9}$$

This expansion is pivotal if and only if $c_{h_C} = a_{h_A} b_{h_B} \neq 0$.

(iv) *If $B(\varepsilon)$, $\varepsilon \in (0, \varepsilon_0]$ is a pivotal (h_B, k_B)-expansion, then there exists $0 < \varepsilon'_0 \leq \varepsilon_0$ such that $B(\varepsilon) \neq 0$, $\varepsilon \in (0, \varepsilon'_0]$, and $C(\varepsilon) = 1/B(\varepsilon)$, $\varepsilon \in (0, \varepsilon'_0]$ is a pivotal (h_C, k_C)-expansion with parameters $h_C = -h_B$, $k_C = k_B - 2h_B$ and coefficients,*

$$c_{h_C} = b_{h_B}^{-1}, \; c_{h_C+r} = -b_{h_B}^{-1} \sum_{1 \le i \le r} b_{h_B+i} c_{h_C+r-i}, r = 1, \ldots, k_C - h_C. \quad (10)$$

(v) *If $A(\varepsilon), \varepsilon \in (0, \varepsilon_0]$ is a (h_A, k_A)-expansion $B(\varepsilon), \varepsilon \in (0, \varepsilon_0]$ is a pivotal (h_B, k_B)-expansion, then, there exists $0 < \varepsilon_0' \le \varepsilon_0$ such that $B(\varepsilon) \ne 0, \varepsilon \in (0, \varepsilon_0']$, and $D(\varepsilon) = A(\varepsilon)/B(\varepsilon), \varepsilon \in (0, \varepsilon_0']$ is a (h_D, k_D)-expansion with parameters $h_D = h_A - h_B, k_D = (k_A - h_B) \wedge (h_A + k_B - 2h_B)$, and coefficients,*

$$d_{h_D+r} = \sum_{0 \le i \le r} c_{h_C+i} a_{h_A+r-i}, r = 0, \ldots, k_D - h_D, \quad (11)$$

where $c_{h_C+j}, j = 0, \ldots, k_C - h_C$ are coefficients of the (h_C, k_C)-expansion $C(\varepsilon) = 1/B(\varepsilon)$ given in the above proposition **(iv)**, *or by formulas,*

$$d_{h_D+r} = b_{h_B}^{-1} \left(a_{h_A+r} - \sum_{1 \le i \le r} b_{h_B+i} d_{h_D+r-i} \right), r = 0, \ldots, k_D - h_D. \quad (12)$$

This expansion is pivotal if and only if $d_{h_D} = a_{h_A} c_{h_C} = a_{h_A}/b_{h_B} \ne 0$.

The following proposition presents "operational" rules for computing parameters of upper bounds for remainders of Laurent asymptotic expansions.

Lemma 10.4 *The above asymptotic expansions have the following operational rules for computing remainders:*

(i) *If $A(\varepsilon), \varepsilon \in (0, \varepsilon_0]$ is a $(h_A, k_A, \delta_A, G_A, \varepsilon_A)$-expansion and c is a constant, then $C(\varepsilon) = cA(\varepsilon), \varepsilon \in (0, \varepsilon_0]$ is a $(h_C, k_C, \delta_C, G_C, \varepsilon_C)$-expansion with parameters $h_C = h_A, k_C = k_A$, coefficients $c_r, r = h_C, \ldots, k_C$ given in proposition* **(i)** *of Lemma 10.3, and parameters $\delta_C, G_C, \varepsilon_C$ given by the following formulas,*

$$\delta_C = \delta_A, \; G_C = |c| G_A, \; \varepsilon_C = \varepsilon_A. \quad (13)$$

(ii) *If $A(\varepsilon), \varepsilon \in (0, \varepsilon_0]$ is a $(h_A, k_A, \delta_A, G_A, \varepsilon_A)$-expansion and $B(\varepsilon), \varepsilon \in (0, \varepsilon_0]$ is a $(h_B, k_B, \delta_B, G_B, \varepsilon_B)$-expansion, then $C(\varepsilon) = A(\varepsilon) + B(\varepsilon), \varepsilon \in (0, \varepsilon_0]$ is a $(h_C, k_C, \delta_C, G_C, \varepsilon_C)$-expansion with parameters $h_C = h_A \wedge h_B, k_C = k_A \wedge k_B$, coefficients $c_r, r = h_C, \ldots, k_C$ given in proposition* **(ii)** *of Lemma 10.3, and parameters $\delta_C, G_C, \varepsilon_C$ given by the following formulas,*

$$\delta_C = \begin{cases} \delta_A & \text{if } k_C = k_A < k_B, \\ \delta_A \wedge \delta_B & \text{if } k_C = k_A = k_B, \\ \delta_B & \text{if } k_C = k_B < k_A, \end{cases}$$

$$\ge \delta_A \wedge \delta_B,$$

$$G_C = G_A \varepsilon_C^{k_A + \delta_A - k_C - \delta_C} + \sum_{k_C < i \leq k_A} |a_i| \varepsilon_C^{i - k_C - \delta_C}$$
$$+ G_B \varepsilon_C^{k_B + \delta_B - k_C - \delta_C} + \sum_{k_C < j \leq k_B} |b_j| \varepsilon_C^{j - k_C - \delta_C},$$

$$\varepsilon_C = \varepsilon_A \wedge \varepsilon_B. \tag{14}$$

(iii) *If* $A(\varepsilon), \varepsilon \in (0, \varepsilon_0]$ *is a* $(h_A, k_A, \delta_A, G_A, \varepsilon_A)$*-expansion and* $B(\varepsilon), \varepsilon \in (0, \varepsilon_0]$ *is a* $(h_B, k_B, \delta_B, G_B, \varepsilon_B)$*-expansion, then* $C(\varepsilon) = A(\varepsilon) \cdot B(\varepsilon), \varepsilon \in (0, \varepsilon_0]$ *is a* $(h_C, k_C, \delta_C, G_C, \varepsilon_C)$*-expansion with parameters* $h_C = h_A + h_B, k_C = (h_A + k_B) \wedge (h_B + k_A)$*, coefficients* $c_r, r = h_C, \ldots, k_C$ *given in proposition* (iii) *of Lemma 10.3, and parameters* $\delta_C, G_C, \varepsilon_C$ *given by the following formulas,*

$$\delta_C = \begin{cases} \delta_A & if \, k_C = h_B + k_A < h_A + k_B, \\ \delta_A \wedge \delta_B & if \, k_C = h_B + k_A = h_A + k_B, \\ \delta_B & if \, k_C = h_A + k_B < h_B + k_A, \end{cases}$$

$$\geq \delta_A \wedge \delta_B,$$

$$G_C = \sum_{k_C < i+j, h_A \leq i \leq k_A, h_B \leq j \leq k_B} |a_i| |b_j| \varepsilon_C^{i+j - k_C - \delta_C}$$
$$+ G_A \sum_{h_B \leq j \leq k_B} |b_j| \varepsilon_C^{j + k_A + \delta_A - k_C - \delta_C}$$
$$+ G_B \sum_{h_A \leq i \leq k_A} |a_i| \varepsilon_C^{i + k_B + \delta_B - k_C - \delta_C}$$
$$+ G_A G_B \varepsilon_C^{k_A + k_B + \delta_A + \delta_B - k_C - \delta_C},$$

$$\varepsilon_C = \varepsilon_A \wedge \varepsilon_B. \tag{15}$$

(iv) *If* $B(\varepsilon), \varepsilon \in (0, \varepsilon_0]$ *is a pivotal* $(h_B, k_B, \delta_B, G_B, \varepsilon_B)$*-expansion, then, there exist* $\varepsilon_C \leq \varepsilon_0' \leq \varepsilon_0$ *such that* $B(\varepsilon) \neq 0, \varepsilon \in (0, \varepsilon_0']$*, and* $C(\varepsilon) = 1/B(\varepsilon), \varepsilon \in (0, \varepsilon_0']$ *is a pivotal* $(h_C, k_C, \delta_C, G_C, \varepsilon_C)$*-expansion with parameters* $h_C = -h_B, k_C = k_B - 2h_B$*, coefficients* $c_r, r = h_C, \ldots, k_C$ *given in proposition* (iv) *of Lemma 10.3, and parameters* $\delta_C, G_C, \varepsilon_C$ *given by the following formulas,*

$$\delta_C = \delta_B,$$

$$G_C = \left(\frac{|b_{h_B}|}{2}\right)^{-1}\left(\sum_{k_B-h_B<i+j,h_B\leq i\leq k_B,h_C\leq j\leq k_C} |b_i||c_j|\varepsilon_C^{i+j-k_B+h_B-\delta_B}\right.$$

$$\left.+G_B\sum_{h_C\leq j\leq k_C} |c_j|\varepsilon_C^{j+h_B}\right),$$

$$\varepsilon_C = \varepsilon_B \wedge \begin{cases} \frac{|b_{h_B}|}{2}\left(\sum_{h_B<i\leq k_B} |b_i|\varepsilon_B^{i-h_B-1}\right. \\ \left.+ G_B\varepsilon_B^{k_B+\delta_B-h_B-1}\right)^{-1} & \text{if } h_B < k_B, \\ \left(\frac{|b_{h_B}|}{2G_B}\right)^{\frac{1}{\delta_B}} & \text{if } h_B = k_B. \end{cases} \qquad (16)$$

(v) *If $A(\varepsilon), \varepsilon \in (0, \varepsilon_0]$ is a $(h_A, k_A, \delta_A, G_A, \varepsilon_A)$-expansion, $B(\varepsilon), \varepsilon \in (0, \varepsilon_0]$ is a pivotal $(h_B, k_B, \delta_B, G_B, \varepsilon_B)$-expansion, then, there exist $\varepsilon_D \leq \varepsilon_0' \leq \varepsilon_0$ such that $B(\varepsilon) \neq 0, \varepsilon \in (0, \varepsilon_0'],$ and $D(\varepsilon) = A(\varepsilon)/B(\varepsilon)$ is a $(h_D, k_D, \delta_D, G_D, \varepsilon_D)$-expansion with parameters $h_D = h_A + h_C = h_A - h_B, k_D = (k_A + h_C) \wedge (h_A + k_C) = (k_A - h_B) \wedge (h_A + k_B - 2h_B),$ coefficients $d_r, r = h_D, \ldots, k_D$ given in proposition (v) of Lemma 10.3, and parameters $\delta_D, G_D, \varepsilon_D$ given by the following formulas,*

$$\delta_D = \begin{cases} \delta_A & \text{if } k_D = h_C + k_A < h_A + k_C \\ \delta_A \wedge \delta_C & \text{if } k_D = h_C + k_A = h_A + k_C, \\ \delta_C & \text{if } k_D = h_A + k_C < h_C + k_A, \end{cases}$$

$$\geq \delta_A \wedge \delta_C = \delta_A \wedge \delta_B,$$

$$G_D = \sum_{k_D<i+j,h_A\leq i\leq k_A,h_C\leq j\leq k_C} |a_i||c_j|\varepsilon_D^{i+j-k_D-\delta_D}$$

$$+ G_A\sum_{h_C\leq j\leq k_C} |c_j|\varepsilon_D^{j+k_A+\delta_A-k_D-\delta_D}$$

$$+ G_C\sum_{h_A\leq i\leq k_A} |a_i|\varepsilon_D^{i+k_C+\delta_C-k_D-\delta_D}$$

$$+ G_AG_C\varepsilon_D^{k_A+k_C+\delta_A+\delta_C-k_D-\delta_D},$$

$$\varepsilon_D = \varepsilon_A \wedge \varepsilon_C, \qquad (17)$$

where coefficients $c_r, r = h_C, \ldots, k_C$ and parameters $h_C, k_C, \delta_C, G_C, \varepsilon_C$ are given for the $(h_C, k_C, \delta_C, G_C, \varepsilon_C)$-expansion of function $C(\varepsilon) = 1/B(\varepsilon)$ in proposition (iv), or by formulas,

$$\delta_D = \begin{cases} \delta_A & \text{if } k_D = k_A - h_B < h_A + k_B - 2h_B, \\ \delta_A \wedge \delta_B & \text{if } k_D = k_A - h_B = h_A + k_B - 2h_B, \\ \delta_B & \text{if } k_D = h_A + k_B - 2h_B < k_A - h_B, \end{cases}$$

$$\geq \delta_A \wedge \delta_B,$$

$$G_D = \left(\frac{|b_{h_B}|}{2}\right)^{-1} \Bigg(\sum_{k_A \wedge (h_A + k_B - h_B) < i \leq k_A} |a_i| \varepsilon_D^{i - h_B - k_D - \delta_D}$$

$$+ \sum_{k_A \wedge (h_A + k_B - h_B) < i + j, h_A \leq i \leq k_A, h_D \leq j \leq k_D} |a_i||d_j| \varepsilon_D^{i + j - k_D - h_B - \delta_D}$$

$$+ G_A \varepsilon_D^{k_A + \delta_A - h_B - k_D - \delta_D} + G_B \sum_{h_D \leq j \leq k_D} |d_j| \varepsilon_D^{j + k_B + \delta_B - h_B - k_D - \delta_D} \Bigg),$$

$$\varepsilon_D = \varepsilon_A \wedge \varepsilon_B \wedge \begin{cases} \frac{|b_{h_B}|}{2} \left(\sum_{h_B < i \leq k_B} |b_i| \varepsilon_B^{i - h_B - 1} \right. \\ \left. + G_B \varepsilon_B^{k_B + \delta_B - h_B - 1} \right)^{-1} & \text{if } h_B < k_B, \\ \left(\frac{|b_{h_B}|}{2G_B} \right)^{\frac{1}{\delta_B}} & \text{if } h_B = k_B. \end{cases} \tag{18}$$

In what follows, the following two lemmas, which present recurrent operational rules for computing coefficients and remainders for multiple summations and multiplications of Laurent asymptotic expansions, will also be used. These lemmas are direct corollaries of Lemmas 10.3 and 10.4.

Let $A_m(\varepsilon) = a_{h_{A_m}, m} \varepsilon^{h_{A_m}} + \cdots + a_{k_{A_m}, m} \varepsilon^{k_{A_m}} + o(\varepsilon^{k_{A_m}}), \varepsilon \in (0, \varepsilon_0]$ be a (h_{A_m}, k_{A_m})-expansion, for $m = 1, \ldots, N$, $B_n(\varepsilon) = A_1(\varepsilon) + \cdots + A_n(\varepsilon), \varepsilon \in (0, \varepsilon_0]$, and $C_n(\varepsilon) = A_1(\varepsilon) \times \cdots \times A_n(\varepsilon), \varepsilon \in (0, \varepsilon_0]$, for $n = 1, \ldots, N$.

The following two lemmas follow, respectively, from Lemmas 10.3 and 10.4 and recurrent relations $B_n(\varepsilon) = B_{n-1}(\varepsilon) + A_n(\varepsilon), \varepsilon \in (0, \varepsilon_0], n = 2, \ldots, N$ and $C_n(\varepsilon) = C_{n-1}(\varepsilon) \cdot A_n(\varepsilon), \varepsilon \in (0, \varepsilon_0], n = 2, \ldots, N$, which hold for any $N \geq 2$.

Lemma 10.5 *The above asymptotic expansions have the following operational rules for computing coefficients:*

(i) *If $A_m(\varepsilon), \varepsilon \in (0, \varepsilon_0]$ is a (h_{A_m}, k_{A_m})-expansion for $m = 1, \ldots, N$ where $N \geq 2$, then $B_n(\varepsilon) = b_{h_{B_n}, n} \varepsilon^{h_{B_n}} + \cdots + b_{k_{B_n}, n} \varepsilon^{k_{B_n}} + o(\varepsilon^{k_{B_n}}), \varepsilon \in (0, \varepsilon_0]$ is a (h_{B_n}, k_{B_n})-expansion, for $n = 1, \ldots, N$, with $h_{B_1} = h_{A_1}, k_{B_1} = k_{A_1}$ and $h_{B_n} = \min(h_{A_1}, \ldots, h_{A_n}) = h_{B_{n-1}} \wedge h_{A_n}, k_{B_n} = \min(k_{A_1}, \ldots, k_{A_n}) = k_{B_{n-1}} \wedge k_{A_n}, n = 2, \ldots, N$ and the coefficients given by formulas $b_{h_{B_1} + l, 1} = a_{h_{A_1} + l, 1}, l = 0, \ldots, k_{B_1} - h_{B_1} = k_{A_1} - h_{A_1}$ and, for $l = 0, \ldots, k_{B_n} - h_{B_n}, n = 2, \ldots, N$, by formulas,*

$$b_{h_{B_n} + l, n} = a_{h_{B_n} + l, 1} + \cdots + a_{h_{B_n} + l, n}, \tag{19}$$

or

$$b_{h_{B_n} + l, n} = b_{h_{B_{n-1}} + l, n-1} + a_{h_{B_n} + l, n}, \tag{20}$$

where $b_{h_{B_{n-1}}+l,n-1} = 0, l = 0, \ldots, h_{B_{n-1}} - h_{B_n}$ and $a_{h_{B_n}+l,m} = 0, l = 0, \ldots$
$h_{A_m} - h_{B_n}, m = 1, \ldots, n$.
Expansions $B_n(\varepsilon), n = 1, \ldots, N$ are pivotal if and only if $b_{h_{B_n},n} = a_{h_{A_1},1} + \cdots + a_{h_{A_n},n} \neq 0, n = 1, \ldots, N$.

(ii) If $A_m(\varepsilon), \varepsilon \in (0, \varepsilon_0]$ is a (h_{A_m}, k_{A_m})-expansion for $m = 1, \ldots, N$ where $N \geq 2$, then $C_n(\varepsilon) = c_{h_{C_n},n}\varepsilon^{h_{C_n}} + \cdots + c_{k_{C_n},n}\varepsilon^{k_{C_n}} + o(\varepsilon^{k_{C_n}}), \varepsilon \in (0, \varepsilon_0]$ is a (h_{C_n}, k_{C_n})-expansion, for $n = 1, \ldots, N$, with $h_{C_1} = h_{A_1}, k_{C_1} = k_{A_1}$ and $h_{C_n} = h_{A_1} + \cdots + h_{A_n} = h_{C_{n-1}} + h_{A_n}, k_{C_n} = \min(k_{A_l} + \sum_{1 \leq r \leq n, r \neq l} h_{A_r}, l = 1, \ldots, n) = (h_{C_{n-1}} + k_{A_n}) \wedge (k_{C_{n-1}} + h_{A_n}), n = 2, \ldots, N$ and coefficients given by formulas, $c_{h_{C_1}+l,1} = a_{h_{A_1}+l,1}, l = 0, \ldots, k_{C_1} - h_{C_1} = k_{A_1} - h_{A_1}$ and, for $l = 0, \ldots, k_{C_n} - h_{C_n}, n = 2, \ldots, N$, by formulas,

$$c_{h_{C_n}+l,n} = \sum_{l_1+\cdots+l_n=l, 0 \leq l_i \leq k_{A_i} - h_{A_i}, i=1,\ldots,n} \prod_{1 \leq i \leq n} a_{h_{A_i}+l_i,i}, \tag{21}$$

or

$$c_{h_{C_n}+l,n} = \sum_{0 \leq l' \leq l} c_{h_{C_{n-1}}+l',n-1} a_{h_{A_n}+l-l',n}. \tag{22}$$

Expansions $C_n(\varepsilon), n = 1, \ldots, N$ are pivotal if and only if $c_{h_{C_n},n} = a_{h_{A_1},1} \times \cdots \times a_{h_{A_n},n} \neq 0, n = 1, \ldots, N$.

(iii) Asymptotic expansions for functions $B_n(\varepsilon) = A_1(\varepsilon) + \cdots + A_n(\varepsilon)$, $n = 1, \ldots, N$ and $C_n(\varepsilon) = A_1(\varepsilon) \times \cdots \times A_n(\varepsilon), n = 1, \ldots, N$ are invariant with respect to any permutation, respectively, of summation and multiplication order in the above formulas (19) and (21).

Lemma 10.6 *The above asymptotic expansions have the following operational rules for computing remainders:*

(i) *If $A_m(\varepsilon), \varepsilon \in (0, \varepsilon_0]$ is a $(h_{A_m}, k_{A_m}, \delta_{A_m}, G_{A_m}, \varepsilon_{A_m})$-expansion for $m = 1, \ldots, N$ where $N \geq 2$, then $B_n(\varepsilon), \varepsilon \in (0, \varepsilon_0]$ is a $(h_{B_n}, k_{B_n}, \delta_{B_n}, G_{b_n}, \varepsilon_{B_n})$-expansion, for $n = 1, \ldots, N$, with parameters $h_{B_1} = h_{A_1}, k_{B_1} = k_{A_1}$ and $h_{B_n} = \min(h_{A_1}, \ldots, h_{A_n}) = h_{B_{n-1}} \wedge h_{A_n}, k_{B_n} = \min(k_{A_1}, \ldots, k_{A_n}) = k_{B_{n-1}} \wedge k_{A_n}, n = 2, \ldots, N$, coefficients $b_{h_{B_n}+l,n}, l = 0, \ldots, k_{B_n} - h_{B_n}, n = 1, \ldots, N$ given in proposition (i) of Lemma 10.5 and parameters $G_{B_n}, \delta_{B_n}, \varepsilon_{B_n}, n = 1, \ldots, N$ given by formulas $\delta_{B_1} = \delta_{A_1} \geq \delta_N^* = \min_{1 \leq m \leq n} \delta_{A_m}, G_{B_1} = G_{A_1}, \varepsilon_{B_1} = \varepsilon_{A_1}$ and, for $n = 2, \ldots, N$, by formulas,*

$$\delta_{B_n} = \min_{m \in \mathbb{K}_n} \delta_{A_m} \geq \delta_N^*,$$

where $\mathbb{K}_n = \{m : 1 \leq m \leq n, k_m = \min(k_1, \ldots, k_n)\}$,

$$G_{B_n} = \sum_{1 \leq i \leq n} \left(G_{A_i}\varepsilon_{B_n}^{k_{A_i}+\delta_{A_i}-k_{B_n}-\delta_{B_n}} + \sum_{k_{B_n} < j \leq k_{A_i}} |a_{A_i,j}|\varepsilon_{B_n}^{j-k_{B_n}-\delta_{B_n}} \right),$$

$$\varepsilon_{B_n} = \min(\varepsilon_{A_1}, \ldots, \varepsilon_{A_n}), \tag{23}$$

or by alternative recurrent formulas,

$$\delta_{B_n} = \min_{m \in \mathbb{K}_n} \delta_{A_m} = \begin{cases} \delta_{B_{n-1}} & \text{if } k_{B_n} = k_{B_{n-1}} < k_{A_n}, \\ \delta_{B_{n-1}} \wedge \delta_{A_n} & \text{if } k_{B_n} = k_{B_{n-1}} = k_{A_n}, \\ \delta_{A_n} & \text{if } k_{B_n} = k_{A_n} < k_{B_{n-1}}, \end{cases}$$

$$\geq \delta_N^*,$$

$$G_{B_n} = G_{B_{n-1}} \varepsilon_{B_n}^{k_{B_{n-1}} + \delta_{B_{n-1}} - k_{B_n} - \delta_{B_n}} + \sum_{k_{B_n} < i \leq k_{B_{n-1}}} |b_{B_{n-1},i}| \varepsilon_{B_n}^{i - k_{B_n} - \delta_{B_n}}$$

$$+ G_{A_n} \varepsilon_{B_n}^{k_{A_n} + \delta_{A_n} - k_{B_n} - \delta_{B_n}} + \sum_{k_{B_n} < j \leq k_{A_n}} |a_{A_n,j}| \varepsilon_{B_n}^{j - k_{B_n} - \delta_{B_n}},$$

$$\varepsilon_{B_n} = \varepsilon_{B_{n-1}} \wedge \varepsilon_{A_n}, \tag{24}$$

(ii) *If $A_m(\varepsilon), \varepsilon \in (0, \varepsilon_0]$ is a $(h_{A_m}, k_{A_m}, \delta_{A_m}, G_{A_m}, \varepsilon_{A_m})$-expansion for $m = 1, \ldots, N$ where $N \geq 2$, then $C_n(\varepsilon), \varepsilon \in (0, \varepsilon_0]$ is a $(h_{C_n}, k_{C_n}, \delta_{C_n}, G_{C_n}, \varepsilon_{C_N})$-expansion, for $n = 1, \ldots, N$, with parameters $h_{C_1} = h_{A_1}, k_{C_1} = k_{A_1}$ and $h_{C_n} = h_{C_{n-1}} + h_{A_n} = h_{A_1} + \cdots + h_{A_n}, k_{C_n} = (h_{C_{n-1}} + k_{A_n}) \wedge (k_{C_{n-1}} + h_{A_n}) = \min_{1 \leq l \leq n}(k_{A_l} + \sum_{1 \leq r \leq n, r \neq l} h_{A_r}), n - 2, \ldots, N$, coefficients $c_{h_{C_n}+l,n}, l - 0, \ldots, k_{C_n} - h_{C_n}, n = 1, \ldots, N$ given in proposition **(ii)** of Lemma 10.5 and parameters $\delta_{C_n}, G_{C_n}, \varepsilon_{C_n}, n = 1, \ldots, N$ given by formulas $\delta_{C_1} = \delta_{A_1} \geq \delta_N^* = \min_{1 \leq m \leq n} \delta_{A_m}, G_{C_1} = G_{A_1}, \varepsilon_{C_1} = \varepsilon_{A_1}$ and, for $n = 2, \ldots, N$, by formulas,*

$$\delta_{C_n} = \min_{m \in \mathbb{L}_n} \delta_{A_m} \geq \delta_N^*,$$

where

$$\mathbb{L}_n = \left\{ m : 1 \leq m \leq n, \left(k_{A_m} + \sum_{1 \leq r \leq n, r \neq m} h_{A_r} \right) \right.$$

$$= \left. \min_{1 \leq l \leq n} \left(k_{A_l} + \sum_{1 \leq r \leq n, r \neq l} h_{A_r} \right) \right\},$$

$$G_{C_n} = \sum_{k_{C_n} < l_1 + \cdots + l_n, h_{A_i} \le l_i \le k_{A_i}, i=1,\ldots,n} \prod_{1 \le i \le n} |a_{A_i, l_i}| \varepsilon_{C_n}^{l_1 + \cdots + l_n - k_{C_n} - \delta_{C_n}}$$

$$+ \sum_{1 \le j \le n} \prod_{1 \le i \le n, i \ne j} \left(\sum_{h_{A_i} \le l \le k_{A_i}} |a_{A_i, l}| \varepsilon_{C_n}^l \right.$$

$$\left. + G_{A_i} \varepsilon_{C_n}^{k_{A_i} + \delta_{A_i}} \right) G_{A_j} \varepsilon_{C_n}^{k_{A_j} + \delta_{A_j} - k_{C_n} - \delta_{C_n}},$$

$$\varepsilon_{C_n} = \min_{1 \le i \le n} \varepsilon_{A_i}. \tag{25}$$

or by alternative recurrent formulas,

$$\delta_{C_n} = \begin{cases} \delta_{C_{n-1}} & \text{if } k_{C_n} = h_{A_n} + k_{C_{n-1}} < h_{C_{n-1}} + k_{A_n}, \\ \delta_{A_n} \wedge \delta_{C_{n-1}} & \text{if } k_{C_n} = h_B + k_A = h_A + k_B, \\ \delta_{A_n} & \text{if } k_{C_n} = h_{C_{n-1}} + k_{A_n} < h_{A_n} + k_{C_{n-1}}, \end{cases}$$

$$\ge \delta_N^*,$$

$$G_{C_n} = \sum_{k_{C_n} < i+j, h_{C_{n-1}} \le i \le k_{C_{n-1}}, h_{A_n} \le j \le k_{A_n}} |c_{C_{n-1}, i}| |a_{A_n, j}| \varepsilon_{C_n}^{i+j-k_{C_n} - \delta_{C_n}}$$

$$+ G_{C_{n-1}} \sum_{h_{A_n} \le j \le k_{A_n}} |a_{A_n, j}| \varepsilon_{C_n}^{j + k_{C_{n-1}} + \delta_{C_{n-1}} - k_{C_n} - \delta_{C_n}}$$

$$+ G_{A_N} \sum_{h_{C_{n-1}} \le i \le k_{C_{n-1}}} |c_{C_{n-1}, i}| \varepsilon_{C_n}^{i + k_{A_n} + \delta_{A_n} - k_{C_n} - \delta_{C_n}}$$

$$+ G_{A_N} G_{C_{n-1}} \varepsilon_{C_n}^{k_{A_n} + k_{C_{n-1}} + \delta_{A_n} + \delta_{C_{n-1}} - k_{C_n} - \delta_{C_n}},$$

$$\varepsilon_{C_n} = \varepsilon_{C_{n-1}} \wedge \varepsilon_{A_n}. \tag{26}$$

(iii) *Parameters* $\delta_{C_n}, G_{C_n}, \varepsilon_{C_n}, n = 1, \ldots, N$ *in upper bounds for remainders in the asymptotic expansions for functions* $B_n(\varepsilon) = A_1(\varepsilon) + \cdots + A_n(\varepsilon), n = 1, \ldots, N$ *and* $C_n(\varepsilon) = A_1(\varepsilon) \times \cdots \times A_n(\varepsilon), n = 1, \ldots, N$ *are invariant with respect to any permutation, respectively, of summation and multiplication order in the above formulas* (23) *and* (25).

It should be noted that formulas (23) and (25) give, in general, the values, which are less or equal than the values for these constants given in alternative formulas, respectively, (24) and (26).

2.3 Proofs of Lemmas 10.1–10.6

The formulas given in Lemmas 10.1 and 10.2 are quite obvious. The same relate to formulas and in propositions **(i)**–**(ii)** (the multiplication by a constant and summation rules) of Lemmas 10.3 and 10.4. They can be obtained by simple accumulation of coefficients for different powers of ε and terms accumulated in the corresponding remainders, as well obvious upper bounds for absolute values of sums of terms accumulated in the corresponding remainders. Lemmas 10.5 and 10.6 are corollaries of Lemmas 10.3 and 10.4.

Let us, therefore, give short proofs of propositions **(iii)**–**(v)** of Lemmas 10.3 and 10.4.

Multiplication of asymptotic expansions $A(\varepsilon)$ and $B(\varepsilon)$ penetrating proposition **(iii)** of Lemma 10.3 and accumulation of coefficients for powers ε^l for $l = h_C, \ldots, k_C$ yields the following relation,

$$
\begin{aligned}
C(\varepsilon) &= A(\varepsilon)B(\varepsilon) \\
&= (a_{h_A}\varepsilon^{h_A} + \cdots + a_{k_A}\varepsilon^{k_A} + o_A(\varepsilon^{k_A}))(b_{h_B}\varepsilon^{h_B} + \cdots + b_{k_B}\varepsilon^{k_B} + o_B(\varepsilon^{k_B})) \\
&= \sum_{h_C \leq l \leq k_C} \sum_{i+j=l, h_A \leq i \leq k_A, h_B \leq j \leq k_B} a_i b_j \varepsilon^l \\
&\quad + \sum_{k_C < i+j, h_A \leq i \leq k_A, h_B \leq j \leq k_B} a_i b_j \varepsilon^{i+j} \\
&\quad + \sum_{h_B \leq j \leq k_B} b_j \varepsilon^j o_A(\varepsilon^{k_A}) + \sum_{h_A \leq i \leq k_A} a_i \varepsilon^i o_B(\varepsilon^{k_B}) + o_A(\varepsilon^{k_A}) o_B(\varepsilon^{k_B}) \\
&= \sum_{h_C \leq l \leq k_C} c_l \varepsilon^l + o_C(\varepsilon^{k_C}),
\end{aligned}
\tag{27}
$$

where

$$
\begin{aligned}
o_C(\varepsilon^{k_C}) &= \sum_{k_C < i+j, h_A \leq i \leq k_A, h_B \leq j \leq k_B} a_i b_j \varepsilon^{i+j} + \sum_{h_B \leq j \leq k_B} b_j \varepsilon^j o_A(\varepsilon^{k_A}) \\
&\quad + \sum_{h_A \leq i \leq k_A} a_i \varepsilon^i o_B(\varepsilon^{k_B}) + o_A(\varepsilon^{k_A}) o_B(\varepsilon^{k_B}).
\end{aligned}
\tag{28}
$$

Obviously,

$$
\frac{o_C(\varepsilon^{k_C})}{\varepsilon^{k_C}} \to 0 \text{ as } 0 < \varepsilon \to 0.
\tag{29}
$$

It should be noted that the accumulation of coefficients for powers ε^l can be made in (27) only up to the maximal value $l = k_C = (h_A + k_B) \wedge (h_B + k_A)$, because of the presence in the expression for the remainder $o_C(\varepsilon^{k_C})$ terms $b_{h_B}\varepsilon^{h_B}o_A(\varepsilon^{k_A})$ and $a_{h_A}\varepsilon^{h_A}o_B(\varepsilon^{k_B})$.

Also, relation (28) readily implies relation (15), which determines parameters $\delta_C, G_C, \varepsilon_C$ in proposition **(iii)** of Lemma 10.4.

The assumptions of proposition **(iv)** in Lemma 10.3 imply that the following relation holds,

$$\varepsilon^{-h_B} B(\varepsilon) \to b_{h_B} \neq 0 \text{ as } 0 < \varepsilon \to 0. \tag{30}$$

This relation implies that there exists $0 < \varepsilon_0' \leq \varepsilon_0$ such that $B(\varepsilon) \neq 0$ for $\varepsilon \in (0, \varepsilon_0']$, and, thus, function $C(\varepsilon) = \frac{1}{B(\varepsilon)}$ is well defined for $\varepsilon \in (0, \varepsilon_0']$.

The assumptions of proposition **(iv)** of Lemma 10.3 also imply that,

$$\varepsilon^{h_B} C(\varepsilon) = \frac{1}{b_{h_B} + b_{h_B+1}\varepsilon \cdots + b_{k_B}\varepsilon^{k_B - h_B} + o_B(\varepsilon^{k_B})\varepsilon^{-h_B}}$$

$$\to \frac{1}{b_{h_B}} = c_{h_C} \text{ as } 0 < \varepsilon \to 0. \tag{31}$$

This relation means that function $\varepsilon^{h_B} C(\varepsilon)$ can be represented in the form $\varepsilon^{h_B} C(\varepsilon) = c_{h_C} + o(1)$, where $c_{h_C} = b_{h_B}^{-1}$, or, equivalently, that the following representation holds,

$$C(\varepsilon) = c_{h_C}\varepsilon^{-h_B} + C_1(\varepsilon), \ \varepsilon \in (0, \varepsilon_0'], \tag{32}$$

where

$$\frac{C_1(\varepsilon)}{\varepsilon^{-h_B}} \to 0 \text{ as } 0 < \varepsilon \to 0. \tag{33}$$

Relations (32) and (33) prove proposition **(iv)** of Lemma 10.3 for the case, where $h_B = k_B$ that is equivalent to the relation $h_C = -h_B = k_C = k_B - 2h_B$.

Note that, in the case $h_B = k_B$, the asymptotic expansion (32) for function $C(\varepsilon)$ can not be extended. Indeed,

$$\varepsilon^{h_B-1} C_1(\varepsilon) = \varepsilon^{h_B-1}(C(\varepsilon) - c_{h_C}\varepsilon^{-h_B})$$

$$= -\frac{c_{h_C}}{b_{h_B} + o_B(\varepsilon^{h_B})\varepsilon^{-h_B}} \frac{o_B(\varepsilon^{h_B})\varepsilon^{-h_B}}{\varepsilon} \tag{34}$$

The term $\frac{o_B(\varepsilon^{h_B})\varepsilon^{-h_B}}{\varepsilon}$ on the right hand side in (34) has an uncertain asymptotic behaviour as $0 < \varepsilon \to 0$.

Let us now assume that $h_B + 1 = k_B$ that is equivalent to the relation $h_C = -h_B = k_C - 1 = k_B - 2h_B - 1$.

In this case, the assumptions of proposition **(iv)** of Lemma 10.3 and relations (32), (33) and (34) imply that

$$\varepsilon^{h_B-1} C_1(\varepsilon) = \varepsilon^{h_B-1}(C(\varepsilon) - c_{h_C}\varepsilon^{-h_B})$$

$$= \frac{-b_{h_B+1}c_{h_C} - o_B(\varepsilon^{h_B+1})\varepsilon^{-h_B-1}c_{h_C}}{b_{h_B} + b_{h_B+1}\varepsilon + o_B(\varepsilon^{h_B+1})\varepsilon^{-h_B}}$$

$$\to \frac{-b_{h_B+1}c_{h_C}}{b_{h_B}} = c_{h_C+1} \text{ as } 0 < \varepsilon \to 0. \tag{35}$$

This relation means that function $\varepsilon^{h_B-1}C_1(\varepsilon)$ can be represented in the form $\varepsilon^{h_B-1}C_1(\varepsilon) = c_{h_C+1} + o(1)$, where $c_{h_C+1} = b_{h_B}^{-1}b_{h_B+1}c_{h_C}$, or, equivalently, that the following representation holds,

$$C(\varepsilon) = c_{h_C}\varepsilon^{-h_B} + c_{h_C+1}\varepsilon^{-h_B+1} + C_2(\varepsilon), \, \varepsilon \in (0, \varepsilon_0'], \tag{36}$$

where

$$\frac{C_2(\varepsilon)}{\varepsilon^{-h_B+1}} \to 0 \text{ as } 0 < \varepsilon \to 0. \tag{37}$$

Relations (36) and (37) yields proposition **(iv)** of Lemma 10.3 for the case, where $h_B + 1 = k_B$.

Note that, in the case $h_B + 1 = k_B$, the asymptotic expansion (36) for function $C(\varepsilon)$ can not be extended. Indeed,

$$\varepsilon^{h_B-2}C_2(\varepsilon) = \varepsilon^{h_B-2}(C(\varepsilon) - c_{h_C}\varepsilon^{-h_B} - c_{h_C+1}\varepsilon^{-h_B+1})$$

$$= -\frac{c_{h_C}}{b_{h_B} + b_{h_B+1}\varepsilon + o_B(\varepsilon^{h_B+1})\varepsilon^{-h_B}} \frac{o_B(\varepsilon^{h_B+1})\varepsilon^{-h_B-1}}{\varepsilon}. \tag{38}$$

The term $\frac{o_B(\varepsilon^{h_B+1})\varepsilon^{-h_B-1}}{\varepsilon}$ on the right hand side in (38) has an uncertain asymptotic behaviour as $0 < \varepsilon \to 0$.

Repeating the above arguments, we can prove that function $C(\varepsilon)$ can be represented in the form of (h_C, k_C)-expansion, with parameters h_C, k_C and coefficients $c_{h_C}, \ldots, c_{k_C}$ given in proposition **(iii)** of Lemma 10.3, for the general case, where $h_B + n = k_B$, or, equivalently, $h_C = -h_B = k_C - n = k_B - 2h_B - n$, for any $n = 0, 1, \ldots$

The (h_C, k_C)-expansion for function $C(\varepsilon) = \frac{1}{B(\varepsilon)}$ can be rewritten in the equivalent form of the following relation,

$$1 = (b_{h_B}\varepsilon^{h_B} + \cdots + b_{k_B}\varepsilon^{k_B} + o_B(\varepsilon^{k_B}))(c_{h_C} + \cdots + c_{h_C}\varepsilon^{k_C} + o_C(\varepsilon^{k_C})). \tag{39}$$

Proposition **(iii)** of Lemma 10.3, applied to the product on the right hand side in (39), permits to represent this product in the form of (h, k)-expansion with parameters $h = h_B + h_C = h_B - h_B = 0$ and $k = (h_B + k_C) \wedge (k_B + h_C) = (k_B - h_B) \wedge (k_B - 2h_B + h_B) = k_B - h_B$.

By canceling coefficient for ε^l on the left and right hand sides in (39), for $l = 0, \ldots, k_B - h_B$ and then solving equation (39) with respect to the remainder $o_C(\varepsilon^{k_C})$ permits to find the following formula for this remainder,

$$o_C(\varepsilon^{kc}) = -\frac{\sum_{k_B-h_B<i+j,h_B\leq i\leq k_B,h_C\leq j\leq k_C} b_i c_j \varepsilon^{i+j} + \sum_{h_C\leq j\leq k_C} c_j \varepsilon^j o_B(\varepsilon^{k_B})}{b_{h_B} e^{h_B} + \cdots + b_{k_B} \varepsilon^{k_B} + o_B(\varepsilon^{k_B})}$$

$$= -\frac{\sum_{k_B-h_B<i+j,h_B\leq i\leq k_B,h_C\leq j\leq k_C} b_i c_j \varepsilon^{i+j-h_B}}{b_{h_B} + \cdots + b_{k_B} \varepsilon^{k_B-h_B} + o_B(\varepsilon^{k_B})\varepsilon^{-h_B}}$$

$$- \frac{\sum_{h_C\leq j\leq k_C} c_j \varepsilon^{j-h_B} o_B(\varepsilon^{k_B})}{b_{h_B} + \cdots + b_{k_B} \varepsilon^{k_B-h_B} + o_B(\varepsilon^{k_B})\varepsilon^{-h_B}}. \tag{40}$$

The assumptions made in proposition **(iv)** of Lemma 10.4, imply that $B(\varepsilon) \neq 0$ and the following inequality holds for $0 < \varepsilon \leq \varepsilon_C$, where ε_C is given in relation (16),

$$|b_{h_B} + \cdots + b_{k_B} \varepsilon^{k_B-h_B} + o_B(\varepsilon^{k_B})\varepsilon^{-h_B}| \geq \frac{|b_{h_B}|}{2} > 0, \tag{41}$$

The assumptions made in proposition **(iv)** of Lemma 10.4 and inequality (41) finally imply that the following inequality holds, for $0 < \varepsilon \leq \varepsilon_C$,

$$|o_C(\varepsilon^{kc})| \leq \varepsilon^{k_B-2h_B+\delta_B} \left(\frac{|b_{h_B}|}{2}\right)^{-1}$$

$$\times \left(\sum_{k_B-h_B<i+j,h_B\leq i\leq k_B,h_C\leq j\leq k_C} |b_i||c_j|\varepsilon_C^{i+j-k_B+h_B-\delta_B} \right.$$

$$\left. +G_B \sum_{h_C\leq j\leq k_C} |c_j|\varepsilon_C^{j+h_B} \right). \tag{42}$$

This inequality proofs the proposition **(iv)** of Lemma 10.4.

The first statement of proposition **(v)** in Lemma 10.3 states that function $D(\varepsilon)$ can be represented as (h_D, k_D)-expansion with parameters h_D, k_D and coefficients $d_{h_d}, \ldots, d_{k_D}$ given in this proposition and relation (11). It is the direct corollary of propositions **(iii)** and **(iv)** of Lemma 10.3, which, just, should be applied to the product $D(\varepsilon) = A(\varepsilon) \cdot \frac{1}{B(\varepsilon)}, \varepsilon \in (0, \varepsilon_0']$.

Note that, in this case, parameters $h_D = h_A + h_C = h_A - h_B$ and $k_D = (k_A + h_C) \wedge (h_A + k_C) = (k_A - h_B) \wedge (h_A + k_B - 2h_B)$.

Now, when it is already proved that $D(\varepsilon)$ is (h_D, k_D)-expansion, its coefficients can be also computed by equalising coefficients for for powers ε^l for $l = h_D, \ldots, k_D$ on the left and right hand sides of relation,

$$A(\varepsilon) = B(\varepsilon)D(\varepsilon)$$

$$= (b_{h_B} e^{h_B} + \cdots + b_{h_B} \varepsilon^{k_B} + o_B(\varepsilon^{k_B}))$$

$$\times (d_{h_D} e^{h_D} + \cdots + d_{h_D} \varepsilon^{k_D} + o_D(\varepsilon^{k_D})). \tag{43}$$

This procedure yields the second statement of proposition **(v)** in Lemma 10.3 and the corresponding formulas given in relation (12).

The first statement of proposition proposition **(v)** in Lemma 10.4 and relations (17) can be obtained by direct application of propositions **(iii)** and **(iv)** and relations (15) and (16) given in Lemma 10.4, to the product $D(\varepsilon) = A(\varepsilon) \cdot \frac{1}{B(\varepsilon)}$.

Proposition **(iii)** of Lemma 10.3, applied to the product on the right hand side in (43), permits to represent this product in the form of (h, k)-expansion with parameters $h = h_B + h_D = h_B + h_A - h_B = h_A$ and $k = (h_B + k_D) \wedge (k_B + h_D) = (h_B + (k_A - h_B)) \wedge (h_A + k_B - 2h_B)) \wedge (k_B + h_A - h_B) = k_A \wedge (k_B + h_A - h_B)$.

By canceling coefficient for ε^l on the left and right hand sides in (43), for $l = h_A, \ldots, k_A \wedge (k_B + h_A - h_B)$ and then solving Eq. (43) with respect to the remainder $o_D(\varepsilon^{k_D})$ yields the following formula for this remainder,

$$
o_D(\varepsilon^{k_D}) = \frac{\sum_{k_A \wedge (k_B + h_A - h_B) < l \leq k_A} a_l \varepsilon^l + o_A(\varepsilon^{k_A})}{b_{h_B} e^{h_B} + \cdots + b_{k_B} \varepsilon^{k_B} + o_B(\varepsilon^{k_B})}
$$
$$
- \frac{\sum_{k_A \wedge (k_B + h_A - h_B) < i+j, h_B \leq i \leq k_B, h_D \leq j \leq k_D} b_i d_j \varepsilon^{i+j}}{b_{h_B} e^{h_B} + \cdots + b_{k_B} \varepsilon^{k_B} + o_B(\varepsilon^{k_B})}
$$
$$
- \frac{\sum_{h_D \leq j \leq k_D} d_j \varepsilon^j o_B(\varepsilon^{k_B})}{b_{h_B} e^{h_B} + \cdots + b_{k_B} \varepsilon^{k_B} + o_B(\varepsilon^{k_B})}
$$

$$
= \frac{\sum_{k_A \wedge (k_B + h_A - h_B) < l \leq k_A} a_l \varepsilon^{l-h_B} + o_A(\varepsilon^{k_A}) \varepsilon^{-h_B}}{b_{h_B} + \cdots + b_{k_B} \varepsilon^{k_B - h_B} + o_B(\varepsilon^{k_B}) \varepsilon^{-h_B}}
$$
$$
- \frac{\sum_{k_A \wedge (k_B + h_A - h_B) < i+j, h_B \leq i \leq k_B, h_D \leq j \leq k_D} b_i d_j \varepsilon^{i+j-h_B}}{b_{h_B} + \cdots + b_{k_B} \varepsilon^{k_B - h_B} + o_B(\varepsilon^{k_B}) \varepsilon^{-h_B}}
$$
$$
- \frac{\sum_{h_D \leq j \leq k_D} d_j \varepsilon^{j-h_B} o_B(\varepsilon^{k_B})}{b_{h_B} + \cdots + b_{k_B} \varepsilon^{k_B - h_B} + o_B(\varepsilon^{k_B}) \varepsilon^{-h_B}} \tag{44}
$$

The assumptions made in proposition **(v)** of Lemma 10.4 and inequality (41) finally imply that the following inequality holds, for $0 < \varepsilon \leq \varepsilon_D$ given in relation (18),

$$
|o_D(\varepsilon^{k_D})| \leq \varepsilon^{k_D + \delta_D} \left(\frac{|b_{h_B}|}{2}\right)^{-1}
$$
$$
\times \left(\sum_{k_A \wedge (k_B + h_A - h_B) < l \leq k_A} |a_l| \varepsilon_D^{l - k_D - h_B - \delta_D} \right.
$$
$$
+ \sum_{k_A \wedge (k_B + h_A - h_B) < i+j, h_B \leq i \leq k_B, h_D \leq j \leq k_D} |b_i||d_j| \varepsilon_D^{i+j - k_D - h_B - \delta_D}
$$
$$
+ G_A \varepsilon_D^{k_A + \delta_A - h_B - k_D - \delta_D}
$$

$$+ G_B \sum_{h_D \leq j \leq k_D} |d_j| \varepsilon_D^{j+k_B+\delta_B-h_B-k_D-\delta_D} \Bigg). \tag{45}$$

2.4 Algebraic and Related Properties of Operational Rules for Laurent Asymptotic Expansions

Let us also introduce parameter $w_A = k_A - h_A$, which is a length of a Laurent asymptotic expansion $A(\varepsilon) = a_h \varepsilon^{h_A} + \cdots + a_k \varepsilon^{k_A} + o_A(\varepsilon^{k_A})$.

The following useful lemma takes place.

Lemma 10.7 *The following relations hold for Laurent asymptotic expansions penetrating Lemma 10.3:*

   (i) *If $C(\varepsilon) = cA(\varepsilon)$, then $w_C = w_A$.*
  (ii) *If $C(\varepsilon) = A(\varepsilon) + B(\varepsilon)$, then $w_A \wedge w_B \leq w_C \leq w_A \vee w_B$.*
 (iii) *If $C(\varepsilon) = A(\varepsilon) \cdot B(\varepsilon)$, then $w_C = w_A \wedge w_B$.*
 (iv) *If $C(\varepsilon) = 1/B(\varepsilon)$, then $w_C = w_B$.*
  (v) *If $D(\varepsilon) = A(\varepsilon)/B(\varepsilon)$, then $w_D = w_A \wedge w_B$.*
 (vi) *If $w_A = w_B = w$ then $w_C = w_D = w$ for all Laurent asymptotic expansions penetrating Lemma 10.3.*

The proof of this simple lemma readily follows from formulas for parameters h and k penetrating propositions **(i)**–**(v)** of Lemma 10.3.

Let us again consider four Laurent asymptotic expansions, $A(\varepsilon) = a_{h_A} \varepsilon^{h_A} + \cdots + a_{k_A} \varepsilon^{k_A} + o_A(\varepsilon^{k_A})$, $B(\varepsilon) = b_{h_B} \varepsilon^{h_B} + \cdots + b_{k_B} \varepsilon^{k_B} + o_B(\varepsilon^{k_B})$, $C(\varepsilon) = c_{h_C} \varepsilon^{h_C} + \cdots + c_{k_C} \varepsilon^{k_C} + o_C(\varepsilon^{k_C})$, and $D(\varepsilon) = d_{h_D} \varepsilon^{h_D} + \cdots + d_{k_D} \varepsilon^{k_D} + o_D(\varepsilon^{k_D})$ defined for $0 < \varepsilon \leq \varepsilon_0$, for some $0 < \varepsilon_0 \leq 1$.

Below, sums $\sum_{l=h}^{k} d_l$ are counted as 0 if $k < h$.

The following lemma is also a corollary of Lemma 10.3.

Lemma 10.8 *The summation and multiplication operations for the Laurent asymptotic expansions penetrating propositions **(ii)** and **(iii)** in Lemma 10.3 possess the following algebraic properties, which should be understood as identities for the corresponding asymptotic expansions:*

  (i) *The summation operation is commutative, i.e., $C(\varepsilon) = A(\varepsilon) + B(\varepsilon) = B(\varepsilon) + A(\varepsilon)$, where $h_C = h_{A+B} = h_{B+A} = h_A \wedge h_B$, $k_C = k_{A+B} = k_{B+A} = k_A \wedge k_B$, and,*

$$C(\varepsilon) = \sum_{l=0}^{k_C - h_C} (a_{h_C+l} + b_{h_C+l}) \varepsilon^{h_C+l} + o_C(\varepsilon^{k_C}), \tag{46}$$

where $a_{h_C+l} = 0$ for $0 \leq l < h_A - h_C$, $b_{h_C+l} = 0$ for $0 \leq l < h_B - h_C$.

(ii) *The summation operation is associative, i.e., $D(\varepsilon) = (A(\varepsilon) + B(\varepsilon)) + C(\varepsilon) = A(\varepsilon) + (B(\varepsilon) + C(\varepsilon)) = A(\varepsilon) + B(\varepsilon) + C(\varepsilon),$ where $h_D = h_{(A+B)+C} = h_{A+(B+C)} = h_{A+B+C} = h_A \wedge h_B \wedge h_C,$ $k_D = k_{(A+B)+C} = k_{A+(B+C)} = k_{A+B+C} = k_A \wedge k_B \wedge k_C,$ and,*

$$D(\varepsilon) = \sum_{l=0}^{k_D-h_D} (a_{h_D+l} + b_{h_D+l} + c_{h_D+l})\varepsilon^{h_D+l} + o_D(\varepsilon^{k_D}), \tag{47}$$

where $a_{h_D+l} = 0$ for $0 \le l < h_A - h_D$, $b_{h_D+l} = 0$ for $0 \le l < h_B - h_D$, $c_{h_D+l} = 0$ for $0 \le l < h_C - h_D$.

(iii) *The multiplication operation is commutative, i.e., $C(\varepsilon) = A(\varepsilon) \cdot B(\varepsilon) = B(\varepsilon) \cdot A(\varepsilon),$ where $h_C = h_{A \cdot B} = h_{B \cdot A} = h_A + h_B,$ $k_C = k_{A \cdot B} = k_{B \cdot A} = (h_A + k_B) \wedge (k_A + h_B),$ and,*

$$C(\varepsilon) = \sum_{l=0}^{k_C-h_C} \left(\sum_{l_1+l_2=l, l_1, l_2 \ge 0} a_{h_A+l_1} b_{h_B+l_2} \right) \varepsilon^{h_C+l} + o_C(\varepsilon^{k_c}). \tag{48}$$

(iv) *The multiplication operation is associative, i.e., $D(\varepsilon) = (A(\varepsilon) \cdot B(\varepsilon)) \cdot C(\varepsilon) = A(\varepsilon) \cdot (B(\varepsilon) \cdot C(\varepsilon)) = A(\varepsilon) \cdot B(\varepsilon) \cdot C(\varepsilon),$ where $h_D = h_{(A \cdot B) \cdot C} = h_{A \cdot (B \cdot C)} = h_{A \cdot B \cdot C} = h_A + h_B + h_C,$ $k_D = k_{(A \cdot B) \cdot C} = k_{A \cdot (B \cdot C)} = k_{A \cdot B \cdot C} = (h_A + h_B + k_C) \wedge (h_A + k_B + h_C) \wedge (k_A + h_B + h_C),$ and,*

$$D(\varepsilon) = \sum_{l=0}^{k_D-h_D} \left(\sum_{l_1+l_2+l_3=l, l_1, l_2, l_3 \ge 0} a_{h_A+l_1} b_{h_B+l_2} c_{h_C+l_3} \right) \varepsilon^{h_D+l} + o_D(\varepsilon^{k_D}). \tag{49}$$

(v) *The summation and multiplication operations possess distributive property, i.e., $D(\varepsilon) = (A(\varepsilon) + B(\varepsilon)) \cdot C(\varepsilon) = A(\varepsilon) \cdot C(\varepsilon) + B(\varepsilon) \cdot C(\varepsilon),$ where $h_D = h_{(A+B) \cdot C} = h_{A \cdot C + B \cdot C} = h_A \wedge h_B + h_C = (h_A + h_C) \wedge (h_B + h_C),$ $k_D = k_{(A+B) \cdot C} = k_{A \cdot C + B \cdot C} = (h_A \wedge h_B + k_C) \wedge (k_A \wedge k_B + h_C) = (h_A + k_C) \wedge (k_A + h_C) \wedge (h_B + k_C) \wedge (k_B + h_C),$ and,*

$$D(\varepsilon) = \sum_{l=0}^{k_D-h_D} \left(\sum_{l_1+l_2=l, l_1, l_2 \ge 0} (a_{h_A \wedge h_B+l_1} \right.$$

$$\left. + b_{h_A \wedge h_B+l_1}) c_{h_C+l_2} \right) \varepsilon^{h_D+l} + o_D(\varepsilon^{k_D})$$

$$= \sum_{l=0}^{k_D-h_A-h_C} \left(\sum_{l_1+l_2=l, l_1, l_2 \ge 0} a_{h_A+l_1} c_{h_C+l_2} \right) \varepsilon^{h_A+h_C+l}$$

$$+ \sum_{l=0}^{k_D-h_B-h_C} \left(\sum_{l_1+l_2=l, l_1, l_2 \ge 0} b_{h_B+l_1} c_{h_C+l_2} \right) \varepsilon^{h_B+h_C+l} + o_D(\varepsilon^{k_D}). \tag{50}$$

where $a_{h_A \wedge h_B+l} = 0.$ for $0 \le l < h_A - h_A \wedge h_B$, $b_{h_A \wedge h_B+l} = 0,$ for $0 \le l < h_B - h_A \wedge h_B$.

The summation and multiplication rules for computing of upper bounds remainders penetrating propositions **(ii)** and **(iii)** in Lemma 10.4 possess the communicative property. This follows from formulas (23) and (25) given in Lemma 10.6.

However, the summation and multiplication rules for computing of upper bounds for remainders presented in propositions (ii) and (iii) of Lemma 10.4 do not possess associative and distributional properties. The question about the form of upper bounds for the corresponding remainders, which would possess these properties, remains open.

As follows from Lemma 10.4, operational rules presented in this lemma possess special property that let one give an effective low bounds for parameter δ_A for any $(h_A, k_A, \delta_A, G_A, \varepsilon_A)$-expansion $A(\varepsilon)$ obtained as the result of a finite sequence of operations (multiplication by a constant, summation, multiplication, and division) performed with $(h_{A_i}, k_{A_i}, \delta_{A_i}, G_{A_i}, \varepsilon_{A_i})$-expansions $A_i(\varepsilon)$, $i = 1, \ldots, N$ from some finite set of such expansions.

The following lemma takes place.

Lemma 10.9 *The operational rules for computing remainders of asymptotic expansions with explicit upper bounds for remainders presented in propositions (ii) and (iii) of Lemma 10.4 possess the following properties:*

(i) *If $C(\varepsilon) = A(\varepsilon) + B(\varepsilon) = B(\varepsilon) + A(\varepsilon)$ then $\delta_C = \delta_{A+B} = \delta_{B+A}$, $G_C = G_{A+B} = G_{B+A}$ and $\varepsilon_C = \varepsilon_{A+B} = \varepsilon_{B+A}$, where parameters δ_C, G_C and ε_C are given by formula (14) in proposition (ii) of Lemma 10.4.*

(ii) *If $C(\varepsilon) = A(\varepsilon) \cdot B(\varepsilon) = B(\varepsilon) \cdot A(\varepsilon)$ then $\delta_C = \delta_{A\cdot B} = \delta_{B\cdot A}$, $G_C = G_{A\cdot B} = G_{B\cdot A}$ and $\varepsilon_C = \varepsilon_{A\cdot B} = \varepsilon_{B\cdot A}$, where parameters δ_C, G_C and ε_C are given by formula (15) in proposition (iii) of Lemma 10.4.*

(iii) *If $A(\varepsilon)$ is $(h_A, k_A, \delta_A, G_A, \varepsilon_A)$-expansion obtained as the result of a finite sequence of operations (multiplication by a constant, summation, multiplication, and quotient) performed with $(h_{A_i}, k_{A_i}, \delta_{A_i}, G_{A_i}, \varepsilon_{A_i})$-expansions $A_i(\varepsilon)$, $i = 1, \ldots, N$ from some finite set of such expansions, then $\delta_A \geq \delta_N^* = \min_{1 \leq i \leq N} \delta_{A_i}$ that makes it possible to rewrite $A(\varepsilon)$ as the $(h_A, k_A, \delta_N^*, G_{A,N}^*, \varepsilon_A)$-expansion, with parameter $G_{A,N}^* = G_A \varepsilon_A^{\delta_A - \delta_N^*}$.*

3 Perturbed Markov Chains and Semi-Markov Processes

In this section, we formulate basic perturbation conditions for Markov chains and semi-Markov processes and give basic formulas for stationary distributions for semi-Markov processes, in particular, formulas connecting stationary distributions with expectations of return times.

3.1 Perturbed Markov Chains

Let $\mathbb{X} = \{1, \ldots, N\}$ and $\eta_n^{(\varepsilon)}$, $n = 0, 1, \ldots$ be, for every $\varepsilon \in (0, \varepsilon_0]$, a homogeneous Markov chain with a phase space $\mathbb{X}$, an initial distribution $p_i^{(\varepsilon)} = \mathsf{P}\{\eta_0^{(\varepsilon)} = i\}$, $i \in \mathbb{X}$ and transition probabilities defined for $i, j \in \mathbb{X}$,

$$p_{ij}(\varepsilon) = \mathsf{P}\{\eta_{n+1}^{(\varepsilon)} = j/\eta_n^{(\varepsilon)} = i\}. \tag{51}$$

Let us assume that the following condition holds:

A: There exist sets $\mathbb{Y}_i \subseteq \mathbb{X}, i \in \mathbb{X}$ and $\varepsilon_0 \in (0, 1]$ such that: **(a)** probabilities $p_{ij}(\varepsilon) > 0, j \in \mathbb{Y}_i, i \in \mathbb{X}$, for $\varepsilon \in (0, \varepsilon_0]$; **(b)** probabilities $p_{ij}(\varepsilon) = 0, j \in \overline{\mathbb{Y}}_i, i \in \mathbb{X}$, for $\varepsilon \in (0, \varepsilon_0]$; **(c)** there exist $n_{ij} \geq 1$ and a chain of states $i = l_{ij,0}, l_{ij,1}, \ldots, l_{ij,n_{ij}} = j$ such that $l_{ij,1} \in \mathbb{Y}_{l_{ij,0}}, \ldots, l_{ij,n_{ij}} \in \mathbb{Y}_{l_{ij,n_{ij}-1}}$, for every pair of states $i, j \in \mathbb{X}$.

We refer to sets $\mathbb{Y}_i, i \in \mathbb{X}$ as transition sets.

Conditions **A** implies that all sets $\mathbb{Y}_i \neq \emptyset, i \in \mathbb{X}$, since matrix $\|p_{ij}(\varepsilon)\|$ is stochastic, for every $\varepsilon \in (0, \varepsilon_0]$.

We now assume that the following perturbation condition holds:

B: $p_{ij}(\varepsilon) = \sum_{l=l_{ij}^-}^{l_{ij}^+} a_{ij}[l]\varepsilon^l + o_{ij}(\varepsilon^{l_{ij}^+})$, where $a_{ij}[l_{ij}^-] > 0$ and $0 \leq l_{ij}^- \leq l_{ij}^+ < \infty$, for $j \in \mathbb{Y}_i, i \in \mathbb{X}$, and $o_{ij}(\varepsilon^{l_{ij}^+})/\varepsilon^{l_{ij}^+} \to 0$ as $\varepsilon \to 0$, for $j \in \mathbb{Y}_i, i \in \mathbb{X}$.

Some additional conditions should be imposed on parameters $\varepsilon_0 \in (0, 1]$ and $l_{ij}^\pm, j \in \mathbb{Y}_i, i \in \mathbb{X}$, and coefficients $a_{ij}[l], l = l_{ij}^-, \ldots, l_{ij}^+, j \in \mathbb{Y}_i, i \in \mathbb{X}$, penetrating the asymptotic expansions condition **B**, in order this condition would be consistent with the model assumption that matrix $\|p_{ij}(\varepsilon)\|$ is stochastic, for every $\varepsilon \in (0, \varepsilon_0]$, and with condition **A**.

Condition **B** implies that there exits $\varepsilon_0 \in (0, 1]$ such that the following relation holds,

$$p_{ij}(\varepsilon) = \sum_{l=l_{ij}^-}^{l_{ij}^+} a_{ij}[l]\varepsilon^l + o_{ij}(\varepsilon^{l_{ij}^+}) > 0, \ j \in \mathbb{Y}_i, \ i \in \mathbb{X}, \ \varepsilon \in (0, \varepsilon_0]. \tag{52}$$

Thus, condition **B** is consistent with condition **A (a)**.

The model assumption that matrix $\|p_{ij}(\varepsilon)\|$ is stochastic is, under conditions **A (a)** and **(b)**, equivalent to the following relation,

$$\sum_{j \in \mathbb{Y}_i} p_{ij}(\varepsilon) = 1, \ j \in \mathbb{Y}_i, i \in \mathbb{X}, \ \varepsilon \in (0, \varepsilon_0]. \tag{53}$$

Condition **B** and proposition **(i)** (the multiple summation rule) of Lemma 10.5 imply that sum $\sum_{j \in \mathbb{Z}} p_{ij}(\varepsilon)$ can, for every subset $\mathbb{Z} \subseteq \mathbb{Y}_i$ and $i \in \mathbb{X}$, be represented in the form of the following asymptotic expansion,

$$\sum_{j \in \mathbb{Z}} p_{ij}(\varepsilon) = \sum_{l=\breve{l}_{i,\mathbb{Z}}^-}^{l_{i,\mathbb{Z}}^+} a_{i,\mathbb{Z}}[l]\varepsilon^l + o_{i,\mathbb{Z}}(\varepsilon^{l_{i,\mathbb{Z}}^+}), \tag{54}$$

where

$$l_{i,\mathbb{Z}}^- = \min_{j \in \mathbb{Z}} l_{ij}^-, \ l_{i,\mathbb{Z}}^+ = \min_{j \in \mathbb{Z}} l_{ij}^+, \tag{55}$$

and

$$a_{i,\mathbb{Z}}[l] = \sum_{j \in \mathbb{Z}} a_{ij}[l], \ l = l_{i,\mathbb{Z}}^-, \dots, l_{i,\mathbb{Z}}^+, \tag{56}$$

where $a_{ij}[l] = 0$, for $0 \le l < l_{ij}^-, j \in \mathbb{Z}$, and

$$o_{i,\mathbb{Z}}(\varepsilon^{l_{i,\mathbb{Z}}^+}) = \sum_{j \in \mathbb{Z}} \left(\sum_{l_{i,\mathbb{Z}}^+ < l \le l_{ij}^+} a_{ij}[l] \varepsilon^l + o_{ij}(\varepsilon^{l_{ij}^l}) \right). \tag{57}$$

In terms of asymptotic expansions, constant 1 can be represented, for every $n = 0, 1, \dots$, in the form of the following pivotal $(0, n)$-expansion,

$$1 = 1 + 0\varepsilon + \cdots 0\varepsilon^n + o_n(\varepsilon^n), \tag{58}$$

where remainders $o_n(\varepsilon^n) \equiv 0, n = 0, 1, \dots$.

Moreover, the above expansion is a $(0, n, 1, G, \varepsilon_0)$-expansion for any $0 < G < \infty$ and $n = 0, 1, \dots$.

Relation (53) permits us apply Lemma 10.2 to the asymptotic expansions given in relations (54) and (58). Not that, in this case, $l_{i,\mathbb{Y}_i}^- = 0$, otherwise the expression on the right hand side in (54) would converge to zero as $\varepsilon \to 0$. Let us take $n = l_{i,\mathbb{Y}_i}^+$ in relation (58). In this case $h_1 = 0$ and $k_1 = l_{i,\mathbb{Y}_i}^+$ in the asymptotic expansion given in relation (58).

Lemma 10.2 and the model stochasticity assumption (53) imply that, in this case, the following condition should hold for the coefficients of asymptotic expansions penetrating condition **B**:

C: (a) $a_{i,\mathbb{Y}_i}[l] = \sum_{j \in \mathbb{Y}_i} a_{ij}[l] = \mathrm{I}(l = 0), \ 0 \le l \le l_{i,\mathbb{Y}_i}^+, \ i \in \mathbb{X}$, where $a_{ij}[l] = 0$, for $0 \le l < l_{ij}^-, j \in \mathbb{Y}_i, i \in \mathbb{X}$; **(b)** $o_{i,\mathbb{Y}_i}(\varepsilon^{l_{i,\mathbb{Y}_i}^+}) \equiv o_{l_{i,\mathbb{Y}_i}^+}(\varepsilon^{l_{i,\mathbb{Y}_i}^+}) \equiv 0, i \in \mathbb{X}$.

Remark 10.1 It is possible to prove that conditions **A**–**C** and the model stochasticity assumption (53) imply that the asymptotic expansion in (54) satisfy, for every $\mathbb{Z} \subseteq \mathbb{Y}_i$ and $i \in \mathbb{X}$, one of the following additional conditions: (a) $l_{i,\mathbb{Z}}^- > 0$; (b) $l_{i,\mathbb{Z}}^- = 0, a_{i,\mathbb{Z}}[0] < 1$; (c) $l_{i,\mathbb{Z}}^- = 0, a_{i,\mathbb{Z}}[0] = 1$ and there exists $0 < l_{i,\mathbb{Z}} \le l_{i,\mathbb{Z}}^+$ such that $a_{i,\mathbb{Z}}[l] = 0, 0 < l < l_{i,\mathbb{Z}}$, but $a_{i,\mathbb{Z}}[l_{i,\mathbb{Z}}] < 0$; or (d) $l_{i,\mathbb{Z}}^- = 0, a_{i,\mathbb{Z}}[0] = 1$ and $a_{i,\mathbb{Z}}[l] = 0, 0 < l \le l_{i,\mathbb{Z}}^+$, but the remainder $o_{i,\mathbb{Z}}(\varepsilon^{l_{i,\mathbb{Z}}^+})$ is a nonpositive function of ε.

The above proposition implies that there exists $\varepsilon_0 \in (0, 1]$ such that $\sum_{l=l_{i,\mathbb{Z}}^-}^{l_{i,\mathbb{Z}}^+} a_{i,\mathbb{Z}}[l] \varepsilon^l + o_{i,\mathbb{Z}}(\varepsilon^{l_{i,\mathbb{Z}}^+}) \le 1, \mathbb{Z} \subseteq \mathbb{Y}_i, i \in \mathbb{X}, \varepsilon \in (0, \varepsilon_0]$. Thus, conditions **A**–**C** are also consistent with the relations $\sum_{j \in \mathbb{Z}} p_{ij}(\varepsilon) \le 1, \mathbb{Z} \subseteq \mathbb{Y}_i, i \in \mathbb{X}, \varepsilon \in (0, \varepsilon_0]$, which follows from the model stochasticity assumption (53).

In the case, where the asymptotic expansions penetrating condition **B** are supposed to be given in the form of asymptotic expansions with explicit upper bounds for remainders, we replace it by the following stronger perturbation condition:

B′: $p_{ij}(\varepsilon) = \sum_{l=l_{ij}^-}^{l_{ij}^+} a_{ij}[l]\varepsilon^l + o_{ij}(\varepsilon^{l_{ij}^+})$, where $a_{ij}[l_{ij}^-] > 0$ and $0 \le l_{ij}^- \le l_{ij}^+ < \infty$, for

$j \in \mathbb{Y}_i, i \in \mathbb{X}$, and $|o_{ij}(\varepsilon^{l_{ij}^+})| \le G_{ij}\varepsilon^{l_{ij}^+ + \delta_{ij}}$, $0 < \varepsilon \le \varepsilon_{ij}$, for $j \in \mathbb{Y}_i, i \in \mathbb{X}$, where
$0 < \delta_{ij} \le 1, 0 < G_{ij} < \infty$ and $0 < \varepsilon_{ij} \le \varepsilon_0$.

3.2 Perturbed Semi-Markov Processes

Let $\mathbb{X} = \{1, \dots, N\}$ and $(\eta_n^{(\varepsilon)}, \kappa_n^{(\varepsilon)})$, $n = 0, 1, \dots$ be, for every $\varepsilon \in (0, 1]$, a Markov
renewal process, i.e., a homogeneous Markov chain with the phase space $\mathbb{X} \times [0, \infty)$,
an initial distribution $p_i^{(\varepsilon)} = \mathsf{P}\{\eta_0^{(\varepsilon)} = i, \kappa_0^{(\varepsilon)} = 0\} = \mathsf{P}\{\eta_0^{(\varepsilon)} = i\}, i \in \mathbb{X}$ and transition probabilities defined for $(i, s), (j, t) \in \mathbb{X} \times [0, \infty)$,

$$Q_{ij}^{(\varepsilon)}(t) = \mathsf{P}\{\eta_{n+1}^{(\varepsilon)} = j, \kappa_{n+1}^{(\varepsilon)} \le t / \eta_n^{(\varepsilon)} = i, \kappa_n^{(\varepsilon)} = s\}. \tag{59}$$

In this case, the random sequence $\eta_n^{(\varepsilon)}, n = 0, 1, \dots$ is also a homogeneous
(embedded) Markov chain with the phase space $\mathbb{X}$ and transition probabilities defined
for $i, j \in \mathbb{X}$,

$$p_{ij}(\varepsilon) = \mathsf{P}\{\eta_{n+1}^{(\varepsilon)} = j / \eta_n^{(\varepsilon)} = i\} = Q_{ij}^{(\varepsilon)}(\infty). \tag{60}$$

We assume that condition **A** holds. This implies that Markov chain $\eta_n^{(\varepsilon)}$ has one
class of communicative states, for every $\varepsilon \in (0, \varepsilon_0]$.

We also assume that the following condition excluding instant transitions holds:

D: $Q_{ij}^{(\varepsilon)}(0) = 0$, $i, j \in \mathbb{X}$, for every $\varepsilon \in (0, \varepsilon_0]$.

Let us now introduce a semi-Markov process,

$$\eta^{(\varepsilon)}(t) = \eta_{\nu^{(\varepsilon)}(t)}^{(\varepsilon)}, \ t \ge 0. \tag{61}$$

where

$$\nu^{(\varepsilon)}(t) = \max(n \ge 0 : \zeta_n^{(\varepsilon)} \le t), \ t \ge 0, \tag{62}$$

is a number of jumps in the time interval $[0, t]$, and

$$\zeta_n^{(\varepsilon)} = \kappa_1^{(\varepsilon)} + \cdots + \kappa_n^{(\varepsilon)}, \ n = 0, 1, \dots, \tag{63}$$

are sequential moments of jumps for the semi-Markov process $\eta^{(\varepsilon)}(t)$.

If $Q_{ij}^{(\varepsilon)}(t) = (1 - e^{-\lambda_i(\varepsilon)t})p_{ij}(\varepsilon), t \ge 0, i, j \in \mathbb{X}$, then $\eta^{(\varepsilon)}(t), t \ge 0$ is a continuous
time homogeneous Markov chain.

If $Q_{ij}^{(\varepsilon)}(t) = I(t \ge 1)p_{ij}(\varepsilon), t \ge 0, i, j \in \mathbb{X}$, then $\eta^{(\varepsilon)}(t) = \eta_{[t]}^{(\varepsilon)}, t \ge 0$ is a discrete
time homogeneous Markov chain embedded in continuous time.

Let us also introduce expectations of sojourn times,

$$e_{ij}(\varepsilon) = \mathsf{E}_i \kappa_1^{(\varepsilon)} I(\eta_1^{(\varepsilon)} = j) = \int_0^\infty t Q_{ij}^{(\varepsilon)}(dt), \ i,j \in \mathbb{X}. \tag{64}$$

We also assume that the following condition holds:

E: $e_{ij}(\varepsilon) < \infty, \ i,j \in \mathbb{X}$, for $\varepsilon \in (0, \varepsilon_0]$.

Here and henceforth, notations P_i and E_i are used for conditional probabilities and expectations under condition $\eta_0^{(\varepsilon)} = i$.
In the case of continuous time Markov chain, $e_{ij}(\varepsilon) = \frac{1}{\lambda_i(\varepsilon)} p_{ij}(\varepsilon), i,j \in \mathbb{X}$.
In the case of discrete time Markov chain, $e_{ij}(\varepsilon) = p_{ij}(\varepsilon), i,j \in \mathbb{X}$.
Conditions **A** and **D** imply that, for every $\varepsilon \in (0, \varepsilon_0]$, expectations $e_{ij}(\varepsilon) > 0$, for $j \in \mathbb{Y}_i, i \in \mathbb{X}$, and $e_{ij}(\varepsilon) = 0$, for $j \in \overline{\mathbb{Y}}_i, i \in \mathbb{X}$.
We now assume that the following perturbation condition holds:

F: $e_{ij}(\varepsilon) = \sum_{l=m_{ij}^-}^{m_{ij}^+} b_{ij}[l] \varepsilon^l + \dot{o}_{ij}(\varepsilon^{m_{ij}^+})$, where $b_{ij}[m_{ij}^-] > 0$ and $-\infty < m_{ij}^- \leq$
$m_{ij}^+ < \infty$, for $j \in \mathbb{Y}_i, i \in \mathbb{X}$ and $\dot{o}_{ij}(\varepsilon^{m_{ij}^+})/\varepsilon^{m_{ij}^+} \to 0$ as $\varepsilon \to 0$, for $j \in \mathbb{Y}_i, i \in \mathbb{X}$.

In particular, in the case of discrete time Markov chain, condition **B** implies condition **F** to hold, since, in this case, expectations $e_{ij}(\varepsilon) = p_{ij}(\varepsilon), j \in \mathbb{Y}_i, i \in \mathbb{X}$.
Condition **F** implies that there exits $\varepsilon_0 \in (0, 1]$ such that the following relation holds,

$$e_{ij}(\varepsilon) > 0, \ j \in \mathbb{Y}_i, \ i \in \mathbb{X}, \ \varepsilon \in (0, \varepsilon_0]. \tag{65}$$

This is consistent with condition **D**.
In the case, where the asymptotic expansions penetration condition **F** are given in the form of asymptotic expansions with explicit upper bounds for remainders, we assume that the following stringer perturbation condition holds:

F': $e_{ij}(\varepsilon) = \sum_{l=m_{ij}^-}^{m_{ij}^+} b_{ij}[l] \varepsilon^l + \dot{o}_{ij}(\varepsilon^{m_{ij}^+})$, where $b_{ij}[l_{ij}^-] > 0$ and $-\infty < m_{ij}^- \leq$
$m_{ij}^+ < \infty$, for $j \in \mathbb{Y}_i, i \in \mathbb{X}$, and $|\dot{o}_{ij}(\varepsilon^{m_{ij}^+})| \leq \dot{G}_{ij} \varepsilon^{m_{ij}^+ + \delta_{ij}}, 0 < \varepsilon \leq \dot{\varepsilon}_{ij}$, for $j \in$
$\mathbb{Y}_i, i \in \mathbb{X}$, where $0 < \delta_{ij} \leq 1, 0 < \dot{G}_{ij} < \infty$ and $0 < \dot{\varepsilon}_{ij} \leq \varepsilon_0$.

It is also worse to note that the perturbation conditions **B** and **F** are independent.
To see this, let us take arbitrary functions $p_{ij}(\varepsilon), j \in \mathbb{Y}_i, i \in \mathbb{X}$ and $e_{ij}(\varepsilon), j \in$
$\mathbb{Y}_i, i \in \mathbb{X}$ satisfying, respectively, conditions **B** and **F**, and, also, relations (52), (53) and (65). Then, there exist semi-Markov transition probabilities $Q_{ij}^{(\varepsilon)}(t), t \geq 0, j \in$
$\mathbb{Y}_i, i \in \mathbb{X}$ such that $Q_{ij}^{(\varepsilon)}(\infty) = p_{ij}(\varepsilon), j \in \mathbb{Y}_i, i \in \mathbb{X}$ and $\int_0^\infty t Q_{ij}^{(\varepsilon)}(dt) = e_{ij}(\varepsilon), j \in$
$\mathbb{Y}_i, i \in \mathbb{X}$, for every $\varepsilon \in (0, \varepsilon_0]$.
It is readily seen that, for example, semi-Markov transition probabilities $Q_{ij}^{(\varepsilon)}(t) =$
$I(t \geq e_{ij}(\varepsilon)/p_{ij}(\varepsilon)) p_{ij}(\varepsilon), t \geq 0, j \in \mathbb{Y}_i, i \in \mathbb{X}$ satisfy the above relations.

3.3   Stationary Distributions for Semi-Markov Processes

Condition **A** guarantees that the phase space $\mathbb{X}$ is one class of communicative states for Markov chain $\eta_n^{(\varepsilon)}$, for every $\varepsilon \in (0, \varepsilon_0]$, i.e., the Markov chain $\eta_n^{(\varepsilon)}$ is ergodic, and, thus, for every $\varepsilon \in (0, \varepsilon_0]$, there exist the unique stationary distribution $\bar{\rho}(\varepsilon) = \langle \rho_1(\varepsilon), \ldots, \rho_N(\varepsilon) \rangle$, which is given by the following ergodic relation, for $i \in \mathbb{X}$,

$$\bar{\mu}_{i,n}^{(\varepsilon)} = \frac{1}{n} \sum_{k=1}^{n} I(\eta_k^{(\varepsilon)} = i) \xrightarrow{\mathsf{P}} \rho_i(\varepsilon) \text{ as } n \to \infty. \tag{66}$$

It is useful to note that the ergodic relation (66) holds for any initial distribution $\bar{p}^{(\varepsilon)} = \langle p_1^{(\varepsilon)}, \ldots p_N^{(\varepsilon)} \rangle$ and the stationary distribution $\bar{\rho}(\varepsilon)$ does not depend on the initial distribution.

As known, $\rho_i(\varepsilon), i \in \mathbb{X}$ is the unique positive solution for the system of linear equations,

$$\begin{cases} \rho_j(\varepsilon) = \sum_{i \in \mathbb{X}} \rho_i(\varepsilon) p_{ij}(\varepsilon), j \in \mathbb{X}, \\ \sum_{i \in \mathbb{X}} \rho_i = 1. \end{cases} \tag{67}$$

Conditions **A**, **D** and **E** imply that, for every $\varepsilon \in (0, \varepsilon_0]$, the semi-Markov process $\eta^{(\varepsilon)}(t)$ is also ergodic and its stationary distribution $\bar{\pi}(\varepsilon) = \langle \pi_1(\varepsilon), \ldots, \pi_N(\varepsilon) \rangle$ is given by the following ergodic relation, for $i \in \mathbb{X}$,

$$\bar{\mu}_i^{(\varepsilon)}(t) = \frac{1}{t} \int_0^t I(\eta^{(\varepsilon)}(s) = i) ds \xrightarrow{\mathsf{P}} \pi_i(\varepsilon) \text{ as } t \to \infty. \tag{68}$$

As in (66), the ergodic relation (68) holds for any initial distribution $\bar{p}^{(\varepsilon)}$ and the stationary distribution $\bar{\pi}(\varepsilon)$ does not depend on the initial distribution.

The stationary distributions for the semi-Markov process $\eta^{(\varepsilon)}(t)$ and the embedded Markov chain $\eta_n^{(\varepsilon)}$ are connected by the following relation,

$$\pi_i(\varepsilon) = \frac{\rho_i(\varepsilon) e_i(\varepsilon)}{\sum_{j \in \mathbb{X}} \rho_j(\varepsilon) e_j(\varepsilon)}, \quad i \in \mathbb{X}, \tag{69}$$

where

$$e_i(\varepsilon) = \mathsf{E}_i \kappa_1^{(\varepsilon)} = \sum_{j \in \mathbb{X}} e_{ij}(\varepsilon), \quad i \in \mathbb{X}. \tag{70}$$

Condition **B** implies that there exist limits,

$$p_{ij}(0) = \lim_{\varepsilon \to 0} p_{ij}(\varepsilon) = \begin{cases} a_{ij}[0] & \text{if } l_{ij}^- = 0, j \in \mathbb{Y}_i, i \in \mathbb{X}, \\ 0 & \text{if } l_{ij}^- > 0, j \in \mathbb{Y}_i, i \in \mathbb{X}, \\ 0 & \text{if } j \in \overline{\mathbb{Y}}_i, i \in \mathbb{X}. \end{cases} \tag{71}$$

Matrix $\|p_{ij}(\varepsilon)\|$ is stochastic, for every $\varepsilon \in (0, \varepsilon_0]$ and, thus, matrix $\|p_{ij}(0)\|$ is also stochastic.

However, it is possible that matrix $\|p_{ij}(0)\|$ has more zero elements than matrices $\|p_{ij}(\varepsilon)\|$.

Therefore, a Markov chain $\eta_n^{(0)}, n = 0, 1, \ldots$, with the phase space $\mathbb{X}$ and the matrix of transition probabilities $\|p_{ij}(0)\|$ can be not ergodic, and its phase space $\mathbb{X}$ can consist of one or several closed classes of communicative states plus, possibly, a class of transient states.

Condition **F** implies that there exist limits,

$$
e_{ij}(0) = \lim_{\varepsilon \to 0} e_{ij}(\varepsilon) = \begin{cases} \infty & \text{if } m_{ij}^- < 0, j \in \mathbb{Y}_i, i \in \mathbb{X}, \\ b_{ij}[0] & \text{if } m_{ij}^- = 0, j \in \mathbb{Y}_i, i \in \mathbb{X}, \\ 0 & \text{if } m_{ij}^- > 0, j \in \mathbb{Y}_i, i \in \mathbb{X}, \\ 0 & \text{if } j \in \overline{\mathbb{Y}}_i, i \in \mathbb{X}. \end{cases} \tag{72}
$$

Out goal is to design an effective algorithm for constructing asymptotic expansions for stationary probabilities $\pi_i(\varepsilon), i \in \mathbb{X}$, under assumption that conditions **A**–**F** hold.

As we shall see, the proposed algorithm, based on a special techniques of sequential phase space reduction, can be applied for models with asymptotically coupled and uncoupled phase spaces and all types of asymptotic behavior of expected sojourn times.

The models of continuous and discrete Markov chains are particular cases.

In particular, asymptotic expansions for stationary probabilities $\rho_i(\varepsilon), i \in \mathbb{X}$ coincide with expansions for stationary probabilities $\pi_i(\varepsilon), i \in \mathbb{X}$, for the discrete time Markov chain, where expectations $e_{ij}(\varepsilon) = p_{ij}(\varepsilon), i, j \in \mathbb{X}$.

3.4 Expected Hitting Times and Stationary Probabilities for Semi-Markov Processes

Let us define hitting times, which are random variables given by the following relation, for $j \in \mathbb{X}$,

$$
\tau_j^{(\varepsilon)} = \sum_{n=1}^{v_j^{(\varepsilon)}} \kappa_n^{(\varepsilon)}, \tag{73}
$$

where

$$
v_j^{(\varepsilon)} = \min(n \geq 1 : \eta_n^{(\varepsilon)} = j). \tag{74}
$$

Let us denote,

$$
E_{ij}(\varepsilon) = \mathsf{E}_i \tau_j^{(\varepsilon)}, \quad i, j \in \mathbb{X}. \tag{75}
$$

As is known, conditions **A**, **D** and **E** imply that, for every $\varepsilon \in (0, \varepsilon_0]$,

$$0 < E_{ij}(\varepsilon) < \infty, \ i, j \in \mathbb{X}. \tag{76}$$

Moreover, under the above conditions, the expectations $E_{ij}(\varepsilon)$, $i \in \mathbb{X}$ are, for every $j \in \mathbb{X}$, the unique solution for the following system of linear equations,

$$\left\{ E_{ij}(\varepsilon) = e_i(\varepsilon) + \sum_{r \neq j} p_{ir}(\varepsilon) E_{rj}(\varepsilon), \ i \in \mathbb{X}. \tag{77} \right.$$

The following relation plays an important role in what follows,

$$\pi_i(\varepsilon) = \frac{e_i(\varepsilon)}{E_{ii}(\varepsilon)}, \ i \in \mathbb{X}. \tag{78}$$

In fact this formula is an alternative form of relation (69). Indeed, as is known, $E_{ii}(\varepsilon) = \sum_{j \in \mathbb{X}} e_j(\varepsilon) f_{ii,j}(\varepsilon)$, where $f_{ii,j}(\varepsilon)$ is the expected number of visits by the Markov chain $\eta_n^{(\varepsilon)}$ the state j between two sequential visits of the state i. As also known, $f_{ii,j}(\varepsilon) = \rho_j(\varepsilon)/\rho_i(\varepsilon)$, $i, j \in \mathbb{X}$.

Formula (78) permits reduce the problem of constructing asymptotic expansions for semi-Markov stationary probabilities $\pi_i(\varepsilon)$ to the problem of constructing Laurent asymptotic expansions for expectation of hitting times $E_{ii}(\varepsilon)$.

4   Semi-Markov Processes with Reduced Phase Spaces

In this section, we present a procedure for one-step procedure of phase space reduction for semi-Markov processes and algorithms for re-calculation of asymptotic expansions for perturbed semi-Markov processes with reduced phase spaces.

4.1   Reduction of a Phase Space for Semi-Markov Process

Let us choose some state r and consider the reduced phase space $_r\mathbb{X} = \{i \in \mathbb{X}, i \neq r\}$, with the state r excluded from the phase space $\mathbb{X}$.

Let us assume that $p_r^{(\varepsilon)} = \mathsf{P}\{\eta_0^{(\varepsilon)} = r\} = 0$ and define the sequential moments of hitting the reduced space $_r\mathbb{X}$ by the embedded Markov chain $\eta_n^{(\varepsilon)}$,

$$_r\xi_n^{(\varepsilon)} = \min(k > {}_r\xi_{n-1}^{(\varepsilon)}, \ \eta_k^{(\varepsilon)} \in {}_r\mathbb{X}), \ n = 1, 2, \ldots, \ {}_r\xi_0^{(\varepsilon)} = 0. \tag{79}$$

Now, let us define the random sequence,

$$({}_r\eta_n^{(\varepsilon)}, {}_r\kappa_n^{(\varepsilon)}) = \begin{cases} (\eta_{{}_r\xi_n^{(\varepsilon)}}^{(\varepsilon)}, \sum_{k={}_r\xi_{n-1}^{(\varepsilon)}+1}^{{}_r\xi_n^{(\varepsilon)}} \kappa_k^{(\varepsilon)}) & \text{for } n = 1, 2, \ldots, \\ (\eta_0^{(\varepsilon)}, 0) & \text{for } n = 0. \end{cases} \tag{80}$$

This sequence is also a Markov renewal process with a phase space $_r\mathbb{X} \times [0, \infty)$, the initial distribution $p_i^{(\varepsilon)} = \mathsf{P}\{\eta_0^{(\varepsilon)} = i\}$, $i \in {}_r\mathbb{X}$ (remind that $p_r^{(\varepsilon)} = 0$), and transition probabilities defined for (i, s), $(j, t) \in {}_r\mathbb{X} \times [0, \infty)$,

$$_rQ_{ij}^{(\varepsilon)}(t) = \mathsf{P}\{_r\eta_{n+1}^{(\varepsilon)} = j, \ _r\kappa_{n+1}^{(\varepsilon)} \le t / \ _r\eta_n^{(\varepsilon)} = i, \ _r\kappa_n^{(\varepsilon)} = s\}. \tag{81}$$

Respectively, one can define the transformed semi-Markov process with the reduced phase space $_r\mathbb{X}$,

$$_r\eta^{(\varepsilon)}(t) = \ _r\eta_{_r\nu^{(\varepsilon)}(t)}^{(\varepsilon)}, \ t \ge 0. \tag{82}$$

where

$$_r\nu^{(\varepsilon)}(t) = \max(n \ge 0 : \ _r\zeta_n^{(\varepsilon)} \le t), \ t \ge 0, \tag{83}$$

is a number of jumps at time interval $[0, t]$, and

$$_r\zeta_n^{(\varepsilon)} = \ _r\kappa_1^{(\varepsilon)} + \cdots + \ _r\kappa_n^{(\varepsilon)}, \ n = 0, 1, \ldots, \tag{84}$$

are sequential moments of jumps for the semi-Markov process $_r\eta^{(\varepsilon)}(t)$.

The transition probabilities $_rQ_{ij}^{(\varepsilon)}(t)$ are expressed via the transition probabilities $Q_{ij}^{(\varepsilon)}(t)$ by the following formula, for $i, j \in {}_r\mathbb{X}, t \ge 0$,

$$_rQ_{ij}^{(\varepsilon)}(t) = Q_{ij}^{(\varepsilon)}(t) + \sum_{n=0}^{\infty} Q_{ir}^{(\varepsilon)}(t) * Q_{rr}^{(\varepsilon)*n}(t) * Q_{rj}^{(\varepsilon)}(t). \tag{85}$$

Here, symbol $*$ is used to denote a convolution of distribution functions (possibly improper) and $Q_{rr}^{(\varepsilon)*n}(t)$ is the n times convolution of the distribution function $Q_{rr}^{(\varepsilon)}(t)$ given by the following recurrent formula, for $r \in \mathbb{X}$,

$$Q_{rr}^{(\varepsilon)*n}(t) = \begin{cases} \int_0^t Q_{rr}^{(\varepsilon)*(n-1)}(t-s)Q_{rr}^{(\varepsilon)}(ds) & \text{for } t \ge 0 \text{ and } n \ge 1, \\ I(t \ge 0) & \text{for } t \ge 0 \text{ and } n = 0. \end{cases} \tag{86}$$

Relation (85) directly implies the following formula for transition probabilities of the embedded Markov chain $_r\eta_n^{(\varepsilon)}$, for $i, j \in {}_r\mathbb{X}$,

$$_rp_{ij}(\varepsilon) = \ _rQ_{ij}^{(\varepsilon)}(\infty)$$

$$= p_{ij}(\varepsilon) + \sum_{n=0}^{\infty} p_{ir}(\varepsilon)p_{rr}(\varepsilon)^n p_{rj}(\varepsilon)$$

$$= p_{ij}(\varepsilon) + p_{ir}(\varepsilon)\frac{p_{rj}(\varepsilon)}{1 - p_{rr}(\varepsilon)}. \tag{87}$$

Let us denote,

$$\mathbb{Y}_{ir}^+ = \{j \in \mathbb{Y}_i : j \neq r\}, \; i, r \in \mathbb{X}. \tag{88}$$

and

$$\mathbb{Y}_{ir}^- = \{j \in {}_r\mathbb{X} : r \in \mathbb{Y}_i, \; j \in \mathbb{Y}_r\}, \; i \in {}_r\mathbb{X}. \tag{89}$$

Condition **A** implies that sets $\mathbb{Y}_{rr}^+ \neq \emptyset, r \in \mathbb{X}$.
Thus, probabilities $1 - p_{rr}(\varepsilon) > 0, r \in \mathbb{X}$, for every $\varepsilon \in (0, \varepsilon_0]$.
That is why,

$$\begin{aligned}
{}_r\mathbb{Y}_i &= \{j \in {}_r\mathbb{X} : {}_r p_{ij}(\varepsilon) > 0, \varepsilon \in (0, \varepsilon_0]\} \\
&= \{j \in {}_r\mathbb{X} : j \in \mathbb{Y}_i\} \cup \{j \in {}_r\mathbb{X} : r \in \mathbb{Y}_i, j \in \mathbb{Y}_r\}, \\
&= \mathbb{Y}_{ir}^+ \cup \mathbb{Y}_{ir}^-, \; i \in {}_r\mathbb{X}.
\end{aligned} \tag{90}$$

Relation (87) and condition **A**, assumed to hold for the Markov chain $\eta_n^{(\varepsilon)}$, imply that condition **A** also holds for the Markov chain ${}_r\eta_n^{(\varepsilon)}$, with the sets ${}_r\mathbb{Y}_i, i \in {}_r\mathbb{X}$.

Indeed, let $i \in {}_r\mathbb{X}$. If $j \in \mathbb{Y}_{ir}^+$ then $p_{ij}(\varepsilon) > 0$ and, thus, ${}_r p_{ij}(\varepsilon) > 0$. If $j \in \mathbb{Y}_{ir}^-$ then $p_{ir}(\varepsilon), p_{rj}(\varepsilon) > 0$ and, again, ${}_r p_{ij}(\varepsilon) > 0$. If $j \notin \mathbb{Y}_{ir}^+ \cup \mathbb{Y}_{ir}^-$ then $p_{ij}(\varepsilon) = 0$ and $p_{ir}(\varepsilon) \cdot p_{rj}(\varepsilon) = 0$. By relation (87), this implies that ${}_r p_{ij}(\varepsilon) = 0$.

Let $i \in {}_r\mathbb{X}$. If $\mathbb{Y}_{ir}^+ \neq \emptyset$ then ${}_r\mathbb{Y}_i \neq \emptyset$. If $\mathbb{Y}_{ir}^+ = \emptyset$ then $r \in \mathbb{Y}_i$ and, thus, $p_{ir}(\varepsilon) > 0$. Then, $\mathbb{Y}_{ir}^- = \{j \in {}_r\mathbb{X} : p_{rj}(\varepsilon) > 0\} = \mathbb{Y}_{rr}^+ \neq \emptyset$. Therefore, sets ${}_r\mathbb{Y}_i \neq \emptyset, i \in {}_r\mathbb{X}$.

Thus, conditions **A (a)** and **(b)** assumed to hold for the Markov chain $\eta_n^{(\varepsilon)}$, imply that these conditions also hold for the Markov chain ${}_r\eta_n^{(\varepsilon)}$, with sets ${}_r\mathbb{Y}_i, i \in {}_r\mathbb{X}$ replacing sets $\mathbb{Y}_i, i \in \mathbb{X}$.

Also, let $i, j \in {}_r\mathbb{X}$ and $i = l_0, l_1, \ldots, l_{n_{ij}} = j$ be a chain of states such that $l_1 \in \mathbb{Y}_{l_0}, \ldots, l_n \in \mathbb{Y}_{l_{n_{ij}-1}}$. As was remarked above, we can always to assume that states $l_1, \ldots, l_{n_{ij}-1}$ are different and that $l_1, \ldots, l_{n_{ij}-1} \neq i, j$. This implies that either $l_1, \ldots, l_{n_{ij}-1} \neq r$ or there exist at most one $1 \leq k \leq n_{ij} - 1$ such that $i_k = r$. In the first case, $l_1 \in {}_r\mathbb{Y}_{l_0}, \ldots, l_{n_{ij}} \in {}_r\mathbb{Y}_{l_{n_{ij}-1}}$. In the second case, $l_1 \in {}_r\mathbb{Y}_{l_0}, \ldots, l_{k-1} \in {}_r\mathbb{Y}_{l_{k-2}}, l_{k-1} \in {}_r\mathbb{Y}_{l_{k+1}}, \ldots, l_{n_{ij}} \in {}_r\mathbb{Y}_{l_{n_{ij}-1}}$.

Thus, condition **A (c)** assumed to hold for the Markov chain $\eta_n^{(\varepsilon)}$, imply that this condition also holds for the Markov chain ${}_r\eta_n^{(\varepsilon)}$.

Let us define distribution functions,

$$F_i^{(\varepsilon)}(t) = \sum_{j \in \mathbb{X}} Q_{ij}^{(\varepsilon)}(t), t \geq 0, \; i, j \in \mathbb{X}. \tag{91}$$

and

$$F_{ij}^{(\varepsilon)}(t) = \begin{cases} Q_{ij}^{(\varepsilon)}(t)/p_{ij}(\varepsilon) & \text{for } t \geq 0 \text{ if } p_{ij}(\varepsilon) > 0, \\ F_i^{(\varepsilon)}(t) & \text{for } t \geq 0 \text{ if } p_{ij}(\varepsilon) = 0. \end{cases} \tag{92}$$

Obviously,

$$\tilde{e}_{ij}(\varepsilon) = \int_0^\infty tF_{ij}^{(\varepsilon)}(dt) = \begin{cases} e_{ij}(\varepsilon)/p_{ij}(\varepsilon) & \text{if } p_{ij}(\varepsilon) > 0, \\ e_i(\varepsilon) & \text{if } p_{ij}(\varepsilon) = 0, \end{cases} \tag{93}$$

and

$$e_i(\varepsilon) = \int_0^\infty tF_i^{(\varepsilon)}(dt), \ i \in \mathbb{X}. \tag{94}$$

Also, let us introduce expectations,

$$_r e_{ij}(\varepsilon) = \int_0^\infty t \, _r Q_{ij}^{(\varepsilon)}(dt), \ i,j \in {}_r\mathbb{X}. \tag{95}$$

Relation (85) directly implies the following formula for expectations $_r e_{ij}(\varepsilon)$, $i, j \in {}_r\mathbb{X}$,

$$_r e_{ij}(\varepsilon) = \tilde{e}_{ij}(\varepsilon)p_{ij}(\varepsilon) + \sum_{n=0}^\infty \big(\tilde{e}_{ir}(\varepsilon) + n\tilde{e}_{rr}(\varepsilon) + \tilde{e}_{rj}(\varepsilon)\big)p_{ir}(\varepsilon)p_{rr}(\varepsilon)^n p_{rj}(\varepsilon)$$

$$= e_{ij}(\varepsilon) + e_{ir}(\varepsilon)\frac{p_{rj}(\varepsilon)}{1 - p_{rr}(\varepsilon)}$$

$$+ e_{rr}(\varepsilon)\frac{p_{ir}(\varepsilon)}{1 - p_{rr}(\varepsilon)}\frac{p_{rj}(\varepsilon)}{1 - p_{rr}(\varepsilon)} + e_{rj}(\varepsilon)\frac{p_{ir}(\varepsilon)}{1 - p_{rr}(\varepsilon)}. \tag{96}$$

Relation (96) implies that conditions **D** and **E**, assumed to hold for the semi-Markov process $\eta^{(\varepsilon)}(t)$, imply that these conditions also hold for the semi-Markov process $_r\eta^{(\varepsilon)}(t)$.

4.2 Hitting Times for Reduced Semi-Markov Processes

The first hitting times to a state $j \neq r$ are connected for Markov chains $\eta_n^{(\varepsilon)}$ and $_r\eta_n^{(\varepsilon)}$ by the following relation,

$$\nu_j^{(\varepsilon)} = \min(n \geq 1 : \eta_n^{(\varepsilon)} = j)$$

$$= \min({}_r\xi_n^{(\varepsilon)} \geq 1 : {}_r\eta_n^{(\varepsilon)} = j) = {}_r\xi_{{}_r\nu_j^{(\varepsilon)}}^{(\varepsilon)}, \tag{97}$$

where

$$_r\nu_j^{(\varepsilon)} = \min(n \geq 1 : {}_r\eta_n^{(\varepsilon)} = j). \tag{98}$$

Relations (97) and (98) imply that the following relation hold for the first hitting times to a state $j \neq r$ for the semi-Markov processes $\eta^{(\varepsilon)}(t)$ and $_r\eta^{(\varepsilon)}(t)$,

$$\tau_j^{(\varepsilon)} = \sum_{n=1}^{v_j^{(\varepsilon)}} \kappa_n^{(\varepsilon)} = \sum_{n=1}^{r\xi_{rv_j^{(\varepsilon)}}^{(\varepsilon)}} \kappa_n^{(\varepsilon)}$$

$$= \sum_{n=1}^{rv_j^{(\varepsilon)}} {}_r\kappa_n^{(\varepsilon)} = {}_r\tau_j^{(\varepsilon)}. \tag{99}$$

Let us summarize the above remarks in the following theorem, which play the key role in what follows.

Theorem 10.1 *Let conditions* **A**, **D** *and* **E** *hold and the initial distribution satisfies the assumption,* $p_r^{(\varepsilon)} = 0$, *for every* $\varepsilon \in (0, \varepsilon_0]$. *Then, for any state* $j \neq r$, *the first hitting times* $\tau_j^{(\varepsilon)}$ *and* ${}_r\tau_j^{(\varepsilon)}$ *to the state* j, *respectively, for semi-Markov processes* $\eta^{(\varepsilon)}(t)$ *and* ${}_r\eta^{(\varepsilon)}(t)$, *coincide.*

4.3 Asymptotic Expansions for Non-absorption Probabilities

As was mentioned above, condition **A** implies that the non-absorption probability $\bar{p}_{ii}(\varepsilon) = 1 - p_{ii}(\varepsilon) > 0$, $i \in \mathbb{X}$, $\varepsilon \in (0, \varepsilon_0]$.

Let us introduce the set,

$$\mathbb{Y} = \{i \in \mathbb{X} : i \in \mathbb{Y}_i\}. \tag{100}$$

Algorithm 1. This is an algorithm for constructing asymptotic expansions for non-absorption probabilities $\bar{p}_{ii}(\varepsilon)$, $i \in \mathbb{X}$.

Case 1: $i \in \mathbb{Y}$.

Let us use the following relation, which holds, for every $i \in \mathbb{Y}$ and $\varepsilon \in (0, \varepsilon_0]$,

$$\bar{p}_{ii}(\varepsilon) = 1 - p_{ii}(\varepsilon) = \sum_{j \in \mathbb{Y}_{ii}^+} p_{ij}(\varepsilon) \tag{101}$$

1.1. To construct the (h_i', k_i')-expansion for the non-absorption probability $\bar{p}_{ii}(\varepsilon) = 1 - p_{ii}(\varepsilon)$ by applying the propositions **(i)** (the multiplication by a constant rule) and **(ii)** (the summation rule) of Lemma 10.3 to the (l_{ii}^-, l_{ii}^+)-expansion for transition probability $p_{ii}(\varepsilon)$ given in condition **B** (first, this expansion is multiplied by constant -1 and, second, is summed with constant 1 represented as $(0, l_{ii}^+)$-expansion given in relation (58)). In this case, parameters $h_i' = 0$, $k_i' = l_{ii}^+$.

1.2. To construct the (h_i'', k_i'')-expansion for the non-absorption probability $\bar{p}_{ii}(\varepsilon) = \sum_{j \in \mathbb{Y}_{ii}^+} p_{ij}(\varepsilon)$ using the corresponding asymptotic expansions for transition probabilities $p_{ij}(\varepsilon)$, $j \in \mathbb{Y}_{ii}^+$ given in condition **B**, and the proposition **(i)** (the multiple summation rule) of Lemma 10.5. In this case, parameters $h_i'' = \min_{j \in \mathbb{Y}_{ii}^+} l_{ij}^-$, $k_i'' = \min_{j \in \mathbb{Y}_{ii}^+} l_{ij}^+$. This asymptotic expansion is pivotal.

1.3. To construct the $(\bar{h}_i, \bar{k}_i)$-expansion for the non-absorption probability $\bar{p}_{ii}(\varepsilon)$ using relation (101), and propositions **(i)–(iv)** of Lemma 10.1. In this case, parameters $\bar{h}_i = 0 \vee h_i'' = h_i''$, $\bar{k}_i = k_i' \vee k_i'' = l_{ii}^+ \vee \min_{j \in \mathbb{Y}_{ii}^+} l_{ij}^+$. This asymptotic expansion is pivotal.

It should be noted that (l_{ii}^-, l_{ii}^+)-expansion for the transition probability $p_{ii}(\varepsilon)$ given in condition **B** and (h_i'', k_i'')-expansion for function $\bar{p}_{ii}(\varepsilon) = \sum_{j \in \mathbb{Y}_{ii}^+} p_{ij}(\varepsilon)$ given in Step 1.2, satisfy, for every $i \in \mathbb{X}$, additional conditions given in Remark 10.1, respectively, for set $\mathbb{Z} = \{i\}$ and set $\mathbb{Z} = \mathbb{Y}_{ii}^+$.

Case 2: $i \in \overline{\mathbb{Y}}$.

1.4. In this case, the non-absorption probability $\bar{p}_{ii}(\varepsilon) \equiv 1$. If necessary, it can be represented in the form of $(0, n)$-expansion given by relation (58), for any $n = 0, 1, \ldots$.

The above remarks can be summarized in the following lemma.

Lemma 10.10 *Let conditions* **A**, **B** *and* **C** *hold. Then, the asymptotic expansions for the non-absorption probabilities* $\bar{p}_{ii}(\varepsilon)$, $i \in \mathbb{X}$ *are given in Algorithm 1.*

Algorithm 2. This is an algorithm for computing upper bounds for remainders of asymptotic expansions for non-absorption probabilities $\bar{p}_{ii}(\varepsilon)$, $i \in \mathbb{X}$.

Case 1: $i \in \mathbb{Y}$.

2.1. To construct $(h_i', k_i', \delta_i', G_i', \varepsilon_i')$-expansion for the non-absorption probability $\bar{p}_{ii}(\varepsilon) = 1 - p_{ii}(\varepsilon)$ by applying the propositions **(i)** (the multiplication by a constant rule) and **(ii** (the summation rule) of Lemma 10.4 to the $(l_{ii}^-, l_{ii}^+, \delta_{ii}, G_{ii}, \varepsilon_{ii})$-expansion for the transition probability $p_{ii}(\varepsilon)$ given in condition **B′** and (first, this expansion is multiplied by constant -1 and, second, is summated with constant 1 represented as $(0, l_{ii}^+, 1, G, \varepsilon_0)$-expansion given in relation (58)), third, constant G can be replaced by 0, since it can be taken an arbitrary small. In this case, parameters $\delta_i' = \delta_{ii}$, $G_i' = G_{ii}$, $\varepsilon_i' = \varepsilon_{ii}$.

2.2. To construct the $(h_i'', k_i'', \delta_i'', G_i'', \varepsilon_i'')$-expansion for the non-absorption probability $\bar{p}_{ii}(\varepsilon) = \sum_{j \in \mathbb{Y}_{ii}^+} p_{ij}(\varepsilon)$ using the $(l_{ij}^-, l_{ij}^+, \delta_{ij}, G_{ij}, \varepsilon_{ij})$-expansions for transition probabilities $p_{ij}(\varepsilon)$, $j \in \mathbb{Y}_{ii}^+$ given in condition **B′**, and the proposition **(i)** (the multiple summation rule) of Lemma 10.6. In this case, parameters δ_i'', G_i'', ε_i'' are given by the corresponding variant of relation (24).

2.3. To construct the $(\bar{h}_i, \bar{k}_i, \bar{\delta}_i, \bar{G}_i, \bar{\varepsilon}_i)$-expansion for the non-absorption probability $\bar{p}_{ii}(\varepsilon)$ using relation (101), and proposition **(i)** of Lemma 10.2. In this case, parameters $\bar{\delta}_i$, $\bar{G}_i$, $\bar{\varepsilon}_i$ are given by the corresponding variant of relation (4).

Case 2: $i \in \overline{\mathbb{Y}}$.

2.4. In this case, the non-absorption probability $\bar{p}_{ii}(\varepsilon) \equiv 1$. If necessary, it can be represented in the form of $(0, n, 1, G, \varepsilon_0)$-expansion given by relation (58), for any $0 < G < \infty$ and $n = 0, 1, \ldots$.

The above remarks can be summarized in the following lemma.

Lemma 10.11 *Let conditions* **A**, **B′** *and* **C** *hold. Then, the asymptotic expansions for the non-absorption probabilities* $\bar{p}_{ii}(\varepsilon)$, $i \in \mathbb{X}$ *with explicit upper bounds for remainders are given in Algorithm 2.*

4.4 Asymptotic Expansions for Transition Probabilities of Reduced Embedded Markov Chains

Relation (87) can be re-written in the following form more convenient for constructing asymptotic expansions for probabilities $_rp_{ij}(\varepsilon), i, j \in {}_r\mathbb{X}$,

$$
{}_rp_{ij}(\varepsilon) = \begin{cases} p_{ij}(\varepsilon) + p_{ir}(\varepsilon)\frac{p_{rj}(\varepsilon)}{1-p_{rr}(\varepsilon)} & \text{if } j \in \mathbb{Y}_{ir}^+ \cap \mathbb{Y}_{ir}^-, \\ p_{ij}(\varepsilon) & \text{if } j \in \mathbb{Y}_{ir}^+ \cap \overline{\mathbb{Y}}_{ir}^-, \\ p_{ir}(\varepsilon)\frac{p_{rj}(\varepsilon)}{1-p_{rr}(\varepsilon)} & \text{if } j \in \overline{\mathbb{Y}}_{ir}^+ \cap \mathbb{Y}_{ir}^-, \\ 0 & \text{if } j \in \overline{\mathbb{Y}}_{ir}^+ \cap \overline{\mathbb{Y}}_{ir}^- = {}_r\overline{\mathbb{Y}}_i. \end{cases} \tag{102}
$$

Algorithm 3. This is an algorithm for constructing asymptotic expansions for transition probabilities $_rp_{ij}(\varepsilon), i, j \in {}_r\mathbb{X}$.

Case 1: $r \in \mathbb{Y}$.

3.1. To construct $(\tilde{h}_{rj}, \tilde{k}_{rj})$-expansions for conditional probabilities $\tilde{p}_{rj}(\varepsilon) = \frac{p_{rj}(\varepsilon)}{1-p_{rr}(\varepsilon)}, j \in \mathbb{Y}_{rr}^+$, using the (l_{rj}^-, l_{rj}^+)-expansions for transition probabilities $p_{rj}(\varepsilon)$ given in condition **B**, the $(\bar{h}_r, \bar{k}_r)$-expansion for the non-absorption probability $\bar{p}_{rr}(\varepsilon) = 1 - p_{rr}(\varepsilon)$ given in Algorithm 1, and the proposition **(v)** (the division rule) of Lemma 10.3. In this case, parameters $\tilde{h}_{rj} = l_{rj}^- - \bar{h}_r, \tilde{k}_{rj} = (l_{rj}^+ - \bar{h}_r) \wedge (l_{rj}^- + \bar{k}_r - 2\bar{h}_r), j \in \mathbb{Y}_{rr}^+$. These asymptotic expansions are pivotal.

3.2. To construct $(\grave{h}_{irj}, \grave{k}_{irj})$-expansions for products $\grave{p}_{irj}(\varepsilon) = p_{ir}(\varepsilon)\tilde{p}_{rj}(\varepsilon) = p_{ir}(\varepsilon) \cdot \frac{p_{rj}(\varepsilon)}{1-p_{rr}(\varepsilon)}, j \in \mathbb{Y}_{ir}^-, i \in {}_r\mathbb{X}$, using the (l_{ir}^-, l_{ir}^+)-expansions for transition probabilities $p_{ir}(\varepsilon)$ given in condition **B**, the $(\tilde{h}_{rj}, \tilde{k}_{rj})$-expansions for conditional probabilities $\tilde{p}_{rj}(\varepsilon)$ given in the above Step 3.1 and the proposition **(iii)** (the multiplication rule) of Lemma 10.3. In this case, parameters $\grave{h}_{irj} = l_{ir}^- + \tilde{h}_{rj}, \grave{k}_{irj} = (l_{ir}^- + \tilde{k}_{rj}) \wedge (l_{ir}^+ + \tilde{h}_{rj}), j \in \mathbb{Y}_{ir}^-, i \in {}_r\mathbb{X}$. These asymptotic expansions are pivotal.

3.3. To construct $(\acute{h}_{irj}, \acute{k}_{irj})$-expansions for sums $\acute{p}_{irj}(\varepsilon) = p_{ij}(\varepsilon) + \grave{p}_{irj}(\varepsilon) = p_{ij}(\varepsilon) + p_{ir}(\varepsilon) \cdot \frac{p_{rj}(\varepsilon)}{1-p_{rr}(\varepsilon)}, j \in \mathbb{Y}_{ir}^+ \cap \mathbb{Y}_{ir}^-, i \in {}_r\mathbb{X}$, using the (l_{ij}^-, l_{ij}^+)-expansions for transition probabilities $p_{ij}(\varepsilon)$ given in condition **B**, the $(\grave{h}_{irj}, \grave{k}_{irj})$-expansions for quantities $\grave{p}_{irj}(\varepsilon)$ given in the above Step 3.2 and the proposition **(ii)** (the summation rule) of Lemma 10.3. In this case, parameters $\acute{h}_{irj} = l_{ij}^- \wedge \grave{h}_{irj}, \acute{k}_{irj} = l_{ij}^+ \wedge \grave{k}_{irj}, j \in \mathbb{Y}_{ir}^-, i \in {}_r\mathbb{X}$. These asymptotic expansions are pivotal.

3.4. To construct $({}_rl_{ij}^-, {}_rl_{ij}^+)$-expansions for transition probabilities $_rp_{ij}(\varepsilon) = \sum_{l={}_rl_{ij}^-}^{{}_rl_{ij}^+} {}_ra_{ij}[l]\varepsilon^l + o(\varepsilon^{{}_rl_{ij}^+}), i, j \in {}_r\mathbb{X}$, using the (l_{ij}^-, l_{ij}^+)-expansions for transition probabilities $p_{ij}(\varepsilon)$ given in condition **B**, the $(\grave{h}_{irj}, \grave{k}_{irj})$-expansions for quantities $\grave{p}_{irj}(\varepsilon)$ and $(\acute{h}_{irj}, \acute{k}_{irj})$-expansions for quantities $\acute{p}_{irj}(\varepsilon)$ given, respectively, in the above Steps 3.2 and 3.3, and the corresponding variants of formulas for transition probabilities $_rp_{ij}(\varepsilon)$ given in relation (102). In this case, parameters $_rl_{ij}^- = \acute{h}_{irj}, {}_rl_{ij}^+ = \acute{k}_{irj}$ if $j \in \mathbb{Y}_{ir}^+ \cap \mathbb{Y}_{ir}^-, i \in {}_r\mathbb{X}$, or $_rl_{ij}^- = l_{ij}^-, {}_rl_{ij}^+ = l_{ij}^+$

if $j \in \mathbb{Y}_{ir}^+ \cap \overline{\mathbb{Y}}_{ir}^-$, $i \in {}_r\mathbb{X}$, or ${}_r l_{ij}^- = \grave{h}_{irj}$, ${}_r l_{ij}^+ = \grave{k}_{irj}$ if $j \in \overline{\mathbb{Y}}_{ir}^+ \cap \mathbb{Y}_{ir}^-$, $i \in {}_r\mathbb{X}$. These asymptotic expansions are pivotal.

Case 2: $r \in \overline{\mathbb{Y}}$.

3.5. The corresponding algorithm is a particular case of the algorithm given in Steps 3.1–3.4. In this case the non-absorption probability $\bar{p}_{rr}(\varepsilon) = 1 - p_{rr}(\varepsilon) \equiv 1$ and, thus, conditional probabilities $\tilde{p}_{rj}(\varepsilon) = \frac{p_{rj}(\varepsilon)}{1-p_{rr}(\varepsilon)} = p_{rj}(\varepsilon)$, $j \in \mathbb{Y}_{rr}^+$. This permits one replace the $(\tilde{h}_{rj}, \tilde{k}_{rj})$-expansions for conditional probabilities $\tilde{p}_{rj}(\varepsilon)$ by the (l_{rj}^-, l_{rj}^+)-expansions for transition probabilities $p_{rj}(\varepsilon)$. This is the only change in the algorithm for construction of asymptotic expansions for transition probabilities ${}_r p_{ij}(\varepsilon)$, $i, j \in {}_r\mathbb{X}$ given in Steps 3.1–3.4, which is required.

The above remarks can be summarized in the following theorem.

Theorem 10.2 *Conditions* **A**, **B** *and* **C** *assumed to hold for the Markov chain $\eta_n^{(\varepsilon)}$, also hold for the reduced Markov chain ${}_r\eta_n^{(\varepsilon)}$, for every $r \in \mathbb{X}$. The asymptotic expansions penetrating condition* **B** *are given for transition probabilities ${}_r p_{ij}(\varepsilon)$, $j \in {}_r\mathbb{Y}_i$, $i \in {}_r\mathbb{X}$, $r \in \mathbb{X}$ in Algorithm 3.*

Algorithm 4. This is an algorithm for computing upper bounds for remainders in asymptotic expansions for transition probabilities ${}_r p_{ij}(\varepsilon)$, $i, j \in {}_r\mathbb{X}$.

4.1. To construct $(\tilde{h}_{rj}, \tilde{k}_{rj}, \tilde{\delta}_{rj}, \tilde{G}_{rj}, \tilde{\varepsilon}_{rj})$-expansions for conditional probabilities $\tilde{p}_{rj}(\varepsilon) = \frac{p_{rj}(\varepsilon)}{1-p_{rr}(\varepsilon)}$, $j \in \mathbb{Y}_{rr}^+$ using the $(l_{rj}^-, l_{rj}^+, \delta_{rj}, G_{rj}, \varepsilon_{rj})$-expansions for transition probabilities $p_{rj}(\varepsilon)$ given in condition **B'**, the $(\bar{h}_r, \bar{k}_r, \bar{\delta}_{rj}, \bar{G}_{rj}, \bar{\varepsilon}_{rj})$-expansion for the non-absorption probability $\bar{p}_{rr}(\varepsilon) = 1 - p_{rr}(\varepsilon)$ given in Algorithm 2, and the proposition (**v**) (the division rule) of Lemma 10.4. In this case, parameters $\tilde{\delta}_{rj}, \tilde{G}_{rj}, \tilde{\varepsilon}_{rj}, j \in \mathbb{Y}_{rr}^+$ are given by the corresponding variants of relation (17).

4.2. To construct $(\grave{h}_{irj}, \grave{k}_{irj}, \grave{\delta}_{irj}, \grave{G}_{irj}, \grave{\varepsilon}_{irj})$-expansions for products $\grave{p}_{irj}(\varepsilon) = p_{ir}(\varepsilon)\tilde{p}_{rj}(\varepsilon) = p_{ir}(\varepsilon) \cdot \frac{p_{rj}(\varepsilon)}{1-p_{rr}(\varepsilon)}$, $j \in \mathbb{Y}_{ir}^+$, $i \in {}_r\mathbb{X}$, using the $(l_{ir}^-, l_{ir}^+, \delta_{ir}, G_{ir}, \varepsilon_{ir})$-expansions for transition probabilities $p_{ir}(\varepsilon)$ given in condition **B'**, the $(\tilde{h}_{rj}, \tilde{k}_{rj}, \tilde{\delta}_{rj}, \tilde{G}_{rj}, \tilde{\varepsilon}_{rj})$-expansions for conditional probabilities $\tilde{p}_{rj}(\varepsilon)$ given in the above Step 4.1 and the proposition (**iii**) (the multiplication rule) of Lemma 10.4. In this case, parameters $\grave{\delta}_{irj}, \grave{G}_{irj}, \grave{\varepsilon}_{irj}, j \in \mathbb{Y}_{ir}^+$, $i \in {}_r\mathbb{X}$ are given by the corresponding variants of relation (15).

4.3. To construct $(\acute{h}_{irj}, \acute{k}_{irj}, \acute{\delta}_{irj}, \acute{G}_{irj}, \acute{\varepsilon}_{irj})$-expansions for sums $\acute{p}_{irj}(\varepsilon) = p_{ij}(\varepsilon) + \grave{p}_{irj}(\varepsilon) = p_{ij}(\varepsilon) + p_{ir}(\varepsilon) \cdot \frac{p_{rj}(\varepsilon)}{1-p_{rr}(\varepsilon)}$, $j \in \mathbb{Y}_{ir}^+ \cap \mathbb{Y}_{ir}^-$, $i \in {}_r\mathbb{X}$, using the $(l_{ij}^-, l_{ij}^+, \delta_{ij}, G_{ij}, \varepsilon_{ij})$-expansions for transition probabilities $p_{ij}(\varepsilon)$ given in condition **B'**, the $(\grave{h}_{irj}, \grave{k}_{irj}, \grave{\delta}_{irj}, \grave{G}_{irj}, \grave{\varepsilon}_{irj})$-expansions for quantities $\grave{p}_{irj}(\varepsilon)$ given in the above Step 4.2 and the proposition (**ii**) (the summation rule) of Lemma 10.4. In this case, parameters $\acute{\delta}_{irj}, \acute{G}_{irj}, \acute{\varepsilon}_{irj}, j \in \mathbb{Y}_{ir}^-$, $i \in {}_r\mathbb{X}$ are given by the corresponding variants of relation (14).

4.4. To construct $({}_r l_{ij}^-, {}_r l_{ij}^+, {}_r \delta_{ij}, {}_r G_{ij}, {}_r \varepsilon_{ij})$-expansions for transition probabilities ${}_r p_{ij}(\varepsilon) = \sum_{l={}_r l_{ij}^-}^{{}_r l_{ij}^+} {}_r a_{ij}[l]\varepsilon^l + o(\varepsilon^{{}_r l_{ij}^+})$, $i, j \in {}_r\mathbb{X}$ using the $(l_{ij}^-, l_{ij}^+, \delta_{ij}, G_{ij}, \varepsilon_{ij})$-expansions for transition probabilities $p_{ij}(\varepsilon)$ given in condition **B'**, the $(\grave{h}_{irj}, \grave{k}_{irj},$

$\grave{\delta}_{irj}, \grave{G}_{irj}, \grave{\varepsilon}_{irj})$-expansions for quantities $\grave{p}_{irj}(\varepsilon)$ and $(\acute{h}_{irj}, \acute{k}_{irj}, \acute{\delta}_{irj}, \acute{G}_{irj}, \acute{\varepsilon}_{irj})$-expansions for quantities $\acute{p}_{irj}(\varepsilon)$ given, respectively, in the above Steps 4.2 and 4.3, and the corresponding variants of formulas for transition probabilities $_r p_{ij}(\varepsilon)$ given in relation (102). In this case, parameters $_r \delta_{ij} = \grave{\delta}_{irj}, {}_r G_{ij} = \grave{G}_{irj}, {}_r \varepsilon_{ij} = \grave{\varepsilon}_{irj}$ if $j \in \mathbb{Y}_{ir}^+ \cap \mathbb{Y}_{ir}^-, i \in {}_r\mathbb{X}$, or $_r \delta_{ij} = \delta_{ij}, {}_r G_{ij} = G_{ij}, {}_r \varepsilon_{ij} = \varepsilon_{ij}$ if $j \in \mathbb{Y}_{ir}^+ \cap \overline{\mathbb{Y}}_{ir}^-, i \in {}_r\mathbb{X}$, or $_r \delta_{ij} = \acute{\delta}_{irj}, {}_r G_{ij} = \acute{G}_{irj}, {}_r \varepsilon_{ij} = \acute{\varepsilon}_{irj}$ if $j \in \overline{\mathbb{Y}}_{ir}^+ \cap \mathbb{Y}_{ir}^-, i \in {}_r\mathbb{X}$.

Case 2: $r \in \overline{\mathbb{Y}}$.

4.5. The corresponding algorithm is a particular case of the algorithm given in Steps 4.1–4.4. In this case the non-absorption probability $\bar{p}_{rr}(\varepsilon) = 1 - p_{rr}(\varepsilon) \equiv 1$ and, thus, conditional probabilities $\tilde{p}_{rj}(\varepsilon) = \frac{p_{rj}(\varepsilon)}{1-p_{rr}(\varepsilon)} = p_{rj}(\varepsilon), i \in \mathbb{Y}_{rr}^+$. This permits one replace the $(\tilde{h}_{rj}, \tilde{k}_{rj}, \tilde{\delta}_{rj}, \tilde{G}_{rj}, \tilde{\varepsilon}_{rj})$-expansions for conditional probabilities $\tilde{p}_{rj}(\varepsilon)$ by the $(l_{rj}^-, l_{rj}^+, \delta_{rj}, G_{rj}, \varepsilon_{rj})$-expansions for transition probabilities $p_{rj}(\varepsilon)$. This is the only change in the algorithm for construction of asymptotic expansions for transition probabilities $_r p_{ij}(\varepsilon), i, j \in {}_r\mathbb{X}$ given in Steps 4.1–4.4, which is required.

The above remarks can be summarized in the following theorem.

Theorem 10.3 *Conditions **A**, **B**′ and **C** assumed to hold for the Markov chain $\eta_n^{(\varepsilon)}$, also hold for the reduced Markov chain $_r\eta_n^{(\varepsilon)}$, for every $r \in \mathbb{X}$. The upper bounds for remainders in asymptotic expansions penetrating condition **B**′ are given for transition probabilities $_r p_{ij}(\varepsilon), j \in {}_r\mathbb{Y}_i, i \in {}_r\mathbb{X}, r \in \mathbb{X}$ in Algorithm 4.*

4.5 Asymptotic Expansions for Expectations of Sojourn Times for Reduced Semi-Markov Processes

Relation (96) can be re-written in the following form more convenient for constructing the corresponding asymptotic expansions probabilities $_r e_{ij}(\varepsilon), i, j \in {}_r\mathbb{X}$,

$$
_r e_{ij}(\varepsilon) = \begin{cases}
\begin{aligned}
& e_{ij}(\varepsilon) + e_{ir}(\varepsilon)\frac{p_{rj}(\varepsilon)}{1-p_{rr}(\varepsilon)} \\
& + e_{rr}(\varepsilon)\frac{p_{ir}(\varepsilon)}{1-p_{rr}(\varepsilon)}\frac{p_{rj}(\varepsilon)}{1-p_{rr}(\varepsilon)} \\
& + e_{rj}(\varepsilon)\frac{p_{ir}(\varepsilon)}{1-p_{rr}(\varepsilon)}
\end{aligned} & \text{if } j \in \mathbb{Y}_{ir}^+ \cap \mathbb{Y}_{ir}^-, \\[4pt]
e_{ij}(\varepsilon) & \text{if } j \in \mathbb{Y}_{ir}^+ \cap \overline{\mathbb{Y}}_{ir}^-, \\[4pt]
\begin{aligned}
& e_{ir}(\varepsilon)\frac{p_{rj}(\varepsilon)}{1-p_{rr}(\varepsilon)} \\
& + e_{rr}(\varepsilon)\frac{p_{ir}(\varepsilon)}{1-p_{rr}(\varepsilon)}\frac{p_{rj}(\varepsilon)}{1-p_{rr}(\varepsilon)} \\
& + e_{rj}(\varepsilon)\frac{p_{ir}(\varepsilon)}{1-p_{rr}(\varepsilon)}
\end{aligned} & \text{if } j \in \overline{\mathbb{Y}}_{ir}^+ \cap \mathbb{Y}_{ir}^-, \\[4pt]
0 & \text{if } j \in \overline{\mathbb{Y}}_{ir}^+ \cap \overline{\mathbb{Y}}_{ir}^-.
\end{cases} \tag{103}
$$

Algorithm 5. This is an algorithm for computing asymptotic expansions for expectations $_r e_{ij}(\varepsilon), i, j \in {}_r\mathbb{X}$.

Case 1: $r \in \mathbb{Y}$.

5.1. To construct $(\hat{h}_{ir}, \hat{k}_{ir})$-expansions for quantities $\hat{p}_{ir}(\varepsilon) = \frac{p_{ir}(\varepsilon)}{\bar{p}_{rr}(\varepsilon)} = \frac{p_{ir}(\varepsilon)}{1-p_{rr}(\varepsilon)}$, $i \in {}_r\mathbb{X}$, using the (l_{ir}^-, l_{ir}^+)-expansions for transition probabilities $p_{ir}(\varepsilon)$ given in condition **B**, the $(\bar{h}_r, \bar{k}_r)$-expansion for the non-absorption probability $\bar{p}_r(\varepsilon) = 1 - p_{rr}(\varepsilon)$ given in Algorithm 1, and proposition **(v)** (the division rule) of Lemma 10.3. In this case, parameters $\hat{h}_{ir} = h_{ir} - \bar{h}_r$, $\hat{k}_{ir} = (h_{ir} + \bar{k}_r - 2\bar{h}_r) \wedge (k_{ir} - \bar{h}_r)$, $i \in {}_r\mathbb{X}$. These asymptotic expansions are pivotal.

5.2. To construct $(\check{h}_{irj}, \check{k}_{irj})$-expansions for products $\check{p}_{irj}(\varepsilon) = \hat{p}_{ir}(\varepsilon)\tilde{p}_{rj}(\varepsilon) = \frac{p_{ir}(\varepsilon)}{1-p_{rr}(\varepsilon)}\frac{p_{rj}(\varepsilon)}{1-p_{rr}(\varepsilon)}$, $j \in \mathbb{Y}_{ir}^-$, $i \in {}_r\mathbb{X}$, using $(\hat{h}_{ir}, \hat{k}_{ir})$-expansions given in the above Step 5.1, the $(\tilde{h}_{rj}, \tilde{k}_{rj})$-expansions for conditional probabilities $\tilde{p}_{rj}(\varepsilon) = \frac{p_{rj}(\varepsilon)}{1-p_{rr}(\varepsilon)}$ given in Step 3.1 of Algorithm 3, and the proposition **(iii)** (the multiplication rule) of Lemma 10.3. In this case, parameters $\check{h}_{irj} = \hat{h}_{ir} + \tilde{h}_{rj}$, $\check{k}_{irj} = (\hat{h}_{ir} + \tilde{k}_{rj}) \wedge (\hat{k}_{ir} + \tilde{h}_{rj})$, $j \in \mathbb{Y}_{ir}^-$, $i \in {}_r\mathbb{X}$. These asymptotic expansions are pivotal.

5.3. To construct $(\dot{\tilde{h}}_{irj}, \dot{\tilde{k}}_{irj})$-expansions for products $\tilde{e}_{irj}(\varepsilon) = e_{ir}(\varepsilon)\tilde{p}_{rj}(\varepsilon) = e_{ir}(\varepsilon)\frac{p_{rj}(\varepsilon)}{1-p_{rr}(\varepsilon)}$, $j \in \mathbb{Y}_{ir}^-$, $i \in {}_r\mathbb{X}$, using the (m_{ir}^-, m_{ir}^+)-expansions for expectations $e_{ir}(\varepsilon)$ given in condition **F**, the $(\tilde{h}_{rj}, \tilde{k}_{rj})$-expansions for conditional probabilities $\tilde{p}_{rj}(\varepsilon) = \frac{p_{rj}(\varepsilon)}{1-p_{rr}(\varepsilon)}$ given in Step 3.1 of Algorithm 3, and the proposition **(iii)** (the multiplication rule) of Lemma 10.3. In this case, parameters $\dot{\tilde{h}}_{irj} = m_{ir}^- + \tilde{h}_{rj}$, $\dot{\tilde{k}}_{irj} = (m_{ir}^- + \tilde{k}_{rj}) \wedge (m_{ir}^+ + \tilde{h}_{rj})$, $j \in \mathbb{Y}_{ir}^-$, $i \in {}_r\mathbb{X}$. These asymptotic expansions are pivotal.

5.4. To construct $(\dot{\check{h}}_{irj}, \dot{\check{k}}_{irj})$-expansions for products $\check{e}_{irj}(\varepsilon) = e_{rr}(\varepsilon) \cdot \check{p}_{irj}(\varepsilon) = e_{rr}(\varepsilon)\frac{p_{ir}(\varepsilon)}{1-p_{rr}(\varepsilon)}\frac{p_{rj}(\varepsilon)}{1-p_{rr}(\varepsilon)}$, $j \in \mathbb{Y}_{ir}^-$, $i \in {}_r\mathbb{X}$, using the (m_{rr}^-, m_{rr}^+)-expansions for expectations $e_{rr}(\varepsilon)$ given in condition **F**, the $(\check{h}_{irj}, \check{k}_{irj})$-expansions for quantities $\check{p}_{irj}(\varepsilon)$ given in the above Step 5.2, and the proposition **(iii)** (the multiplication rule) of Lemma 10.3. In this case, parameters $\dot{\check{h}}_{irj} = m_{rr}^- + \check{h}_{irj}$, $\dot{\check{k}}_{irj} = (m_{rr}^- + \check{k}_{irj}) \wedge (m_{rr}^+ + \check{h}_{irj})$, $j \in \mathbb{Y}_{ir}^-$, $i \in {}_r\mathbb{X}$. These asymptotic expansions are pivotal.

5.5. To construct $(\dot{\hat{h}}_{irj}, \dot{\hat{k}}_{irj})$-expansions for products $\hat{e}_{irj}(\varepsilon) = e_{rj}(\varepsilon) \cdot \hat{p}_{ir}(\varepsilon) = e_{rj}(\varepsilon)\frac{p_{ir}(\varepsilon)}{1-p_{rr}(\varepsilon)}$, $j \in \mathbb{Y}_{ir}^-$, $i \in {}_r\mathbb{X}$, using the (m_{rj}^-, m_{rj}^+)-expansions for expectations $e_{rj}(\varepsilon)$ given in condition **F**, the $(\hat{h}_{ir}, \hat{k}_{ir})$-expansions for quantities $\hat{p}_{ir}(\varepsilon)$ given in the above Step 5.1, and the proposition **(iii)** (the multiplication rule) of Lemma 10.3. In this case, parameters $\dot{\hat{h}}_{irj} = m_{rj}^- + \hat{h}_{ir}$, $\dot{\hat{k}}_{irj} = (m_{rj}^- + \hat{k}_{ir}) \wedge (m_{rj}^+ + \hat{h}_{ir})$, $j \in \mathbb{Y}_{ir}^-$, $i \in {}_r\mathbb{X}$. These asymptotic expansions are pivotal.

5.6. To construct $(\ddot{h}_{irj}, \ddot{k}_{irj})$-expansions for sums $\ddot{e}_{irj}(\varepsilon) = \tilde{e}_{irj}(\varepsilon) + \check{e}_{irj}(\varepsilon) + \hat{e}_{irj}(\varepsilon) = e_{ir}(\varepsilon)\frac{p_{rj}(\varepsilon)}{1-p_{rr}(\varepsilon)} + e_{rr}(\varepsilon)\frac{p_{ir}(\varepsilon)}{1-p_{rr}(\varepsilon)}\frac{p_{rj}(\varepsilon)}{1-p_{rr}(\varepsilon)} + e_{rj}(\varepsilon)\frac{p_{ir}(\varepsilon)}{1-p_{rr}(\varepsilon)}$, using the $(\dot{\tilde{h}}_{irj}, \dot{\tilde{k}}_{irj})$-expansions for quantities $\tilde{e}_{irj}(\varepsilon)$, the $(\dot{\check{h}}_{irj}, \dot{\check{k}}_{irj})$-expansions for quantities $\check{e}_{irj}(\varepsilon)$ and the $(\dot{\hat{h}}_{irj}, \dot{\hat{k}}_{irj})$-expansions for quantities $\hat{e}_{irj}(\varepsilon)$ given, respectively, in the above Steps 5.3, 5.4 and 5.5, and the proposition **(i)** (the summation rule) of Lemma 10.5. In this case, parameters $\ddot{h}_{irj} = \dot{\tilde{h}}_{irj} \wedge \dot{\check{h}}_{irj} \wedge \dot{\hat{h}}_{irj}$, $\ddot{k}_{irj} = \dot{\tilde{k}}_{irj} \wedge \dot{\check{k}}_{irj} \wedge \dot{\hat{k}}_{irj}$, $j \in \mathbb{Y}_{ir}^-$, $i \in {}_r\mathbb{X}$. These asymptotic expansions are pivotal.

5.7. To construct $(\dddot{h}_{irj}, \dddot{k}_{irj})$-expansions for sums $\dddot{e}_{irj}(\varepsilon) = e_{ij}(\varepsilon) + \ddot{e}_{irj}(\varepsilon) = e_{ij}(\varepsilon) + e_{ir}(\varepsilon)\frac{p_{rj}(\varepsilon)}{1-p_{rr}(\varepsilon)} + e_{rr}(\varepsilon)\frac{p_{ir}(\varepsilon)}{1-p_{rr}(\varepsilon)}\frac{p_{rj}(\varepsilon)}{1-p_{rr}(\varepsilon)} + e_{rj}(\varepsilon)\frac{p_{ir}(\varepsilon)}{1-p_{rr}(\varepsilon)}$, using (m_{ij}^-, m_{ij}^+)-expansions for expectations $e_{ij}(\varepsilon)$ given in condition **F**, the $(\ddot{h}_{irj}, \ddot{k}_{irj})$-expansions-expansions for quantities $\ddot{e}_{irj}(\varepsilon)$ given in the above Step 5.6, and the proposition **(ii)** (the summation rule) of Lemma 10.3. In this case, parameters $\dddot{h}_{irj} = m_{ij}^- \wedge \ddot{h}_{irj}$, $\dddot{k}_{irj} = m_{ij}^+ \wedge \ddot{k}_{irj}$, $j \in \mathbb{Y}_{ir}^-$, $i \in {}_r\mathbb{X}$. These asymptotic expansions are pivotal.

5.8. To construct $({}_rm_{ij}^-, {}_rm_{ij}^+)$-expansions for probabilities ${}_re_{ij}(\varepsilon) = \sum_{l={}_rm_{ij}^-}^{{}_rm_{ij}^+} {}_rb_{ij}[l] \times \varepsilon^l + o(\varepsilon^{{}_rm_{ij}^+})$, $i, j \in {}_r\mathbb{X}$ using the asymptotic expansions for expectations $e_{ij}(\varepsilon)$ given in condition **F**, the $(\ddot{h}_{irj}, \ddot{k}_{irj})$-expansions for quantities $\ddot{e}_{irj}(\varepsilon)$ and $(\dddot{h}_{irj}, \dddot{k}_{irj})$-expansions for quantities $\dddot{e}_{irj}(\varepsilon)$ given, respectively, in Steps 5.6 and 5.7, and the corresponding variants of formulas for expectations ${}_re_{ij}(\varepsilon)$ given in relation (103). In this case, parameters ${}_rm_{ij}^- = \dddot{h}_{irj}$, ${}_rm_{ij}^+ = \dddot{k}_{irj}$ if $j \in \mathbb{Y}_{ir}^+ \cap \mathbb{Y}_{ir}^-$, $i \in {}_r\mathbb{X}$, or ${}_rm_{ij}^- = m_{ij}^-$, ${}_rm_{ij}^+ = m_{ij}^+$ if $j \in \mathbb{Y}_{ir}^+ \cap \overline{\mathbb{Y}}_{ir}^-$, $i \in {}_r\mathbb{X}$, or ${}_rm_{ij}^- = \ddot{h}_{irj}$, ${}_rm_{ij}^+ = \ddot{k}_{irj}$ if $j \in \overline{\mathbb{Y}}_{ir}^+ \cap \mathbb{Y}_{ir}^-$, $i \in {}_r\mathbb{X}$. These asymptotic expansions are pivotal.

Case 2: $r \in \mathbb{Y}$.

5.9. The corresponding algorithm is a particular case of the algorithm given in Steps 5.1–5.8. In this case the non-absorption probability $\bar{p}_{rr}(\varepsilon) = 1 - p_{rr}(\varepsilon) \equiv 1$ and, thus, conditional probabilities $\tilde{p}_{rj}(\varepsilon) = \frac{p_{rj}(\varepsilon)}{1-p_{rr}(\varepsilon)} = p_{rj}(\varepsilon)$, $j \in \mathbb{Y}_{rr}^+$ and quantities $\hat{p}_{ir}(\varepsilon) = \frac{p_{ir}(\varepsilon)}{1-p_{rr}(\varepsilon)} = p_{ir}(\varepsilon)$, $i \in {}_r\mathbb{X}$. This permits one replace the $(\tilde{h}_{rj}, \tilde{k}_{rj})$-expansions for conditional probabilities $\tilde{p}_{rj}(\varepsilon)$ by the (l_{rj}^-, l_{rj}^+)-expansions for transition probabilities $p_{rj}(\varepsilon)$ and the $(\hat{h}_{rj}, \hat{k}_{rj})$-expansions for quantities $\hat{p}_{rj}(\varepsilon)$ by the (l_{ir}^-, l_{ir}^+)-expansions for transition probabilities $p_{ir}(\varepsilon)$. These are the only changes in the algorithm for construction of asymptotic expansions for expectations ${}_re_{ij}(\varepsilon)$, $i, j \in {}_r\mathbb{X}$ given in Steps 5.1–5.8, which are required.

The above remarks can be summarized in the following theorem.

Theorem 10.4 *Conditions **A**–**F** assumed to hold for the semi-Markov process $\eta^{(\varepsilon)}(t)$, also hold for the reduced semi-Markov process ${}_r\eta^{(\varepsilon)}(t)$, for every $r \in \mathbb{X}$. The asymptotic expansions penetrating conditions **B** and **F** are given for transition probabilities ${}_rp_{ij}(\varepsilon), j \in {}_r\mathbb{Y}_i$, $i \in {}_r\mathbb{X}, r \in \mathbb{X}$ and expectations ${}_re_{ij}(\varepsilon), j \in {}_r\mathbb{Y}_i$, $i \in {}_r\mathbb{X}, r \in \mathbb{X}$ in Algorithms 3 and 5.*

Algorithm 6. This is an algorithm for computing upper bounds for remainders in asymptotic expansions for expectations ${}_re_{ij}(\varepsilon)$, $i, j \in {}_r\mathbb{X}$.

6.1. To construct $(\hat{h}_{ir}, \hat{k}_{ir}, \hat{\delta}_{ir}, \hat{G}_{ir}, \hat{\varepsilon}_{ir})$-expansions for quantities $\hat{p}_{ir}(\varepsilon) = \frac{p_{ir}(\varepsilon)}{\bar{p}_{rr}(\varepsilon)} = \frac{p_{ir}(\varepsilon)}{1-p_{rr}(\varepsilon)}$, $i \in {}_r\mathbb{X}$, using the $(l_{ir}^-, l_{ir}^+, \delta_{ij}, G_{ij}, \varepsilon_{ij})$-expansions for transition probabilities $p_{ir}(\varepsilon)$ given in condition **B'**, the $(\bar{h}_r, \bar{k}_r, \bar{\delta}_{ir}, \bar{G}_{ir}, \bar{\varepsilon}_{ir})$-expansion for the non-absorption probability $\bar{p}_r(\varepsilon) = 1 - p_{rr}(\varepsilon)$ given in Algorithm 2, and proposition **(v)** (the division rule) of Lemma 10.4. In this case, parameters $\hat{\delta}_{ir}, \hat{G}_{ir}, \hat{\varepsilon}_{ir}, i \in {}_r\mathbb{X}$ are given by the corresponding variants of relation (17).

6.2. To construct $(\check{h}_{irj}, \check{k}_{irj}, \check{\delta}_{irj}, \check{G}_{irj}, \check{\varepsilon}_{irj})$-expansions for products $\check{p}_{irj}(\varepsilon) = \hat{p}_{ir}(\varepsilon)$ $\tilde{p}_{rj}(\varepsilon) = \frac{p_{ir}(\varepsilon)}{1-p_{rr}(\varepsilon)}\frac{p_{rj}(\varepsilon)}{1-p_{rr}(\varepsilon)}$, $j \in \mathbb{Y}_{ir}^-$, $i \in {}_r\mathbb{X}$, using $(\hat{h}_{ir}, \hat{k}_{ir}, \hat{\delta}_{ir}, \hat{G}_{ir}, \hat{\varepsilon}_{ir})$-

expansions given in the above Step 6.1, the $(\tilde{h}_{rj}, \tilde{k}_{rj}, \tilde{\delta}_{rj}, \tilde{G}_{rj}, \tilde{\varepsilon}_{rj})$-expansions for conditional probabilities $\tilde{p}_{rj}(\varepsilon) = \frac{p_{rj}(\varepsilon)}{1-p_{rr}(\varepsilon)}$ given in Step 4.1 of Algorithm 4, and the proposition **(iii)** (the multiplication rule) of Lemma 10.4. In this case, parameters $\check{\delta}_{irj}, \check{G}_{irj}, \check{\varepsilon}_{irj}, j \in \mathbb{Y}_{ir}^{-}, i \in {}_{r}\mathbb{X}$ are given by the corresponding variants of relation (15).

6.3. To construct $(\dot{\tilde{h}}_{irj}, \dot{\tilde{k}}_{irj}, \dot{\tilde{\delta}}_{irj}, \dot{\tilde{G}}_{irj}, \dot{\tilde{\varepsilon}}_{irj})$-expansions for products $\tilde{e}_{irj}(\varepsilon) = e_{ir}(\varepsilon)$ $\tilde{p}_{rj}(\varepsilon) = e_{ir}(\varepsilon)\frac{p_{rj}(\varepsilon)}{1-p_{rr}(\varepsilon)}$, $j \in \mathbb{Y}_{ir}^{-}$, $i \in {}_{r}\mathbb{X}$, using the $(m_{ir}^{-}, m_{ir}^{+}, \delta_{ir}, G_{ir}, \varepsilon_{ir})$-expansions for expectations $e_{ir}(\varepsilon)$ given in condition $\mathbf{F'}$, the $(\tilde{h}_{rj}, \tilde{k}_{rj}, \tilde{\delta}_{rj}, \tilde{G}_{rj}, \tilde{\varepsilon}_{rj})$-expansions for conditional probabilities $\tilde{p}_{rj}(\varepsilon) = \frac{p_{rj}(\varepsilon)}{1-p_{rr}(\varepsilon)}$ given in Step 4.1 of Algorithm 4, and the proposition **(iii)** (the multiplication rule) of Lemma 10.4. In this case, parameters $\dot{\tilde{\delta}}_{irj}, \dot{\tilde{G}}_{irj}, \dot{\tilde{\varepsilon}}_{irj}, j \in \mathbb{Y}_{ir}^{-}, i \in {}_{r}\mathbb{X}$ are given by the corresponding variants of relation (15).

6.4. To construct $(\dot{\tilde{h}}_{irj}, \dot{\tilde{k}}_{irj}, \dot{\tilde{\delta}}_{irj}, \dot{\tilde{G}}_{irj}, \dot{\tilde{\varepsilon}}_{irj})$-expansions for products $\check{e}_{irj}(\varepsilon) = e_{rr}(\varepsilon) \cdot \check{p}_{irj}(\varepsilon) = e_{rr}(\varepsilon)\frac{p_{ir}(\varepsilon)}{1-p_{rr}(\varepsilon)}\frac{p_{rj}(\varepsilon)}{1-p_{rr}(\varepsilon)}$, $j \in \mathbb{Y}_{ir}^{-}$, $i \in {}_{r}\mathbb{X}$, using the $(m_{rr}^{-}, m_{rr}^{+}, \delta_{rr}, G_{rr}, \varepsilon_{rr})$-expansions for expectations $e_{rr}(\varepsilon)$ given in condition $\mathbf{F'}$, the $(\check{h}_{irj}, \check{k}_{irj}, \check{\delta}_{irj}, \check{G}_{irj}, \check{\varepsilon}_{irj})$-expansions for quantities $\check{p}_{irj}(\varepsilon)$ given in the above Step 6.2, and the proposition **(iii)** (the multiplication rule) of Lemma 10.4. In this case, parameters $\dot{\tilde{\delta}}_{irj}, \dot{\tilde{G}}_{irj}, \dot{\tilde{\varepsilon}}_{irj}), j \in \mathbb{Y}_{ir}^{-}, i \in {}_{r}\mathbb{X}$ are given by the corresponding variants of relation (15).

6.5. To construct $(\hat{\tilde{h}}_{irj}, \hat{\tilde{k}}_{irj}, \hat{\tilde{\delta}}_{irj}, \hat{\tilde{G}}_{irj}, \hat{\tilde{\varepsilon}}_{irj})$-expansions for products $\hat{e}_{irj}(\varepsilon) = e_{rj}(\varepsilon) \cdot \hat{p}_{ir}(\varepsilon) = e_{rj}(\varepsilon)\frac{p_{ir}(\varepsilon)}{1-p_{rr}(\varepsilon)}$, $j \in \mathbb{Y}_{ir}^{-}$, $i \in {}_{r}\mathbb{X}$, using the $(m_{rj}^{-}, m_{rj}^{+}, \delta_{rj}, G_{rj}, \varepsilon_{rj})$-expansions for expectations $e_{rj}(\varepsilon)$ given in condition $\mathbf{F'}$, the $(\hat{h}_{ir}, \hat{k}_{ir}, \hat{\delta}_{ir}, \hat{G}_{ir}, \hat{\varepsilon}_{ir})$-expansions for quantities $\hat{p}_{ir}(\varepsilon)$ given in the above Step 6.1, and the proposition **(iii)** (the multiplication rule) of Lemma 10.4. In this case, parameters $\hat{\tilde{\delta}}_{irj}, \hat{\tilde{G}}_{irj}, \hat{\tilde{\varepsilon}}_{irj}), j \in \mathbb{Y}_{ir}^{-}, i \in {}_{r}\mathbb{X}$ are given by the corresponding variants of relation (15).

6.6. To construct $(\ddot{h}_{irj}, \ddot{k}_{irj}, \ddot{\delta}_{irj}, \ddot{G}_{irj}, \ddot{\varepsilon}_{irj})$-expansions for sums $\ddot{e}_{irj}(\varepsilon) = \tilde{e}_{irj}(\varepsilon) + \check{e}_{irj}(\varepsilon) + \hat{e}_{irj}(\varepsilon) = e_{ir}(\varepsilon)\frac{p_{rj}(\varepsilon)}{1-p_{rr}(\varepsilon)} + e_{rr}(\varepsilon)\frac{p_{ir}(\varepsilon)}{1-p_{rr}(\varepsilon)}\frac{p_{rj}(\varepsilon)}{1-p_{rr}(\varepsilon)} + e_{rj}(\varepsilon)\frac{p_{ir}(\varepsilon)}{1-p_{rr}(\varepsilon)}$, using the $(\dot{\tilde{h}}_{irj}, \dot{\tilde{k}}_{irj}, \dot{\tilde{\delta}}_{irj}, \dot{\tilde{G}}_{irj}, \dot{\tilde{\varepsilon}}_{irj})$-expansions for quantities $\tilde{e}_{irj}(\varepsilon)$, the $(\dot{\tilde{h}}_{irj}, \dot{\tilde{k}}_{irj}, \dot{\tilde{\delta}}_{irj}, \dot{\tilde{G}}_{irj}, \dot{\tilde{\varepsilon}}_{irj})$-expansions for quantities $\check{e}_{irj}(\varepsilon)$ and the $(\hat{\tilde{h}}_{irj}, \hat{\tilde{k}}_{irj}, \hat{\tilde{\delta}}_{irj}, \hat{\tilde{G}}_{irj}, \hat{\tilde{\varepsilon}}_{irj})$-expansions for quantities $\hat{e}_{irj}(\varepsilon)$ given, respectively, in the above Steps 6.3, 6.4 and 6.5, and the proposition **(i)** (the summation rule) of Lemma 10.6. In this case, parameters $\ddot{\delta}_{irj}, \ddot{G}_{irj}, \ddot{\varepsilon}_{irj}, j \in \mathbb{Y}_{ir}^{-}, i \in {}_{r}\mathbb{X}$ are given by the corresponding variants of relation (24).

6.7. To construct $(\dddot{h}_{irj}, \dddot{k}_{irj}, \dddot{\delta}_{irj}, \dddot{G}_{irj}, \dddot{\varepsilon}_{irj})$-expansions for sums $\dddot{e}_{irj}(\varepsilon) = e_{ij}(\varepsilon) + \ddot{e}_{irj}(\varepsilon) = e_{ij}(\varepsilon) + e_{ir}(\varepsilon)\frac{p_{rj}(\varepsilon)}{1-p_{rr}(\varepsilon)} + e_{rr}(\varepsilon)\frac{p_{ir}(\varepsilon)}{1-p_{rr}(\varepsilon)}\frac{p_{rj}(\varepsilon)}{1-p_{rr}(\varepsilon)} + e_{rj}(\varepsilon)\frac{p_{ir}(\varepsilon)}{1-p_{rr}(\varepsilon)}$, using $(m_{ij}^{-}, m_{ij}^{+}, \delta_{ij}, G_{ij}, \varepsilon_{ij})$-expansions for expectations $e_{ij}(\varepsilon)$ given in condition $\mathbf{F'}$, the $(\ddot{h}_{irj}, \ddot{k}_{irj}, \ddot{\delta}_{irj}, \ddot{G}_{irj}, \ddot{\varepsilon}_{irj})$-expansions for quantities $\ddot{e}_{irj}(\varepsilon)$ given in the above Step 6.6, and the proposition **(ii)** (the summation rule) of Lemma 10.4. In this case, parameters $\dddot{\delta}_{irj}, \dddot{G}_{irj}, \dddot{\varepsilon}_{irj}, j \in \mathbb{Y}_{ir}^{-}, i \in {}_{r}\mathbb{X}$ are given by the corresponding variants of relation (14).

6.8. To construct $({}_rm_{ij}^-, {}_rm_{ij}^+, {}_r\dot\delta_{ij}, {}_r\dot G_{ij}, {}_r\dot\varepsilon_{ij})$-expansions expansions for expectations ${}_re_{ij}(\varepsilon) = \sum_{l={}_rm_{ij}^-}^{{}_rm_{ij}^+} {}_rb_{ij}[l]\varepsilon^l + o(\varepsilon^{{}_rm_{ij}^+}), i,j \in {}_r\mathbb{X}$, using the $(m_{ij}^-, m_{ij}^+, \dot\delta_{ij}, \dot G_{ij}, \dot\varepsilon_{ij})$-expansions for expectations $e_{ij}(\varepsilon)$ given in condition **F'**, the $(\ddot h_{irj}, \ddot k_{irj}, \ddot\delta_{irj}, \ddot G_{irj}, \ddot\varepsilon_{irj})$-expansions for quantities $\ddot e_{irj}(\varepsilon)$ and the $(\dddot h_{irj}, \dddot k_{irj}, \dddot\delta_{irj}, \dddot G_{irj}, \dddot\varepsilon_{irj})$-expansions for quantities $\dddot e_{irj}(\varepsilon)$ given, respectively, in Steps 6.6 and 6.7, and the corresponding variants of formulas for expectations ${}_re_{ij}(\varepsilon)$ given in relation (103). In this case, parameters ${}_r\dot\delta_{ij} = \dddot\delta_{irj}, {}_r\dot G_{ij} = \dddot G_{irj}, {}_r\dot\varepsilon_{ij} = \dddot\varepsilon_{irj}$ if $j \in \mathbb{Y}_{ir}^+ \cap \mathbb{Y}_{ir}^-, i \in {}_r\mathbb{X}$, or ${}_r\dot\delta_{ij} = \delta_{ij}, {}_r\dot G_{ij} = G_{ij}, {}_r\dot\varepsilon_{ij} = \varepsilon_{ij}$ if $j \in \mathbb{Y}_{ir}^+ \cap \overline{\mathbb{Y}}_{ir}^-, i \in {}_r\mathbb{X}$, or ${}_r\dot\delta_{ij} = \ddot\delta_{irj}, {}_r\dot G_{ij} = \ddot G_{irj}, {}_r\dot\varepsilon_{ij} = \ddot\varepsilon_{irj}$ if $j \in \overline{\mathbb{Y}}_{ir}^+ \cap \mathbb{Y}_{ir}^-, i \in {}_r\mathbb{X}$.

Case 2: $r \in \overline{\mathbb{Y}}$.

6.9. The corresponding algorithm is a particular case of the algorithm given in Steps 5.1–5.8. In this case the non-absorption probability $\bar p_{rr}(\varepsilon) = 1 - p_{rr}(\varepsilon) \equiv 1$ and, thus, conditional probabilities $\tilde p_{rj}(\varepsilon) = \frac{p_{rj}(\varepsilon)}{1-p_{rr}(\varepsilon)} = p_{rj}(\varepsilon), j \in \mathbb{Y}_{rr}^+$ and quantities $\hat p_{ir}(\varepsilon) = \frac{p_{ir}(\varepsilon)}{1-p_{rr}(\varepsilon)} = p_{ir}(\varepsilon), i \in {}_r\mathbb{X}$. This permits one replace the $(\tilde h_{rj}, \tilde k_{rj}, \tilde\delta_{rj}, \tilde G_{rj}, \tilde\varepsilon_{rj})$-expansions for conditional probabilities $\tilde p_{rj}(\varepsilon)$ by the $(l_{rj}^-, l_{rj}^+, \delta_{rj}, G_{rj}, \varepsilon_{rj})$-expansions for transition probabilities $p_{rj}(\varepsilon)$ and the $(\hat h_{rj}, \hat k_{rj}, \hat\delta_{rj}, \hat G_{rj}, \hat\varepsilon_{rj})$-expansions for quantities $\hat p_{rj}(\varepsilon)$ by the $(l_{ir}^-, l_{ir}^+, \delta_{ir}, G_{ir}, \varepsilon_{ir})$-expansions for transition probabilities $p_{ir}(\varepsilon)$. These are the only changes in the algorithm for construction of asymptotic expansions for expectations ${}_re_{ij}(\varepsilon), i,j \in {}_r\mathbb{X}$ given in Steps 6.1–6.8, which are required.

The above remarks can be summarized in the following theorem.

Theorem 10.5 *Conditions* **A**, **B'**, **C–E**, **F'** *assumed to hold for the semi-Markov process* $\eta^{(\varepsilon)}(t)$, *also hold for the reduced semi-Markov process* ${}_r\eta^{(\varepsilon)}(t)$, *for every* $r \in \mathbb{X}$. *The upper bounds for remainders in expansions penetrating conditions* **B'** *and* **F'** *are given for transition probabilities* ${}_rp_{ij}(\varepsilon), j \in {}_r\mathbb{Y}_i, i \in {}_r\mathbb{X}, r \in \mathbb{X}$ *and expectations* ${}_re_{ij}(\varepsilon), j \in {}_r\mathbb{Y}_i, i \in {}_r\mathbb{X}, r \in \mathbb{X}$ *in Algorithms 4 and 6.*

5 Sequential Reduction of Phase Space for Semi-Markov Processes

In this section, we present algorithms of sequential reduction of phase spaces for semi-Markov processes.

5.1 Algorithms of Sequential Reduction of Phase Spaces for Semi-Markov Processes

Let $\eta^{(\varepsilon)}(t)$ be a semi-Markov process with the phase space $\mathbb{X} = \{1, \ldots, N\}$, which satisfy conditions **A–F**.

Let $\langle r_1, \ldots, r_N \rangle$ is a permutation of the sequence $\langle 1, \ldots, N \rangle$, and $\bar{r}_n = \langle r_1, \ldots, r_n \rangle$, $n = 1, \ldots, N$ is the corresponding sequence of growing chains of states from space $\mathbb{X}$.

Let us choose state $i \in \mathbb{X}$, and a permutation $\langle r_1, \ldots, r_N \rangle$ such that $r_N = i$.

Let us also assume that initial distribution $\bar{p}(\varepsilon)$ is concentrated in the state i, i.e., $p_i^{(\varepsilon)} = 1$.

Algorithm 7. This is an algorithm for sequential reduction of the phase space for the semi-Markov process $\eta^{(\varepsilon)}(t)$ and constructing asymptotic expansions for transition probabilities and expectation of sojourn times for semi-Markov processes with reduced phase spaces.

7.1. Let $_{\bar{r}_1}\eta^{(\varepsilon)}(t) = {}_{r_1}\eta^{(\varepsilon)}(t)$ be the reduced semi-Markov process which is the result of reduction of state r_1 for the semi-Markov process $\eta^{(\varepsilon)}(t)$. This semi-Markov process has the phase space $_{\bar{r}_1}\mathbb{X} = \mathbb{X} \setminus \{r_1\}$, transition probabilities of the embedded Markov chain $_{\bar{r}_1}p_{i'j'}(\varepsilon)$, $i',j' \in {}_{r_1}\mathbb{X}$ and expectations of transition times $_{\bar{r}_1}e_{i'j'}(\varepsilon)$, $i',j' \in {}_{r_1}\mathbb{X}$, which are determined by the transition probabilities and the expectations of transition times for the process $\eta^{(\varepsilon)}(t)$ via relations (87) and (96). According Theorem 10.1, the expectations of hitting times $E_{i'j'}(\varepsilon)$, $i',j' \in {}_{\bar{r}_1}\mathbb{X}$ coincide for the semi-Markov processes $\eta^{(\varepsilon)}(t)$ and $_{\bar{r}_1}\eta^{(\varepsilon)}(t)$. According Theorems 10.2 and 10.4, the semi-Markov process $_{\bar{r}_1}\eta^{(\varepsilon)}(t)$ satisfy conditions **A–F**. The transition sets $_{\bar{r}_1}\mathbb{Y}_{i'} = {}_{r_1}\mathbb{Y}_{i'}$, $i' \in {}_{\bar{r}_1}\mathbb{X}$ are determined for the process $_{\bar{r}_1}\eta^{(\varepsilon)}(t)$ by condition **A** and relation (90). Therefore, the $(_{\bar{r}_1}l_{i'j'}^-, {}_{\bar{r}_1}l_{i'j'}^+)$-expansions for transition probabilities $_{\bar{r}_1}p_{i'j'}(\varepsilon)$, $j' \in {}_{\bar{r}_1}\mathbb{Y}_{i'}$, $i' \in {}_{\bar{r}_1}\mathbb{X}$ and $(_{\bar{r}_1}m_{i'j'}^-, {}_{\bar{r}_1}m_{i'j'}^+)$-expansions for expectations $_{\bar{r}_1}e_{i'j'}(\varepsilon)$, $j' \in {}_{\bar{r}_1}\mathbb{Y}_{i'}$, $i' \in {}_{\bar{r}_1}\mathbb{X}$ can be constructed by applying Algorithms 1, 3 and 5 to the $(l_{i'j'}^-, l_{i'j'}^+)$-expansions for transition probabilities $p_{i'j'}(\varepsilon)$, $j' \in \mathbb{Y}_{i'}$, $i' \in \mathbb{X}$ and $(m_{i'j'}^-, m_{i'j'}^+)$-expansions for expectations $e_{i'j'}(\varepsilon)$, $j' \in \mathbb{Y}_{i'}$, $i' \in \mathbb{X}$. These expansions are pivotal.

7.2. Let $_{\bar{r}_2}\eta^{(\varepsilon)}(t)$ be the reduced semi-Markov process which is the result of reduction of state r_2 for the semi-Markov process $_{\bar{r}_1}\eta^{(\varepsilon)}(t)$. This semi-Markov process has the phase space $_{\bar{r}_2}\mathbb{X} = \mathbb{X} \setminus \{r_1, r_2\}$, the transition probabilities of the embedded Markov chain $_{\bar{r}_2}p_{i'j'}(\varepsilon)$, $i',j' \in {}_{\bar{r}_1}\mathbb{X}$ and the expectations of transition times $_{\bar{r}_2}e_{i'j'}(\varepsilon)$, $i',j' \in {}_{\bar{r}_2}\mathbb{X}$, which are determined by the transition probabilities and the expectations of transition times for the process $_{\bar{r}_1}\eta^{(\varepsilon)}(t)$ via relations (87) and (96). According to Theorem 10.1, the expectations of hitting times $E_{i'j'}(\varepsilon)$, $i',j' \in {}_{\bar{r}_2}\mathbb{X}$ coincide for the semi-Markov processes $\eta^{(\varepsilon)}(t)$, $_{\bar{r}_1}\eta^{(\varepsilon)}(t)$ and $_{\bar{r}_2}\eta^{(\varepsilon)}(t)$. According Theorems 10.2 and 10.4, the transition probabilities of the embedded Markov chain $_{\bar{r}_2}p_{i'j'}(\varepsilon)$, $i',j' \in {}_{\bar{r}_2}\mathbb{X}$ and the expectations of transition times $_{\bar{r}_2}e_{i'j'}(\varepsilon)$, $i',j' \in {}_{\bar{r}_2}\mathbb{X}$ satisfy conditions **A–F**. The transition sets $_{\bar{r}_2}\mathbb{Y}_{i'}$, $i' \in {}_{\bar{r}_2}\mathbb{X}$ are determined for the process $_{\bar{r}_2}\eta^{(\varepsilon)}(t)$ by condition **A** and relation (90) in the same way as the transition sets $_{\bar{r}_1}\mathbb{Y}_{i'}$, $i' \in {}_{r_1}\mathbb{X}$ are determined by condition **A** and relation (90) for the process $_{\bar{r}_1}\eta^{(\varepsilon)}(t)$. Therefore, the $(_{\bar{r}_2}l^-_{i'j'}, _{\bar{r}_2}l^+_{i'j'})$-expansions for transition probabilities $_{\bar{r}_2}p_{i'j'}(\varepsilon), j' \in _{\bar{r}_2}\mathbb{Y}_{i'}, i' \in _{\bar{r}_2}\mathbb{X}$ and $(_{\bar{r}_2}m^-_{i'j'}, _{\bar{r}_2}m^+_{i'j'})$-expansions for expectations $_{\bar{r}_2}e_{i'j'}(\varepsilon), j' \in _{\bar{r}_2}\mathbb{Y}_{i'}, i' \in _{\bar{r}_2}\mathbb{X}$ can be constructed by applying Algorithms 1, 3 and 5 to the $(_{\bar{r}_1}l^-_{i'j'}, _{\bar{r}_1}l^+_{i'j'})$-expansions for transition probabilities $_{\bar{r}_1}p_{i'j'}(\varepsilon), j' \in _{\bar{r}_1}\mathbb{Y}_{i'}, i' \in _{\bar{r}_1}\mathbb{X}$ and $(_{\bar{r}_1}m^-_{i'j'}, _{\bar{r}_1}m^+_{i'j'})$-expansions for expectations $_{\bar{r}_1}e_{i'j'}(\varepsilon), j' \in _{\bar{r}_1}\mathbb{Y}_{i'}, i' \in _{\bar{r}_1}\mathbb{X}$. These expansions ara pivotal.

7.3. By continuing the above procedure of phase space reduction for states $r_3, \ldots, r_{N-1}$, we construct the semi-Markov process $_{\bar{r}_{N-1}}\eta^{(\varepsilon)}(t)$ with the phase space $_{\bar{r}_{N-1}}\mathbb{X} = \mathbb{X} \setminus \{r_1, r_2, \ldots, r_{N-1}\} = \{i\}$ (which is a one-point set), the transition probabilities of the embedded Markov chain $_{\bar{r}_{N-1}}p_{ii}(\varepsilon) = 1$, and the expectations of transition times $_{\bar{r}_{N-1}}e_{ii}(\varepsilon)$, which are determined by the transition probabilities and the expectations of transition times of the process $_{\bar{r}_{N-2}}\eta^{(\varepsilon)}(t)$ via relations (87) and (96). According to Theorem 10.1, the expectations of hitting times $E_{ii}(\varepsilon)$ for the semi-Markov processes $\eta^{(\varepsilon)}(t)$, $_{\bar{r}_1}\eta^{(\varepsilon)}(t), \ldots, _{\bar{r}_{N-1}}\eta^{(\varepsilon)}(t)$ coincide. According to Theorems 10.2 and 10.4, the transition probabilities of the embedded Markov chain $_{\bar{r}_{N-1}}p_{ii}(\varepsilon) = 1$ and the expectations of transition times $_{\bar{r}_{N-1}}e_{ii}(\varepsilon)$ satisfy conditions **A–F**. In this case, the transition set $_{\bar{r}_{N-1}}\mathbb{Y}_i = \{i\}$, for every $i \in \mathbb{X}$. Therefore, the $(_{\bar{r}_{N-1}}l^-_{i'j'}, _{\bar{r}_{N-1}}l^+_{i'j'})$-expansions for transition probabilities $_{\bar{r}_{N-1}}p_{i'j'}(\varepsilon) = 1, j' \in _{\bar{r}_{N-1}}\mathbb{Y}_{i'}, i' \in _{\bar{r}_{N-1}}\mathbb{X}$ (which take the form of relation (58)) and $(_{\bar{r}_{N-1}}m^-_{i'j'}, _{\bar{r}_{N-1}}m^+_{i'j'})$-expansions for expectations $_{\bar{r}_{N-1}}e_{i'j'}(\varepsilon), j' \in _{\bar{r}_{N-1}}\mathbb{Y}_{i'}, i' \in _{\bar{r}_{N-1}}\mathbb{X}$ can be constructed by applying Algorithms 1, 3 and 5 to the $(_{\bar{r}_{N-2}}l^-_{i'j'}, _{\bar{r}_{N-2}}l^+_{i'j'})$ -expansions for transition probabilities $_{\bar{r}_{N-2}}p_{i'j'}(\varepsilon), j' \in _{\bar{r}_{N-2}}\mathbb{Y}_{i'}, i' \in _{\bar{r}_{N-2}}\mathbb{X}$ and $(_{\bar{r}_{N-2}}m^-_{i'j'}, _{\bar{r}_{N-2}}m^+_{i'j'})$-expansions for expectations $_{\bar{r}_{N-2}}e_{i'j'}(\varepsilon), j' \in _{\bar{r}_{N-2}}\mathbb{Y}_{i'}, i' \in _{\bar{r}_{N-2}}\mathbb{X}$. These expansions ara pivotal.

7.4. The semi-Markov process $_{\bar{r}_{N-1}}\eta^{(\varepsilon)}(t)$ has the one-point phase space $_{\bar{r}_{N-1}}\mathbb{X} = \{i\}$ and, thus, the transition probability $_{\bar{r}_{N-1}}p_{ii}(\varepsilon) \equiv 1$, while the expectation of transition time $_{\bar{r}_{N-1}}e_{ii}(\varepsilon) = E_{ii}(\varepsilon)$. The above algorithm of sequential reduction of phase space should be repeated for every $i \in \mathbb{X}$. In this way, the Laurent asymptotic expansions for quantities $E_{ii}(\varepsilon), i \in \mathbb{X}$ can be written down. These asymptotic expansions have the following form,

$$E_{ii}(\varepsilon) = \sum_{l=M^-_{ii}}^{M^+_{ii}} B_{ii}[l]\varepsilon^l + \hat{o}_i(\varepsilon^{M^+_{ii}}), i \in \mathbb{X}, \tag{104}$$

where parameters $M^\pm_{ii} = _{\bar{r}_{N-1}}m^\pm_{ii}, i \in \mathbb{X}$ and the coefficients $B_{ii}[l] = _{\bar{r}_{N-1}}b_{ii}[l], l = M^-_{ii}, \ldots, M^+_{ii}, i \in \mathbb{X}$, where $_{\bar{r}_{N-1}}b_{ii}[l]$ are coefficients in the corresponding $(_{\bar{r}_{N-1}}m^-_{i'j'}, _{\bar{r}_{N-1}}m^+_{i'j'})$-expansions for expectations $_{\bar{r}_{N-1}}e_{i'j'}(\varepsilon), j' \in _{\bar{r}_{N-1}}\mathbb{Y}_{i'}, i' \in _{\bar{r}_{N-1}}\mathbb{X}$. These expansions are pivotal.

It should be noted that, for every $n = 1, \ldots, N-1$, the reduced semi-Markov process $_{\bar{r}_n}\eta^{(\varepsilon)}(t)$ is invariant with respect to any permutation $\bar{r}'_n = (r'_1, \ldots, r'_n)$ of the the sequence $\bar{r}_n = (r_1, \ldots, r_n)$.

Indeed, for every such permutation $\bar{r}'_n = (r'_1, \ldots, r'_n)$, the corresponding reduced semi-Markov process $_{\bar{r}'_n}\eta^{(\varepsilon)}(t)$ is constructed from the initial semi-Markov process $\eta^{(\varepsilon)}(t)$, as the sequence of its states at sequential moment of hitting into the same reduced phase space $_{\bar{r}_n'}\mathbb{X} = \mathbb{X} \setminus \{r_1', \ldots, r_n'\} = _{\bar{r}_n}\mathbb{X} = \mathbb{X} \setminus \{r_1, \ldots, r_n\}$ and times between sequential jumps of the reduced semi-Markov process $_{\bar{r}_n'}\eta^{(\varepsilon)}(t)$ which are times between sequential hitting of the above reduced space by the initial semi-Markov process $\eta^{(\varepsilon)}(t)$.

This implies that the expectation of transition time $_{\bar{r}_n}e_{i'j'}(\varepsilon)$ is, for every $i', j' \in {}_{\bar{r}_n}\mathbb{X}$ and $n = 1, \ldots, N - 1$, invariant with respect to any permutation $\bar{r}_n' = (r_1', \ldots, r_n')$ of the sequence $\bar{r}_n = (r_1, \ldots, r_n)$.

Moreover, as follows from the Algorithms 1–7, the expectation of transition time $_{\bar{r}_n}e_{i'j'}(\varepsilon)$ is a rational function of initial transition probabilities $p_{ij}(\varepsilon), j \in \mathbb{Y}_i, i \in \mathbb{X}$ and expectations $e_{ij}(\varepsilon), j \in \mathbb{Y}_i, i \in \mathbb{X}$ (a quotient of two sums of products of some of these probabilities and expectations), which, according the above remarks, is invariant with respect to any permutation $\bar{r}_n' = (r_1', \ldots, r_n')$ of the sequence $\bar{r}_n = (r_1, \ldots, r_n)$.

By using identical arithmetical transformations (disclosure of brackets, imposition of a common factor out of the brackets, bringing a fractional expression to a common denominator, permutation of summands or multipliers, elimination of expression with equal absolute values and opposite signs in the sums and elimination of equal expressions in the quotients, etc.) the rational function $_{\bar{r}_n'}e_{i'j'}(\varepsilon)$ given by Algorithm 7 can be transformed in the rational function $_{\bar{r}_n}e_{i'j'}(\varepsilon)$ given by Algorithm 7 and vice versa.

By Lemma 10.8, these transformations do not affect the corresponding asymptotic expansions for expectation $_{\bar{r}_n}e_{i'j'}(\varepsilon)$ given by Algorithm 7, and, thus, these asymptotic expansions are invariant with respect to any permutation $\bar{r}_n' = (r_1', \ldots, r_n')$ of the sequence $\bar{r}_n = (r_1, \ldots, r_n)$.

The above remarks can be summarized in the following theorem.

Theorem 10.6 *Let conditions **A–F** hold for semi-Markov processes $\eta^{(\varepsilon)}(t)$. Then, for every $i \in \mathbb{X}$, the Laurent asymptotic expansion (104) for the expectation of hitting times $E_{ii}(\varepsilon)$ given by Algorithm 7 can be written down. This expansion is invariant with respect to the choice of permutation $\langle r_1, \ldots, r_{N-1}, i \rangle$ of sequence $\langle 1, \ldots, N \rangle$, in the above algorithm.*

Let us now assume that conditions **A, B', C–E, F'** hold for the semi-Markov process $\eta^{(\varepsilon)}(t)$.

Algorithm 8. This is an algorithm for computing upper bounds for remainders in asymptotic expansions for transition probabilities and expectation of sojourn times for semi-Markov processes with reduced phase spaces.

8.1. Let $_{\bar{r}_1}\eta^{(\varepsilon)}(t) = _{r_1}\eta^{(\varepsilon)}(t)$ be be the reduced semi-Markov process, which is constructed as this is described in Step 7.1 of Algorithm 7. According to Theorems 10.3 and 10.5, the semi-Markov process $_{\bar{r}_1}\eta^{(\varepsilon)}(t)$ satisfies conditions **A, B', C–E, F'**. Therefore, $(_{\bar{r}_1}l_{i'j'}^-, {}_{\bar{r}_1}l_{i'j'}^+, {}_{\bar{r}_1}\delta_{i'j'}, {}_{\bar{r}_1}G_{i'j'}, {}_{\bar{r}_1}\varepsilon_{i'j'})$-expansions for transition probabilities $_{\bar{r}_1}p_{i'j'}(\varepsilon), j' \in {}_{\bar{r}_1}\mathbb{Y}_{i'}, i' \in {}_{\bar{r}_1}\mathbb{X}$ and $(_{\bar{r}_1}m_{i'j'}^-, {}_{\bar{r}_1}m_{i'j'}^+, {}_{\bar{r}_1}\dot{\delta}_{i'j'}, {}_{\bar{r}_1}\dot{G}_{i'j'}, {}_{\bar{r}_1}\dot{\varepsilon}_{i'j'})$-expansions for expectations $_{\bar{r}_1}e_{i'j'}(\varepsilon), j' \in {}_{\bar{r}_1}\mathbb{Y}_{i'}, i' \in {}_{\bar{r}_1}\mathbb{X}$ can be constructed by applying Algorithms 1–5 to the $(l_{i'j'}^-, l_{i'j'}^+, \delta_{i'j'}, G_{i'j'}, \varepsilon_{i'j'})$-expansions for transition probabilities $p_{i'j'}(\varepsilon), j' \in \mathbb{Y}_{i'}, i' \in \mathbb{X}$ and $(m_{i'j'}^-, m_{i'j'}^+, \dot{\delta}_{i'j'}, \dot{G}_{i'j'}, \dot{\varepsilon}_{i'j'})$-expansions for expectations $e_{i'j'}(\varepsilon), j' \in \mathbb{Y}_{i'}, i' \in \mathbb{X}$.

8.2. Let $_{\bar{r}_2}\eta^{(\varepsilon)}(t)$ be the reduced semi-Markov process, which is constructed as this is described in Step 7.2 of Algorithm 7. According to Theorems 10.3 and 10.5, the semi-Markov process $_{\bar{r}_2}\eta^{(\varepsilon)}(t)$ satisfies conditions **A**, **B**′, **C–E**, **F**′. Therefore, $(_{\bar{r}_2}l^-_{i'j'}, {}_{\bar{r}_2}l^+_{i'j'}, {}_{\bar{r}_2}\delta_{i'j'}, {}_{\bar{r}_2}G_{i'j'}, {}_{\bar{r}_2}\varepsilon_{i'j'})$-expansions for transition probabilities $_{\bar{r}_2}p_{i'j'}(\varepsilon), j' \in {}_{\bar{r}_2}\mathbb{Y}_{i'}, i' \in {}_{\bar{r}_2}\mathbb{X}$ and $(_{\bar{r}_2}m^-_{i'j'}, {}_{\bar{r}_2}m^+_{i'j'}, {}_{\bar{r}_2}\dot{\delta}_{i'j'}, {}_{\bar{r}_2}\dot{G}_{i'j'}, {}_{\bar{r}_2}\dot{\varepsilon}_{i'j'})$-expansions for expectations $_{\bar{r}_2}e_{i'j'}(\varepsilon), j' \in {}_{\bar{r}_2}\mathbb{Y}_{i'}, i' \in {}_{\bar{r}_2}\mathbb{X}$ can be constructed by applying Algorithms 1–5 to the $(_{\bar{r}_1}l^-_{i'j'}, {}_{\bar{r}_1}l^+_{i'j'}, {}_{\bar{r}_1}\delta_{i'j'}, {}_{\bar{r}_1}G_{i'j'}, {}_{\bar{r}_1}\varepsilon_{i'j'})$-expansions for transition probabilities $p_{i'j'}(\varepsilon), j' \in \mathbb{Y}_{i'}, i' \in {}_{\bar{r}_1}\mathbb{X}$ and $(_{\bar{r}_1}m^-_{i'j'}, {}_{\bar{r}_1}m^+_{i'j'}, {}_{\bar{r}_1}\dot{\delta}_{i'j'}, {}_{\bar{r}_1}\dot{G}_{i'j'}, {}_{\bar{r}_1}\dot{\varepsilon}_{i'j'})$-expansions for expectations $e_{i'j'}(\varepsilon), j' \in {}_{\bar{r}_1}\mathbb{Y}_{i'}, i' \in {}_{\bar{r}_1}\mathbb{X}$.

8.3. Finally, let $_{\bar{r}_{N-1}}\eta^{(\varepsilon)}(t)$ be the reduced semi-Markov process, which is constructed as this is described in Step 7.2 of Algorithm 7. According to Theorems 10.3 and 10.5, the semi-Markov process $_{\bar{r}_{N-1}}\eta^{(\varepsilon)}(t)$ satisfies conditions **A**, **B**′, **C–E**, **F**′. Therefore, $(_{\bar{r}_{N-1}}l^-_{i'j'}, {}_{\bar{r}_{N-1}}l^+_{i'j'}, {}_{\bar{r}_{N-1}}\delta_{i'j'}, {}_{\bar{r}_{N-1}}G_{i'j'}, {}_{\bar{r}_{N-1}}\varepsilon_{i'j'})$-expansions for transition probabilities $_{\bar{r}_{N-1}}p_{i'j'}(\varepsilon), j' \in {}_{\bar{r}_{N-1}}\mathbb{Y}_{i'}, i' \in {}_{\bar{r}_{N-1}}\mathbb{X}$ and $(_{\bar{r}_{N-1}}m^-_{i'j'}, {}_{\bar{r}_{N-1}}m^+_{i'j'}, {}_{\bar{r}_{N-1}}\dot{\delta}_{i'j'},$ $_{\bar{r}_{N-1}}\dot{G}_{i'j'}, {}_{\bar{r}_{N-1}}\dot{\varepsilon}_{i'j'})$-expansions for expectations $_{\bar{r}_{N-1}}e_{i'j'}(\varepsilon), j' \in {}_{\bar{r}_{N-1}}\mathbb{Y}_{i'}, i' \in {}_{\bar{r}_{N-1}}\mathbb{X}$ can be constructed by applying Algorithms 1–5 to the $(_{\bar{r}_{N-2}}l^-_{i'j'}, {}_{\bar{r}_{N-2}}l^+_{i'j'}, {}_{\bar{r}_{N-2}}\delta_{i'j'}, {}_{\bar{r}_{N-2}}G_{i'j'},$ $_{\bar{r}_{N-2}}\varepsilon_{i'j'})$-expansions for transition probabilities $p_{i'j'}(\varepsilon), j' \in \mathbb{Y}_{i'}, i' \in {}_{\bar{r}_{N-2}}\mathbb{X}$ and $(_{\bar{r}_{N-2}}m^-_{i'j'}, {}_{\bar{r}_{N-2}}m^+_{i'j'}, {}_{\bar{r}_{N-2}}\dot{\delta}_{i'j'}, {}_{\bar{r}_{N-2}}\dot{G}_{i'j'}, {}_{\bar{r}_{N-2}}\dot{\varepsilon}_{i'j'})$-expansions for expectations $e_{i'j'}(\varepsilon), j' \in {}_{\bar{r}_{N-2}}\mathbb{Y}_{i'}, i' \in {}_{\bar{r}_{N-2}}\mathbb{X}$.

8.4. Finally, due to equalities $_{\bar{r}_{N-1}}e_{ii}(\varepsilon) = E_{ii}(\varepsilon), i \in \mathbb{X}$, we get that the asymptotic expansion (104) for expectations $E_{ii}(\varepsilon), i \in \mathbb{X}$, given in the Step 7.4 of Algorithm 7, is a $(M^-_{ii}, M^+_{ii}, \delta^\circ_{ii}, G^\circ_{ii}, \varepsilon^\circ_{ii})$-expansion with parameters $M^-_{ii} = {}_{\bar{r}_{N-1}}m^-_{ii}, M^+_{ii} = {}_{\bar{r}_{N-1}}m^+_{ii}, \delta^\circ_{ii} = {}_{\bar{r}_{N-1}}\dot{\delta}_{ii}, G^\circ_{ii} = {}_{\bar{r}_{N-1}}\dot{G}_{ii}, \varepsilon^\circ_{ii} = {}_{\bar{r}_{N-1}}\dot{\varepsilon}_{ii}$.

In this case, the invariance of explicit upper bounds for remainders given by Algorithm 8, with respect to the choice of any permutation $\langle r_1, \ldots, r_{N-1}, i \rangle$ of sequence $\langle 1, \ldots, N \rangle$, can not be guaranteed.

However, Lemma 10.9 guarantees that the following inequalities hold for the parameters $\delta^\circ_{ii}, i \in \mathbb{X}$,

$$\delta^\circ_{ii} \geq \delta^\circ = \min_{j \in \mathbb{Y}_i, i \in \mathbb{X}} (\delta_{ij} \wedge \dot{\delta}_{ij}). \tag{105}$$

The following theorem takes place.

Theorem 10.7 *Let conditions* **A**, **B**′, **C–E**, **F**′ *hold for semi-Markov processes* $\eta^{(\varepsilon)}(t)$. *Then, for every* $i \in \mathbb{X}$, *the* (M^-_{ii}, M^+_{ii})-*expansion* (104) *for the expectations of hitting times* $E_{ii}(\varepsilon)$, *given by Algorithm 7, is a* $(M^-_{ii}, M^+_{ii}, \delta^\circ_{ii}, G^\circ_{ii}, \varepsilon^\circ_{ii})$-*expansion, with parameters* $\delta^\circ_{ii}, G^\circ_{ii}, \varepsilon^\circ_{ii}$ *given in Algorithm 8. The inequality* (105) *holds for parameters* $\delta^\circ_{ii}, i \in \mathbb{X}$.

5.2 Laurent Asymptotic Expansions for Expectations of Hitting Times

Algorithms presented above yields the Laurent asymptotic expansions for expectations of hitting times $E_{ij}(\varepsilon), i, j \in \mathbb{X}$. Indeed, let choose two states $i, j \in \mathbb{X}$ and a chain of states $\bar{r}_{N-2} = (r_1, \ldots, r_{N-2}), r_1, \ldots, r_{N-2} \neq i, j$.

According to Theorem 10.1 and Algorithm 8 the expectations $E_{ij}(\varepsilon)$ coincides for the initial semi-Markov process $\eta^{(\varepsilon)}(t)$ and the semi-Markov process $\bar{r}_{N-2}\eta^{(\varepsilon)}(t)$. The semi-Markov process $\bar{r}_{N-2}\eta^{(\varepsilon)}(t)$ has a two-points phase space $\bar{r}_{N-2}\mathbb{X} = \{i, j\}$. The expectations of hitting times $E_{i'j'}(\varepsilon), i' \in \{i, j\}$ can be found by solving, for every $j' \in \{i, j\}$, the system of (two, in this case) linear equations (77) that yields the following formulas, for every $j' \in \{i, j\}$,

$$\begin{cases} E_{i'j'}(\varepsilon) = \bar{r}_{N-2}e_{i'}(\varepsilon) \cdot \frac{1}{\bar{r}_{N-2}p_{i'j'}(\varepsilon)}, \\ E_{j'j'}(\varepsilon) = \bar{r}_{N-2}e_{j'}(\varepsilon) + \bar{r}_{N-2}e_{i'}(\varepsilon) \cdot \frac{\bar{r}_{N-2}p_{j'i'}(\varepsilon)}{\bar{r}_{N-2}p_{i'j'}(\varepsilon)}, \end{cases} \tag{106}$$

where $i' \neq j'$ in both equations in (106) and,

$$\bar{r}_{N-2}e_{i'}(\varepsilon) = \bar{r}_{N-2}e_{i'i}(\varepsilon) + \bar{r}_{N-2}e_{i'j}(\varepsilon), \ i' \in \{i, j\}. \tag{107}$$

The corresponding asymptotic expansions for $E_{ij}(\varepsilon)$ can be constructing by using the asymptotic expansions for transition probabilities $p_{j'i'}(\varepsilon)$ and expectations $\bar{r}_{N-2}e_{i'}(\varepsilon)$ given in Algorithms 7 and 8 and the operational rules for Laurent asymptotic expansions presented in Lemmas 10.1–10.9.

6 Asymptotic Expansions for Stationary Distributions

In this section, we present algorithms for construction of asymptotic expansions for stationary distributions of nonlinearly perturbed semi-Markov processes.

6.1 Asymptotic Expansions for Stationary Probabilities of Perturbed Semi-Markov Processes

Let us recall relation (78) for stationary probabilities of the semi-Markov process $\eta^{(\varepsilon)}(t)$,

$$\pi_i(\varepsilon) = \frac{e_i(\varepsilon)}{E_{ii}(\varepsilon)}, \ i \in \mathbb{X}. \tag{108}$$

Algorithm 9. This is an algorithm for constructing asymptotic expansions for stationary probabilities of perturbed semi-Markov processes.

9.1. Conditions **A–F** and proposition **(i)** (the multiple summation rule) of Lemma 10.5, permits one can construct (m_i^-, m_i^+)-expansions for expectations $e_i(\varepsilon), i \in \mathbb{X}$, which take the following forms,

$$
\begin{aligned}
e_i(\varepsilon) &= \sum_{j \in \mathbb{Y}_i} e_{ij}(\varepsilon) \\
&= \sum_{j \in \mathbb{Y}_i} \left(\sum_{l=m_{ij}^-}^{m_i^+} b_{ij}[l]\varepsilon^l + \dot{o}_{ij}(\varepsilon^{m_{ij}^+}) \right) \\
&= \sum_{l=m_i^-}^{m_i^+} b_i[l]\varepsilon^l + \dot{o}_i(\varepsilon^{m_i^+}), \ i \in \mathbb{X},
\end{aligned}
\tag{109}
$$

where

$$
m_i^- = \min_{j \in \mathbb{Y}_i} m_{ij}^-, \ m_i^+ = \min_{j \in \mathbb{Y}_i} m_{ij}^+, \ i \in \mathbb{X},
\tag{110}
$$

and

$$
b_i[m_i^- + l] = \sum_{j \in Y_i} b_{ij}[m_i^- + l], \ l = 0, \ldots, m_i^+ - m_i^-, \ i \in \mathbb{X},
\tag{111}
$$

where $b_{ij}[m_i^- + l] = 0$, for $0 \leq l < m_{ij}^- - m_i^-, j \in Y_i, \ i \in \mathbb{X}$.

The above asymptotic expansions are pivotal for all $i \in \mathbb{X}$.

9.2. Conditions **A–F**, relation (108) and proposition **(v)** (the division rule) of Lemma 10.3, permits us construct (n_i^-, n_i^+)-expansions for stationary probabilities $\pi_i(\varepsilon), i \in \mathbb{X}$, which take the following forms,

$$
\pi_i(\varepsilon) = \sum_{l=n_i^-}^{n_i^+} c_i[l]\varepsilon^l + o_i(\varepsilon^{n_i^+}), \ i \in \mathbb{X},
\tag{112}
$$

where

$$
n_i^- = m_i^- - M_{ii}^-, \ n_i^+ = (m_i^+ - M_{ii}^-) \wedge (m_i^- + M_{ii}^+ - 2M_{ii}^-), \ i \in \mathbb{X},
\tag{113}
$$

and

$$
c_i[n_i^- + l] = \frac{b_i[m_i^- + l] - \sum_{1 \leq l' \leq l} B_{ii}[M_{ii}^- + l]c_i[n_i^- + l - l']}{B_{ii}[M_{ii}^-]},
$$
$$
l = 0, \ldots, n_i^+ - n_i^-, \ i \in \mathbb{X}.
\tag{114}
$$

Since $\pi_i(\varepsilon) > 0, i \in \mathbb{X}, \varepsilon \in (0, \varepsilon_0]$, the asymptotic expansions (109) are pivotal, i.e., coefficients,

$$c_i[n_i^-] = b_i[m_i^-]/B_{ii}[M_{ii}^-] > 0, \ i \in \mathbb{X}. \tag{115}$$

By the definition, $e_i(\varepsilon) \leq E_{ii}(\varepsilon), \ i \in \mathbb{X}, \varepsilon \in (0, \varepsilon_0]$. This implies that parameters $M_i^- \leq m_i^-, i \in \mathbb{X}$ and thus, parameters

$$n_i^- \geq 0, \ i \in \mathbb{X}. \tag{116}$$

Moreover, since $\sum_{i \in \mathbb{X}} \pi_i(\varepsilon) = 1$, for every $\varepsilon \in (0, \varepsilon_0]$, the parameters $n_i^\pm, i \in \mathbb{X}$ and coefficients $c_i[l], l = n_i^-, \ldots, n_i^+, i \in \mathbb{X}$ satisfies the following relations,

$$n^- = \min_{i \in \mathbb{X}} n_i^- = 0, \tag{117}$$

and

$$c[l] = \sum_{i \in \mathbb{X}} c_i[l] = \begin{cases} 1 & \text{for } l = 0, \\ 0 & \text{for } 0 < l \leq n^+ = \min_{i \in \mathbb{X}} n_i^+. \end{cases} \tag{118}$$

Let us introduce sets,

$$\mathbb{X}_0 = \{i \in \mathbb{X} : n_i^- = 0\}.$$

By the above remarks, the following relation takes place,

$$\pi_i(0) = \lim_{\varepsilon \to 0} \pi_i(\varepsilon) = \begin{cases} c_i[0] > 0 & \text{if } i \in \mathbb{X}_0, \\ 0 & \text{if } i \notin \mathbb{X}_0. \end{cases} \tag{119}$$

Theorem 10.8 *Let conditions **A–F** hold for semi-Markov processes $\eta^{(\varepsilon)}(t)$. Then, the (n_i^-, n_i^+)-expansions (112), for the stationary probabilities $\pi_i(\varepsilon), i \in \mathbb{X}$ given by Algorithm 9, can be written down. This expansion is invariant with respect to the choice of permutation $\langle r_1, \ldots, r_{N-1}, i \rangle$ of sequence $\langle 1, \ldots, N \rangle$, in the above algorithm. Relations (115)–(119) hold for these expansions.*

6.2 Asymptotic Expansions for Stationary Probabilities of Perturbed Semi-Markov Processes with Explicit Upper Bounds for Remainders

Algorithm 10. This is an algorithm for computing upper bounds for remainders in asymptotic expansions for stationary probabilities of perturbed semi-Markov processes.

10.1. Conditions **A, B′, C–E, F′** and the proposition **(i)** (the multiple summation rule) of Lemma 10.6 imply that the (m_i^-, m_i^+)-expansions for expectations $e_i(\varepsilon), i \in$

$\mathbb{X}$ are $(m_i^-, m_i^+, \dot{\delta}_i, \dot{G}_i, \dot{\varepsilon}_i)$-expansions, with parameters $\dot{\delta}_i, \dot{G}_i, \dot{\varepsilon}_i, i \in \mathbb{X}$ given by the following formulas,

$$\dot{\delta}_i = \min_{j \in \mathbb{Y}_i, \, m_{ij}^+ = m_i^+} \dot{\delta}_{ij},$$

$$\dot{G}_i = \sum_{j \in \mathbb{Y}_i} \left(\dot{G}_{ij} \dot{\varepsilon}_i^{m_{ij}^+ + \dot{\delta}_{ij} - m_i^+ - \dot{\delta}_i} + \sum_{m_i^+ < j \leq m_{ij}^+} |b_{ij}| \dot{\varepsilon}_i^{j - m_i^+ - \dot{\delta}_i} \right),$$

$$\dot{\varepsilon}_i = \min_{j \in \mathbb{Y}_i} \dot{\varepsilon}_{ij}. \tag{120}$$

10.2. Conditions **A**, **B′**, **C–E**, **F′** and the propositions **(iv)** (the reciprocal rule) and **(v)** (the division rule) of Lemma 10.6 imply that the (n_i^-, n_i^+)-expansions for expectations $\pi_i(\varepsilon), i \in \mathbb{X}$ are $(n_i^-, n_i^+, \delta_i^*, G_i^*, \varepsilon_i^*)$-expansions, with parameters $\delta_i^*, G_i^*, \varepsilon_i^*, i \in \mathbb{X}$ given by the following formulas,

$$\delta_i^* = \begin{cases} \dot{\delta}_i & \text{if } n_i^+ = m_i^+ - M_{ii}^- < n_i^- + M_{ii}^+ - 2M_{ii}^-, \\ \dot{\delta}_i \wedge \delta_{ii}^\circ & \text{if } n_i^+ = m_i^+ - M_{ii}^- < n_i^- + M_{ii}^+ - 2M_{ii}^-, \\ \delta_B & \text{if } n_i^+ = n_i^- + M_{ii}^+ - 2M_{ii}^- < m_i^+ - M_{ii}^-, \end{cases}$$

$$G_i^* = \left(\frac{B_{ii}[M_{ii}^-]}{2} \right)^{-1} \left(\sum_{m_i^+ \wedge (m_i^- + M_{ii}^+ - M_{ii}^-) < l \leq m_i^+} |b_i[l]| (\varepsilon_i^*)^{l - n_i^+ - M_{ii}^- - \delta_i^*} \right.$$

$$+ \sum_{m_i^+ \wedge (m_i^- + M_{ii}^+ - M_{ii}^-) < l + k, \, m_i^- \leq l \leq m_i^+, \, M_{ii}^- \leq k \leq M_{ii}^+} |b_i[l]| |c_i[k]| (\varepsilon_i^*)^{l + k - n_i^+ - M_{ii}^- - \delta_i^*}$$

$$+ \dot{G}_i (\varepsilon_i^*)^{m_i^+ + \dot{\delta}_i - n_i^+ - M_{ii}^- - \delta_i^*}$$

$$\left. + G_{ii}^\circ \sum_{n_i^- \leq k \leq n_i^+} |c_i[k]| (\varepsilon_i^*)^{k + M_{ii}^+ + \delta_i^\circ - n_i^+ - M_{ii}^- - \delta_i^*} \right),$$

$$\varepsilon_i^* = \dot{\varepsilon}_i \wedge \varepsilon_i^\circ \wedge \begin{cases} \frac{B_{ii}[M_{ii}^-]}{2} \left(\sum_{M_{ii}^- < l \leq M_{ii}^+} |B_{ii}[l]| (\varepsilon_i^\circ)^{l - M_{ii}^- - 1} \right. \\ \left. + G_{ii}^\circ (\varepsilon_i^\circ)^{M_{ii}^+ + \delta_i^\circ - M_{ii}^- - 1} \right)^{-1} & \text{if } M_{ii}^- < M_{ii}^+, \\ \left(\frac{B_{ii}[M_{ii}^-]}{2 G_i^\circ} \right)^{\frac{1}{\delta_i^\circ}} & \text{if } M_{ii}^- = M_{ii}^+. \end{cases} \tag{121}$$

Theorem 10.9 *Let conditions **A**, **B′**, **C–E**, **F′** hold for semi-Markov processes $\eta^{(\varepsilon)}(t)$. Then, the $(n_i^-, n_i^+, \delta_i^*, G_i^*, \varepsilon_i^*)$-expansions (112) for the stationary probabilities $\pi_i(\varepsilon), i \in \mathbb{X}$ given by Algorithms 9 and 10 can be written down. The inequalities $\delta_i^* \geq \delta^\circ, i \in \mathbb{X}$ hold, where parameter δ° is given in relation (105).*

7 Future Studies and Bibliographical Remarks

In this section, we present some directions for future studies and short bibliographical remarks concerned works in the area.

7.1 Directions for Future Studies

The method of sequential reduction of a phase space presented in the paper can also be applied for getting asymptotic expansions for high order power and exponential moments of hitting times, for nonlinearly perturbed semi-Markov processes.

In the present paper, we consider the model, where the pre-limiting perturbed semi-Markov processes have a phase space which is one class of communicative states, while the limiting unperturbed semi-Markov process has a phase space which consists of one or several classes of communicative states and possibly a class of transient states. However, the method of sequential reduction of the phase space can also be applied to nonlinearly perturbed semi-Markov processes with absorption and, therefore, to the model, where the pre-limiting semi-Markov processes also have a phase space, which consists of several classes of communicative states and a class of transient states.

We are quite sure that combination of results in the above two directions with the methods of asymptotic analysis for nonlinearly perturbed regenerative processes developed in Silvestrov [301, 304, 305] and Gyllenberg and Silvestrov [99, 100, 102, 104] will make it possible to expand results concerned asymptotic expansions for quasi-stationary distributions and other characteristics for nonlinearly perturbed semi-Markov processes with absorption, where the limiting semi-Markov process has a phase space which consists of one class of communicative states and a class of transient states, to a general case, where the limiting semi-Markov process has a phase space, which consists of several classes of communicative states and a class of transient states. Some additional results and examples can be found in the recent paper by Silvestrov, D. and Silvestrov, S. [321].

The problems of aggregation of steps in the time-space screening procedures for semi-Markov processes, tracing pivotal orders for different groups of states as well as getting explicit matrix formulas, for coefficients and parameters of upper bounds for remainders in the corresponding asymptotic expansions for stationary distributions and moments of hitting times, do require additional studies. It can be expected that such formulas can be obtained, for example, for birth-death type semi-Markov processes, for which the proposed algorithms of reduction of a phase space preserve the birth-death structure for reduced semi-Markov processes. Some initial results in this direction are presented in the recent paper by Silvestrov, Petersson and Hössjer [319].

We are going to present results concerned Laurent asymptotic expansions for power and exponential moments of hitting times, quasi-stationary distributions and explicit formulas for coefficients and parameters of upper bounds for remainders for some specific classes of semi-Markov models, as well as applications to some models of population genetics, information networks and queuing systems, in future publications.

7.2 Bibliographical Remarks

Note first of all that the model of perturbed discrete time Markov chains, at least, in the most difficult case of so-called singularly perturbed Markov chains and semi-Markov processes with absorption and asymptotically uncoupled phase spaces, attracted attention of researchers in the mid of the 20th century.

The methods used for construction of asymptotic expansions for stationary distributions and related functionals such as moments of hitting times can be split in three groups.

(1) The first works related to asymptotical problems for the above models are Meshalkin [221], Simon and Ando [323], Hanen [106–109], Kingman [169], Darroch and Seneta [65, 66], Keilson [160, 161], Seneta [273–276], Schweitzer [265], Korolyuk [177], Silvestrov [287–293], Anisimov [11–15], Korolyuk and Turbin [193, 194], Gusak and Korolyuk [96], Turbin [344, 345], Korolyuk, Penev and Turbin [189], Kovalenko [199, 200], Poliščuk and Turbin [256], Korolyuk, Brodi and Turbin [179], Pervozvanskiĭ and Smirnov [247], Courtois [56] and Gaĭtsgori and Pervozvanskiĭ [88].

(2) Convergence results, for distributions and moments of hitting times, eigenvalues, eigenvectors, stationary and quasi-stationary distributions, Perron roots, coefficients of ergodicity, etc. have been studied in works by Meshalkin [221], Hanen, [106–109], Kingman [169], Darroch and Seneta [65, 66], Keilson [160, 161], Seneta [273–276, 282], Schweitzer [265, 266], Korolyuk [177, 178], Silvestrov [287–289, 291, 293–297, 300, 303, 309], Anisimov [11–19], Korolyuk and Turbin [193–196], Gusak and Korolyuk [96], Turbin [344], Korolyuk, Penev and Turbin [189], Masol and Silvestrov [222], Zakusilo [360, 361], Kovalenko [199, 200], Korolyuk, Brodi and Turbin [179], Gaĭtsgori and Pervozvanskiy [88, 89], Allen, Anderssen and Seneta [8], Kaplan [146, 147], Korolyuk, Turbin and Tomusjak [197], Shurenkov [285, 286], Anisimov and Chernyak [20], Anisimov, Voĭna and Lebedev [21], Coderch, Willsky, Sastry and Castañon [53], Korolyuk, D. [174], Korolyuk, D. and Silvestrov [175, 176], Stewart [329], Koury, McAllister and Stewart [198], McAllister, Stewart and Stewart, W. [220], Cao and Stewart [51], Kartashov [152, 153, 155], Korolyuk and Tadzhiev [192], Burnley [49], Gibson and Seneta [90], Haviv [113], Haviv, Ritov and Rothblum [121], Rohlichek [259], Rohlicek and Willsky [260, 261], Silvestrov and Velikii [322], Alimov and Shurenkov [6, 7], Hunter [138], Latouche [208], Pollett and Stewart [257], Motsa and Silvestrov [223], Hoppensteadt, Salehi and Skorokhod [127], Kalashnikov [143], Korolyuk and Limnios [181, 187], Marek and Mayer [216], Yin and Zhang [354–357], Craven [64], Hernández-Lerma and Lasserre [125], Yin, Zhang and Badowski [358], Silvestrov and Drozdenko [313, 314], Kupsa and Lacroix [204], Drozdenko [70–72], Barbour and Pollett [38, 39], Glynn [91], Benois, Landim and Mourragui [42], Serlet [283] and Meyer [227].

(3) Rates of convergence, errors of approximation, sensitivity and related stability theorems for Markov chains and related models of stochastic processes have been studied in works by Schweitzer [265, 267–269], Silvestrov [287, 290, 292], Seneta [276–282], Courtois [56, 58], Gaĭtsgori and Pervozvanskiy [88, 89], Kovalenko [200], Kalashnikov [142–144], Latouche and Louchard [209] (1978), Berman and Plemmons [43], Meyer [225, 228], Kalashnikov and Anichkin [145], Bobrova [47], Louchard and Latouche [214, 215], Stewart [327–329, 331–337], Courtois and Semal [60–63], Haviv and Rothblum [123], Haviv and Van der Heyden [124], Koury, McAllister and Stewart [198], McAllister, Stewart, G. and Stewart, W. [220], Funderlic and Meyer [87], Kartashov [148–150, 152–155, 157], Vantilborgh [347], Haviv [112, 115, 117], Haviv and Ritov [118, 120], Rohlichek [259], Rohlicek and Willsky [260, 261], Stewart and Sun [339], Hunter [137–139, 141], Stewart and Zhang [340], Hassin and Haviv [111], Barlow [40], Meyn and Tweedie [230], Lasserre [206], Pollett and Stewart [257], Stewart, G., Stewart, W. and McAllister [338], Borovkov [48], Yin and Zhang [354–357], Li, Yin, G., Yin, K. and Zhang [213], Craven [64], Kontoyiannis and Meyn [173], Mitrophanov [231–234], Zhang and Yin [362], Mitrophanov, Lomsadze and Borodovsky [235], Guo [95] and Sirl, Zhang and Pollett [324].

(4) Asymptotic expansions for distributions of hitting times, moments of hitting times, resolvents, eigenvalues, eigenvectors, stationary and quasi-stationary distributions, Perron roots, etc., have been studied in works by Turbin [345], Poliščuk and Turbin [256], Koroljuk, Brodi and Turbin [179], Pervozvanskiĭ and Smirnov [247], Courtois and Louchard [59], Korolyuk and Turbin [195, 196], Courtois [57], Latouche and Louchard [209], Kokotović, Phillips and Javid [170], Korolyuk, Penev and Turbin [190], Phillips and Kokotović [253], Delebecque [67], Abadov [1], Kartashov [151, 155], Haviv [112], Korolyuk [178], Stewart and Sun [339], Silvestrov and Abadov [311, 312], Haviv, Ritov and Rothblum [122], Haviv and Ritov [119], Schweitzer and Stewart [272], (1993), Silvestrov [301, 304, 305], Englund and Silvestrov [77], Gyllenberg and Silvestrov [99, 100, 102, 104], Korolyuk and Limnios [181–187], Stewart [335, 336], Yin and Zhang [354, 356, 357], Avrachenkov [26, 27], Avrachenkov and Lasserre [34], Korolyuk, V.S. and Korolyuk, V.V. [180], Englund [75, 76], Yin, G., Zhang, Yang and Yin, K. [359], Avrachenkov and Haviv [31, 32], Craven [64], Avrachenkov, Filar and Howlett [30], Petersson [248–252], Silvestrov, D. and Silvestrov, S. [320, 321] and Silvestrov, Petersson and Hössjer [319].

(5) Asymptotic expansions for other characteristics of Markov type processes are presented in works by Nagaev [236, 237], Leadbetter [211], Poliščuk and Turbin [256], Quadrat [258], Abadov [1], Silvestrov and Abadov [310–312], Stewart and Sun [339], Kartashov [155, 158], Khasminskii, Yin and Zhang [165, 166], Wentzell [350, 351], Cao [50], Gyllenberg and Silvestrov, [99, 100, 102, 104], Fuh and Lai [86], Kontoyiannis and Meyn [173], Fuh [84, 85], Samoĭlenko [263, 264], Silvestrov [301, 304, 305], Ni [238–242]. Ni, Silvestrov and Malyarenko [243], Albeverio, Koroliuk

and Samoilenko [3], (2009) and Avrachenkov, Filar and Howlett [30], Petersson [248] and Silvestrov and Petersson [318].

(6) We would like especially to mention books including problems on perturbed Markov chains, semi-Markov processes and related problems. These are Seneta [276, 282], Silvestrov [293], Korolyuk and Turbin [194, 195], Courtois [57], Kalashnikov [142, 144], Anisimov [18, 19], Stewart and Sun [339], Korolyuk and Swishchuk [191], Meyn and Tweedie [230], Kartashov [155], Borovkov [48], Stewart [335, 336], Yin and Zhang [355–357], Korolyuk, V.S. and Korolyuk, V.V. [180], Bini, Latouche and Meini [46], Koroliuk and Limnios [187], Gyllenberg and Silvestrov [104] and Avrachenkov, Filar and Howlett [30].

(7) General results of perturbation theory of matrices and linear operators are presented in works by Vishik and Lyusternik [349], Wilkinson [353], Stewart [325–327, 330, 335, 336], Plotkin and Turbin [254, 255], Korolyuk and Turbin [195, 196], Berman and Plemmons [43], Wentzell and Freidlin [352], Haviv [114], Meyer and Stewart [229], Bielecki and Stettner [44], Delebecque [68], Stewart and Sun [339], Hunter [139, 141], Haviv and Ritov [120], Lasserre [206], Kartashov [155], Hoppensteadt, Salehi and Skorokhod [129], Avrachenkov [26], Korolyuk, V.S. and Korolyuk, V.V. [180], Li and Stewart [212], Avrachenkov, Haviv and Howlett [33], Howlett and Avrachenkov [134], Howlett, Pearce and Torokhti [136], Torokhti, Howlett and Pearce [343], Verhulst [348], Howlett, Avrachenkov, Pearce and Ejov [135], Howlett, Albrecht and Pearce [133], Albrecht, Howlett and Pearce [4, 5] and Avrachenkov and Lasserre [35]. In particular, we would like to mention some books, which contains materials on general perturbation matrix and operator theory. These are Erdélyi [81], Kato [159], Cole [54], Korolyuk and Turbin [195, 196], Wentzell and Freidlin [352], Kevorkian and Cole [163, 164], Baumgärtel [41], Stewart [335, 336], Korolyuk, V.S. and Korolyuk, V.V. [180], Konstantinov, Gu, Mehrmann and Petkov [171], Verhulst [348], Gyllenberg and Silvestrov [104] and Avrachenkov, Filar and Howlett [30].

(8) Applications of results on perturbed Markov type processes to the control theory, decision processes, Internet, queuing theory, mathematical genetics, population dynamics and epidemic models, insurance and financial mathematics are presented in works by Simon and Ando [323], Kovalenko [200, 201], Courtois [57], Kalashnikov [142, 144], Delebecque and Quadrat [69], Kovalenko and Kuznetsov [202], Quadrat [258], Gut and Holst [97], Schweitzer [266], Anisimov, Zakusilo and Donchenko [22], Latouche [207], Pervozvanskii and Gaitsgori [246], Meyer [224], Ho and Cao [126], Asmussen [23, 24], Gyllenberg and Silvestrov [98, 101, 103, 104], Pollett and Stewart [257], Abbad and Filar [2], Hoppensteadt, Salehi and Skorokhod [128], Kijima [167], Kovalenko, Kuznetsov, and Pegg [203], Borovkov [48], Yin and Zhang [354–357], Englund [73, 74], Yin, G., Zhang, Yang and Yin, K. [359], Avrachenkov, Filar and Haviv [29], Altman, Avrachenkov and Núñez-Queija [9], Langville and Meyer [205], Silvestrov and Drozdenko [314], Avrachenkov, Litvak, and Son Pham [36, 37], Drozdenko [70, 72], Andersson and Silvestrov, S. [10], Anisimov [19], Konstantinov and Petkov [172], Avrachenkov, Borkar and Nemirovsky [28], Barbour and Pollett [38, 39], Blanchet and Zwart [45], Hössjer [130, 131], Engström and Silvestrov, S. [78–80], Hössjer and Ryman [132], Ni [242], Petersson [249, 252], Silvestrov [306–308] and Silvestrov, Petersson and Hössjer [319]. In particular, we

would like to mention some books in this area that are Kovalenko [200], Kalashnikov [142, 144], Anisimov, Zakusilo and Donchenko [22], Pervozvanskiĭ and Gaitsgori [246], Kijima [167], Kovalenko, Kuznetsov, and Pegg [203], Asmussen [24], Anisimov [19], Gyllenberg and Silvestrov [104], Koroliuk and Limnios [188], Asmussen and Albrecher [25], Avrachenkov, Filar and Howlett [30], Silvestrov [307, 308].

(9) Exact and related approximative computational methods for stationary and quasi-stationary distributions of Markov chains and semi-Markov processes and related problems are presented in works by Romanovskiĭ [262], Feller [83], Kemeny and Snell [162], Golub and Seneta [92], Seneta [276, 282], Paige, Styan and Wachter [245], Silvestrov [298, 299, 302], Chatelin and Miranker [52], Harrod and Plemmons [110], Schweitzer [266, 269], Grassman, Taksar and Heyman [94], Schweitzer, Puterman and Kindle [271], Sheskin [284], Hunter [137, 138], Schweitzer and Kindle [270], Feinberg and Chiu [82], Haviv [113, 115], Haviv, Ritov and Rothblum [121], Sumita and Reiders [342], Mattingly and Meyer [219], Stewart, W. [341], Kim and Smith [168], Stewart [335, 336], Latouche and Ramaswami [210], Kartashov [156], Meyer [226], Häggström [105], Bini, Latouche and Meini [46], Golub and Van Loan [93], Silvestrov, Manca and Silvestrova [317], Van Doorn and Pollett [346] and Silvestrov and Manca [315, 316]. In particular, we would like to mention some related books that are Romanovskiĭ [262], Feller [83], Kemeny and Snell [162], Golub and Seneta [92], Seneta [276, 282], Berman and Plemmons [43], Silvestrov [298], Meyer [226], Häggström [105], Bini, Latouche and Meini [46], Meyn and Tweedie [230], Hernández-Lerma and Lasserre, [125], Gyllenberg and Silvestrov [104], Nåsell [244], Avrachenkov, Filar and Howlett [30] and Collet, Martínez and San Martín [55].

References

1. Abadov, Z.A.: Asymptotical expansions with explicit estimation of constants for exponential moments of sums of random variables defined on a markov chain and their applications to limit theorems for first hitting times. Candidate of Science dissertation, Kiev State University (1984)

2. Abbad, M., Filar, J.A.: Algorithms for singularly perturbed Markov control problems: a survey. In: Leondes, C.T. (ed.) Techniques in Discrete-Time Stochastic Control Systems. Control and Dynamic Systems, vol. 73, pp. 257–289. Academic Press, New York (1995)

3. Albeverio, S., Koroliuk, V.S., Samoilenko, I.V.: Asymptotic expansion of semi-Markov random evolutions. Stochastics **81**(5), 477–502 (2009)

4. Albrecht, A.R., Howlett, P.G., Pearce, C.E.M.: Necessary and sufficient conditions for the inversion of linearly-perturbed bounded linear operators on Banach space using Laurent series. J. Math. Anal. Appl. **383**(1), 95–110 (2011)

5. Albrecht, A.R., Howlett, P.G., Pearce, C.E.M.: The fundamental equations for inversion of operator pencils on Banach space. J. Math. Anal. Appl. **413**(1), 411–421 (2014)

6. Alimov, D., Shurenkov, V.M.: Markov renewal theorems in triangular array model. Ukr. Mat. Zh. **42**, 1443–1448 (1990) (English translation in Ukr. Math. J. **42**, 1283–1288)

7. Alimov, D., Shurenkov, V.M.: Asymptotic behavior of terminating Markov processes that are close to ergodic. Ukr. Mat. Zh. **42**, 1701–1703 (1990) (English translation in Ukr. Math. J. **42** 1535–1538)

8. Allen, B., Anderssen, R.S., Seneta, E.: Computation of stationary measures for infinite Markov chains. In: Neuts, M.F. (ed.) Algorithmic Methods in Probability. Studies in the Management Sciences, vol. 7, pp. 13–23. North-Holland, Amsterdam (1977)
9. Altman, E., Avrachenkov, K.E., Núñez-Queija, R.: Perturbation analysis for denumerable Markov chains with application to queueing models. Adv. Appl. Probab. **36**(3), 839–853 (2004)
10. Andersson, F., Silvestrov, S.: The mathematics of Internet search engines. Acta Appl. Math. **104**, 211–242 (2008)
11. Anisimov, V.V. Limit theorems for semi-Markov processes. I, II. Teor. Veroyatn. Mat. Stat. **2**, I: 3–12; II: 13–21 (1970)
12. Anisimov, V.V.: Limit theorems for sums of random variables on a Markov chain, connected with the exit from a set that forms a single class in the limit. Teor. Veroyatn. Mat. Stat. **4**, 3–17 (1971) (English translation in Theory Probab. Math. Statist. **4**, 1–13)
13. Anisimov, V.V.: Limit theorems for sums of random variables that are given on a subset of states of a Markov chain up to the moment of exit, in a series scheme. Teor. Veroyatn. Mat. Stat. **4**, 18–26 (1971) (English translation in Theory Probab. Math. Statist. **4**, 15–22)
14. Anisimov, V.V.: Limit theorems for sums of random variables that are given on a countable subset of the states of a Markov chain up to the first exit time. Teor. Veroyatn. Mat. Stat. **8**, 3–13 (1973)
15. Anisimov, V.V.: Asymptotical consolidation of states for random processes. Kibernetika (3), 109–117 (1973)
16. Anisimov, V.V.: Limit theorems for processes admitting the asymptotical consolidation of states. Theor. Veroyatn. Mat. Stat. **22**, 3–15 (1980) (English translation in Theory Probab. Math. Statist. **22**, 1–13)
17. Anisimov, V.V.: Approximation of Markov processes that can be asymptotically lumped. Teor. Veroyatn. Mat. Stat. **34**, 1–12 (1986) (English translation in Theory Probab. Math. Statist. **34**, 1–11)
18. Anisimov, V.V.: Random Processes with Discrete Components, 183 pp. Vysshaya Shkola and Izdatel'stvo Kievskogo Universiteta, Kiev (1988)
19. Anisimov, V.V.: Switching Processes in Queueing Models. Applied Stochastic Methods. ISTE, London and Wiley, Hoboken, NJ (2008). 345 pp
20. Anisimov, V.V., Chernyak, A.V.: Limit theorems for certain rare functionals on Markov chains and semi-Markov processes. Teor. Veroyatn. Mat. Stat. **26**, 3–8 (1982) (English translation in Theory Probab. Math. Statist. **26**, 1–6)
21. Anisimov, V.V., Voĭna, A.A., Lebedev, E.A.: Asymptotic estimation of integral functionals and consolidation of stochastic systems. Vestnik Kiev. Univ. Model. Optim. Slozhn. Sist. (2), 41–50 (1983)
22. Anisimov, V.V., Zakusilo, O.K., Donchenko, V.S.: Elements of Queueing and Asymptotical Analysis of Systems, 248 pp. Lybid', Kiev (1987)
23. Asmussen, S.: Busy period analysis, rare events and transient behavior in fluid flow models. J. Appl. Math. Stoch. Anal. **7**(3), 269–299 (1994)
24. Asmussen, S.: Applied Probability and Queues. Second edition, Applications of Mathematics, Stochastic Modelling and Applied Probability, vol. 51, xii+438 pp. Springer, New York (2003)
25. Asmussen, S., Albrecher, H.: Ruin Probabilities. Second edition, Advanced Series on Statistical Science & Applied Probability, vol. 14, xviii+602 pp. World Scientific, Hackensack, NJ (2010)
26. Avrachenkov, K.E.: Analytic perturbation theory and its applications. Ph.D. thesis, University of South Australia (1999)
27. Avrachenkov, K.E.: Singularly perturbed finite Markov chains with general ergodic structure. In: Boel, R., Stremersch, G. (eds.) Discrete Event Systems. Analysis and Control. Kluwer International Series in Engineering and Computer Science, vol. 569, pp. 429–432. Kluwer, Boston (2000)
28. Avrachenkov, K., Borkar, V., Nemirovsky, D.: Quasi-stationary distributions as centrality measures for the giant strongly connected component of a reducible graph. J. Comput. Appl. Math. **234**(11), 3075–3090 (2010)

29. Avrachenkov, K.E., Filar, J., Haviv, M.: Singular perturbations of Markov chains and decision processes. In: Feinberg, E.A., Shwartz, A. (eds.) Handbook of Markov Decision Processes. Methods and Applications. International Series in Operations Research & Management Science, vol. 40, pp. 113–150. Kluwer, Boston (2002)

30. Avrachenkov, K.E., Filar, J.A., Howlett, P.G.: Analytic Perturbation Theory and Its Applications, xii+372 pp. SIAM, Philadelphia, PA (2013)

31. Avrachenkov, K.E., Haviv, M.: Perturbation of null spaces with application to the eigenvalue problem and generalized inverses. Linear Algebr. Appl. **369**, 1–25 (2003)

32. Avrachenkov, K.E., Haviv, M.: The first Laurent series coefficients for singularly perturbed stochastic matrices. Linear Algebr. Appl. **386**, 243–259 (2004)

33. Avrachenkov, K.E., Haviv, M., Howlett, P.G.: Inversion of analytic matrix functions that are singular at the origin. SIAM J. Matrix Anal. Appl. **22**(4), 1175–1189 (2001)

34. Avrachenkov, K.E., Lasserre, J.B.: The fundamental matrix of singularly perturbed Markov chains. Adv. Appl. Probab. **31**(3), 679–697 (1999)

35. Avrachenkov, K.E., Lasserre, J.B.: Analytic perturbation of generalized inverses. Linear Algebr. Appl. **438**(4), 1793–1813 (2013)

36. Avrachenkov, K., Litvak, N., Son Pham, K.: Distribution of PageRank mass among principle components of the web. In: Chung, F.R.K., Bonato, A. (eds.) Algorithms and Models for the Web-Graph. Lecture Notes in Computer Science, vol. 4863, pp. 16–28. Springer, Berlin (2007)

37. Avrachenkov, K., Litvak, N., Son Pham, K.: A singular perturbation approach for choosing the PageRank damping factor. Internet Math. **5**(1–2), 47–69 (2008)

38. Barbour, A.D., Pollett, P.K.: Total variation approximation for quasi-stationary distributions. J. Appl. Probab. **47**(4), 934–946 (2010)

39. Barbour, A.D., Pollett, P.K.: Total variation approximation for quasi-equilibrium distributions II. Stoch. Process. Appl. **122**(11), 3740–3756 (2012)

40. Barlow, J.: Perturbation results for nearly uncoupled Markov chains with applications to iterative methods. Numer. Math. **65**(1), 51–62 (1993)

41. Baumgärtel, H.: Analytic Perturbation Theory for Matrices and Operators. Operator Theory: Advances and Applications, vol. 15, 427 pp. Birkhäuser, Basel (1985)

42. Benois, O., Landim, C., Mourragui, M.: Hitting times of rare events in Markov chains. J. Stat. Phys. **153**(6), 967–990 (2013)

43. Berman, A., Plemmons, R.J.: Nonnegative Matrices in the Mathematical Sciences.Classics in Applied Mathematics, vol. 9, xx+340 pp. SIAM, Philadelphia (1994). (A revised reprint ofNonnegative Matrices in the Mathematical Sciences. Computer Science and Applied Mathematics, xviii+316 pp. Academic Press, New York, 1979)

44. Bielecki, T., Stettner, Ł.: On ergodic control problems for singularly perturbed Markov processes. Appl. Math. Optim. **20**(2), 131–161 (1989)

45. Blanchet, J., Zwart, B.: Asymptotic expansions of defective renewal equations with applications to perturbed risk models and processor sharing queues. Math. Methods Oper. Res. **72**, 311–326 (2010)

46. Bini, D.A., Latouche, G., Meini, B.: Numerical Methods for Structured Markov Chains. Numerical Mathematics and Scientific Computation, p. xii+327. Oxford Science Publications, Oxford University Press, New York (2005)

47. Bobrova, A.F.: Estimates of accuracy of an asymptotic consolidation of countable Markov chains. Stability Problems for Stochastic Models, pp. 16–24. Trudy Seminara, VNIISI, Moscow (1983)

48. Borovkov, A.A.: Ergodicity and Stability of Stochastic Processes. Wiley Series in Probability and Statistics, vol. 314, p. xxiv+585. Wiley, Chichester (1998) (Translation from the 1994 Russian original)

49. Burnley, C.: Perturbation of Markov chains. Math. Mag. **60**(1), 21–30 (1987)

50. Cao, X.R.: The Maclaurin series for performance functions of Markov chains. Adv. Appl. Probab. **30**, 676–692 (1998)

51. Cao, W.L., Stewart, W.J.: Iterative aggregation/disaggregation techniques for nearly uncoupled Markov chains. J. Assoc. Comput. Mach. **32**, 702–719 (1985)
52. Chatelin, F., Miranker, W.L.: Aggregation/disaggregation for eigenvalue problems. SIAM J. Numer. Anal. **21**(3), 567–582 (1984)
53. Coderch, M., Willsky, A.S., Sastry, S.S., Castañon, D.A.: Hierarchical aggregation of singularly perturbed finite state Markov processes. Stochastics **8**, 259–289 (1983)
54. Cole, J.D.: Perturbation Methods in Applied Mathematics, p. vi+260. Blaisdell, Waltham, Mass (1968)
55. Collet, P., Martínez, S., San Martín, J.: Quasi-Stationary Distributions. Markov Chains, Diffusions and Dynamical Systems. Probability and its Applications, p. xvi+280. Springer, Heidelberg (2013)
56. Courtois, P.J.: Error analysis in nearly-completely decomposable stochastic systems. Econometrica **43**(4), 691–709 (1975)
57. Courtois, P.J.: Decomposability: Queueing and Computer System Applications. ACM Monograph Series, p. xiii+201. Academic Press, New York (1977)
58. Courtois, P.J.: Error minimization in decomposable stochastic models. In: Ralph, L.D., Teunis, J.O. (eds.) Applied Probability - Computer Science: the Interface. Progress in Computer Science, 2, vol. I, pp. 189–210. Birkhäuser, Boston (1982)
59. Courtois, P.J., Louchard, G.: Approximation of eigen characteristics in nearly-completely decomposable stochastic systems. Stoch. Process. Appl. **4**, 283–296 (1976)
60. Courtois, P.J., Semal, P.: Error bounds for the analysis by decomposition of non-negative matrices. In: Iazeolla, G., Courtois, P.J., Hordijk, A. (eds.) Mathematical Computer Performance and Reliability, pp. 209–224. North-Holland, Amsterdam (1984)
61. Courtois, P.J., Semal, P.: Block decomposition and iteration in stochastic matrices. Philips J. Res. **39**(4–5), 178–194 (1984)
62. Courtois, P.J., Semal, P.: Bounds for the positive eigenvectors of nonnegative matrices and for their approximations by decomposition. J. Assoc. Comput. Mach. **31**(4), 804–825 (1984)
63. Courtois, P.J., Semal, P. Bounds for transient characteristics of large or infinite Markov chains. In: Stewart, W.J. (ed.) Numerical Solution of Markov Chains. Probability: Pure and Applied, vol. 8, pp. 413–434. Marcel Dekker, New York (1991)
64. Craven, B.D.: Perturbed Markov processes. Stoch. Models **19**(2), 269–285 (2003)
65. Darroch, J., Seneta, E.: On quasi-stationary distributions in absorbing discrete-time finite Markov chains. J. Appl. Probab. **2**, 88–100 (1965)
66. Darroch, J., Seneta, E.: On quasi-stationary distributions in absorbing continuous-time finite Markov chains. J. Appl. Probab. **4**, 192–196 (1967)
67. Delebecque, F.: A reduction process for perturbed Markov chains. SIAM J. Appl. Math. **43**, 325–350 (1983)
68. Delebecque, F.: On the resolvent approach to the spectral decomposition of a regular matrix pencil. Linear Algebr. Appl. **129**, 63–75 (1990)
69. Delebecque, F., Quadrat, J.P.: Optimal control of Markov chains admitting strong and weak interactions. Automatica **17**(2), 281–296 (1981)
70. Drozdenko, M.: Weak convergence of first-rare-event times for semi-Markov processes I. Theory Stoch. Process. **13**(4), 29–63 (2007)
71. Drozdenko, M.: Weak convergence of first-rare-event times for semi-Markov processes, vol. 49. Doctoral dissertation, Mälardalen University, Västerås (2007)
72. Drozdenko, M.: Weak convergence of first-rare-event times for semi-Markov processes II. Theory Stoch. Process. **15**(2), 99–118 (2009)
73. Englund, E.: Perturbed renewal equations with application to M/M queueing systems. 1. Teor. Ĭmovirn. Mat. Stat. **60**, 31–37 (1999) (Also in Theory Probab. Math. Stat. **60**, 35–42)
74. Englund, E.: Perturbed renewal equations with application to M/M queueing systems. 2. Teor. Ĭmovirn. Mat. Stat. **61**, 21–32 (1999) (Also in Theory Probab. Math. Stat. **61**, 21–32)
75. Englund, E.: Nonlinearly perturbed renewal equations with applications to a random walk. In: Silvestrov, D., Yadrenko, M., Olenko A., Zinchenko, N. (eds.) Proceedings of the Third International School on Applied Statistics, Financial and Actuarial Mathematics, Feodosiya (2000). (Theory Stoch. Process. **6**(22)(3–4), 33–60)

76. Englund, E.: Nonlinearly perturbed renewal equations with applications. Doctoral dissertation, Umeå University (2001)
77. Englund, E., Silvestrov, D.S.: Mixed large deviation and ergodic theorems for regenerative processes with discrete time. In: Jagers, P., Kulldorff, G., Portenko, N., Silvestrov, D. (eds.) Proceedings of the Second Scandinavian–Ukrainian Conference in Mathematical Statistics, Vol. I, Umeå, 1997. (Theory Stoch. Process. **3**(19)(1–2), 164–176)
78. Engström, C., Silvestrov, S.: Generalisation of the damping factor in PageRank for weighted networks. In: Silvestrov, D., Martin-Löf, A. (eds.) Modern Problems in Insurance Mathematics. Chap. 19. EAA series, pp. 313–334. Springer, Cham (2014)
79. Engström, C., Silvestrov, S.: PageRank, a look at small changes in a line of nodes and the complete graph. In: Silvestrov S., Rančić M. (eds.) Engineering Mathematics II. Algebraic, Stochastic and Analysis Structures for Networks, Data Classification and Optimization. Springer, Heidelberg (2016)
80. Engström, C., Silvestrov, S.: PageRank, connecting a line of nodes with a complete graph. In: Silvestrov S., Rančić M. (eds.) Engineering Mathematics II. Algebraic, Stochastic and Analysis Structures for Networks, Data Classification and Optimization. Springer, Heidelberg (2016)
81. Erdélyi, A.: Asymptotic Expansions, vi+108 pp. Dover, New York (1956)
82. Feinberg, B.N., Chiu, S.S.: A method to calculate steady-state distributions of large Markov chains by aggregating states. Oper. Res. **35**(2), 282–290 (1987)
83. Feller, W.: An Introduction to Probability Theory and Its Applications, vol. I, xviii+509 pp. Wiley, New York (1968). (3rd edition of An Introduction to Probability Theory and Its Applications, vol. I, xii+419 pp. Wiley, New York, 1950)
84. Fuh, C.D.: Uniform Markov renewal theory and ruin probabilities in Markov random walks. Ann. Appl. Probab. **14**(3), 1202–1241 (2004)
85. Fuh, C.D.: Asymptotic expansions on moments of the first ladder height in Markov random walks with small drift. Adv. Appl. Probab. **39**, 826–852 (2007)
86. Fuh, C.D., Lai, T.L.: Asymptotic expansions in multidimensional Markov renewal theory and first passage times for Markov random walks. Adv. Appl. Probab. **33**, 652–673 (2001)
87. Funderlic, R.E., Meyer, C.D., Jr.: Sensitivity of the stationary distribution vector for an ergodic Markov chain. Linear Algebr. Appl. **76**, 1–17 (1985)
88. Gaĭtsgori, V.G., Pervozvanskiĭ, A.A.: Aggregation of states in a Markov chain with weak interaction. Kibernetika (3), 91–98 (1975). (English translation in Cybernetics **11**(3), 441–450)
89. Gaĭtsgori, V.G., Pervozvanskiy, A.A.: Decomposition and aggregation in problems with a small parameter. Izv. Akad. Nauk SSSR, Tekhn. Kibernet. (1), 33–46 (1983) (English translation in Eng. Cybern. **2**(1), 26–38)
90. Gibson, D., Seneta, E.: Augmented truncations of infinite stochastic matrices. J. Appl. Probab. **24**(3), 600–608 (1987)
91. Glynn, P.: On exponential limit laws for hitting times of rare sets for Harris chains and processes. In: Glynn, P., Mikosch, T., Rolski, T. (eds.) New Frontiers in Applied Probability: a Festschrift for Søren Asmussen. vol. **48A**, 319–326 (2011). (J. Appl. Probab. Spec.)
92. Golub, G.H., Seneta, E.: Computation of the stationary distribution of an infinite Markov matrix. Bull. Aust. Math. Soc. **8**, 333–341 (1973)
93. Golub, G.H., Van Loan, C.F: Matrix Computations. Johns Hopkins Studies in the Mathematical Sciences, xiv+756 pp. Johns Hopkins University Press, Baltimore, MD (2013). (4th edition of Matrix Computations. Johns Hopkins Series in the Mathematical Sciences, vol. 3, xvi+476 pp. Johns Hopkins University Press, Baltimore, MD, 1983)
94. Grassman, W.K., Taksar, M.I., Heyman, D.P.: Regenerative analysis and steady state distributions for Markov chains. Oper. Res. **33**, 1107–1116 (1985)
95. Guo, D.Z.: On the sensitivity of the solution of nearly uncoupled Markov chains. SIAM J. Matrix Anal. Appl. **14**(4), 1112–1123 (2006)
96. Gusak, D.V., Korolyuk, V.S.. Asymptotic behaviour of semi-Markov processes with a decomposable set of states. Teor. Veroyatn. Mat. Stat. **5**, 43–50 (1971) (English translation in Theory Probab. Math. Stat. **5**, 43–51)

97. Gut, A., Holst, L.: On the waiting time in a generalized roulette game. Stat. Probab. Lett. **2**(4), 229–239 (1984)
98. Gyllenberg, M., Silvestrov, D.S.: Quasi-stationary distributions of a stochastic metapopulation model. J. Math. Biol. **33**, 35–70 (1994)
99. Gyllenberg, M., Silvestrov, D.S.: Quasi-stationary phenomena in semi-Markov models. In: Janssen, J., Limnios, N. (eds.) Proceedings of the Second International Symposium on Semi-Markov Models: Theory and Applications, Compiègne, pp. 87–93 (1998)
100. Gyllenberg, M., Silvestrov, D.S.: Quasi-stationary phenomena for semi-Markov processes. In: Janssen, J., Limnios, N. (eds.) Semi-Markov Models and Applications, pp. 33–60. Kluwer, Dordrecht (1999)
101. Gyllenberg, M., Silvestrov, D.S.: Cramér-Lundberg and diffusion approximations for non-linearly perturbed risk processes including numerical computation of ruin probabilities. In: Silvestrov, D., Yadrenko, M., Borisenko, O., Zinchenko, N. (eds.) Proceedings of the Second International School on Actuarial and Financial Mathematics, Kiev (1999). (Theory Stoch. Process., **5**(**21**)(1–2), 6–21)
102. Gyllenberg, M., Silvestrov, D.S.: Nonlinearly perturbed regenerative processes and pseudo-stationary phenomena for stochastic systems. Stoch. Process. Appl. **86**, 1–27 (2000)
103. Gyllenberg, M., Silvestrov, D.S.: Cramér-Lundberg approximation for nonlinearly perturbed risk processes. Insur. Math. Econ. **26**, 75–90 (2000)
104. Gyllenberg, M., Silvestrov, D.S.: Quasi-Stationary Phenomena in Nonlinearly Perturbed Stochastic Systems. De Gruyter Expositions in Mathematics, vol. 44, ix+579 pp. Walter de Gruyter, Berlin (2008)
105. Häggström, O.: Finite Markov Chains and Algorithmic Applications. London Mathematical Society Student Texts, vol. 52, 126 pp. Cambridge University Press, Cambridge (2002)
106. Hanen, A.: Probème central limite dans le cas Markovien fini. La matrice limite n'a qu'une seule classe ergodique et pas d'état transitoire. C. R. Acad. Sci. Paris **256**, 68–70 (1963)
107. Hanen, A.: Problème central limite dans le cas Markovien fini. II. La matrice limite a plusieurs classes ergodiques et pas d'états transitoires. C. R. Acad. Sci. Paris **256**, 362–364 (1963)
108. Hanen, A.: Problème central limite dans le cas Markovien fini. Cas général. C. R. Acad. Sci. Paris **256**, 575–577 (1963)
109. Hanen, A.: Théorèmes limites pour une suite de chaînes de Markov. Ann. Inst. H. Poincaré **18**, 197–301 (1963)
110. Harrod, W.J., Plemmons, R.J.: Comparison of some direct methods for computing stationary distributions of Markov chains. SIAM J. Sci. Stat. Comput. **5**, 453–469 (1984)
111. Hassin, R., Haviv, M.: Mean passage times and nearly uncoupled Markov chains. SIAM J. Discret. Math. **5**, 386–397 (1992)
112. Haviv, M.: An approximation to the stationary distribution of a nearly completely decomposable Markov chain and its error analysis. SIAM J. Algebr. Discret. Methods **7**(4), 589–593 (1986)
113. Haviv, M.: Aggregation/disaggregation methods for computing the stationary distribution of a Markov chain. SIAM J. Numer. Anal. **24**(4), 952–966 (1987)
114. Haviv, M.: Error bounds on an approximation to the dominant eigenvector of a nonnegative matrix. Linear Multilinear Algebr. **23**(2), 159–163 (1988)
115. Haviv, M.: An aggregation/disaggregation algorithm for computing the stationary distribution of a large Markov chain. Commun. Stat. Stoch. Models **8**(3), 565–575 (1992)
116. Haviv, M.: On censored Markov chains, best augmentations and aggregation/disaggregation procedures. Aggregation and disaggregation in operations research. Comput. Oper. Res. **26**(10–11), 1125–1132 (1999)
117. Haviv, M.: More on Rayleigh-Ritz refinement technique for nearly uncoupled stochastic matrices. SIAM J. Matrix Anal. Appl. **10**(3), 287–293 (2006)
118. Haviv, M., Ritov, Y.: An approximation to the stationary distribution of a nearly completely decomposable Markov chain and its error bound. SIAM J. Algebr. Discret. Methods **7**(4), 583–588 (1986)

119. Haviv, M., Ritov, Y.: On series expansions and stochastic matrices. SIAM J. Matrix Anal. Appl. **14**(3), 670–676 (1993)
120. Haviv, M., Ritov, Y.: Bounds on the error of an approximate invariant subspace for non-self-adjoint matrices. Numer. Math. **67**(4), 491–500 (1994)
121. Haviv, M., Ritov, Y., Rothblum, U.G.: Iterative methods for approximating the subdominant modulus of an eigenvalue of a nonnegative matrix. Linear Algebr. Appl. **87**, 61–75 (1987)
122. Haviv, M., Ritov, Y., Rothblum, U.G.: Taylor expansions of eigenvalues of perturbed matrices with applications to spectral radii of nonnegative matrices. Linear Algebr. Appl. **168**, 159–188 (1992)
123. Haviv, M., Rothblum, U.G.: Bounds on distances between eigenvalues. Linear Algebr. Appl. **63**, 101–118 (1984)
124. Haviv, M., Van der Heyden, L.: Perturbation bounds for the stationary probabilities of a finite Markov chain. Adv. Appl. Probab. **16**, 804–818 (1984)
125. Hernández-Lerma, O. Lasserre, J.B.: Markov Chains and Invariant Probabilities. Progress in Mathematics, vol. 211, xvi+205 pp. Birkhäuser, Basel (2003)
126. Ho, Y.C., Cao, X.R.: Perturbation Analysis of Discrete Event Dynamic Systems. The Springer International Series in Engineering and Computer Science, 433 pp. Springer, New York (1991)
127. Hoppensteadt, F., Salehi, H., Skorokhod, A.: On the asymptotic behavior of Markov chains with small random perturbations of transition probabilities. In: Gupta, A.K. (ed.) Multidimensional Statistical Analysis and Theory of Random Matrices: Proceedings of the Sixth Eugene Lukacs Symposium. Bowling Green, OH, 1996, pp. 93–100. VSP, Utrecht (1996)
128. Hoppensteadt, F., Salehi, H., Skorokhod, A.: Markov chain with small random perturbations with applications to bacterial genetics. Random Oper. Stoch. Equ. **4**(3), 205–227 (1996)
129. Hoppensteadt, F., Salehi, H., Skorokhod, A.: Discrete time semigroup transformations with random perturbations. J. Dyn. Differ. Equ. **9**(3), 463–505 (1997)
130. Hössjer, O.: Coalescence theory for a general class of structured populations with fast migration. Adv. Appl. Probab. **43**(4), 1027–1047 (2011)
131. Hössjer, O.: Spatial autocorrelation for subdivided populations with invariant migration schemes. Methodol. Comput. Appl. Probab. **16**(4), 777–810 (2014)
132. Hössjer, O., Ryman, N.: Quasi equilibrium, variance effective size and fixation index for populations with substructure. J. Math. Biol. **69**(5), 1057–1128 (2014)
133. Howlett, P., Albrecht, A., Pearce, C.: Laurent series for inversion of linearly perturbed bounded linear operators on Banach space. J. Math. Anal. Appl. **366**(1), 112–123 (2010)
134. Howlett, P., Avrachenkov, K.: Laurent series for the inversion of perturbed linear operators on Hilbert space. In: Glover, B.M., Rubinov, A.M. (eds.) Optimization and Related Topics. Applied Optimisation, vol. 47, pp. 325–342. Kluwer, Dordrecht (2001)
135. Howlett, P., Avrachenkov, K., Pearce, C., Ejov, V.: Inversion of analytically perturbed linear operators that are singular at the origin. J. Math. Anal. Appl. **353**(1), 68–84 (2009)
136. Howlett, P., Pearce, C., Torokhti, A.: On nonlinear operator approximation with preassigned accuracy. J. Comput. Anal. Appl. **5**(3), 273–297 (2003)
137. Hunter, J.J.: Stationary distributions of perturbed Markov chains. Linear Algebr. Appl. **82**, 201–214 (1986)
138. Hunter, J.J.: The computation of stationary distributions of Markov chains through perturbations. J. Appl. Math. Stoch. Anal. **4**(1), 29–46 (1991)
139. Hunter, J.J.: A survey of generalized inverses and their use in applied probability. Math. Chron. **20**, 13–26 (1991)
140. Hunter, J.J.: Stationary distributions and mean first passage times of perturbed Markov chains. Linear Algebr. Appl. **410**, 217–243 (2005)
141. Hunter, J.J.: Generalized inverses of Markovian kernels in terms of properties of the Markov chain. Linear Algebr. Appl. **447**, 38–55 (2014)
142. Kalashnikov, V.V.: Qualitative Analysis of the Behaviour of Complex Systems by the Method of Test Functions. Theory and Methods of Systems Analysis, 247 pp. Nauka, Moscow (1978)
143. Kalashnikov, V.V.: Solution of the problem of approximating a denumerable Markov chain. Eng. Cybern. (3), 92–95 (1997)

144. Kalashnikov, V.V.: Geometric Sums: Bounds for Rare Events with Applications. Mathematics and its Applications, vol. 413, ix+265 pp. Kluwer, Dordrecht (1997)
145. Kalashnikov, V.V., Anichkin, S.A.: Continuity of random sequences and approximation of Markov chains. Adv. Appl. Probab. **13**(2), 402–414 (1981)
146. Kaplan, E.I.: Limit theorems for exit times of random sequences with mixing. Teor. Veroyatn. Mat. Stat. **21**, 53–59 (1979). (English translation in Theory Probab. Math. Stat. **21**, 59–65)
147. Kaplan, E.I.: Limit Theorems for Sum of Switching Random Variables with an Arbitrary Phase Space of Switching Component. Candidate of Science dissertation, Kiev State University (1980)
148. Kartashov, N.V.: Inequalities in stability and ergodicity theorems for Markov chains with a general phase space. I. Teor. Veroyatn. Primen. vol. 30, 230–240 (1985). (English translation in Theory Probab. Appl. **30**, 247–259)
149. Kartashov, N.V.: Inequalities in stability and ergodicity theorems for Markov chains with a general phase space. II. Teor. Veroyatn. Primen. **30**, 478–485 (1985). (English translation in Theory Probab. Appl. **30**, 507–515)
150. Kartashov, N.V.: Asymptotic representations in an ergodic theorem for general Markov chains and their applications. Teor. Veroyatn. Mat. Stat. **32**, 113–121 (1985). (English translation in Theory Probab. Math. Stat. **32**, 131–139)
151. Kartashov, N.V.: Asymptotic expansions and inequalities in stability theorems for general Markov chains under relatively bounded perturbations. In: Stability Problems for Stochastic Models, Varna, 1985, pp. 75–85. VNIISI, Moscow (1985). (English translation in J. Soviet Math. **40**(4), 509–518)
152. Kartashov, N.V.: Inequalities in theorems of consolidation of Markov chains. Theor. Veroyatn. Mat. Stat. **34**, 62–73 (1986). (English translation in Theory Probab. Math. Stat. **34**, 67–80)
153. Kartashov, N.V.: Estimates for the geometric asymptotics of Markov times on homogeneous chains. Teor. Veroyatn. Mat. Stat. **37**, 66–77 (1987). (English translation in Theory Probab. Math. Stat. **37**, 75 88)
154. Kartashov, M.V.: Computation and estimation of the exponential ergodicity exponent for general Markov processes and chains with recurrent kernels. Teor. Ĭmovirn. Mat. Stat. **54**, 47–57 (1996). (English translation in Theory Probab. Math. Stat. **54**, 49–60)
155. Kartashov, M.V.: Strong Stable Markov Chains, 138 pp. VSP, Utrecht and TBiMC, Kiev (1996)
156. Kartashov, M.V.: Calculation of the spectral ergodicity exponent for the birth and death process. Ukr. Mat. Zh. **52**, 889–897 (2000) (English translation in Ukr. Math. J. **52**, 1018–1028)
157. Kartashov, M.V.: Ergodicity and stability of quasihomogeneous Markov semigroups of operators. Teor. Ĭmovirn. Mat. Stat. **72**, 54–62 (2005). (English translation in Theory Probab. Math. Stat. **72**, 59–68)
158. Kartashov, M.V.: Quantitative and qualitative limits for exponential asymptotics of hitting times for birth-and-death chains in a scheme of series. Teor. Imovirn. Mat. Stat. **89**, 40–50 (2013). (English translation in Theory Probab. Math. Stat. **89**, 45–56 (2014))
159. Kato, T.: Perturbation Theory for Linear Operators. Classics of Mathematics, 623 pp. Springer, Berlin (2013). (2nd edition of Perturbation Theory for Linear Operators, xix+592 pp. Springer, New York)
160. Keilson, J.: A limit theorem for passage times in ergodic regenerative processes. Ann. Math. Stat. **37**, 866–870 (1966)
161. Keilson, J.: Markov Chain Models – Rarity and Exponentiality. Applied Mathematical Sciences, vol. 28, xiii+184 pp. Springer, New York (1979)
162. Kemeny, J.G., Snell, J.L.: Finite Markov Chains. The University Series in Undergraduate Mathematics, viii+210 pp. D. Van Nostrand, Princeton, NJ (1960)
163. Kevorkian, J., Cole, J.D.: Perturbation Methods in Applied Mathematics. Applied Mathematical Sciences, vol. 34, x+558 pp. Springer, New York (1981)
164. Kevorkian, J., Cole, J.D.: Multiple Scale and Singular Perturbation Methods. Applied Mathematical Sciences, vol. 114, viii+632 pp. Springer, New York (1996)

165. Khasminskii, R.Z., Yin, G., Zhang, Q.: Singularly perturbed Markov chains: quasi-stationary distribution and asymptotic expansion. In: Proceedings of Dynamic Systems and Applications, vol. 2, pp. 301–308. Atlanta, GA, 1995. Dynamic, Atlanta, GA (1996)
166. Khasminskii, R.Z., Yin, G., Zhang, Q.: Asymptotic expansions of singularly perturbed systems involving rapidly fluctuating Markov chains. SIAM J. Appl. Math. **56**(1), 277–293 (1996)
167. Kijima, M.: Markov Processes for Stochastic Modelling. Stochastic Modeling Series, x+341 pp. Chapman & Hall, London (1997)
168. Kim, D.S., Smith, R.L.: An exact aggregation/disaggregation algorithm for large scale Markov chains. Nav. Res. Logist. **42**(7), 1115–1128 (1995)
169. Kingman, J.F.: The exponential decay of Markovian transition probabilities. Proc. Lond. Math. Soc. **13**, 337–358 (1963)
170. Kokotović, P.V., Phillips, R.G., Javid, S.H.: Singular perturbation modeling of Markov processes. In: Bensoussan, A., Lions, J.L. (eds.) Analysis and optimization of systems: Proceedings of the Fourth International Conference on Analysis and Optimization. Lecture Notes in Control and Information Science, vol. 28, pp. 3–15. Springer, Berlin (1980)
171. Konstantinov, M. Gu, D.W., Mehrmann, V., Petkov, P.: Perturbation Theory for Matrix Equations. Studies in Computational Mathematics, vol. 9, xii+429 pp. North-Holland, Amsterdam (2003)
172. Konstantinov, M.M., Petkov, P.H.: Perturbation methods in linear algebra and control. Appl. Comput. Math. **7**(2), 141–161 (2008)
173. Kontoyiannis, I., Meyn, S.P.: Spectral theory and limit theorems for geometrically ergodic Markov processes. Ann. Appl. Probab. **13**(1), 304–362 (2003)
174. Korolyuk, D.V.: Limit theorems for hitting time type functionals defined on processes with semi-Markov switchings. Candidate of Science Dissertation, Kiev State University (1983)
175. Korolyuk, D.V., Silvestrov D.S.: Entry times into asymptotically receding domains for ergodic Markov chains. Teor. Veroyatn. Primen. **28**, 410–420 (1983). (English translation in Theory Probab. Appl. **28**, 432–442)
176. Korolyuk, D.V., Silvestrov D.S.: Entry times into asymptotically receding regions for processes with semi-Markov switchings. Teor. Veroyatn. Primen. **29**, 539–544 (1984). (English translation in Theory Probab. Appl. **29**, 558–563)
177. Korolyuk, V.S.: On asymptotical estimate for time of a semi-Markov process being in the set of states. Ukr. Mat. Zh. **21**, 842–845 (1969). (English translation in Ukr. Math. J. **21**, 705–710)
178. Korolyuk, V.S.: Stochastic Models of Systems, 208 pp. Naukova Dumka, Kiev (1989)
179. Korolyuk, V.S., Brodi, S.M., Turbin, A.F.: Semi-Markov processes and their application. Probability Theory. Mathematical Statistics. Theoretical Cybernetics, vol. 11, pp. 47–97. VINTI, Moscow (1974)
180. Korolyuk, V.S., Korolyuk, V.V.: Stochastic Models of Systems. Mathematics and its Applications, vol. 469, xii+185 pp. Kluwer, Dordrecht (1999)
181. Korolyuk, V.S., Limnios, N.: Diffusion approximation of integral functionals in merging and averaging scheme. Teor. Ĭmovirn. Mat. Stat. **59**, 99–105 (1998). (English translation in Theory Probab. Math. Stat. **59**, 101–107)
182. Korolyuk, V.S., Limnios, N.: Diffusion approximation for integral functionals in the double merging and averaging scheme. Teor. Ĭmovirn. Mat. Stat. **60**, 77–84 (1999). (English translation in Theory Probab. Math. Stat. **60**, 87–94)
183. Korolyuk, V.S., Limnios, N.: Evolutionary systems in an asymptotic split phase space. In: Limnios, N., Nikulin, M. (eds.) Recent Advances in Reliability Theory: Methodology, Practice and Inference, pp. 145–161. Birkhäuser, Boston (2000)
184. Korolyuk, V.S., Limnios, N.: Markov additive processes in a phase merging scheme. In: Korolyuk, V., Prokhorov, Yu., Khokhlov, V., Klesov, O. (eds.) Proceedings of the Conference Dedicated to the 90th Anniversary of Boris Vladimirovich Gnedenko, Kiev (2002). (Theory Stoch. Process., **8**(3–4), 213–225)
185. Korolyuk, V.S., Limnios, N.: Average and diffusion approximation of stochastic evolutionary systems in an asymptotic split state space. Ann. Appl. Probab. **14**, 489–516 (2004)

186. Korolyuk, V.S., Limnios, N.: Diffusion approximation of evolutionary systems with equilibrium in asymptotic split phase space. Teor. Ĭmovirn. Mat. Stat. **70**, 63–73 (2004). (English translation in Theory Probab. Math. Stat. **70**, 71–82)

187. Koroliuk, V.S., Limnios, N.: Stochastic Systems in Merging Phase Space, xv+331 pp. World Scientific, Singapore (2005)

188. Koroliuk, V.S., Limnios, N.: Reliability of semi-Markov systems with asymptotic merging phase space. In: Rykov, V.V., Balakrishnan, N., Nikulin, M.S. (eds.) Mathematical and Statistical Models and Methods in Reliability, pp. 3–18. Birkhäuser, Boston (2010)

189. Korolyuk, V.S., Penev, I.P., Turbin, A.F.: The asymptotic behavior of the distribution of the absorption time of a Markov chain. Kibernetika (2), 20–22 (1972)

190. Korolyuk, V.S., Penev, I.P., Turbin, A.F.: Asymptotic expansion for the distribution of the absorption time of a weakly inhomogeneous Markov chain. In: Korolyuk, V.S. (ed.) Analytic Methods of Investigation in Probability Theory, pp. 97–105. Akad. Nauk Ukr. SSR, Inst. Mat., Kiev (1981)

191. Korolyuk, V., Swishchuk, A.: Semi-Markov Random Evolutions, 254 pp. Naukova Dumka, Kiev (1992). (English revised edition of Semi-Markov Random Evolutions. Mathematics and its Applications, vol. 308, x+310 pp. Kluwer, Dordrecht, 1995)

192. Korolyuk, V.V., Tadzhiev, A.: Asymptotic behavior of Markov evolutions prior to the time of absorption. Ukr. Mat. Zh. **38**, 248–251 (1986). (English translation in Ukr. Math. J. **38**, 219–222)

193. Korolyuk, V.S., Turbin, A.F.: On the asymptotic behaviour of the occupation time of a semi-Markov process in a reducible subset of states. Teor. Veroyatn. Mat. Stat. **2**, 133–143 (1970). (English translation in Theory Probab. Math. Stat. **2**, 133–143)

194. Korolyuk, V.S., Turbin, A.F.: A certain method of proving limit theorems for certain functionals of semi-Markov processes. Ukr. Mat. Zh. **24**, 234–240 (1972)

195. Korolyuk, V.S., Turbin, A.F.: Semi-Markov Processes and its Applications, 184 pp. Naukova Dumka, Kiev (1976)

196. Korolyuk, V.S., Turbin, A.F.: Mathematical Foundations of the State Lumping of Large Systems, 218 pp.. Naukova Dumka, Kiev (1978). (English edition: Mathematical Foundations of the State Lumping of Large Systems. Mathematics and its Applications, vol. 264, x+278 pp. Kluwer, Dordrecht, 1993)

197. Korolyuk, V.S., Turbin, A.F., Tomusjak, A.A.: Sojourn time of a semi-Markov process in a decomposing set of states. Analytical Methods of Probability Theory, vol. 152, pp. 69–79. Naukova Dumka, Kiev (1979)

198. Koury, J.R., McAllister, D.F., Stewart, W.J.: Iterative methods for computing stationary distributions of nearly completely decomposable Markov chains. SIAM J. Algebr. Discret. Methods **5**, 164–186 (1984)

199. Kovalenko, I.N.: An algorithm of asymptotic analysis of a sojourn time of Markov chain in a set of states. Dokl. Acad. Nauk Ukr. SSR, Ser. A **6**, 422–426 (1973)

200. Kovalenko, I.N.: Studies in the Reliability Analysis of Complex Systems, 210 pp. Naukova Dumka, Kiev (1975)

201. Kovalenko, I.N.: Rare events in queuing theory - a survey. Queuing Syst. Theory Appl. **16**(1–2), 1–49 (1994)

202. Kovalenko, I.N., Kuznetsov, M.Ju.: Renewal process and rare events limit theorems for essentially multidimensional queueing processes. Math. Oper. Stat. Ser. Stat. **12**(2), 211–224 (1981)

203. Kovalenko, I.N., Kuznetsov, N.Y., Pegg, P.A.: Mathematical Theory of Reliability of Time Dependent Systems with Practical Applications. Wiley Series in Probability and Statistics, 316 pp. Wiley, New York (1997)

204. Kupsa, M., Lacroix, Y.: Asymptotics for hitting times. Ann. Probab. **33**(2), 610–619 (2005)

205. Langville, A.N., Meyer, C.D.: Updating Markov chains with an eye on Google's PageRank. SIAM J. Matrix Anal. Appl. **27**(4), 968–987 (2006)

206. Lasserre, J.B.: A formula for singular perturbations of Markov chains. J. Appl. Probab. **31**, 829–833 (1994)

207. Latouche, G.: Perturbation analysis of a phase-type queue with weakly correlated arrivals. Adv. Appl. Probab. **20**, 896–912 (1988)
208. Latouche, G.: First passage times in nearly decomposable Markov chains. In: Stewart, W.J. (ed.) Numerical Solution of Markov Chains. Probability: Pure and Applied, vol. 8, pp. 401–411. Marcel Dekker, New York (1991)
209. Latouche, G., Louchard, G.: Return times in nearly decomposable stochastic processes. J. Appl. Probab. **15**, 251–267 (1978)
210. Latouche, G., Ramaswami, V.: Introduction to Matrix Analytic Methods in Stochastic Modeling. ASA-SIAM Series on Statistics and Applied Probability, xiv+334 pp. SIAM, Philadelphia, PA and ASA, Alexandria, VA (1999)
211. Leadbetter, M.R.: On series expansion for the renewal moments. Biometrika **50**(1–2), 75–80 (1963)
212. Li, R.C., Stewart, G.W.: A new relative perturbation theorem for singular subspaces. Linear Algebr. Appl. **313**(1–3), 41–51 (2000)
213. Li, X., Yin, G., Yin, K., Zhang, Q.: A numerical study of singularly perturbed Markov chains: quasi-equilibrium distributions and scaled occupation measures. Dyn. Contin. Discret. Impuls. Syst. **5**(1–4), 295–304 (1999)
214. Louchard, G., Latouche, G.: Random times in nearly-completely decomposable, transient Markov chains. Cahiers Centre Études Rech. Opér. **24**(2–4), 321–352 (1982)
215. Louchard, G., Latouche, G.: Geometric bounds on iterative approximations for nearly completely decomposable Markov chains. J. Appl. Probab. **27**(3), 521–529 (1990)
216. Marek, I., Mayer, P.: Convergence analysis of an iterative aggregation/disaggregation method for computing stationary probability vectors of stochastic matrices. Numer. Linear Algebr. Appl. **5**(4), 253–274 (1998)
217. Marek, I., Mayer, P., Pultarová, I.: Convergence issues in the theory and practice of iterative aggregation/disaggregation methods. Electron. Trans. Numer. Anal. **35**, 185–200 (2009)
218. Marek, I., Pultarová, I.: A note on local and global convergence analysis of iterative aggregation-disaggregation methods. Linear Algebr. Appl. **413**(2–3), 327–341 (2006)
219. Mattingly, R.B., Meyer, C.D.: Computing the stationary distribution vector of an irreducible Markov chain on a shared-memory multiprocessor. In: Stewart, W.J. (ed.) Numerical Solution of Markov Chains. Probability: Pure and Applied, vol. 8, pp. 491–510. Marcel Dekker, New York (1991)
220. McAllister, D.F., Stewart, G.W., Stewart, W.J.: On a Rayleigh-Ritz refinement technique for nearly uncoupled stochastic matrices. Linear Algebr. Appl. **60**, 1–25 (1984)
221. Meshalkin, L.D.: Limit theorems for Markov chains with a finite number of states. Teor. Veroyatn. Primen. **3**, 361–385 (1958). (English translation in Theory Probab. Appl. **3**, 335–357)
222. Masol, V.I., Silvestrov, D.S.: Record values of the occupation time of a semi-Markov process. Visnik Kiev. Univ. Ser. Mat. Meh. **14**, 81–89 (1972)
223. Motsa, A.I., Silvestrov, D.S.: Asymptotics of extremal statistics and functionals of additive type for Markov chains. In: Klesov, O., Korolyuk, V., Kulldorff, G., Silvestrov, D. (eds.) Proceedings of the First Ukrainian–Scandinavian Conference on Stochastic Dynamical Systems, Uzhgorod, 1995 (1996). (Theory Stoch. Proces., **2(18)**(1–2), 217–224)
224. Meyer, C.D.: Stochastic complementation, uncoupling Markov chains, and the theory of nearly reducible systems. SIAM Rev. **31**(2), 240–272 (1989)
225. Meyer, C.D.: Sensitivity of the stationary distribution of a Markov chain. SIAM J. Matrix Anal. Appl. **15**(3), 715–728 (1994)
226. Meyer, C.D. (2000). *Matrix Analysis and Applied Linear Algebra*. SIAM, Philadelphia, PA, xii+718 pp
227. Meyer, C.D.: Continuity of the Perron root. Linear Multilinear Algebr. **63**(7), 1332–1336 (2015)
228. Meyer Jr., C.D.: The condition of a finite Markov chain and perturbation bounds for the limiting probabilities. SIAM J. Algebr. Discret. Methods **1**(3), 273–283 (1980)

229. Meyer, C.D., Stewart, G.W.: Derivatives and perturbations of eigenvectors. SIAM J. Numer. Anal. **25**(3), 679–691 (1988)
230. Meyn, S.P., Tweedie, R.L.: Markov Chains and Stochastic Stability, xxviii+594 pp. Cambridge University Press, Cambridge (2009). (2nd edition of Markov Chains and Stochastic Stability. Communications and Control Engineering Series, xvi+ 548 pp. Springer, London, 1993)
231. Mitrophanov, A.Y.: Stability and exponential convergence of continuous-time Markov chains. J. Appl. Probab. **40**, 970–979 (2003)
232. Mitrophanov, A.Y.: Sensitivity and convergence of uniformly ergodic Markov chains. J. Appl. Probab. **42**, 1003–1014 (2005)
233. Mitrophanov, A.Y.: Ergodicity coefficient and perturbation bounds for continuous-time Markov chains. Math. Inequal. Appl. **8**(1), 159–168 (2005)
234. Mitrophanov, A.Y.: Stability estimates for finite homogeneous continuous-time Markov chains. Theory Probab. Appl. **50**(2), 319–326 (2006)
235. Mitrophanov, A.Y., Lomsadze, A., Borodovsky, M.: Sensitivity of hidden Markov models. J. Appl. Probab. **42**, 632–642 (2005)
236. Nagaev, S.V.: Some limit theorems for stationary Markov chains. Teor. Veroyatn. Primen. **2**, 389–416 (1957). (English translation in Theory Probab. Appl. **2**, 378–406)
237. Nagaev, S.V.: A refinement of limit theorems for homogeneous Markov chains. Teor. Veroyatn. Primen. **6**, 67–86 (1961). (English translation in Theory Probab. Appl. **6**, 62–81)
238. Ni, Y.: Perturbed renewal equations with multivariate nonpolynomial perturbations. In: Frenkel, I., Gertsbakh, I., Khvatskin, L., Laslo, Z., Lisnianski, A. (eds.) Proceedings of the International Symposium on Stochastic Models in Reliability Engineering, Life Science and Operations Management, pp. 754–763. Beer Sheva, Israel (2010)
239. Ni, Y.: Analytical and numerical studies of perturbed renewal equations with multivariate non-polynomial perturbations. J. Appl. Quant. Methods **5**(3), 498–515 (2010)
240. Ni, Y.: Nonlinearly perturbed renewal equations: asymptotic results and applications. Doctoral Dissertation, **106**, Mälardalen University, Västerås (2011)
241. Ni, Y.: Nonlinearly perturbed renewal equations: the non-polynomial case. Teor. İmovirn. Mat. Stat. **84**, 111–122 (2012). (Also in Theory Probab. Math. Stat. **84**, 117–129)
242. Ni, Y.: Exponential asymptotical expansions for ruin probability in a classical risk process with non-polynomial perturbations. In: Silvestrov, D., Martin-Löf, A. (eds.) Modern Problems in Insurance Mathematics. Chap. 6, EAA series, pp. 67–91. Springer, Cham (2014)
243. Ni, Y., Silvestrov, D., Malyarenko, A.: Exponential asymptotics for nonlinearly perturbed renewal equation with non-polynomial perturbations. J. Numer. Appl. Math. **1**(96), 173–197 (2008)
244. Nåsell, I.: Extinction and Quasi-Stationarity in the Stochastic Logistic SIS Model. Lecture Notes in Mathematics, Mathematical Biosciences Subseries, vol. 2022, xii+199 pp. Springer, Heidelberg (2011)
245. Paige, C.C., Styan, G.P.H., Wachter, P.G.: Computation of the stationary distribution of a Markov chain. J. Stat. Comput. Simul. **4**, 173–186 (1975)
246. Pervozvanskiĭ, A.A., Gaitsgori, V.G.: Theory of Suboptimal Decisions Decomposition and Aggregation. Mathematics and Its Applications (Soviet Series), vol. 12, xviii+384 pp. Kluwer, Dordrecht (1988)
247. Pervozvanskiĭ, A.A., Smirnov, I.N.: An estimate of the steady state of a complex system with slowly varying constraints. Kibernetika (4), 45–51 (1974). (English translation in Cybernetics **10**(4), 603–611)
248. Petersson, M.: Quasi-stationary distributions for perturbed discrete time regenerative processes. Teor. İmovirn. Mat. Stat. **89**, 140–155 (2013). (Also in Theor. Probab. Math. Stat. **89**, 153–168)
249. Petersson, M.: Asymptotics of ruin probabilities for perturbed discrete time risk processes. In: Silvestrov, D., Martin-Löf, A. (eds.) Modern Problems in Insurance Mathematics. Chapter 7, EAA series, pp. 93–110. Springer, Cham (2014)
250. Petersson, M.: Asymptotic expansions for moment functionals of perturbed discrete time semi-Markov processes. In: Silvestrov, S., Rančić, M. (eds.) Engineering Mathematics II.

Algebraic, Stochastic and Analysis Structures for Networks, Data Classification and Optimization. Springer, Heidelberg (2016)

251. Petersson, M.: Asymptotic for quasi-stationary distributions of perturbed discrete time semi-Markov processes. In: Silvestrov, S., Rančić, M. (eds.) Engineering Mathematics II. Algebraic, Stochastic and Analysis Structures for Networks, Data Classification and Optimization. Springer, Heidelberg (2016)

252. Petersson, M.: Perturbed discrete time stochastic models. Doctoral Dissertation, Stockholm University (2016)

253. Phillips, R.G., Kokotović, P.V.: A singular perturbation approach to modeling and control of Markov chains. IEEE Trans. Autom. Control 26, 1087–1094 (1981)

254. Plotkin, J.D., Turbin, A.F.: Inversion of linear operators that are perturbed on the spectrum. Ukr. Mat. Zh. 23, 168–176 (1971)

255. Plotkin, J.D., Turbin, A.F.: Inversion of normally solvable linear operators that are perturbed on the spectrum. Ukr. Mat. Zh. 27(4), 477–486 (1975)

256. Poliščuk, L.I., Turbin, A.F.: Asymptotic expansions for certain characteristics of semi-Markov processes. Teor. Veroyatn. Mat. Stat. 8, 122–127 (1973). (English translation in Theory Probab. Math. Stat. 8, 121–126)

257. Pollett, P.K., Stewart, D.E.: An efficient procedure for computing quasi-stationary distributions of Markov chains with sparse transition structure. Adv. Appl. Probab. 26, 68–79 (1994)

258. Quadrat, J.P.: Optimal control of perturbed Markov chains: the multitime scale case. In: Ardema, M.D. (ed.) Singular Perturbations in Systems and Control. CISM Courses and Lectures, vol. 280, pp. 215–239. Springer, Vienna (1983)

259. Rohlichek, J.R.: Aggregation and time scale analysis of perturbed markov systems. Ph.D. thesis, Massachusetts Inst. Tech., Cambridge, MA (1987)

260. Rohlicek, J.R., Willsky, A.S.: Multiple time scale decomposition of discrete time Markov chains. Syst. Control Lett. 11(4), 309–314 (1988)

261. Rohlicek, J.R., Willsky, A.S.: The reduction of perturbed Markov generators: an algorithm exposing the role of transient states. J. Assoc. Comput. Mach. 35(3), 675–696 (1988)

262. Romanovskiĭ, V.I.: Discrete Markov Chains, 436 pp. Gostehizdat, Moscow-Leningrad (1949)

263. Samoĭlenko, Ĭ.V.: Asymptotic expansion of Markov random evolution. Ukr. Mat. Visn. 3(3), 394–407 (2006). (English translation in Ukr. Math. Bull. 3(3), 381–394)

264. Samoĭlenko, Ĭ.V.: Asymptotic expansion of semi-Markov random evolution. Ukr. Mat. Zh. 58, 1234–1248 (2006). (English translation in Ukr. Math. J. 58, 1396–1414)

265. Schweitzer, P.J.: Perturbation theory and finite Markov chains. J. Appl. Probab. 5, 401–413 (1968)

266. Schweitzer, P.J.: Aggregation methods for large Markov chains. In: Iazeolla, G., Courtois, P.J., Hordijk, A. (eds.) Mathematical Computer Performance and Reliability, pp. 275–286. North-Holland, Amsterdam (1984)

267. Schweitzer, P.J.: Posterior bounds on the equilibrium distribution of a finite Markov chain. Commun. Stat. Stoch. Models 2(3), 323–338 (1986)

268. Schweitzer, P.J.: Dual bounds on the equilibrium distribution of a finite Markov chain. J. Math. Anal. Appl. 126(2), 478–482 (1987)

269. Schweitzer, P.J.: A survey of aggregation-disaggregation in large Markov chains. In: Stewart, W.J. (ed.) Numerical solution of Markov chains. Probability: Pure and Applied, vol. 8, pp. 63–88. Marcel Dekker, New York (1991)

270. Schweitzer, P.J., Kindle, K.W.: Iterative aggregation for solving undiscounted semi-Markovian reward processes. Commun. Stat. Stoch. Models 2(1), 1–41 (1986)

271. Schweitzer, P.J., Puterman, M.L., Kindle, K.W.: Iterative aggregation-disaggregation procedures for discounted semi-Markov reward processes. Oper. Res. 33(3), 589–605 (1985)

272. Schweitzer, P., Stewart, G.W.: The Laurent expansion of pencils that are singular at the origin. Linear Algebr. Appl. 183, 237–254 (1993)

273. Seneta, E.: Finite approximations to infinite non-negative matrices. Proc. Camb. Philos. Soc. 63, 983–992 (1967)

274. Seneta, E.: Finite approximations to infinite non-negative matrices. II. Refinements and applications. Proc. Camb. Philos. Soc. **64**, 465–470 (1968)
275. Seneta, E.: The principle of truncations in applied probability. Comment. Math. Univ. Carolinae **9**, 237–242 (1968)
276. Seneta, E.: Nonnegative Matrices. An Introduction to Theory and Applications, x+214 pp. Wiley, New York (1973)
277. Seneta, E.: Iterative aggregation: convergence rate. Econ. Lett. **14**(4), 357–361 (1984)
278. Seneta, E.: Sensitivity to perturbation of the stationary distribution: some refinements. Linear Algebr. Appl. **108**, 121–126 (1988)
279. Seneta, E.: Perturbation of the stationary distribution measured by ergodicity coefficients. Adv. Appl. Probab. **20**, 228–230 (1988)
280. Seneta, E.: Sensitivity analysis, ergodicity coefficients, and rank-one updates for finite Markov chains. In: Stewart, W.J. (ed.) Numerical Solution of Markov Chains. Probability: Pure and Applied, vol. 8, pp. 121–129. Marcel Dekker, New York (1991)
281. Seneta, E.: Sensitivity of finite Markov chains under perturbation. Stat. Probab. Lett. **17**(2), 163–168 (1993)
282. Seneta, E.: Nonnegative Matrices and Markov chains. Springer Series in Statistics, xvi+287 pp. Springer, New York (2006). (A revised reprint of 2nd edition of Nonnegative Matrices and Markov Chains. Springer Series in Statistics, xiii+279 pp. Springer, New York, 1981)
283. Serlet, L.: Hitting times for the perturbed reflecting random walk. Stoch. Process. Appl. **123**(1), 110–130 (2013)
284. Sheskin, T.J.: A Markov chain partitioning algorithm for computing steady state probabilities. Oper. Res. **33**, 228–235 (1985)
285. Shurenkov, V.M.: Transition phenomena of the renewal theory in asymptotical problems of theory of random processes 1. Mat. Sbornik, **112**, 115–132 (1980). (English translation in Math. USSR: Sbornik, **40**(1), 107–123)
286. Shurenkov, V.M.: Transition phenomena of the renewal theory in asymptotical problems of theory of random processes 2. Mat. Sbornik **112**, 226–241 (1980). (English translation in Math. USSR: Sbornik, **40**(2), 211–225)
287. Silvestrov, D.S.: Limit theorems for a non-recurrent walk connected with a Markov chain. Ukr. Mat. Zh. **21**, 790–804 (1969). (English translation in Ukr. Math. J. **21**, 657–669)
288. Silvestrov, D.S.: Limit theorems for semi-Markov processes and their applications. 1, 2. Teor. Veroyatn. Mat. Stat. **3**, 155–172, 173–194 (1970). (English translation in Theory Probab. Math. Stat. **3**, 159–176, 177–198)
289. Silvestrov, D.S.: Limit theorems for semi-Markov summation schemes. 1. Teor. Veroyatn. Mat. Stat. **4**, 153–170 (1971). (English translation in Theory Probab. Math. Stat. **4**, 141–157)
290. Silvestrov, D.S.: Uniform estimates of the rate of convergencxe for sums of random variables defined on a finite homogeneous Markov chain with absorption. Theor. Veroyatn. Mat. Stat. **5**, 116–127 (1971). (English translation in Theory Probab. Math. Stat. **5**, 123–135)
291. Silvestrov, D.S.: Limit distributions for sums of random variables that are defined on a countable Markov chain with absorption. Dokl. Acad. Nauk Ukr. SSR, Ser. A (4), 337–340 (1972)
292. Silvestrov, D.S.: Estimation of the rate of convergence for sums of random variables that are defined on a countable Markov chain with absorption. Dokl. Acad. Nauk Ukr. SSR, Ser. A **5**, 436–438 (1972)
293. Silvestrov, D.S.: Limit Theorems for Composite Random Functions, 318 pp. Vysshaya Shkola and Izdatel'stvo Kievskogo Universiteta, Kiev (1974)
294. Silvestrov, D.S.: A generalization of the renewal theorem. Dokl. Akad. Nauk Ukr. SSR, Ser. A **11**, 978–982 (1976)
295. Silvestrov, D.S.: The renewal theorem in a series scheme. 1. Teor. Veroyatn. Mat. Stat. **18**, 144–161 (1978). (English translation in Theory Probab. Math. Stat. **18**, 155–172)
296. Silvestrov, D.S.: The renewal theorem in a series scheme 2. Teor. Veroyatn. Mat. Stat. **20**, 97–116 (1979). (English translation in Theory Probab. Math. Stat. **20**, 113–130)
297. Silvestrov, D.S.: A remark on limit distributions for times of attainment for asymptotically recurrent Markov chains. Theory Stoch. Process. **7**, 106–109 (1979)

298. Silvestrov, D.S.: Semi-Markov Processes with a Discrete State Space, 272 pp. Library for the Engineer in Reliability, Sovetskoe Radio, Moscow (1980)
299. Silvestrov, D.S.: Mean hitting times for semi-Markov processes, and queueing networks. Elektron. Infor. Kybern. **16**, 399–415 (1980)
300. Silvestrov, D.S.: Theorems of large deviations type for entry times of a sequence with mixing. Teor. Veroyatn. Mat. Stat. **24**, 129–135 (1981). (English translation in Theory Probab. Math. Stat. **24**, 145–151)
301. Silvestrov, D.S.: Exponential asymptotic for perturbed renewal equations. Teor. Imovirn. Mat. Stat. **52**, 143–153 (1995). (English translation in Theory Probab. Math. Stat. **52**, 153–162)
302. Silvestrov, D.S.: Recurrence relations for generalised hitting times for semi-Markov processes. Ann. Appl. Probab. **6**, 617–649 (1996)
303. Silvestrov, D.S.: Nonlinearly perturbed Markov chains and large deviations for lifetime functionals. In: Limnios, N., Nikulin, M. (eds.) Recent Advances in Reliability Theory: Methodology, Practice and Inference, pp. 135–144. Birkhäuser, Boston (2000)
304. Silvestrov, D.S.: Asymptotic expansions for quasi-stationary distributions of nonlinearly perturbed semi-Markov processes. Theory Stoch. Process. **13**(1–2), 267–271 (2007)
305. Silvestrov D.S.: Nonlinearly perturbed stochastic processes and systems. In: Rykov, V., Balakrishnan, N., Nikulin, M. (eds.) Mathematical and Statistical Models and Methods in Reliability, Chapter 2, pp. 19–38. Birkhäuser (2010)
306. Silvestrov, D.S.: Improved asymptotics for ruin probabilities. In: Silvestrov, D., Martin-Löf, A. (eds.) Modern Problems in Insurance Mathematics, Chap. 5. EAA series, pp. 93–110. Springer, Cham (2014)
307. Silvestrov D.S.: American-Type Options. Stochastic Approximation Methods, Volume 1. De Gruyter Studies in Mathematics, vol. 56, x+509 pp. Walter de Gruyter, Berlin (2014)
308. Silvestrov D.S.: American-Type Options. Stochastic Approximation Methods, Volume 2. De Gruyter Studies in Mathematics, vol. 57, xi+558 pp. Walter de Gruyter, Berlin (2015)
309. Silvestrov D.: Necessary and sufficient conditions for convergence of first-rare-event times for perturbed semi-Markov processes. Research Report 2016:4, Department of Mathematics, Stockholm University, 39 pp. (2016). arXiv:1603.04344
310. Silvestrov, D.S., Abadov, Z.A.: Asymptotic behaviour for exponential moments of sums of random variables defined on exponentially ergodic Markov chains. Dokl. Acad. Nauk Ukr. SSR, Ser. A (4), 23–25 (1984)
311. Silvestrov, D.S., Abadov, Z.A.: Uniform asymptotic expansions for exponential moments of sums of random variables defined on a Markov chain and distributions of passage times. 1. Teor. Veroyatn. Mat. Stat. **45**, 108–127 (1991). (English translation in Theory Probab. Math. Stat. **45**, 105–120)
312. Silvestrov, D.S., Abadov, Z.A.: Uniform representations of exponential moments of sums of random variables defined on a Markov chain and of distributions of passage times. 2. Teor. Veroyatn. Mat. Stat. **48**, 175–183 (1993). (English translation in Theory Probab. Math. Stat. **48**, 125–130)
313. Silvestrov, D.S., Drozdenko, M.O.: Necessary and sufficient conditions for the weak convergence of the first-rare-event times for semi-Markov processes. Dopov. Nac. Akad. Nauk Ukr., Mat. Prirodozn. Tekh Nauki (11), 25–28 (2005)
314. Silvestrov, D.S., Drozdenko, M.O.: Necessary and sufficient conditions for weak convergence of first-rare-event times for semi-Markov processes. Theory Stoch. Process., 12(28), no. 3–4, Part I: 151–186. Part II, 187–202 (2006)
315. Silvestrov, D., Manca, R.: Reward algorithms for semi-Markov processes. Research Report 2015:16, Department of Mathematics, Stockholm University, 23 pp. (2015). arXiv:1603.05693
316. Silvestrov, D., Manca, R.: Reward algorithms for exponential moments of hitting times for semi-Markov processes. Research Report 2016:1, Department of Mathematics, Stockholm University, 23 pp. (2016)
317. Silvestrov, D., Manca, R., Silvestrova, E.: Computational algorithms for moments of accumulated Markov and semi-Markov rewards. Commun. Stat. Theory. Methods **43**(7), 1453–1469 (2014)

318. Silvestrov, D.S., Petersson, M.: Exponential expansions for perturbed discrete time renewal equations. In: Karagrigoriou, A., Lisnianski, A., Kleyner, A., Frenkel, I. (eds.) Applied Reliability Engineering and Risk Analysis. Probabilistic Models and Statistical Inference. Chapter 23, pp. 349–362. Wiley, New York (2013)
319. Silvestrov, D.S., Petersson, M., Hössjer, O.: Nonlinearly perturbed birth-death-type models. Research Report 2016:6, Department of Mathematics, Stockholm University, 63 pp. (2016). arXiv:1604.02295
320. Silvestrov, D., Silvestrov, S.: Asymptotic expansions for stationary distributions of perturbed semi-Markov processes. Research Report 2015:9, Department of Mathematics, Stockholm University, 75 pp. (2015). arXiv:1603.03891
321. Silvestrov, D., Silvestrov, S.: Asymptotic expansions for stationary distributions of nonlinearly perturbed semi-Markov processes. I, II. (2016). Part I: arXiv:1603.04734, 30 pp., Part II: arXiv:1603.04743, 33 pp
322. Silvestrov, D.S., Velikii, Y.A.: Necessary and sufficient conditions for convergence of attainment times. In: Zolotarev, V.M., Kalashnikov, V.V. (eds.) Stability Problems for Stochastic Models. Trudy Seminara, VNIISI, Moscow, 129–137 (1988). (English translation in J. Soviet. Math. **57**, 3317–3324)
323. Simon, H.A., Ando, A.: Aggregation of variables in dynamic systems. Econometrica **29**, 111–138 (1961)
324. Sirl, D., Zhang, H., Pollett, P.: Computable bounds for the decay parameter of a birth-death process. J. Appl. Probab. **44**(2), 476–491 (2007)
325. Stewart, G.W.: On the continuity of the generalized inverse. SIAM J. Appl. Math. **17**, 33–45 (1969)
326. Stewart, G.W.: Error and perturbation bounds for subspaces associated with certain eigenvalue problems. SIAM Rev. **15**, 727–764 (1973)
327. Stewart, G.W.: Perturbation bounds for the definite generalized eigenvalue problem. Linear Algebr. Appl. **23**, 69 85 (1979)
328. Stewart, G.W.: Computable error bounds for aggregated Markov chains. J. Assoc. Comput. Mach. **30**(2), 271–285 (1983)
329. Stewart, G.W.: On the structure of nearly uncoupled Markov chains. In: Iazeolla, G., Courtois, P.J., Hordijk, A. (eds.) Mathematical Computer Performance and Reliability, pp. 287–302. North-Holland, Amsterdam (1984)
330. Stewart, G.W.: A second order perturbation expansion for small singular values. Linear Algebr. Appl. **56**, 231–235 (1984)
331. Stewart, G.W.: Stochastic perturbation theory. SIAM Rev. **32**(4), 579–610 (1990)
332. Stewart, G.W.: On the sensitivity of nearly uncoupled Markov chains. In: Stewart, W.J. (ed.) Numerical Solution of Markov Chains. Probability: Pure and Applied, vol. 8, pp. 105–119. Marcel Dekker, New York (1991)
333. Stewart, G.W.: Gaussian elimination, perturbation theory, and Markov chains. In: Meyer, C.D., Plemmons, R.J. (eds). Linear Algebra, Markov Chains, and Queueing Models. IMA Volumes in Mathematics and its Applications, vol. 48, pp. 59–69. Springer, New York (1993)
334. Stewart, G.W.: On the perturbation of Markov chains with nearly transient states. Numer. Math. **65**(1), 135–141 (1993)
335. Stewart, G.W.: Matrix Algorithms. Vol. I. Basic Decompositions, xx+458 pp. SIAM, Philadelphia, PA (1998)
336. Stewart, G.W.: Matrix Algorithms. Vol. II. Eigensystems, xx+469 pp. SIAM, Philadelphia, PA (2001)
337. Stewart, G.W.: On the powers of a matrix with perturbations. Numer. Math. **96**(2), 363–376 (2003)
338. Stewart, G.W., Stewart, W.J., McAllister, D.F.: A two-stage iteration for solving nearly completely decomposable Markov chains. In: Golub, G., Greenbaum, A., Luskin, M. (eds.) Recent Advances in Iterative Methods. IMA Volumes in Mathematics and its Applications, vol. 60, pp. 201–216. Springer, New York (1994)

339. Stewart, G.W., Sun, J.G.: Matrix Perturbation Theory. Computer Science and Scientific Computing, xvi+365 pp. Academic Press, Boston (1990)

340. Stewart, G.W., Zhang, G.: On a direct method for the solution of nearly uncoupled Markov chains. Numer. Math. **59**(1), 1–11 (1991)

341. Stewart, W.J.: Introduction to the Numerical Solution of Markov Chains, xx+539 pp. Princeton University Press, Princeton, NJ (1994)

342. Sumita, U., Reiders, M.: A new algorithm for computing the ergodic probability vector for large Markov chains: Replacement process approach. Probab. Eng. Inf. Sci. **4**, 89–116 (1988)

343. Torokhti, A., Howlett, P., Pearce, C.: Method of best successive approximations for nonlinear operators. J. Comput. Anal. Appl. **5**(3), 299–312 (2003)

344. Turbin, A.F.: On asymptotic behavior of time of a semi-Markov process being in a reducible set of states. Linear case. Teor. Veroyatn. Mat. Stat. **4**, 179–194 (1971). (English translation in Theory Probab. Math. Stat. **4**, 167–182)

345. Turbin, A.F.: An application of the theory of perturbations of linear operators to the solution of certain problems that are connected with Markov chains and semi-Markov processes. Teor. Veroyatn. Mat. Stat. **6**, 118–128 (1972). (English translation in Theory Probab. Math. Stat. **6**, 119–130)

346. Van Doorn, E.A., Pollett, P.K.: Quasi-stationary distributions for discrete-state models. Eur. J. Oper. Res. **230**, 1–14 (2013)

347. Vantilborgh, H.: Aggregation with an error of $O(\varepsilon^2)$. J. Assoc. Comput. Mach. **32**(1), 162–190 (1985)

348. Verhulst, F.: Methods and Applications of Singular Perturbations: Boundary Layers and Multiple Timescale Dynamics. Texts in Applied Mathematics, vol. 50, xvi+324 pp. Springer, New York (2005)

349. Vishik, M.I., Lyusternik, L.A.: The solution of some perturbation problems in the case of matrices and self-adjoint and non-self-adjoint differential equations. Uspehi Mat. Nauk **15**, 3–80 (1960)

350. Wentzell, A.D.: Asymptotic expansions in limit theorems for stochastic processes I. Probab. Theory Relat. Fields **106**(3), 331–350 (1996)

351. Wentzell, A.D.: Asymptotic expansions in limit theorems for stochastic processes II. Probab. Theory Relat. Fields **113**(2), 255–271 (1999)

352. Wentzell, A.D., Freidlin, M.I. (1979). *Fluctuations in Dynamical Systems Subject to Small Random Perturbations*. Probability Theory and Mathematical Statistics, Nauka, Moscow, 424 pp. (English edition: *Random Perturbations of Dynamical Systems*. Fundamental Principles of Mathematical Sciences, **260**, Springer, New York (1998, 2012), xxviii+458 pp)

353. Wilkinson, J.H.: Error analysis of direct method of matrix inversion. J. Assoc. Comput. Mach. **8**, 281–330 (1961)

354. Yin, G.G., Zhang, Q.: Continuous-time Markov Chains and Applications. A Singular Perturbation Approach. Applications of Mathematics, vol. 37, ivx+349 pp. Springer, New York (1998)

355. Yin, G., Zhang, Q.: Discrete-time singularly perturbed Markov chains. In: Yao, D.D., Zhang, H., Zhou, X.Y. (eds.) Stochastic Modeling and Optimization, pp. 1–42. Springer, New York (2003)

356. Yin, G.G., Zhang, Q.: Discrete-time Markov chains. Two-time-scale methods and applications. Stochastic Modelling and Applied Probability, xix+348 pp. Springer, New York (2005)

357. Yin, G.G., Zhang, Q.: Continuous-Time Markov Chains and Applications. A Two-Time-Scale Approach. Stochastic Modelling and Applied Probability, vol. 37, xxii+427 pp. Springer, New York (2013). (2nd revised edition of Continuous-Time Markov Chains and Applications. A Singular Perturbation Approach. Applications of Mathematics, vol. 37, xvi+349 pp. Springer, New York, 1998)

358. Yin, G., Zhang, Q., Badowski, G.: Discrete-time singularly perturbed Markov chains: aggregation, occupation measures, and switching diffusion limit. Adv. Appl. Probab. **35**(2), 449–476 (2003)

359. Yin, G., Zhang, Q., Yang, H., Yin, K.: Discrete-time dynamic systems arising from singularly perturbed Markov chains. In: Proceedings of the Third World Congress of Nonlinear Analysts, Part 7, Catania, 2000 (2001). (Nonlinear Anal., **47**(7), 4763–4774)
360. Zakusilo, O.K.: Thinning semi-Markov processes. Teor. Veroyatn. Mat. Stat. **6**, 54–59 (1972). (English translation in Theory Probab. Math. Stat. **6**, 53–58)
361. Zakusilo, O.K.: Necessary conditions for convergence of semi-Markov processes that thin. Teor. Veroyatn. Mat. Stat. **7**, 65–69 (1972). (English translation in Theory Probab. Math. Stat. **7**, 63–66)
362. Zhang, Q., Yin, G.: Exponential bounds for discrete-time singularly perturbed Markov chains. J. Math. Anal. Appl. **293**(2), 645–662 (2004)

PageRank, a Look at Small Changes
in a Line of Nodes and the Complete Graph

Christopher Engström and Sergei Silvestrov

Abstract In this article we will look at the PageRank algorithm used as part of the ranking process of different Internet pages in search engines by for example Google. This article has its main focus in the understanding of the behavior of PageRank as the system dynamically changes either by contracting or expanding such as when adding or subtracting nodes or links or groups of nodes or links. In particular we will take a look at link structures consisting of a line of nodes or a complete graph where every node links to all others. We will look at PageRank as the solution of a linear system of equations and do our examination in both the ordinary normalized version of PageRank as well as the non-normalized version found by solving corresponding linear system. We will show that using two different methods we can find explicit formulas for the PageRank of some simple link structures.

Keywords PageRank · Graph · Random walk · Block matrix

1 Introduction

PageRank is a method in which we can rank nodes in different link structures such as Internet pages on the Web in order of "importance" given the link structure of the complete system. It is important that the method is extremely fast since there is a huge number of Internet pages. It is also important that the algorithm returns the most relevant results first since very few people will look through more than a couple of pages when doing a search in a search engine, [6].

While PageRank was originally constructed for use in search engines, there are other uses of PageRank or similar methods, for example in the EigenTrust

C. Engström (✉) · S. Silvestrov
Division of Applied Mathematics, School of Education,
Culture and Communication, Mälardalen University, Box 883, 721 23 Västerås, Sweden
e-mail: christopher.engstrom@mdh.se

S. Silvestrov
e-mail: sergei.silvestrov@mdh.se

© Springer International Publishing Switzerland 2016
S. Silvestrov and M. Rančić (eds.), *Engineering Mathematics II*,
Springer Proceedings in Mathematics & Statistics 179,
DOI 10.1007/978-3-319-42105-6_11

algorithm for reputation management to decrease distribution of unauthentic files in P2P networks, [14].

Calculating PageRank is usually done using the Power method which can be implemented very efficiently, even for very large systems. The convergence speed of the Power method and it's dependence on certain parameters have been studied to some extent. For example the Power method on a graph structure such as that created by the Web will converge with a convergence rate of c, where c is one of the parameters used in the definition [11], and the problem is well conditioned unless c is very close to 1 [13]. However since the number of pages on the Web is huge, extensive work has been done in trying to improve the computation time of PageRank even further. One example is by aggregating webpages that are "close" and are expected to have a similar PageRank as in [12]. Another method used to speed up calculations is found in [18] where they do not compute the PageRank of pages that have already converged in every iteration. Other methods to speed up calculations include removing "dangling nodes" before computing PageRank and then calculate them at the end or explore other methods such as using a power series formulation of PageRank [2].

There are also work done on the large scale using PageRank and other measures in order to learn more about the Web, for example looking at the distribution of PageRank both theoretically and experimentally such as in [8].

While the theory behind PageRank is well understood from Perron–Frobenius theory for non-negative irreducible matrices [3, 10, 15] and the study of Markov chains [16, 17], how PageRank is affected from changes in the the system or parameters is not as well known.

In this article we start by giving a short introduction on PageRank and some notation and definitions used throughout the article. We will look at PageRank as the solution to a linear system of equations and what we can learn using this representation. Looking at some common graph structures we want to gain a better understanding of the changes in PageRank as the graph structure changes. This could for example be used in finding good approximations of PageRank of certain structures in order to speed up calculations further.

We will look at both the "ordinary" normalized version of PageRank as well as a non-normalized version we get by solving the linear system. We will see how this non-normalized version corresponds to the probabilities of a random walk through the graph and how we can use this to find the PageRank of some systems using this perspective rather than solving the system or computing the dominant eigenvector.

Mainly two different structures, first a simple line in Sect. 5 and later a complete graph in Sect. 6 will be examined. In both cases we will see that we can find explicit expressions for the PageRank depending on the number of nodes. In both cases of the "ordinary" PageRank as well as a non-normalized version expressions for the PageRank will be found for both the structure itself as well as the PageRank after doing some simple modifications.

2   Calculating PageRank

Starting with a number of nodes (Internet pages) and the non-negative matrix A with every element $a_{ij} \neq 0$ corresponding to a link from node i to node j. The value of element $a_{ij} = 1/n$ where n is the number of outgoing links from node i. An example of a graph and corresponding matrix can be seen in Fig. 1.

By convention we do not allow any loops (nodes linking to themselves). We also need that no nodes have zero outgoing links (dangling nodes) resulting in a row with all zeros. For now we assume that none of these dangling nodes are present in the link matrix. This means that every row will sum to one in the link matrix A.

The PageRank vector $\mathbf{R}$ we want for ranking the nodes (pages) is the eigenvector corresponding to the dominant eigenvalue with value one of matrix M:

$$\mathsf{M} = c\mathsf{A}^\top + (1 - c)\mathbf{u}\mathbf{e}^\top,$$

where $0 < c < 1$, usually $c \approx 0.85$, A is the link matrix, $\mathbf{e}$ is a column vector of the same length as the number of nodes (n) filled with ones and $\mathbf{u}$ is a column vector of the same length with elements u_i, $0 \leq u_i \leq 1$ such that $||\mathbf{u}||_1 = 1$. For $\mathbf{u}$ we will usually use the uniform vector (all elements equal) with $u_i = 1/n$ where n is the number of nodes. The result after calculating the PageRank of the example matrix for the system in Fig. 1 can be seen below:

$$\mathbf{R} \approx \begin{bmatrix} 0.3328 \\ 0.3763 \\ 0.1974 \\ 0.0934 \end{bmatrix}.$$

This can be seen as a random walk where we start in a random node depending on the weight vector $\mathbf{u}$. Then with a probability c we go to any of the nodes linked to from that node and with a probability $1 - c$ we instead go to a random (in the case of uniform $\mathbf{u}$) new node. The PageRank vector can be seen as the probability that you after a long time is located in the node in question [2]. More on why an eigenvector with eigenvalue 1 always exists can be seen in for example [7].

Fig. 1 Directed graph and corresponding matrix system matrix A

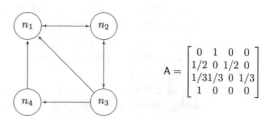

$$A = \begin{bmatrix} 0 & 1 & 0 & 0 \\ 1/2 & 0 & 1/2 & 0 \\ 1/3 & 1/3 & 0 & 1/3 \\ 1 & 0 & 0 & 0 \end{bmatrix}$$

Role of c.

Looking at the formula it is not immediately obvious why we demand $0 < c < 1$ and what role c holds. We can easily see what happens at the limits, if $c = 0$ the PageRank is decided only by the initial weights $\mathbf{u}$. However if $c = 1$ the weights have no role and the algorithm used for calculating PageRank might not even converge. As c increases, nodes further and further away have an impact on the PageRank of individual nodes. And the opposite for low c, the lower c is the more important is the immediate surrounding of a node in deciding its PageRank. The parameter c is also a very important factor in how fast the algorithms used to calculate PageRank converges, the higher c is the slower the algorithm will converge.

Handling of dangling nodes.

If $\mathbf{A}$ contains dangling nodes, corresponding row no longer sums to one and there therefor will probably not be any eigenvector with eigenvalue equal to one. The method we use in order to fix this is to instead assume that the dangling nodes link to all nodes equally (or some other distribution over the nodes). This gives us: $\mathbf{T} = \mathbf{A} + \mathbf{g}\mathbf{w}^{\top}$, where $\mathbf{g}$ is a column vector with elements equal to one for a dangling node and zero for all other nodes. Here $\mathbf{w}$ is the distribution according to how we make the dangling nodes link to other nodes (usually uniform or equal to $\mathbf{u}$). In this work we always use $\mathbf{w} = \mathbf{u}$ to simplify calculations.

There are other ways to handle dangling nodes, for example by adding one new node linking only to itself and let all dangling nodes link to this node. Assuming $\mathbf{w} = \mathbf{u}$ these methods should be essentially the same apart from implementation [5].

3   Notation and Definitions

Here we give some notes on the notation used through the rest of the article in order to clarify which variation of PageRank is used as well as some overall notation and the definition of some common important link structures. We will repeatedly use the L^1 norm in comparing the size of different vectors or (parts of) matrices.

First some overall notation:

- S_G: The system of nodes and links for which we want to calculate PageRank, contains the system matrix $\mathbf{A}_G$ as well as a weight vector $\mathbf{v}_G$. Subindex G can be either a capital letter or a number in the case of multiple systems.
- n_G: The number of nodes in system S_G.
- $\mathbf{A}_G$: System matrix where a zero element a_{ij} means there is no link from node i to node j. Non-zero elements are equal to $1/r_i$ where r_i is the number of links from node i. Size $n_G \times n_G$.
- $\mathbf{v}_G$: Non-negative weight vector, not necessary with sum one. Size $n_G \times 1$.
- $\mathbf{u}_G$: The weight vector $\mathbf{v}_G$ normalized such that $\|\mathbf{u}_G\|_1 = 1$. We note that $\mathbf{u}_G$ is proportional to $\mathbf{v}_G$ ($\mathbf{u}_G \propto \mathbf{v}_G$). Size $n_G \times 1$.
- c: Parameter $0 < c < 1$ for calculating PageRank, usually $c = 0.85$.
- $\mathbf{g}_G$: Vector with elements equal to one for dangling nodes and zero for all other in S_G. Size $n_G \times 1$.

- M_G: Modified system matrix, $\mathsf{M}_G = c(\mathsf{A}_G + \mathbf{g}_G\mathbf{u}_G^\top)^\top + (1-c)\mathbf{u}_G\mathbf{e}^\top$ used to calculate PageRank, where $\mathbf{e}$ is the unit vector. Size $n_G \times n_G$.

In the cases where there is only one possible system the subindex G will often be omitted.

From earlier we saw how we could calculate PageRank for a system S, we also make the assumption that $\mathbf{w} = \mathbf{u}$ both since it simplifies calculations as well as having no obvious disadvantages since both vectors play largely the same role in that they can be used to penalize or promote certain certain nodes.

We will use two different ways to define different versions of PageRank using the notation $\mathbf{R}_G^{(t)}$ where t is the type of PageRank used, G is the graph or part of graph for which $\mathbf{R}$ is the PageRank. Often G is the whole graph in which case the subindex is usually omitted $\mathbf{R}^{(t)}$.

We will sometimes give the formula for a specific node j in this case it will be noted as $\mathbf{R}_j^{(t)}$. When normalizing the resulting elements such that their sum equal to one we get the traditional PageRank:

Defnition 1 $\mathbf{R}_G^{(1)}$ for system S_G is defined as the eigenvector with eigenvalue one to the matrix $\mathsf{M}_G = c(\mathsf{A}_G + \mathbf{g}_G\mathbf{u}_G^\top)^\top + (1-c)\mathbf{u}_G\mathbf{e}^\top$.

Note that we always have $||\mathbf{R}^{(1)}||_1 = 1$ and that non-zero elements in $\mathbf{R}_G^{(1)}$ are all positive. The fact that $||\mathbf{R}^{(1)}||_1 = 1$ is generally not the case in other versions of PageRank. When instead setting up the resulting equation system and solving it we get the second definition, the result is multiplied with n_G in order to get multiplication with the one vector in case of uniform $\mathbf{u}_G$.

Defnition 2 $\mathbf{R}_G^{(2)}$ for system S_G is defined as $\mathbf{R}_G^{(2)} = (\mathsf{I} - c\mathsf{A}_G^\top)^{-1}n_G\mathbf{u}_G$

We note that generally $||\mathbf{R}^{(2)}||_1 \neq 1$ as well as $\mathbf{R}_G^{(2)} \neq n_G\mathbf{R}_G^{(1)}$ unless there are no dangling nodes in the system. However the two versions of PageRank are proportional to each other ($\mathbf{R}_G^{(2)} \propto \mathbf{R}_G^{(1)}$).

Defnition 3 A simple line is a graph with n_L nodes where node n_L links to node n_{L-1} which in turn links to node n_{L-2} all the way until node n_2 link to node n_1.

The link matrix A_L and graph for system S_L consisting of a simple line with 5 nodes can be seen in Fig. 2.

Defnition 4 A complete graph is a group of nodes in which all nodes in the group links to all other nodes in the group.

The link matrix A_G for system S_G consisting of a complete graph with 5 nodes can be seen in Fig. 3.

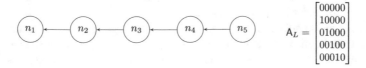

$$\mathsf{A}_L = \begin{bmatrix} 0&0&0&0&0 \\ 1&0&0&0&0 \\ 0&1&0&0&0 \\ 0&0&1&0&0 \\ 0&0&0&1&0 \end{bmatrix}$$

Fig. 2 The simple line with 5 nodes and corresponding system matrix

Fig. 3 A complete graph
with five nodes and
corresponding system matrix

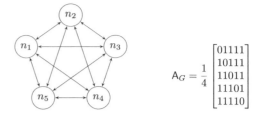

$$A_G = \frac{1}{4} \begin{bmatrix} 01111 \\ 10111 \\ 11011 \\ 11101 \\ 11110 \end{bmatrix}$$

4 Calculating Non-normalized PageRank

While ordinary normalized PageRank $\mathbf{R}^{(1)}$ is usually calculated using the Power
method or some other similar iterative method, in order to find nice analytic forms
using non-normalized PageRank we will use a number of different ways to calculate
it. From now we will assume uniform $\mathbf{u}$ which simplifies calculations significantly.

In this article we will look at two methods two calculate PageRank ($\mathbf{R}^{(2)}$), while
neither method is especially useful for calculating PageRank of large systems, they
give exact answers as compared to the usual iterative methods. The goal is to use
these in order to learn something about the behavior of some common typical graph
structures within a system. From earlier we have:

$$\mathbf{R}^{(1)} = \mathbf{M}\mathbf{R}^{(1)} = (c(\mathbf{A} + \mathbf{g}\mathbf{u}^\top)^\top + (1 - c)\mathbf{u}\mathbf{e}^\top)\mathbf{R}^{(1)}. \tag{1}$$

Calculating the dominant eigenvector $\mathbf{R}^{(1)}$ is the same as solving the linear system:

$$\mathbf{R}^{(1)} = \mathbf{M}\mathbf{R}^{(1)} \Leftrightarrow (c\mathbf{A}^\top - \mathsf{I})\mathbf{R}^{(1)} = -(c\mathbf{u}\mathbf{g}^\top + (1 - c)\mathbf{u}\mathbf{e}^\top)\mathbf{R}^{(1)}. \tag{2}$$

Since every column of $\mathbf{u}\mathbf{g}^\top$ is either equal to $\mathbf{u}$ or zero and all columns equal to $\mathbf{u}$
for $\mathbf{u}\mathbf{e}^\top$ we can see that $-(c\mathbf{u}\mathbf{g}^\top + (1 - c)\mathbf{u}\mathbf{e}^\top)\mathbf{R}^{(1)}$ will be proportional to $\mathbf{u}$. This
can be written as: $(c\mathbf{A}^\top - \mathsf{I})\mathbf{R}^{(1)} = k\mathbf{u}$.

We choose $k = -n$ in order to get $k\mathbf{u}$ equal to the one vector in the case of uniform
$\mathbf{u}$, the minus sign to get positive rank and solving the system we get:

$$\mathbf{R}^{(2)} = (\mathsf{I} - c\mathbf{A}^\top)^{-1}n\mathbf{u}. \tag{3}$$

To get the rank to sum to one it is a simple matter of normalizing the result.
$\mathbf{R}^{(1)} = \mathbf{R}^{(2)}/\|\mathbf{R}^{(2)}\|_1$ [5]. We note the similarity with this formulation of PageRank
(solution to $\mathbf{R}^{(2)} = c\mathbf{A}^T\mathbf{R}^{(2)} + n\mathbf{u}$) with the one for the potential of a Markov chain
with a discounted cost (solution to $\mathbf{R}^{(2)} = \alpha\mathbf{A}\mathbf{R}^{(2)} + c$), where $0 < \alpha < 1$ is the
discount factor and c is a cost vector, [16].

Note that we do not need to take any care of the dangling nodes when calculating
the PageRank in this way but obviously a lot slower than using the Power method or
other conventional methods of calculating PageRank since we need to invert a large
sparse matrix. Although we do not need to change $\mathbf{A}$ for dangling nodes, the result

when doing so is changed (but still proportional to $\mathbf{R}^{(1)}$). We will never change $\mathbf{A}$ for dangling nodes when solving the linear system and only use the version defined above. Note that while solving the equation system is slow it is possible to get to this or similar non-normalized version of PageRank using another PageRank algorithm, such as using a power series formulation as in [1] or by first calculating the ordinary normalized PageRank and then scale it appropriately [9].

The following theorem explains how PageRank ($\mathbf{R}^{(2)}$) can be computed and how it can be interpret from a probabilistic viewpoint using random walks on a graph and the hitting probabilities of said random walks.

Theorem 1 *Consider a random walk on a graph described by $c\mathbf{A}$ described as before. We walk to a new node with probability c and stop with probability $1 - c$. PageRank $\mathbf{R}^{(2)}$ of a node when using uniform $\mathbf{u}$ can be written:*

$$\mathbf{R}_j^{(2)} = \left(\sum_{e_i \in S, e_i \neq e_j} P(e_i \to e_j) + 1 \right) \left(\sum_{k=0}^{\infty} (P(e_j \to e_j))^k \right), \qquad (4)$$

where $P(e_i \to e_j)$ is the probability to hit node e_j in a random walk starting in node e_i described as above. This can be seen as the expected number of visits to e_j if we do multiple random walks, starting in every node once.

Proof $(c\mathbf{A}^\top)_{ij}^k$ is the probability to be in node e_i starting in node e_j after k steps. Multiplying with the unit vector $\mathbf{e}$ (vector with all elements equal to one) therefor gives the sum of all the probabilities to be in node e_i after k steps starting in every node once. The expected total number of visits is the sum of all probabilities to be in node e_i for every step starting in every node:

$$\mathbf{R}_j^{(2)} = \left(\left(\sum_{k=0}^{\infty} (c\mathbf{A}^\top)^k \right) \mathbf{e} \right)_j . \qquad (5)$$

$\sum_{k=0}^{\infty} (c\mathbf{A}^\top)^k$ is the Neumann series of $(\mathbf{I} - c\mathbf{A}^\top)^{-1}$ which is guaranteed to converge since $c\mathbf{A}^\top$ is non-negative and have column sum < 1. If $\mathbf{u}$ is uniform we get by the definition:

$$\mathbf{R}^{(2)} = (\mathbf{I} - c\mathbf{A}^\top)^{-1} n\mathbf{u} = (\mathbf{I} - c\mathbf{A}^\top)^{-1}\mathbf{e} = \left(\sum_{k=0}^{\infty} (c\mathbf{A}^\top)^k \right) \mathbf{e} \qquad (6)$$

$$\Rightarrow \mathbf{R}_j^{(2)} = \left(\sum_{e_i \in S, e_i \neq e_j} P(e_i \to e_j) + 1 \right) \left(\sum_{k=0}^{\infty} (P(e_j \to e_j))^k \right). \qquad (7)$$

$\square$

5 Changes in the Simple Line

Using the simple line as defined earlier we recall that we had the link matrix with an image of the system in Fig. 2

$$A = \begin{bmatrix} 0\,0\,0\,0\,0 \\ 1\,0\,0\,0\,0 \\ 0\,1\,0\,0\,0 \\ 0\,0\,1\,0\,0 \\ 0\,0\,0\,1\,0 \end{bmatrix}.$$

By setting up the system of equations we get the inverse $(I - cA^\top)^{-1}$ as:

$$(I - cA^\top)^{-1} = \begin{bmatrix} 1 & c & c^2 & c^3 & c^4 \\ 0 & 1 & c & c^2 & c^3 \\ 0 & 0 & 1 & c & c^2 \\ 0 & 0 & 0 & 1 & c \\ 0 & 0 & 0 & 0 & 1 \end{bmatrix}.$$

Note that this needs only to be multiplied with $n\mathbf{u}$ or a multiple of $\mathbf{u}$ for us to get a meaningful ranking. This gives us $\mathbf{R}^{(2)}$ (for uniform $\mathbf{u}$):

$$\mathbf{R}^{(2)} = [1 + c + c^2 + c^3 + c^4, 1 + c + c^2 + c^3, 1 + c + c^2, 1 + c, 1]^\top.$$

If wanted to get the common normalized ranking $\mathbf{R}^{(1)}$ we need to normalize the result to sum to one. Looking at the elements a_{ij} of $(I - cA^\top)^{-1}$ and considering the example with a random walk through the graph, we can see the value of every element a_{ij} as the probability to get from node e_j to node e_j. In the case where the link matrix contain nodes with paths back to itself we will later see that it is actually not the probability to get there but the sum of all probabilities to get from e_j to e_i corresponding to Theorem 1. We can motivate this further by looking at the same line but adding a link back from the first node to the second node.

5.1 The Simple Line with Node One Linking to Node Two

Letting node one link to node two in the earlier example gives us the graph in Fig. 4. The resulting inverse can be written as

$$(I - cA^\top)^{-1} = \begin{bmatrix} s & sc & sc^2 & sc^3 & sc^4 \\ sc & s & sc & sc^2 & sc^3 \\ 0 & 0 & 1 & c & c^2 \\ 0 & 0 & 0 & 1 & c \\ 0 & 0 & 0 & 0 & 1 \end{bmatrix},$$

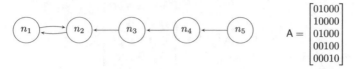

Fig. 4 Simple line where the first node links to the second and corresponding system matrix

where $s = \sum_{k=0}^{\infty} c^{2k} = \frac{1}{1-c^2}$ is the sum of all the probabilities of getting from node 1 or 2 back to itself.

From this we can see that the following observations seem to be true.

- The sum of a column c_j is at most $\sum_{k=0}^{\infty} c^k = \frac{1}{1-c}$ when using uniform **u**, with equality if there are no paths to any dangling node from node j and node j is not a dangling node itself.
- A diagonal element is equal to one if the node have no paths leading back to itself.
- Setting one element in u_i to zero only effects the influence of a random walk starting in the corresponding node.
- Every non zero element in the same row can be written as the diagonal element on the same line times the sum of probabilities of getting from all other nodes to the node corresponding to the current line.
- Each element e_{ij} of $(I - cA^{\top})^{-1}$ contains the sum of probabilities of all paths starting in node j and ending in node i. When doing a random walk by choosing a random link with probability c and stopping with probability $1 - c$.

This is consistent with the statement that the normalized PageRank $\mathbf{R}_j^{(1)}$ of a node is the probability that a surfer that starts in a random node (page) and keeps clicking links with probability c or starts at a new random page with probability (1-c) is in a given node. However here we can explicitly see all the probabilities and their influence on the ranking, [7].

5.2 Removing a Link Between Two Nodes

When removing a link between two nodes in the simple line we end up with two smaller disjoint lines instead. We note that these could be calculated separately and we would still have the same relation between them. This is interesting since when using the "Power method" or straight calculating $\mathbf{R}^{(1)}$ this is not possible since more nodes in a system obviously means a lower mean rank since we in that case normalize the result to one.

Especially in the inverse $(I - cA^{\top})^{-1}$ we see that when we remove one link, we remove all the elements in the upper right corresponding to paths from nodes above the removed link to all the ones below it. An example of what the new inverse looks like when removing the link between the third node and the second node in Fig. 2 can be seen below:

$$(I - c\mathbf{A}^\top)^{-1} = \begin{bmatrix} 1 & c & 0 & 0 & 0 \\ 0 & 1 & 0 & 0 & 0 \\ 0 & 0 & 1 & c & c^2 \\ 0 & 0 & 0 & 1 & c \\ 0 & 0 & 0 & 0 & 1 \end{bmatrix},$$

with PageRank: $\mathbf{R}^{(2)} = [1 + c, 1, 1 + c + c^2, 1 + c, 1]$ and normalizing constant $N = 5 + 3c + c^2$, when using a uniform $\mathbf{u}$.

5.3 Adding a New Node Pointing at One Node in the Simple Line

A more interesting example is when looking at what happens when we add a single new node, linking to one other node in the simple line. Since we make no changes in the line that part of the inverse will stay the same. We will however add a new row and column. The non diagonal element of the new column can be found immediately as c times the column corresponding to the node our new node links to. This since we got the probability c to get to that node instead of 1 when we start in it. At last we need to add the one at the new element in the diagonal. An example of what the inverse looks like after adding a new node pointing at node 3 in Fig. 2 can be seen below

$$(I - c\mathbf{A}^\top)^{-1} = \begin{bmatrix} 1 & c & c^2 & c^3 & c^4 & c^3 \\ 0 & 1 & c & c^2 & c^3 & c^2 \\ 0 & 0 & 1 & c & c^2 & c \\ 0 & 0 & 0 & 1 & c & 0 \\ 0 & 0 & 0 & 0 & 1 & 0 \\ 0 & 0 & 0 & 0 & 0 & 1 \end{bmatrix}.$$

From this easy example we can immediately get an expression for the PageRank of a simple line with one or more added nodes linking to any of the nodes in the simple line.

Theorem 2 *The PageRank of a node e_i belonging to the line in a system containing a simple line with one outside node linking to one of the nodes in the line when using uniform weight vector $\mathbf{u}$ can be written:*

$$\mathbf{R}_i^{(2)} = \sum_{k=0}^{n_L - i} c^k + b_{ij} = \frac{1 - c^{n_L - i + 1}}{1 - c} + b_{ij} \tag{8}$$

$$b_{ij} = \begin{cases} c^{j+1-i}, & j \geq i \\ 0, & j < i \end{cases} \tag{9}$$

where n_L is the number of nodes in the line and the new node link to node j. The new node has rank 1. After normalization we get the PageRank of node i as:

$$\mathbf{R}_i^{(1)} = \frac{\frac{1-c^{n_L-i+1}}{1-c} + b_{ij}}{n_L + 1 + (n_L - 1)c + (n_L - 2)c^2 + \cdots + c^{n_L-1} + \frac{1-c^j}{1-c}} \quad (10)$$

where $\mathbf{R}_i^{(1)}$, $\mathbf{R}_i^{(2)}$ is the PageRank of one of the nodes in the original line, L is the number of nodes on the line, j is the number of the node linked to be the new node.

Additionally adding new nodes linking to the line means adding additional b_{ij} parts and adding the corresponding part $\frac{1-c^j}{1-c}$ to the normalizing constant.

Proof From Theorem 1 PageRank for a node when using uniform $\mathbf{u}$ can be written as:

$$\mathbf{R}_i^{(2)} = \left(\sum_{e_j \in S, e_j \neq e_i} P(e_j \to e_i) + 1 \right) \left(\sum_{k=0}^{\infty} (P(e_i \to e_i))^k \right),$$

where $P(e_j \to e_i)$ is the probability to hit node e_i starting in node e_j. When we consider a random walk on a graph given by $c\mathbf{A}$ described as before. We walk to a new node with probability c and stop with probability $1 - c$.

The probability of getting to any node e_i in the line from any other node e_j in the line once is:

$$P(e_j \to e_i) = c^{j-i}, \quad j > i, \quad (11)$$

and zero otherwise. Summation over all $j > i$ gives

$$\sum_{e_j \in S, e_j \neq e_i} P(e_j \to e_i) + 1 = \sum_{k=1}^{n_L-i} c^k + 1 = \frac{1 - c^{n_L-i+1}}{1 - c}, \quad (12)$$

where L is the number of nodes in the line. With the first part shown we need to show that the single outside node linking to node e_j adds $b_{ij} = c^{j+1-i}$, $j \geq i$. We get this probability in the same way by instead looking at the line created by the first j nodes plus the extra node added linking to node j. We get the probability to reach node e_j as c and then c^2 for the next and so on. If $i > j$, e_i does not belong to this line, and we obviously cannot reach it from e_j hence $b_{ij} = 0$, $i > j$.

Last the PageRank of the "outside" node linking to a node in the line is obviously 1 since no node links to it. The normalized PageRank is found by dividing $\mathbf{R}^{(2)}$ with $\|\mathbf{R}^{(2)}\|_1$. $\qquad\square$

We also give a proof using matrices but first we will need the following lemma for blockwise inversion used repeatedly throughout the article. We note that we label the blocks from $\mathbf{B}$ to $\mathbf{E}$ rather than from $\mathbf{A}$ to $\mathbf{D}$ in order to avoid confusion with the system matrix $\mathbf{A}$.

Lemma 1

$$\begin{bmatrix} B & C \\ D & E \end{bmatrix}^{-1} = \begin{bmatrix} (B - CE^{-1}D)^{-1} & -(B - CE^{-1}D)^{-1}CE^{-1} \\ -E^{-1}D(B - CE^{-1}D)^{-1} & E^{-1} + E^{-1}D(B - CE^{-1}D)^{-1}CE^{-1} \end{bmatrix},$$

$$\tag{13}$$

where B, E *is square and* E, $(B - CE^{-1}D)$ *are nonsingular.*

Proof To prove the Lemma it is enough to show that:

$$\begin{bmatrix} B & C \\ D & E \end{bmatrix} \begin{bmatrix} (B - CE^{-1}D)^{-1} & -(B - CE^{-1}D)^{-1}CE^{-1} \\ -E^{-1}D(B - CE^{-1}D)^{-1} & E^{-1} + E^{-1}D(B - CE^{-1}D)^{-1}CE^{-1} \end{bmatrix} = I.$$

$$\tag{14}$$

Looking at the result blockwise we get:

$$B(B - CE^{-1}D)^{-1} - CE^{-1}D(B - CE^{-1}D)^{-1} = \tag{15}$$
$$= (B - CE^{-1}D)(B - CE^{-1}D)^{-1} = I,$$

$$- B(B - CE^{-1}D)^{-1}CE^{-1} + C(E^{-1} + E^{-1}D(B - CE^{-1}D)^{-1}CE^{-1}) = \tag{16}$$
$$= CE^{-1} - (B - CE^{-1}D)(B - CE^{-1}D)^{-1}CE^{-1} = CE^{-1} - ICE^{-1} = 0,$$

$$D(B - CE^{-1}D)^{-1} - EE^{-1}D(B - CE^{-1}D)^{-1} = \tag{17}$$
$$= D(B - CE^{-1}D)^{-1} - D(B - CE^{-1}D)^{-1} = 0,$$

$$- D(B - CE^{-1}D)^{-1}CE^{-1} + E(E^{-1} + E^{-1}D(B - CE^{-1}D)^{-1}CE^{-1}) = \tag{18}$$
$$= -D(B - CE^{-1}D)^{-1}CE^{-1} + I + D(B - CE^{-1}D)^{-1}CE^{-1} = I.$$

This gives:

$$\begin{bmatrix} B & C \\ D & E \end{bmatrix} \begin{bmatrix} (B - CE^{-1}D)^{-1} & -(B - CE^{-1}D)^{-1}CE^{-1} \\ -E^{-1}D(B - CE^{-1}D)^{-1} & E^{-1} + E^{-1}D(B - CE^{-1}D)^{-1}CE^{-1} \end{bmatrix} = \tag{19}$$

$$= \begin{bmatrix} I & 0 \\ 0 & I \end{bmatrix} = I.$$

Furthermore we need that E and $(B - CE^{-1}D)$ is nonsingular in order for the matrix to be invertible, [4]. $\qquad\square$

When using Lemma 1 we will denote the individual blocks off the inverse matrix as described in Definition 5.

Defnition 5 Given a block matrix $\mathbf{M}$ we denote the inverse as:

$$\mathbf{M} = \begin{bmatrix} \mathbf{M}_{1,1} & \mathbf{M}_{1,2} & \cdots & \mathbf{M}_{1,n} \\ \mathbf{M}_{2,1} & \mathbf{M}_{2,2} & \cdots & \mathbf{M}_{2,n} \\ \vdots & \vdots & \ddots & \vdots \\ \mathbf{M}_{m,1} & \mathbf{M}_{m,2} & \cdots & \mathbf{M}_{m,n} \end{bmatrix}, \mathbf{M}^{-1} = \begin{bmatrix} \mathbf{M}_{1,1}^{\text{inv}} & \mathbf{M}_{1,2}^{\text{inv}} & \cdots & \mathbf{M}_{1,n}^{\text{inv}} \\ \mathbf{M}_{2,1}^{\text{inv}} & \mathbf{M}_{2,2}^{\text{inv}} & \cdots & \mathbf{M}_{2,n}^{\text{inv}} \\ \vdots & \vdots & \ddots & \vdots \\ \mathbf{M}_{m,1}^{\text{inv}} & \mathbf{M}_{m,2}^{\text{inv}} & \cdots & \mathbf{M}_{m,n}^{\text{inv}} \end{bmatrix}. \tag{20}$$

We can now give a matrix proof of Theorem 2 as well.

Proof (*Proof of Theorem 2*) We let $\mathbf{B}$ be the part of the matrix $(\mathbf{I} - c\mathbf{A}^\top)$ corresponding to the nodes in the line which gives:

$$(\mathbf{I} - c\mathbf{A}^\top) = \begin{bmatrix} \mathbf{B} & \mathbf{C} \\ \mathbf{0} & 1 \end{bmatrix}. \tag{21}$$

We write

$$(\mathbf{I} - c\mathbf{A}^\top)^{-1} = \begin{bmatrix} \mathbf{B}^{\text{inv}} & \mathbf{C}^{\text{inv}} \\ \mathbf{D}^{\text{inv}} & \mathbf{E}^{\text{inv}} \end{bmatrix}. \tag{22}$$

Using Lemma 1 for blockwise inverse we get $\mathbf{B}^{\text{inv}} = (\mathbf{B} - \mathbf{C}\mathbf{E}^{-1}\mathbf{D})^{-1} = \mathbf{B}^{-1}$. Since $\mathbf{B}$ is the matrix for the simple line found earlier we get:

$$\mathbf{B}^{\text{inv}} = (\mathbf{I} - c\mathbf{A}^\top)^{-1} = \begin{bmatrix} 1 & c & c^2 & \cdots & c^{L-1} \\ 0 & 1 & c & \cdots & c^{L-2} \\ 0 & 0 & 1 & \cdots & c^{L-3} \\ \vdots & \vdots & \ddots & \ddots & \vdots \\ 0 & \cdots & \cdots & 0 & 1 \end{bmatrix}, \tag{23}$$

where L is the total number of nodes in the line. $\mathbf{C} = [0 \ \cdots \ c \ 0 \cdots \ 0]^\top$ where the non-zero element c is at position j gives:

$$\mathbf{C}^{\text{inv}} = -\mathbf{B}^{\text{inv}}\mathbf{C}\mathbf{E}^{-1} = -\mathbf{B}^{\text{inv}}\mathbf{C} = [c^j \ c^{j-1} \ \cdots \ c \ 0 \cdots \ 0]^\top. \tag{24}$$

Last, since $\mathbf{D} = 0$ we get $\mathbf{D}^{\text{inv}} = 0$, $\mathbf{E}^{\text{inv}} = 1$. Since the weight vector $\mathbf{u}$ is uniform we get the PageRank of a node as the sum of corresponding row in $(\mathbf{I} - c\mathbf{A}^\top)^{-1}$. For the nodes in the line we get PageRank:

$$\mathbf{R}_i^{(2)} = \sum_{k=0}^{n_L-i} c^k + b_{ij} = \frac{1-c^{n_L-i+1}}{1-c} + b_{ij}, \tag{25}$$

$$b_{ij} = \begin{cases} c^{j+1-i}, & j \geq i \\ 0, & j < i \end{cases} \tag{26}$$

where the sum is the sum of the first n_L values and b_{ij} is the value on the last column. For the last row we obviously get sum 1.

We get the normalized PageRank $\mathbf{R}^{(1)}$ by dividing $\mathbf{R}^{(1)} = \mathbf{R}^{(2)}/||\mathbf{R}^{(2)}||_1$. $\qquad\square$

6   Changes in a Complete Graph

Complete graphs or similar structures are common both as parts of a site and as a way between different sites to try and gain a better rank. An image of a complete graph with five nodes can be seen in Fig. 3. We recall that the system matrix for this system is:

$$
A = \frac{1}{4}
\begin{bmatrix}
0 & 1 & 1 & 1 & 1 \\
1 & 0 & 1 & 1 & 1 \\
1 & 1 & 0 & 1 & 1 \\
1 & 1 & 1 & 0 & 1 \\
1 & 1 & 1 & 1 & 0
\end{bmatrix}.
$$

Using this we get the inverse of this system as:

$$
(I - cA^\top)^{-1} =
\begin{bmatrix}
\frac{3c-4}{c^2+3c-4} & \frac{-c}{c^2+3c-4} & \frac{-c}{c^2+3c-4} & \frac{-c}{c^2+3c-4} & \frac{-c}{c^2+3c-4} \\
\frac{-c}{c^2+3c-4} & \frac{3c-4}{c^2+3c-4} & \frac{-c}{c^2+3c-4} & \frac{-c}{c^2+3c-4} & \frac{-c}{c^2+3c-4} \\
\frac{-c}{c^2+3c-4} & \frac{-c}{c^2+3c-4} & \frac{3c-4}{c^2+3c-4} & \frac{-c}{c^2+3c-4} & \frac{-c}{c^2+3c-4} \\
\frac{-c}{c^2+3c-4} & \frac{-c}{c^2+3c-4} & \frac{-c}{c^2+3c-4} & \frac{3c-4}{c^2+3c-4} & \frac{-c}{c^2+3c-4} \\
\frac{-c}{c^2+3c-4} & \frac{-c}{c^2+3c-4} & \frac{-c}{c^2+3c-4} & \frac{-c}{c^2+3c-4} & \frac{3c-4}{c^2+3c-4}
\end{bmatrix}.
$$

After normalization we will obviously end up with $R_i^{(1)} = 1/5$ as PageRank for every node i. However since there is not any dangling nodes in the complete graph all the nodes will have maximum influence on the PageRank of the system. Additionally since they only point to each other they will not share any of it with the outside in the case of a bigger link matrix with a part of it being a complete graph. This makes a complete graph similar to a dangling node in that it will not increase the rank of anyone else, but with the addition of having a higher rank in itself since it can increase its own rank to a certain extent.

Trying to find an expression for the elements in the inverse $(I - cA^\top)^{-1}$ for the complete graph we formulate the following lemma:

Lemma 2 *The diagonal element a_d of the inverse $(I - cA^\top)^{-1}$ of the complete graph with n nodes is:*

$$
a_d = \frac{(n-1) - c(n-2)}{(n-1) - c(n-2) - c^2}. \tag{27}
$$

The non diagonal elements a_{ij} can be written as:

$$
a_{ij} = \frac{c}{(n-1) - c(n-2) - c^2}. \tag{28}
$$

Proof The diagonal element is the sum of the probabilities of all paths to node e_d from itself. This can be written as a geometric sum: $a_d = \sum_{k=0}^{\infty} P(e_d \to e_d)^k$, where

$P(e_d \rightarrow e_d)$ is the probability of getting from node e_d to node e_d. The probability $P(e_d \rightarrow e_d)$ can be written as:

$$P(e_d \rightarrow e_d) = \frac{c^2}{n-1} + \frac{c^3(n-2)}{(n-1)^2} + \frac{c^4(n-2)^2}{(n-1)^3} + \cdots \tag{29}$$

$$= \frac{c^2}{n-1} \sum_{k=0}^{\infty} \left(\frac{c(n-2)}{n-1} \right)^k = \frac{c^2}{(n-1) - c(n-2)}.$$

This gives:

$$a_d = \sum_{k=0}^{\infty} \left(\frac{c^2}{(n-1) - c(n-2)} \right)^k = \frac{(n-1) - c(n-2)}{(n-1) - c(n-2) - c^2}. \tag{30}$$

For non-diagonal elements e_{ij} we get $e_{ij} = P(e_i \rightarrow e_j)a_d$, where $P(e_i \rightarrow e_j)$ is the probability of getting from node e_i to node e_j where $e_i \neq e_j$. This probability can be written as:

$$P(e_i \rightarrow e_j) = \frac{c}{n-1} + \frac{c^2(n-2)}{(n-1)^2} + \frac{c^3(n-2)^2}{(n-1)^3} + \cdots \tag{31}$$

$$= \frac{c}{n-1} \sum_{k=0}^{\infty} \left(\frac{c(n-2)}{n-1} \right)^k = \frac{c}{(n-1) - c(n-2)}.$$

This gives:

$$a_{ij} = \frac{c}{(n-1) - c(n-2)} \frac{(n-1) - c(n-2)}{(n-1) - c(n-2) - c^2} = \frac{c}{(n-1) - c(n-2) - c^2}. \tag{32}$$

$\square$

We give a matrix proof of Lemma 2 as well:

Proof (Proof of Lemma 2) We consider a general matrix $\mathbf{A}$ of the form:

$$\mathbf{A} = \begin{bmatrix} 1 & a & a & \dots & a \\ a & 1 & a & \dots & a \\ a & a & 1 & \ddots & a \\ \vdots & \vdots & \ddots & \ddots & \vdots \\ a & a & a & \dots & 1 \end{bmatrix}.$$

We use Gauss-Jordan elimination to find the inverse $\mathbf{A}^{-1}$:

$$
\begin{bmatrix}
1 & a & a & \dots & a & 1 & 0 & 0 & \dots & 0 \\
a & 1 & a & \dots & a & 0 & 1 & 0 & \dots & 0 \\
a & a & 1 & \ddots & a & 0 & 0 & 1 & \ddots & 0 \\
\vdots & \vdots & \ddots & \ddots & \vdots & \vdots & \vdots & \ddots & \ddots & 0 \\
a & a & a & \dots & 1 & 0 & 0 & 0 & \dots & 1
\end{bmatrix}.
$$

We add $-ar_1$ where r_1 is the first row to every other row to eliminate the elements below 1 on the first column:

$$
\begin{bmatrix}
1 & a & a & \dots & a & 1 & 0 & 0 & \dots & 0 \\
0 & 1-a^2 & a-a^2 & \dots & a-a^2 & -a & 1 & 0 & \dots & 0 \\
0 & a-a^2 & 1-a^2 & \ddots & a-a^2 & -a & 0 & 1 & \ddots & 0 \\
\vdots & \vdots & & \ddots & \ddots & \vdots & \vdots & \vdots & \ddots & \ddots & 0 \\
0 & a-a^2 & a-a^2 & \dots & 1-a^2 & -a & 0 & 0 & \dots & 1
\end{bmatrix}.
$$

Next we eliminate the values to the right of the 1 on the first row. We add $-k\sum_{i=2}^{n} r_i$, where r_i is row i to the first row giving the equation:

$$
a = -k(1 - a^2 + (n-2)(a - a^2)) \tag{33}
$$
$$
\rightarrow k - \frac{-a}{(1 - a^2 + (n-2)(a - a^2))}.
$$

This gives:

$$
\begin{bmatrix}
1 & 0 & 0 & \dots & 0 & 1-(n-1)ak & k & k & \dots & k \\
0 & 1-a^2 & a-a^2 & \dots & a-a^2 & -a & 1 & 0 & \dots & 0 \\
0 & a-a^2 & 1-a^2 & \ddots & a-a^2 & -a & 0 & 1 & \ddots & 0 \\
\vdots & \vdots & \ddots & \ddots & \vdots & \vdots & & \vdots & \ddots & \ddots & 0 \\
0 & a-a^2 & a-a^2 & \dots & 1-a^2 & -a & 0 & 0 & \dots & 1
\end{bmatrix}.
$$

We are now done calculating the first row of the inverse A^{-1}. We get the other rows using the same calculations if we start with another pivot element. For the inverse matrix we get diagonal elements $d = 1 - (n-1)ak$ and for all other elements $e = k$, where n is the total number of rows giving a inverse like below:

$$
A^{-1} =
\begin{bmatrix}
1-(n-1)ak & k & k & \dots & k \\
k & 1-(n-1)ak & k & \dots & k \\
k & k & 1-(n-1)ak & \ddots & \vdots \\
\vdots & \vdots & & \ddots & k \\
k & k & \dots & k & 1-(n-1)ak
\end{bmatrix}.
$$

Calculating for $a = -c/(n-1)$ as for a complete graph gives:

$$k = \frac{-a}{(1 - a^2 + (n-2)(a - a^2))} = \frac{c}{(n-1) - (n-2)c - c^2}, \quad (34)$$

$$d = 1 - (n-1)ak = \frac{(n-1) - (n-2)c - c^2 - (n-1)(-c)/(n-1)c}{(n-1) - (n-2)c - c^2} \quad (35)$$

$$= \frac{(n-1) - (n-2)c}{(n-1) - (n-2)c - c^2}.$$

And the proof is complete. □

Using this we immediately get the PageRank (before normalization) of elements in a complete graph with uniform **u**:

Theorem 3 *Given a complete graph with $n > 1$ nodes, PageRank $\mathbf{R}^{(2)}$ before normalization can be written as:*

$$\mathbf{R}_i^{(2)} = \frac{1}{1-c}. \quad (36)$$

Proof From Lemma 2 We already have the inverse $(\mathsf{I} - c\mathbf{A}^\top)^{-1}$, We then find the PageRank by summation of any row of the matrix (since all rows have equal sum).

$$\mathbf{R}_i^{(2)} = a_d + (n-1)a_{ij}, \quad i \neq j, \quad (37)$$

$$= \frac{(n-1) - c(n-2) + c(n-1)}{(n-1) - c(n-2) - c^2} = \frac{c + (n-1)}{(n-1) - c(n-2) - c^2} = \frac{1}{1-c}.$$

□

We do note that since we have no dangling nodes all the probability from a node in the complete graph is distributed within the complete graph. Also the size of the graph is irrelevant for the individual nodes as long as none are linked to from outside sources and it consists of at least two nodes. In the $\mathbf{R}^{(1)}$ sense the size obviously changes the result since we would increase the overall number of nodes in the system by increasing the size of the complete graph. Two things is important to note however: The higher ones own PageRank before joining the complete graph (probability of getting there from outside nodes) the more gain there is by joining a small complete graph in order to maximize the probability of returning to itself. In the same way if a node have a very low rank it gains much by joining a large complete graph of nodes with higher rank than itself.

6.1 Adding a Link Out of a Complete Graph

If we want to see how the complete graph changes when adding one link from one node (node one) out of the complete graph we end up with the following system matrix for the nodes in the complete graph:

$$
(I - cA^\top) = \begin{bmatrix}
1 & -c/4 & -c/4 & -c/4 & -c/4 \\
-c/5 & 1 & -c/4 & -c/4 & -c/4 \\
-c/5 & -c/4 & 1 & -c/4 & -c/4 \\
-c/5 & -c/4 & -c/4 & 1 & -c/4 \\
-c/5 & -c/4 & -c/4 & -c/4 & 1
\end{bmatrix}.
$$

After taking the inverse and multiplying with -1 we get:

$$(I - cA^\top)^{-1} =$$

$$
\begin{bmatrix}
\frac{15c-20}{s} & \frac{-5c}{s} & \frac{-5c}{s} & \frac{-5c}{s} & \frac{-5c}{s} \\
\frac{-4c}{s} & \frac{12c^2+40c-80}{(c+4)s} & -\frac{4c(5+c)}{(c+4)s} & -\frac{4c(5+c)}{(c+4)s} & -\frac{4c(5+c)}{(c+4)s} \\
\frac{-4c}{s} & -\frac{4c(5+c)}{(c+4)s} & \frac{12c^2+40c-80}{(c+4)s} & -\frac{4c(5+c)}{(c+4)s} & -\frac{4c(5+c)}{(c+4)s} \\
\frac{-4c}{s} & -\frac{4c(5+c)}{(c+4)s} & -\frac{4c(5+c)}{(c+4)s} & \frac{12c^2+40c-80}{(c+4)s} & -\frac{4c(5+c)}{(c+4)s} \\
\frac{-4c}{s} & -\frac{4c(5+c)}{(c+4)s} & -\frac{4c(5+c)}{(c+4)s} & -\frac{4c(5+c)}{(c+4)s} & \frac{12c^2+40c-80}{(c+4)s}
\end{bmatrix},
$$

where $s = 4c^2 + 15c - 20$.

We find the expression for the PageRank in a complete graph with one node linking out to be the following assuming uniform **u**.

Theorem 4 *The PageRank of the nodes in a complete graph with the first node linking out of the complete graph, the PageRank can be written as:*

$$
\mathbf{R}_1^{(2)} = \frac{n(n-1) + nc}{n(n-1) - (n-1)c^2 - n(n-2)c}, \tag{38}
$$

$$
\mathbf{R}_i^{(2)} = \frac{(c+n)(n-1)}{n(n-1) - (n-1)c^2 - n(n-2)c}, \quad n \geq i > 1, \tag{39}
$$

where n is the number of nodes in the complete graph and node one links out of the complete graph.

Proof We start by looking at the PageRank as a probability, we let e_1 be the node linking out. The probability to get from e_1 back to itself is:

$$P(e_1 \to e_1) = \frac{c(n-1)}{n} \frac{c}{n-1} + \frac{c(n-1)}{n} \frac{c}{n-1} \frac{c(n-2)}{n-1} \qquad (40)$$

$$+ \frac{c(n-1)}{n} \frac{c}{n-1} \left(\frac{c(n-2)}{n-1}\right)^2 + \cdots$$

$$= \frac{c^2}{n} \sum_{k=0}^{\infty} \left(\frac{c(n-2)}{n-1}\right)^k = \frac{c^2}{n} \frac{n-1}{(n-1) - c(n-2)}.$$

And we get the sum of all probabilities from e_1 back to itself as:

$$\sum_{k=0}^{\infty} (P(e_1 \to e_1))^k = \sum_{k=0}^{\infty} \left(\frac{c^2}{n} \frac{n-1}{(n-1) - c(n-2)}\right)^k \qquad (41)$$

$$= \frac{n((n-1) - c(n-2))}{n((n-1) - c(n-2)) - c^2(n-1)} = B^{inv}.$$

We remember that on the diagonal of $(I - cA^{\top})$, we have the sums of probabilities of nodes going back to themselves. So if we divide the matrix $(I - cA^{\top})$ in blocks:

$$(I - cA^{\top}) = \begin{bmatrix} B & C \\ D & E \end{bmatrix},$$

and inverse matrix:

$$(I - cA^{\top})^{-1} = \begin{bmatrix} B^{inv} & C^{inv} \\ D^{inv} & E^{inv} \end{bmatrix}.$$

We note that B^{inv} is not the inverse of B but the part of the inverse $(I - cA^{\top})^{-1}$ corresponding to block B. We let $B = [1]$ corresponding to the node linking out and we get B^{inv} as above.

For the elements C_i^{inv}, $i \neq 1$ of C^{inv} we find them as

$$C_i^{inv} = \sum_{k=0}^{\infty} (P(e_i \to e_1))^k \sum_{k=0}^{\infty} (P(e_1 \to e_1))^k \qquad (42)$$

$$= \frac{c}{n-1} \sum_{k=0}^{\infty} \left(\frac{c(n-2)}{n-1}\right)^k B^{inv} = \frac{cn}{n((n-1) - c(n-2)) - c^2(n-1)}.$$

Since E and $DB^{-1}C$ are both symmetric and have every non-diagonal element equal as well as all diagonal elements equal, the inverse $E^{inv} = (E - DB^{-1}C)^{-1}$ should be the same as well. Especially every row and column have the same sum. From Lemma 1 for blockwise inversion we get:

$$C_i^{inv} = -\frac{-c}{n-1} \sum_{k=1}^{n-1} E_{ki}^{inv}, \qquad (43)$$

$$D_i^{inv} = -\frac{c}{n} \sum_{k=1}^{n-1} E_{ik}^{inv} = -\frac{c}{n} \sum_{k=1}^{n-1} E_{ki}^{inv}, \tag{44}$$

$$\Rightarrow \begin{cases} D_i^{inv} = \dfrac{(n-1)C_i^{inv}}{n}, \\[2mm] \displaystyle\sum_{k=1}^{n-1} E_{ik}^{inv} = \dfrac{(n-1)C_i^{inv}}{c}. \end{cases} \tag{45}$$

We get the PageRank as:

$$\mathbf{R}_1^{(2)} = \mathsf{B}^{inv} + (n-1)\mathsf{C}^{inv} = \frac{n((n-1) - c(n-2))}{n((n-1) - c(n-2)) - c^2(n-1)} \tag{46}$$
$$+ \frac{(n-1)cn}{n((n-1) - c(n-2)) - c^2(n-1)} = \frac{n(n-1) + nc}{n(n-1) - (n-1)c^2 - n(n-2)c},$$

$$\mathbf{R}_i^{(2)} = D^{inv} + \sum_{k=1}^{n-1} E_{ik}^{inv} = \frac{(n-1)C_i nv_i}{n} + \frac{(n-1)C_i^{inv}}{c} \tag{47}$$
$$= \frac{(n-1)C_i^{inv}(c+n)}{nc} = \frac{(c+n)(n-1)}{n(n-1) - (n-1)c^2 - n(n-2)c}.$$

And the proof is complete.　　　　　　　　　　　　　　　　　　　　　　□

We give a matrix proof of Theorem 4 as well:

Proof (Proof of Theorem 4) We consider the square matrix A with n rows:

$$\mathsf{A} = \begin{bmatrix} 1 & a & a & \dots & a \\ b & 1 & a & \dots & a \\ b & a & 1 & \ddots & a \\ \vdots & \vdots & \ddots & \ddots & \vdots \\ b & a & a & \dots & a \end{bmatrix},$$

where $a = -c/(n-1)$, $b = -c/(n)$. We divide the matrix in blocks:

$$\mathsf{A} = \begin{bmatrix} \mathsf{B} & \mathsf{C} \\ \mathsf{D} & \mathsf{E} \end{bmatrix},$$

where $\mathsf{B} = [1]$, $\mathsf{C} = [a\ a\ \dots\ a]$, $\mathsf{D} = [b\ b\ \dots\ b]^\top$ and E looks like the matrix for a complete graph but of size $(n-1) \times (n-1)$:

$$E = \begin{bmatrix} 1 & a & a & \ldots & a \\ a & 1 & a & \ldots & a \\ a & a & 1 & \ddots & a \\ \vdots & \vdots & \ddots & \ddots & \vdots \\ a & a & a & \ldots & a \end{bmatrix}.$$

In the same way as in the proof of Lemma 2 we find the elements of B, C by choosing the top left element as pivot element. This gives

$$k_A = \frac{-a}{(1 - ab) + (n - 2)(a - ab)}. \tag{48}$$

We write A^{-1} as blocks:

$$A^{-1} = \begin{bmatrix} B^{inv} & C^{inv} \\ D^{inv} & E^{inv} \end{bmatrix},$$

and get: $B^{inv} = 1 - (n - 1)bk_A$ and $C_i^{inv} = k_A$.

From the matrix proof of Lemma 2 we get the non-diagonal elements E_e and diagonal elements E_d of E^{-1} as

$$E_e = k_D = \frac{-a}{(1 - a^2) + (n - 3)(a - a^2)} = \frac{(n - 1)c}{(n - 1)^2 - (n - 3)(n - 1)c - (n - 2)c^2}, \tag{49}$$

$$E_d = 1 - (n - 2)ak_D = \frac{(n - 1)^2 - (n - 3)(n - 1)c}{(n - 1)^2 - (n - 3)(n - 1)c - (n - 2)c^2}. \tag{50}$$

From Lemma 1 we then get:

$$B^{inv} = (B - CE^{-1}D)^{-1} = 1 - (n - 1)bk_A, \tag{51}$$

$$C^{inv} = -(B - CE^{-1}D)^{-1}CE^{-1} \tag{52}$$
$$\Rightarrow C_i^{inv} = -(B - CE^{-1}D)^{-1}b(E_d + (n - 2)E_e) = k_A,$$

$$D^{inv} = -E^{-1}D(B - CE^{-1}D)^{-1} \tag{53}$$
$$\Rightarrow D_i^{inv} = -a(E_d + (n - 2)E_e)(B - CE^{-1}D)^{-1} = \frac{bk_A}{a},$$

$$E^{inv} = E^{-1} + E^{-1}D(B - CE^{-1}D)^{-1}CE^{-1}, \tag{54}$$

$$\Rightarrow \begin{cases} E_d^{inv} = E_d + b(E_d + (n-2)E_e)(\mathsf{B} - \mathsf{CE}^{-1}\mathsf{D})^{-1}a(E_d + (n-2)E_e) \\ \qquad = E_d - b(E_d + (n-2)E_e)k_A \\ E_e^{inv} = E_e + b(E_d + (n-2)E_e)(\mathsf{B} - \mathsf{CE}^{-1}\mathsf{D})^{-1}a(E_d + (n-2)E_e) \\ \qquad = E_e - b(E_d + (n-2)E_e)k_A \end{cases} \tag{55}$$

We replace $a = -c/(n-1)$ and $b = -c/n$ as for our complete graph and get inverse:

$$(\mathsf{I} - c\mathsf{A}^\top)^{-1} =$$

$$= \begin{bmatrix} 1 - (n-1)bk_A & k_A & k_A & \cdots & k_A \\ \frac{bk_A}{a} & 1 - (n-2)ak_D & k_D & \cdots & k_D \\ \frac{bk_A}{a} & k_D & 1 - (n-2)ak_D & \ddots & k_D \\ \vdots & \vdots & & \ddots & \vdots \\ \frac{bk_A}{a} & k_D & k_D & \cdots & 1 - (n-2)ak_D \end{bmatrix}.$$

For the PageRank of the node linking out we get:

$$\mathbf{R}_1^{(2)} = \mathsf{B}^{inv} + (n-1)C_i^{inv} = \tag{56}$$
$$= 1 - (n-1)bk_A + (n-1)k_A = 1 - (n-1)(b-1)k_A$$
$$= \frac{(1-ab) + (n-2)(a-ab) + (n-1)(b-1)a}{(1-ab) + (n-2)(a-ab)}$$
$$= \frac{(1-ab) - (a-ab) - (n-1)a}{(1-ab) + (n-2)(a-ab)} = \frac{n(n-1) + cn}{n(n-1) - n(n-2)c - (n-1)c^2}.$$

For all other nodes we get PageRank:

$$\mathbf{R}_i^{(2)} = D_i^{inv} + E_d^{inv} + (n-2)E_e^{inv} = \tag{57}$$
$$= \frac{bk_A}{a} + E_d - b(E_d + (n-2)E_e)k_A$$
$$\quad + (n-2)E_e - (n-2)b(E_d + (n-2)E_e)k_A$$
$$= E_d + (n-2)E_e - (n-1)b(E_d + (n-2)E_e)k_A + (b/a)k_A$$
$$= \frac{1-a}{(1-a^2) + (n-3)(a-a^2)} + \frac{-b}{(1-ab) + (n-2)(a-ab)}$$
$$\quad + \frac{(n-1)ab(1-a)}{((1-a^2) + (n-3)(a-a^2))((1-ab) + (n-2)(a-ab))}$$
$$= \frac{1-b}{1 - ab + (n-2)(a-ab)} = \frac{(n-1)(n+c)}{n(n-1) - n(n-2)c - (n-1)c^2}.$$

And the proof is complete.   □

Just looking at the expression it is hard to say how the PageRank changes after linking out. We can however note a couple of things: First of all the PageRank is lower than for the complete graph (since we now have a chance to escape the graph). But more interesting, when comparing the node that links out with the others we formulate the following theorem:

Theorem 5 *In a complete graph not linked to from the outside but with one link out, the node that links out will have the highest PageRank in the complete graph.*

Proof Using the expression for PageRank in a complete graph with one link out we want to prove $\mathbf{R}_k^{(2)} > \mathbf{R}_i^{(2)}$ where $\mathbf{R}_k^{(2)}$ is the PageRank for the node linking out and $\mathbf{R}_i^{(2)}$ is the PageRank of all the other nodes.

$$\mathbf{R}_k^{(2)} > \mathbf{R}_i^{(2)}$$

$$\Leftrightarrow \frac{n(n-1)+nc}{n(n-1)-(n-1)c^2-n(n-2)c} > \frac{(c+n)(n-1)}{n(n-1)-(n-1)c^2-n(n-2)c}$$

$$\Leftrightarrow n(n-1)+nc > (c+n)(n-1)$$

$$\Leftrightarrow n^2+nc-n > n^2+nc-n-c$$

(58)

where $0 < c < 1$ and $n > 1$ is the number of nodes in the complete graph. This is obviously true and the proof is complete. $\qquad\square$

Apart from the knowledge that it is the node that links out of a complete graph that loses the least from it we can also see that as the number of nodes in the complete graph increases the difference between them decreases since we have a factor n^2 in the denominator compared to only a difference of c in the nominator.

6.2 Effects of Linking to a Complete Graph

In the case of a link to a complete graph without a link back from the complete graph we can easily guess the result. From earlier we know that for a node linking to one other node in a link matrix with no change of getting back to itself the column corresponding to the node linking out is c times the column of the node it links to. Additionally we need to add a one to the diagonal element for that column.

The fact that there is no probability (or a very low if it is only close to complete) to escape the complete graph and give any advantage to other parts of the system means the complete graph as a whole get maximum benefit from the links to it. Looking at how the additional probability $c/(1-c) = c + c^2 + c^3 + \cdots + c^\infty$ get distributed within the complete graph we realize that the node linked to gains all of the initial c^1, then loses a part c^2 distributed among all other nodes in the complete graph, after that the rest is close to evenly distributed between all the nodes in the complete graph. As such we see that the node linked to is the node which gains the most from the link (which is what we would expect).

7   Conclusions

We have seen that we can solve the resulting equation system instead of using the definition directly or using the Power method. While this method is significantly slower it has made it possible to get a bigger understanding of the different roles of the link matrix **A** and the weight vector **u**. We have seen how PageRank changes when doing some small changes in a couple of simple systems. For these systems we also found explicit expressions for the PageRank and in particular two ways to find these. Either by solving the equation system itself or by using a probabilistic perspective and calculate:

$$
\mathbf{R}_g^{(2)} = \left(\sum_{e_i \in S,\ e_i \neq e_g} P(e_i \to e_g) + 1 \right) \left(\sum_{k=0}^{\infty} (P(e_g \to e_g))^k \right),
$$

where $P(e_i \to e_g)$ is the sum of probability of all paths from node e_i to node e_g and the weight vector **u** is uniform.

One of the main advantages in using non-normalized PageRank over the ordinary normalized version is that it is possible to split a large system into multiple disjoint systems and calculate $\mathbf{R}^{(2)}$ for every subsystem separately, something which cannot be done as easily.

Acknowledgements This research was supported in part by the Swedish Research Council (621-2007-6338), Swedish Foundation for International Cooperation in Research and Higher Education (STINT), Royal Swedish Academy of Sciences, Royal Physiographic Society in Lund and Crafoord Foundation.

References

1. Andersson, F.: Estimation of the quality of hyperlinked documents using a series formulation of pagerank. Master's thesis, Mathematics, Centre for Mathematical sciences, Lund Institute of Technology, Lund University (2006:E22). LUTFMA-3132-2006
2. Andersson, F., Silvestrov, S.: The mathematics of internet search engines. Acta Appl. Math. **104**, 211–242 (2008)
3. Berman, A., Plemmons, R.: Nonnegative Matrices in the Mathematical Sciences. No. Del 11 in Classics in Applied Mathematics. Society for Industrial and Applied Mathematics. Academic Press, New York (1994)
4. Bernstein, D.: Matrix Mathematics. Princeton University Press, Princeton (2005)
5. Bianchini, M., Gori, M., Scarselli, F.: Inside pagerank. ACM Trans. Int. Technol. **5**(1), 92–128 (2005)
6. Brin, S., Page, L.: The anatomy of a large-scale hypertextual web search engine. Comput. Netw. ISDN Syst. **30**(1–7), 107–117 (1998) (Proceedings of the Seventh International World Wide Web Conference)
7. Bryan, K., Leise, T.: The $ 25,000,000,000 eigenvector: the linear algebra behind google. SIAM Rev. **48**(3), 569–581 (2006)

8. Dhyani, D., Bhowmick, S.S., Ng, W.K.: Deriving and verifying statistical distribution of a hyperlink-based web page quality metric. Data Knowl. Eng. **46**(3), 291–315 (2003)
9. Engström, C., Silvestrov, S.: Non-normalized pagerank and random walks on n-partite graphs. In: Proceedings of 3rd Stochastic Modeling Techniques and Data Analysis, pp. 192–202 (2015)
10. Gantmacher, F.: The Theory of Matrices. Chelsea, New York (1959)
11. Haveliwala, T., Kamvar, S.: The second eigenvalue of the google matrix. Technical Report 2003–20, Stanford InfoLab (2003)
12. Ishii, H., Tempo, R., Bai, E.W., Dabbene, F.: Distributed randomized pagerank computation based on web aggregation. In: Proceedings of the 48th IEEE Conference on Decision and Control, 2009 Held Jointly with the 2009 28th Chinese Control Conference. CDC/CCC 2009, pp. 3026–3031 (2009)
13. Kamvar, S., Haveliwala, T.: The condition number of the pagerank problem. Technical Report 2003–36, Stanford InfoLab (2003)
14. Kamvar, S.D., Schlosser, M.T., Garcia-Molina, H.: The eigentrust algorithm for reputation management in P2P networks. In: Proceedings of the 12th International Conference on World Wide Web. WWW '03, pp. 640–651. ACM, New York (2003)
15. Lancaster, P.: Theory of Matrices. Academic Press, New York (1969)
16. Norris, J.R.: Markov Chains. Cambridge University Press, New York (2009)
17. Rydén, T., Lindgren, G.: Markovprocesser. Lund University, Lund (2000)
18. Sepandar, K., Taher, H., Gene, G.: Adaptive methods for the computation of pagerank. Linear Algebra Appl. **386**(0), 51–65 (2004) (Special Issue on the Conference on the Numerical Solution of Markov Chains 2003)

PageRank, Connecting a Line of Nodes with a Complete Graph

Christopher Engström and Sergei Silvestrov

Abstract The focus of this article is the PageRank algorithm originally defined by S. Brin and L. Page as the stationary distribution of a certain random walk on a graph used to rank homepages on the Internet. We will attempt to get a better understanding of how PageRank changes after you make some changes to the graph such as adding or removing edge between otherwise disjoint subgraphs. In particular we will take a look at link structures consisting of a line of nodes or a complete graph where every node links to all others and different ways to combine the two. Both the ordinary normalized version of PageRank as well as a non-normalized version of PageRank found by solving corresponding linear system will be considered. We will see that it is possible to find explicit formulas for the PageRank in some simple link structures and using these formulas take a more in-depth look at the behavior of the ranking as the system changes.

Keywords PageRank · Graph · Random walk · Block matrix

1 Introduction

PageRank was initially used to rank homepages (nodes) based on the structure of links between these pages. This is important in order to return the most relevant results in for example search engines. Since the number of pages on the Internet is huge and ever increasing it is important that the method is extremely fast but there is also a heavy requirement on the quality of the ranking since very few people will look through more than a couple of the highest ranked pages when looking for something, [4].

C. Engström (✉) · S. Silvestrov
Division of Applied Mathematics, School of Education,
Culture and Communication, Mälardalen University, Box 883, 721 23 Västerås, Sweden
e-mail: christopher.engstrom@mdh.se

S. Silvestrov
e-mail: sergei.silvestrov@mdh.se

© Springer International Publishing Switzerland 2016
S. Silvestrov and M. Rančić (eds.), *Engineering Mathematics II*,
Springer Proceedings in Mathematics & Statistics 179,
DOI 10.1007/978-3-319-42105-6_12

249

Some other applications of PageRank or similar methods include the EigenTrust algorithm used for reputation management in P2P networks [12] and DebtRank for evaluating risk in financial networks [1].

Calculating PageRank is usually calculated using the Power method which can be implemented very efficiently, even for very large systems. The convergence speed of the Power method and it's dependence on certain parameters have been studied to some extent. For example the Power method on a graph structure such as that created by the Web will converge with a convergence rate of c, where c is one of the parameters used in the definition [9], and the problem is well conditioned unless c is very close to 1 [11]. Many methods have been developed in order to speed up the calculations of PageRank such as by aggregating webpages that are "close" and are expected to have a similar PageRank as in [10] or by partitioning the graph into different components as in [6].

There are also work done on the large scale using PageRank and other measures in order to learn more about the Web, for example looking at the distribution of PageRank both theoretically and experimentally such as in [5].

While the theory behind PageRank is well understood from Perron–Frobenius theory for non-negative irreducible matrices [2, 8, 13] and the study of Markov chains [14, 15], how PageRank is affected from changes in the system or parameters is not as well known. We will start by giving some necessary definitions as well as describing the notation used throughout the article. Before continuing to the main part of the article we will also give some previous results described in [7] which will be needed throughout the rest of the article. As in said previous work we will consider PageRank as the solution to a linear system of equations as well as probabilities of a random walk through the graph and see what we can learn from this. Both ordinary normalized PageRank as well as non-normalized PageRank will be considered and we will highlight some differences between the two as parameter c or the size of the graph changes. In this article we will look at how PageRank changes as we combine a line of vertices with edges in one direction with a complete graph in different ways in Sect. 3. And after that in Sects. 4 and 5 we will take a closer look at the found formulas for some of the examples mainly by looking at partial derivatives of the PageRank. We will see one of the possible reasons why c is usually chosen to be around $c \approx 0.85$. PageRank for some nodes increases extremely fast while for some other nodes decreases extremely fast for larger c, while for lower c the difference in PageRank between nodes is smaller the lower c gets and the initial weight vector have a much larger influence on the final ranking.

2   Notation, Definitions and Previous Results

We will start by describing the notation used throughout the article as well as describing some common link structures which will be used. At the end of this section we will give a couple of lemmas and theorems without proofs summarizing previous results.

First some overall notation:

- S_G: The system of nodes and links for which we want to calculate PageRank, contains the system matrix A_G as well as a weight vector $\mathbf{v}_G$. Subindex G can be either a capital letter or a number in the case of multiple systems.
- n_G: The number of nodes in system S_G.
- A_G: System matrix where a zero element a_{ij} means there is no link from node i to node j. Non-zero elements are equal to $1/r_i$ where r_i is the number of links from node i. Size $n_G \times n_G$.
- $\mathbf{v}_G$: Non-negative weight vector, not necessary with sum one. Size $n_G \times 1$.
- $\mathbf{u}_G$: The weight vector $\mathbf{v}_G$ normalized such that $||\mathbf{u}_G||_1 = 1$. We note that $\mathbf{u}_G$ is proportional to $\mathbf{v}_G$ ($\mathbf{u}_G \propto \mathbf{v}_G$). Size $n_G \times 1$.
- c: Parameter $0 < c < 1$ for calculating PageRank, usually $c = 0.85$.
- $\mathbf{g}_G$: Vector with elements equal to one for dangling nodes and zero for all other in S_G. Size $n_G \times 1$.
- M_G: Modified system matrix, $M_G = c(A_G + \mathbf{g}_G \mathbf{u}_G^\top)^\top + (1 - c)\mathbf{u}_G \mathbf{e}^\top$ used to calculate PageRank, where $\mathbf{e}$ is the unit vector. Size $n_G \times n_G$.
- S: Global system made up of multiple disjoint subsystems $S = S_1 \cup S_2 \ldots \cup S_N$, where N is the number of subsystems.
- V: Global weight vector for system S, $\mathbf{V} = [\mathbf{v}_1^\top \ \mathbf{v}_2^\top \ \ldots \ \mathbf{v}_N^\top]^\top$, where N is the number of subsystems.

In the cases where there is only one possible system the subindex G will often be omitted. For the systems making up S we define disjoint systems in the following way.

Definition 1 Two systems S_1, S_2 are disjoint if there are no paths from any nodes in S_1 to S_2 or from any nodes in S_2 to S_1.

Different versions of PageRank will be denoted as follows

$$\mathbf{R}_G^{(t)}[S_H \to S_I, S_J \to S_K \ldots],$$

where t is the type of PageRank used, $S_G \subseteq S$ is the nodes in the global system S for which $\mathbf{R}$ is the PageRank. Often $S_G = S$ and we write it as $\mathbf{R}_S^{(t)}$. In the last part within brackets we write possible connections between otherwise disjoint subsystems in S, for example an arrow to the right means there are links from the left system to the right system. How many and what type of links however needs to be specified for every individual case.

We will sometimes give the formula for a specific node j in this case it will be noted as $\mathbf{R}_{G,j}^{(t)}[S_H \to S_I, S_J \to S_K \ldots]$. When it is obvious which system to use (for example when only one is specified) and there are no connections between systems S_G as well as the brackets with connections between systems will usually be omitted resulting in $\mathbf{R}_j^{(t)}$. It should be obvious when this is the case. When normalizing the resulting elements such that their sum is equal to one we get the traditional normalized PageRank.

Definition 2 $\mathbf{R}_G^{(1)}$ for system S_G is defined as the eigenvector with eigenvalue one to the matrix $\mathsf{M}_G = c(\mathsf{A}_G + \mathbf{g}_G\mathbf{u}_G^\top)^\top + (1 - c)\mathbf{u}_G\mathbf{e}^\top$.

Note that M is a stochastic matrix and therefor PageRank itself can be seen as the stationary distribution of a Markov chain describing a random walk on a graph described by A (with some correction for vertices with no outgoing edges) and a small random chance $(1 - c)$ to move to a random node depending on the distribution described by $\mathbf{u}$. By convention PageRank is normalized such that $||\mathbf{R}^{(1)}||_1 = 1$ to get said stationary distribution and it is usually reasonable to assume M to be irreducible and primitive hence $\mathbf{R}_G^{(1)}$ will be a positive vector easily shown using Perron-Frobenius theory for non-negative irreducible matrices [2, 8, 13]. The fact that $||\mathbf{R}^{(1)}||_1 = 1$ is generally not the case in our other version of PageRank. If we instead set up the resulting equation system and solving it we get the second definition, the result is multiplied with n_G in order to get multiplication with the one vector in case of uniform $\mathbf{u}_G$.

Definition 3 $\mathbf{R}_G^{(2)}$ for system S_G is defined as $\mathbf{R}_G^{(2)} = (\mathsf{I} - c\mathsf{A}_G^\top)^{-1}n_G\mathbf{u}_G$

We note that generally $||\mathbf{R}^{(2)}||_1 \neq 1$ as well as $\mathbf{R}_G^{(2)} \neq n_G\mathbf{R}_G^{(1)}$ unless there are no dangling nodes in the system. However the two versions of PageRank are proportional to each other ($\mathbf{R}_G^{(2)} \propto \mathbf{R}_G^{(1)}$).

Another way to calculate PageRank using this second definition is from a probabilistic perspective. For a proof of the theorem we refer to our earlier work [7].

Theorem 1 *Consider a random walk on a graph described by $c\mathsf{A}$ described as before. We walk to a new node with probability c and stop with probability $1 - c$. PageRank $\mathbf{R}^{(2)}$ of a node when using uniform $\mathbf{u}$ can be written:*

$$\mathbf{R}_j^{(2)} = \left(\sum_{e_i \in S, e_i \neq e_j} P(e_i \to e_j) + 1 \right) \left(\sum_{k=0}^{\infty} (P(e_j \to e_j))^k \right), \tag{1}$$

where $P(e_i \to e_j)$ is the probability to hit node e_j in a random walk starting in node e_i described as above. This can be seen as the expected number of visits to e_j if we do multiple random walks, starting in every node once.

Two small graph-structures that will be used are the simple line and complete graph.

Definition 4 A simple line is a graph with n_L nodes where node n_L links to node n_{L-1} which in turn links to node n_{L-2} all the way until node n_2 link to node n_1.

Definition 5 A complete graph is a group of nodes in which all nodes in the group links to all other nodes in the group.

The following well known lemma for blockwise inversion easily verified by calculating the matrix with the inverse according to the lemma [3, 7].

Lemma 1

$$
\begin{bmatrix} B & C \\ D & E \end{bmatrix}^{-1} = \begin{bmatrix} (B - CE^{-1}D)^{-1} & -(B - CE^{-1}D)^{-1}CE^{-1} \\ -E^{-1}D(B - CE^{-1}D)^{-1} & E^{-1} + E^{-1}D(B - CE^{-1}D)^{-1}CE^{-1} \end{bmatrix},
$$
(2)

where B, E is square and E, $(B - CE^{-1}D)$ are nonsingular.

Below follows two previous results regarding PageRank for the simple line and complete graph by themselves, because of size considerations we refer to [7] for proofs of both theorems.

Theorem 2 *The PageRank of a node e_i belonging to the line in a system containing a simple line with one outside node linking to one of the nodes in the line when using uniform weight vector $\mathbf{u}$ can be written:*

$$
\mathbf{R}_i^{(2)} = \sum_{k=0}^{n_L - i} c^k + b_{ij} = \frac{1 - c^{n_L - i + 1}}{1 - c} + b_{ij},
$$
(3)

$$
b_{ij} = \begin{cases} c^{j+1-i}, & j \geq i, \\ 0, & j < i, \end{cases}
$$
(4)

where n_L is the number of nodes in the line and the new node link to node j. The new node has rank 1.

Theorem 3 *The PageRank of the nodes in a complete graph with the first node linking out of the complete graph, the PageRank can be written as:*

$$
\mathbf{R}_1^{(2)} = \frac{n(n-1) + nc}{n(n-1) - (n-1)c^2 - n(n-2)c},
$$
(5)

$$
\mathbf{R}_i^{(2)} = \frac{(c+n)(n-1)}{n(n-1) - (n-1)c^2 - n(n-2)c}, \quad n \geq i > 1,
$$
(6)

where n is the number of nodes in the complete graph and node one links out of the complete graph.

3 Changes in PageRank When Connecting the Simple Line with the Complete Graph

Looking at some simple structures and how PageRank changes as we change them, the goal is to learn something in how and why the rank changes as it does. This in an attempt to answer questions such as: How do I connect my two sites or within my one site in such a way that I won't get any undesired results? In all these examples we will assume uniform $\mathbf{u}$ (which means we can multiply the inverse $(I - cA^\top)^{-1}$ with the one vector in order to get $\mathbf{R}^{(2)}$).

Here we will look at what happens when we connect a complete graph with a simple line in various ways. This way we can get some information on what type of structure is most effective in getting a high PageRank and see how they interact with each other.

3.1 Connecting the Simple Line with a Link from a Node in the Complete Graph to a Node in the Line

Looking at the system where we let one node in a complete graph link to one node in a simple line we get a system similar to the case where we added a single node to the line (complete graph with one node). An example of what the system could look like can be seen in Fig. 1. We have the two systems S_L, S_G as the original systems for the simple line and complete graph respectively. We want to find the new PageRank of these nodes after creating our new system S by adding a link from the first node in the complete graph $e_{G,1}$ to node $e_{L,j}$ in the simple line. When using $n_L = 5, n_G = 5, j = 3$ we get the system with $(I - c\mathbf{A}^\top)$ seen in Fig. 1.

Assuming uniform $\mathbf{u}$ the PageRank in the simple line after adding the link from the complete graph $\mathbf{R}_L^{(2)}[S_G \to S_L]$ can still be written in about the same way:

Theorem 4 *Observing the nodes in a system S made up of two systems, a simple line S_L with n_L nodes and a complete graph S_G with n_G nodes where we add one link from node e_g in the complete graph to node e_j in the simple line. Assuming uniform $\mathbf{u}$ we get the PageRank $\mathbf{R}_{L,i}^{(2)}[S_G \to S_L]$ for the nodes in the line after the new link and $\mathbf{R}_{G,i}^{(2)}[S_G \to S_L]$ for the nodes in the complete graph after the new link as:*

Fig. 1 Simple line with one link from a complete graph to one node in the line

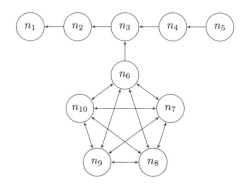

$$\mathbf{R}_{L,i}^{(2)}[S_G \to S_L] = \sum_{k=0}^{n_L-i} c^k + b_{ij} = \frac{1 - c^{n_L-i+1}}{1 - c} + b_{ij}, \tag{7}$$

$$b_{ij} = -c^{j+1-i} \frac{c + (n_G - 1)}{(n_G - 1)c^2 + n_G(n_G - 2)c - n_G(n_G - 1)}, \quad j \geq i,$$

$$b_{ij} = 0, \quad j < i.$$

For the nodes in the complete graph we get:

$$\mathbf{R}_{G,1}^{(2)}[S_G \to S_L] = -\frac{n_G(n_G - 1) + n_G c}{(n_G - 1)c^2 + n_G(n_G - 2)c - n_G(n_G - 1)}, \tag{8}$$

$$\mathbf{R}_{G,j}^{(2)}[S_G \to S_L] = \frac{(c + n_G)(n_G - 1)}{n_G(n_G - 1) - (n_G - 1)c^2 - n_G(n_G - 2)c}, \tag{9}$$

where $\mathbf{R}_{G,1}^{(2)}[S_G \to S_L]$ is PageRank for the node in the complete graph linking to the line and $\mathbf{R}_{G,j}^{(2)}[S_G \to S_L]$ is the PageRank of the other nodes in the complete graph.

Proof For the nodes in the complete graph we get the PageRank immediately from Theorem 3.

For the nodes in the line we get a similar result as when adding a link from a single node to the line in Theorem 2. We get the same PageRank for the nodes we can not reach from the complete graph ($b_{ij} = 0$, $j < i$). For the nodes we can reach we need to modify b_{ij}. The sum of all probability to reach the node in the complete graph linking to the line is found in (5) in Theorem 3.

$$\mathbf{R}_{G,1}^{(2)}[S_G \to S_L] = -\frac{n_G(n_G - 1) + n_G c}{(n_G - 1)c^2 + n_G(n_G - 2)c - n_G(n_G - 1)}.$$

The probability to reach the linked to node in the line e_j is then

$$\left(\frac{c}{n_G}\right) \mathbf{R}_{e_i \in S_G}^{(2)}[S_G \to S_L],$$

and for any further node in the line we need to multiply with c for every extra step. This gives:

$$b_{ij} = -c^{j-i} \frac{c}{n_G} \frac{n_G(n_G - 1) + n_G c}{(n_G - 1)c^2 + n_G(n_G - 2)c - n_G(n_G - 1)} \tag{10}$$

$$= -c^{j+1-i} \frac{c + (n_G - 1)}{(n_G - 1)c^2 + n_G(n_G - 2)c - n_G(n_G - 1)}, \quad j \geq i,$$

and the proof is complete. □

Another way to prove the theorem is by setting up the linear system and using the lemma for blockwise inversion (Lemma 1). If we want to know the common normalized PageRank we find the normalizing constant as the sum of the PageRank of all the nodes:

$$N = \frac{n_L}{1-c} - \frac{c\left(1 - c^{n_L}\right)}{(1-c)^2} \tag{11}$$

$$+ \frac{c\left(1 - c^{n_L - i + 2}\right)(c + n_G - 1)}{(1-c)\left((n_G - 1)c^2 + n_G(n_G - 2)c - n_G(n_G - 1)\right)}$$

$$+ n_G(n_G - 1) + n_G c(n_G - 1)c^2 + n_G(n_G - 2)c - n_G(n_G - 1)$$

$$+ n_G\left((n_G - 1)c^2 + (2n_G(n_G - 1) - 1)c + (n_G(n_G - 1))^2\right)$$

$$\left(c\left((n_G - 1)c^2 + n_G(n_G - 2)c - n_G(n_G - 1)\right)\right.$$

$$\left. + n_G\left((n_G - 1)c^2 + n_G(n_G - 2)c - n_G(n_G - 1)\right) - 1\right)^{-1},$$

which can be used to get the normalized PageRank:

$$\mathbf{R}_i^{(1)}[S_G \to S_L] = \mathbf{R}_i^{(2)}[S_G \to S_L]/N. \tag{12}$$

3.2 Connecting the Simple Line with a Complete Graph by Adding a Link from a Node in the Line to a Node in the Complete Graph

When we instead let one node e_j in the simple line link to one node in the complete graph we get a system that could look like the system in Fig. 2.

For the PageRank we formulate the following:

Theorem 5 *Observing the nodes in a system S made up of two systems, a simple line S_L with n_L nodes and a complete graph S_G with n_G nodes where we add one*

Fig. 2 Simple line with one node in the line linking to a node in a complete graph

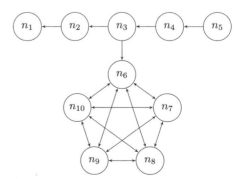

*link from node e_j in the line to node e_g in the complete graph. Assuming uniform **u** we get the PageRank $\mathbf{R}_{L,i}^{(2)}[S_L \to S_G]$ for the nodes in the line after the new link and $\mathbf{R}_{G,i}^{(2)}[S_L \to S_G]$ for the nodes in the complete graph after the new link as:*

$$\mathbf{R}_{L,i}^{(2)}[S_L \to S_G] = \frac{1 - c^{n_L+1-i}}{1 - c}, \quad i \geq j, \tag{13}$$

$$\mathbf{R}_{G,g}^{(2)}[S_L \to S_G] = \left(\frac{c(1 - c^{n_L+1-j})}{2(1 - c)} \right) \left(\frac{((n_G - 1) - c(n_G - 2))}{((n_G - 1) - c(n_G - 2)) - c^2} \right) + \frac{1}{1 - c}, \tag{14}$$

$$\mathbf{R}_{G,i}^{(2)}[S_L \to S_G] = \left(\frac{c^2(1 - c^{n_L+1-j})}{2(1 - c)} \right) \left(\frac{1}{((n_G - 1) - c(n_G - 2)) - c^2} \right) + \frac{1}{1 - c}, \tag{15}$$

$$\mathbf{R}_{L,i}^{(2)}[S_L \to S_G] = \frac{1 - c^{j-i}}{1 - c} + \left(\frac{c^{j-i}}{2} \right) \frac{1 - c^{n_L-j+1}}{1 - c}, \quad i < j. \tag{16}$$

Proof We divide the matrix $(\mathsf{I} - c\mathbf{A}^\top)$ in blocks:

$$(\mathsf{I} - c\mathbf{A}^\top) = \begin{bmatrix} \mathsf{B} & \mathsf{C} \\ \mathsf{D} & \mathsf{E} \end{bmatrix},$$

where B is the part corresponding to the line, C is a zero matrix (since we have no links from nodes in the complete graph to the line). D is a zero matrix except for one element $D_{g,j} = -c/2$, where e_j is j:th the node in the line linking to the complete graph and e_g is the g:th node in the graph linked to by node e_j. We note that j, g are the internal number for the complete graph and line respectively and not their "number" in the combined graph. E is the part corresponding to the complete graph.

In the same way we divide the inverse in blocks:

$$(\mathsf{I} - c\mathbf{A}^\top)^{-1} = \begin{bmatrix} \mathsf{B}^{inv} & \mathsf{C}^{inv} \\ \mathsf{D}^{inv} & \mathsf{E}^{inv} \end{bmatrix}.$$

Using Lemma 1 for blockwise inversion we get:

$$\mathsf{B}^{inv} = (\mathsf{B} - \mathsf{D}\mathsf{E}^{-1}\mathsf{C})^{-1} = \mathsf{B}^{-1}, \tag{17}$$
$$\mathsf{C}^{inv} = -(\mathsf{B} - \mathsf{D}\mathsf{E}^{-1}\mathsf{C})^{-1}\mathsf{C}\mathsf{E}^{-1} = 0,$$
$$\mathsf{D}^{inv} = -\mathsf{E}^{-1}\mathsf{D}(\mathsf{B} - \mathsf{D}\mathsf{E}^{-1}\mathsf{C})^{-1} = \mathsf{E}^{-1}\mathsf{D}\mathsf{B}^{-1},$$
$$\mathsf{E}^{inv} = \mathsf{E}^{-1} + \mathsf{E}^{-1}\mathsf{D}(\mathsf{B} - \mathsf{D}\mathsf{E}^{-1}\mathsf{C})^{-1}\mathsf{C}\mathsf{E}^{-1} = \mathsf{E}^{-1}.$$

Since one of the nodes in the line links out we get B divided in blocks:

$$\mathsf{B} = \begin{bmatrix} \mathsf{B}_B & \mathsf{B}_C \\ \mathsf{B}_D & \mathsf{B}_E \end{bmatrix},$$

$$
B_B = \begin{bmatrix} 1 & -c & 0 & \cdots & 0 \\ 0 & 1 & -c & \ddots & \vdots \\ 0 & 0 & 1 & \ddots & 0 \\ \vdots & \ddots & \ddots & \ddots & -c \\ 0 & \cdots & 0 & 0 & 1 \end{bmatrix}, \quad B_C = \begin{bmatrix} 0 & & \cdots\cdots & & 0 \\ \vdots & & \ddots & \ddots & \vdots \\ 0 & & & \ddots & \vdots \\ -c/2 & 0 & & \cdots & 0 \end{bmatrix},
$$

where B_D is a zero matrix and B_E looks the same as B_B although possibly with a different size. The size of the blocks are: $B_B : (j-1) \times (j-1), B_C : (j-1) \times (n_L - j+1), B_D : (n_L - j+1) \times (j-1)$ and $B_E : (n_L - j+1) \times (n_L - j+1)$, where n_L is the total number of nodes in the line.

For the blocks of the inverse we get:

$$
\begin{aligned}
B_B^{inv} &= B_B^{-1}, \\
B_C^{inv} &= -B_B^{-1} B_C B_E^{-1}, \\
B_D^{inv} &= 0, \\
B_E^{inv} &= E^{-1},
\end{aligned} \tag{18}
$$

B_B^{inv} and B_E^{inv} are found as the inverse for the simple line, leaving B_C^{inv} to be computed. The only difference compared to a simple line is that the only non-zero element in B_C is $-c/2$ rather than $-c$. In other words B^{-1} is exactly as it would have been for a simple line, except block corresponding to B_C^{inv} which is multiplied with 0.5.

We can now find the PageRank of the nodes in the line:

$$
R_{L,i}^{(2)}[S_L \to S_G] = \sum_{k=1}^{n_L} B_{i,k}^{-1} = \sum_{k=1}^{j-1} B_{i,k}^{-1} + \sum_{k=j}^{n_L} B_{i,k}^{-1} \tag{19}
$$

$$
= \sum_{k=0}^{j-i-1} c^k + \sum_{k=j-i}^{n_L-i} \frac{c^k}{2} = \frac{1 - c^{j-i}}{1-c} + \left(\frac{c^{j-i}}{2} \right) \frac{1 - c^{n_L-j+1}}{1-c}.
$$

For the nodes in the complete graph we first need to find D^{inv}, to do so we start by calculating DB^{-1}. Since only one element D_{gj} of D is non-zero, only row g of DB^{-1} can be non-zero. We get row g as:

$$
(DB^{-1})_{\text{row}_g} = \frac{-c}{2} [0 \ \cdots \ 1 \ c \ c^2 \ \cdots \ c^L - j],
$$

where there are $j-1$ zeros before the 1, (B^{-1} upper triangular). Multiplying this with E^{-1} gives:

$$-E^{-1}DB^{-1} = \frac{c}{2} \begin{bmatrix} 0 \ldots 0 \ s \ cs \ \ldots \ c^{n_L-j}s \\ \vdots \ \vdots \ \vdots \ \vdots \ \vdots \ \vdots \ \vdots \\ 0 \ldots 0 \ s \ cs \ \ldots \ c^{n_L-j}s \\ 0 \ldots 0 \ d \ cd \ \ldots \ c^{n_L-j}d \\ 0 \ldots 0 \ s \ cs \ \ldots \ c^{n_L-j}s \\ \vdots \ \vdots \ \vdots \ \vdots \ \vdots \ \vdots \ \vdots \\ 0 \ldots 0 \ s \ cs \ \ldots \ c^{n_L-j}s \end{bmatrix},$$

$$s = \frac{c}{(n_G - 1) - c(n_G - 2) - c^2}, \quad d = \frac{(n_G - 1) - c(n_G - 2)}{(n_G - 1) - c(n_G - 2) - c^2}.$$

E^{-1} was calculated in one of our previous works [7, Lemma 2, p. 14-17] where we considered the problem of a complete graph without any additional connections, E^{-1} can be seen at the end of this proof.

We can now find the PageRank of the nodes in the complete graph by summation of corresponding row:

$$\mathbf{R}_{G,i}^{(2)}[S_L \to S_G] = \sum_{k=1}^{n_L} D_{ik}^{inv} + \sum_{k=1}^{n_L} E_{ik}^{inv} = \sum_{k=1}^{n_L} D_{ik}^{inv} + \frac{1}{1-c}. \tag{20}$$

We separate between the node in the complete graph linked to from the line and the other nodes in the complete graph.

$$\mathbf{R}_{G,i}^{(2)}[S_L \to S_G] = \frac{c}{2} \sum_{k=0}^{n_L-j} c^k s + \frac{1}{1-c} \tag{21}$$

$$= \left(\frac{c(1 - c^{n_L-j+1})}{2(1-c)} \right) \left(\frac{c}{(n_G - 1) - c(n_G - 2) - c^2} \right) + \frac{1}{1-c}, \quad i \neq g,$$

$$\mathbf{R}_{G,g}^{(2)}[S_L \to S_G] = \frac{c}{2} \sum_{k=0}^{n_L-j} c^k d + \frac{1}{1-c} \tag{22}$$

$$= \left(\frac{c(1 - c^{n_L+1-j})}{2(1-c)} \right) \left(\frac{((n_G - 1) - c(n_G - 2))}{((n_G - 1) - c(n_G - 2)) - c^2} \right) + \frac{1}{1-c}.$$

And the proof is complete. For completeness we include the complete inverse as well:

$$(I - cA^\top)^{-1} = \begin{bmatrix} B^{inv} & C^{inv} \\ D^{inv} & E^{inv} \end{bmatrix}, \quad B^{-1} \begin{bmatrix} 1 \ c \ c^2 \ \ldots \ c^{n_L-1}/2 \\ 0 \ 1 \ c \ \ldots \ c^{n_L-2}/2 \\ 0 \ 0 \ 1 \ \ddots \ \vdots \\ \vdots \ \vdots \ \ddots \ \ddots \ c/2 \\ 0 \ 0 \ldots \ 0 \ 1 \end{bmatrix},$$

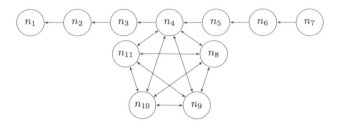

Fig. 3 Simple line with one node in the line being a part of a complete graph

$$
C^{-1} = \begin{bmatrix} 0 \dots 0 \\ \vdots \ddots \vdots \\ 0 \dots 0 \end{bmatrix}, \quad D^{-1} = -E^{-1}DB^{-1}\text{(seen above)},
$$

$$
E^{-1} = \begin{bmatrix}
1 - (n_G - 1)ak & k & k & \dots & k \\
k & 1 - (n_G - 1)ak & k & \dots & k \\
k & k & 1 - (n_G - 1)ak & \ddots & \vdots \\
\vdots & \vdots & & \ddots & k \\
k & k & \dots & k & 1 - (n_G - 1)ak
\end{bmatrix},
$$

$$
a = -c/(n-1), \quad k = \frac{-a}{(1 - a^2 + (n - 2)(a - a^2))}.
$$

$\square$

This could also have been showed using a similar method as the one used to prove Theorem 4. The normalizing constant can then be found by summation of the individual PageRank of all the nodes in order to get the normalized PageRank $\mathbf{R}^{(1)}$.

We note that while the node in the line that links to the complete graph does not lose anything from the new link, the nodes below it in the line do lose quite a lot because of it. Likewise the PageRank of the node thats linked to gains more from the new link than the others in the complete graph.

3.3 Connecting the Simple Line with a Complete Graph by Letting One Node in the Line Be Part of the Complete Graph

If we instead let one node in the line be part of the complete graph we get another interesting example to look at. An example of what the system could look like can be seen in Fig. 3.

We formulate the following theorem for the PageRank of the given example:

Theorem 6 *The PageRank of the nodes in system S_L made up of a simple line and system S_G made up of a complete graph after one of the nodes $e_j \in S_L$ becomes part of the complete graph assuming uniform $\mathbf{u}$ can be written:*

$$\mathbf{R}_{L,i}^{(2)}[S_L \leftrightarrow S_G] = \frac{1 - c^{n_L+1-i}}{1-c}, \quad i > j, \tag{23}$$

$$\mathbf{R}_{L,j}^{(2)}[S_L \leftrightarrow S_G] = \left(\frac{1 - c^{n_L+1-j}}{1-c} + \frac{c(n_G - 1)}{(n_G - 1) - c(n_G - 2)} \right) \tag{24}$$
$$\left(\frac{n_G((n_G - 1) - c(n_G - 2))}{n_G((n_G - 1) - c(n_G - 2)) - c^2(n_G - 1)} \right),$$

$$\mathbf{R}_{L,i}^{(2)}[S_L \leftrightarrow S_G] = \frac{c^{j-i}\mathbf{R}_{L,j}^{(2)}}{n_G} + \frac{1 - c^{j-i}}{1-c}, i < j, \tag{25}$$

$$\mathbf{R}_{G,i}^{(2)}[S_L \leftrightarrow S_G] = \frac{(c + n_G)(n_G - 1)(1 - c) + (n_G - 1)c^2(1 - c^{n_L-j})}{(1 - c)(n_G(n_G - 1) - (n_G - 1)c^2 - n_G(n_G - 2)c)}, \tag{26}$$

where $\mathbf{R}_{G,i}^{(2)}[S_L \leftrightarrow S_G]$ is the PageRank for the nodes in the complete graph (except the node also being a part of the line) and $\mathbf{R}_{L,i}^{(2)}[S_L \leftrightarrow S_G]$ is the PageRank of nodes in the line. n_G, n_L is the number of nodes in the complete graph and simple line respectively after making one node in the line part of the complete graph.

Proof For the proof of the nodes $e_i \in S_L$, $i > j$ we get the PageRank for a simple line. In order to find $\mathbf{R}_{L,j}^{(2)}[S_L \leftrightarrow S_G]$ we first use Theorem 1 to write it as:

$$\mathbf{R}_{L,j}^{(2)}[S_L \leftrightarrow S_G] = \left(\sum_{e_i \in S, e_i \neq e_j} P(e_i \to e_j) + 1 \right) \left(\sum_{k=0}^{\infty} (P(e_j \to e_j))^k \right), \tag{27}$$

where $P(e_i \to e_j)$ is the probability of getting from node e_i to node e_j.

$$\sum_{e_i \in S, e_i \neq e_j} P(e_i \to e_j) + 1 = \sum_{k=0}^{n_L+1-j} c^k + (n - 1)\frac{c}{n_G - 1} \sum_{k=0}^{\infty} \left(\frac{c(n_G - 2)}{(n_G - 1)} \right)^k \tag{28}$$
$$= \frac{1 - c^{n_L+1-j}}{1-c} + \frac{c(n_G - 1)}{(n_G - 1) - c(n_G - 2)},$$

$$P(e_j \rightarrow e_j) = \frac{c(n_G - 1)}{n_G} \frac{c}{n_G - 1} + \frac{c(n_G - 1)}{n_G} \frac{c(n_G - 2)}{n_G - 1} \frac{c}{n_G - 1} \tag{29}$$

$$+ \frac{c(n_G - 1)}{n_G} \frac{c^2(n_G - 2)^2}{(n_G - 1)^2} \frac{c}{n_G - 1} + \cdots$$

$$= \frac{c^2}{n_G} \sum_{k=0}^{\infty} \left(\frac{c(n_G - 2)}{n - 1} \right)^k = \frac{c^2(n_G - 1)}{n_G((n_G - 1) - c(n_G - 2))},$$

$$\sum_{k=0}^{\infty} (P(e_j \rightarrow e_j))^k = \frac{n_G((n_G - 1) - c(n_G - 2))}{n_G((n_G - 1) - c(n_G - 2)) - c^2(n_G - 1)}. \tag{30}$$

Multiplication of the results from Eqs. (29) and (30) gives

$$\mathbf{R}_{L,j}^{(2)}[S_L \leftrightarrow S_G] = \left(\frac{1 - c^{n_L + 1 - j}}{1 - c} + \frac{c(n_G - 1)}{(n_G - 1) - c(n_G - 2)} \right) \tag{31}$$

$$\left(\frac{n_G((n_G - 1) - c(n_G - 2))}{n_G((n_G - 1) - c(n_G - 2)) - c^2(n_G - 1)} \right).$$

For the nodes below the one in the complete graph we can write the PageRank as:

$$\mathbf{R}_{L,i}^{(2)}[S_L \leftrightarrow S_G] = \mathbf{R}_{L,j}^{(2)}[S_L \leftrightarrow S_G]P(e_j \rightarrow e_i) + \sum_{k=0}^{j-i-1} c^k, \quad i < j, \tag{32}$$

$$P(e_j \rightarrow e_i) = \frac{c^{j-i}}{n_G}. \tag{33}$$

This gives:

$$\mathbf{R}_{L,i}^{(2)}[S_L \leftrightarrow S_G] = \frac{c^{j-i}\mathbf{R}_{L,j}^{(2)}[S_L \leftrightarrow S_G]}{n_G} + \frac{1 - c^{j-i}}{1 - c}, i < j. \tag{34}$$

Left to prove we have the formula for the nodes in the complete graph not directly connected to the line $\mathbf{R}_{G,i}^{(2)}[S_L \leftrightarrow S_G]$. We do not need to consider the part of the line following the complete graph, since we can not get from this part of the graph back to the complete graph. We already have the PageRank for the nodes not linking out in the complete graph in the case where we have no line of nodes linking to the complete graph from Theorem 3.

Since all paths $P(e_k \rightarrow e_i)$ where $e_k \in S_L$ need to go through node e_j we can write these as a product of the probability to get to node e_j times the probability to from there get to node e_i for which we want to calculate PageRank.

$$\mathbf{R}_{G,i}^{(2)}[S_L \leftrightarrow S_G] = \left(\sum_{k=0}^{\infty} (P(e_i \to e_i))^k \right) \qquad (35)$$

$$\left(\left(\sum_{\substack{e_k \in S_L \\ e_k \neq e_j}} P(e_k \to e_j) + 1 \right) P(e_j \to e_i) + 1 + \sum_{\substack{e_k \in S_G \\ e_k \neq e_j, e_i}} P(e_k \to e_i) \right).$$

Looking at the part depending on e_j we get

$$P(e_j \to e_i) \left(\sum_{k=0}^{\infty} (P(e_i \to e_i))^k \right) = \frac{c(n_G - 1)}{n_G(n_G - 1) - n_G(n_G - 2)c - (n_G - 1)c^2}. \qquad (36)$$

This can be either in the matrix proof of Theorem 3 in [7] or by calculating corresponding hitting probabilities. Using this we can decompose the PageRank into two parts

$$\mathbf{R}_{G,i}^{(2)}[S_L \leftrightarrow S_G] = + \left(1 + \sum_{\substack{e_k \in S_G \\ e_k \neq e_j, e_i}} P(e_k \to e_i) \right) \left(\sum_{k=0}^{\infty} (P(e_i \to e_i))^k \right) \qquad (37)$$

$$\left(\sum_{\substack{e_k \in S_L \\ e_k \neq e_j}} P(e_k \to e_j) + 1 \right) P(e_j \to e_i) \left(\sum_{k=0}^{\infty} (P(e_i \to e_i))^k \right).$$

Decomposing the PageRank for a complete graph with links out and the PageRank for a simple line for the part corresponding to the line we get

$$\mathbf{R}_{G,i}^{(2)}[S_L \leftrightarrow S_G] = \frac{(n_G - 1)(n_G + c)}{n_G(n_G - 1) - n_G(n_G - 2)c - (n_G - 1)c^2} \qquad (38)$$

$$+ \left(\frac{c(1 - c^{n_L - j})}{1 - c} \right) \frac{c(n_G - 1)}{n_G(n_G - 1) - n_G(n_G - 2)c - (n_G - 1)c^2}$$

$$= \frac{(c + n_G)(n_G - 1)(1 - c) + (n_G - 1)c^2(1 - c^{n_L - j})}{(1 - c)(n_G(n_G - 1) - (n_G - 1)c^2 - n_G(n_G - 2)c)}.$$

□

Theorem 7 *The normalizing constant N for the simple line with one node being part of a complete graph using uniform $\mathbf{u}$ can be written as:*

$$N = (n_G - 1)\, \mathbf{R}^{(2)}_{G,i \notin L}[S_L \leftrightarrow S_G] + \mathbf{R}^{(2)}_{L,j}[S_L \leftrightarrow S_G] + \frac{n_L - 1}{1 - c} \tag{39}$$

$$-\frac{c\left(1 - c^{n_L - j}\right)}{(1 - c)^2} - \frac{c\left(1 - c^{j-1}\right)}{(1 - c)^2} + \frac{c\left(1 - c^{j-1}\right) \mathbf{R}^{(2)}_{L,j}[S_L \leftrightarrow S_G]}{n_G\,(1 - c)}$$

where:

- $\mathbf{R}^{(2)}_{G,i}[S_L \leftrightarrow S_G]$ *is the PageRank of nodes in the complete graph,*
- $\mathbf{R}^{(2)}_{L,j}[S_L \leftrightarrow S_G]$ *is for the node in both the line and complete graph, and*
- $\mathbf{R}^{(2)}_{L,i}[S_L \leftrightarrow S_G]$ *is for the nodes in the line.*

Proof The normalizing constant is equal to the sum of the non-normalized PageRank of all nodes.

We got n_G nodes in the complete graph, $(n - 1)$ not directly connected to the line and one connected to the line. This gives:

$$N = (n - 1)\mathbf{R}^{(2)}_{G,i}[S_L \leftrightarrow S_G] + \mathbf{R}^{(2)}_{L,j}[S_L \leftrightarrow S_G] + \sum_{i \neq j} \mathbf{R}^{(2)}_{L,j}[S_L \leftrightarrow S_G], \tag{40}$$

where $\mathbf{R}^{(2)}_{L,j}[S_L \leftrightarrow S_G]$ is the PageRank of individual nodes in the line except for the node node j in the line for which we have $\mathbf{R}^{(2)}_{L,j}[S_L \leftrightarrow S_G]$. For those nodes we got PageRank:

$$\mathbf{R}^{(2)}_{L,j}[S_L \leftrightarrow S_G] = \begin{cases} \dfrac{1 - c^{n_L + 1 - i}}{1 - c}, & i > j, \\[3mm] \dfrac{c^{j-i} \mathbf{R}^{(2)}_{L,j}}{n_G} + \dfrac{1 - c^{j-i}}{1 - c}, & i < j. \end{cases} \tag{41}$$

The sum of all nodes for which $i > j$ can be written:

$$\sum_{i=j+1}^{n_L} \mathbf{R}^{(2)}_{L,j}[S_L \leftrightarrow S_G] = \frac{n_L - j}{1 - c} - \frac{c\left(1 - c^{n_L - j}\right)}{(1 - c)^2}, \tag{42}$$

where we use that the second part $\sum_{i=j}^{n_L} \frac{-c^{n_L + 1 - i}}{1 - c}$ is a geometric sum. Calculating the sum for $i < j$ in the same way we get:

$$\sum_{i=1}^{j-1} \mathbf{R}^{(2)}_{L,j}[S_L \leftrightarrow S_G] = \frac{j - 1}{1 - c} - \frac{c\left(1 - c^{j-1}\right)}{(1 - c)^2} + \frac{c\left(1 - c^{j-1}\right) \mathbf{R}^{(2)}_{L,j}}{n_G\,(1 - c)}. \tag{43}$$

Summation of all individual parts completes the proof.

$\square$

Now that we have an explicit formula for this example we can look at what happens when we change various parameters like c or the size of either the line or complete graph.

4 A Closer Look at the Formulas for PageRank in Our Examples

Now that we have formulas for the PageRank of a couple different graph structures we are going to take a short look at what happens when we change some parameters. We will also take a look at the partial derivative with respect to c.

4.1 Partial Derivatives with Respect to C

In the case of the simple line with formula as seen earlier we get the derivative with respect to c as:

$$\frac{\partial}{\partial c} \mathbf{R}_i^{(2)}[S_L] = (1-c)^{-2} - \frac{c^{n_L-i+1}(n_L-i+1)}{c(1-c)} - \frac{c^{n_L-i+1}}{(1-c)^2}. \tag{44}$$

Rewriting it and looking to see if it is positive we get:

$$\frac{1 + c^{n_L-i}(i-n_L)(1-c)}{(-1+c)^2} \geq 0 \Leftrightarrow c^{n_L-i}((i-n_L)(1-c) + \frac{1}{c^{n_L-i}}) \geq 0, \tag{45}$$

$$\Leftrightarrow \frac{1}{c^{n_L-i}} \geq (n_L-i)(1-c) \Leftrightarrow \frac{1}{1-c} \geq (n_L-i)c^{n_L-i} \Leftrightarrow \sum_{k=0}^{\infty} c^k \geq (n_L-i)c^{n_L-i}. \tag{46}$$

Since we have $0 < c < 1$, $n_L \geq i$ we have that $c^k > c^{k+1}$ the first $n_L - i$ elements of the left sum is at least as large as c^{n_L-i}, this gives:

$$\sum_{k=0}^{\infty} c^k \geq \sum_{k=1}^{n_L-i} c^k \geq (n_L-i)c^{n_L-i}. \tag{47}$$

For our case with a line connected to a complete graph by letting one node in the complete graph be part of the line we get the following derivative with respect to c:

$$\frac{\partial}{\partial c} \mathbf{R}_{L,j}^{(2)}[S_L \leftrightarrow S_G] \tag{48}$$

$$= \frac{\left(\left((-1+c)n_G^2 + (-1+c)^2 n_G - c^2\right)\left((-1+c)n_G - 2c + 1\right)\frac{\partial}{\partial c}G(c)\right)n_G}{\left((-1+c)n_G^2 + (-1+c)^2 n_G - c^2\right)^2}$$

$$\frac{-\left((n_G-1)\left(c\left((-2+c)n_G + 2 - 2c\right)G(c) - (n_G-1)\left(n_G + c^2\right)\right)\right)n_G}{\left((-1+c)n_G^2 + (-1+c)^2 n_G - c^2\right)^2},$$

$$G(c) = \frac{1 - c^{n_L - j + 1}}{1 - c},$$

$$\frac{\partial}{\partial c} G(c) = (1 - c)^{-2} - \frac{c^{n_L - j + 1} (n_L - j + 1)}{c(1 - c)} - \frac{c^{L - j + 1}}{(1 - c)^2}.$$

The derivatives have about the same shape as the original function. As c gets large so does the derivative and as n_G increases the slope gets steeper for large c.

Looking at the other nodes in the complete graph we get the derivative with respect to c as:

$$\frac{\partial}{\partial c} \mathbf{R}_{G,i}^{(2)} [S_L \leftrightarrow S_G] \qquad (49)$$

$$= \frac{(n_G - 1)(1 - c) - (c - n_G)(n_G - 1) + 2(n_G - 1)c\left(1 - c^{n_L - j}\right)}{(1 - c)\left(n_G(n_G - 1) - (n_G - 1)c^2 - n_G(n_G - 2)c\right)}$$

$$- \frac{(n_G - 1)c^{1 + n_L - j}(n_L - j)}{(1 - c)\left(n_G(n_G - 1) - (n_G - 1)c^2 - n_G(n_G - 2)c\right)}$$

$$+ \frac{(c + n_G)(n_G - 1)(1 - c) + (n_G - 1)c^2\left(1 - c^{n_L - j}\right)}{(1 - c)^2\left(n_G(n_G - 1) - (n_G - 1)c^2 - n_G(n_G - 2)c\right)}$$

$$- \frac{\left((c + n_G)(n_G - 1)(1 - c) + (n_G - 1)c^2\left(1 - c^{n_L - j}\right)\right)(2c + (2 - 2c - n_G)n_G)}{(1 - c)\left(n_G(n_G - 1) - (n_G - 1)c^2 - n_G(n_G - 2)c\right)^2}.$$

4.1.1 Changes in the Size of the Complete Graph for Our Last Example

When we change the size of the complete graph we can see for example what size would be the most effective for increasing ones PageRank. In all these examples we will use $n_L = 10, j = 6, c = 0.85$ and n_G will wary between 1 and 50. First we note that the part above the complete graph is unaffected by the change of n_G. It is obvious however that as n_G increases the normalizing constant in the normalized PageRank will likely get larger resulting in a lower PageRank as long as it is part of a small system.

For the nodes in the complete graph we get the result in Fig. 4.

Here we see something curious, the nodes seems to be gaining rank in the beginning while starting to fall after a while and possibly converging towards a single value depending on if it links out of the complete graph or not. The reason we get a local maximum is the fact that for a moderately large n_G we maximize the probability of $\mathbf{R}_{L,j}^{(2)}$ getting back to itself while keeping the complete graph lare enough to keep most probability for itself. Here we can see that it is not always a good idea for an individual node to join a complete graph. If the node in question already have larger PageRank than the other nodes in the complete graph it actually might lose PageRank from joining it.

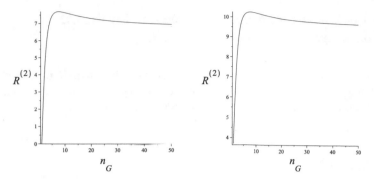

Fig. 4 $\mathbf{R}^{(2)}$ of the nodes in the complete graph not part of the line (*left*) and part of the line (*right*) as a function of n_G

The result for the node below the complete graph we get the result in Fig. 5.

Here we see the great loser as n_G increases. Since the chance of escaping the complete graph depends on $\mathbf{R}^{(2)}_{S_G,j}[S_L \leftrightarrow S_G]/n_G$ as n_G increases so does this nodes PageRank as well. From this we see a clear example of the effects of complete graphs on its surrounding nodes. A complete graph can be seen as a type of sink, all links to the complete graph will be used to maximum effect within the complete graph. And even worse, even if the complete graph have some nodes that point out of it their influence will be very small since the nodes in the complete graph having a large number of links the chance of escaping is low.

Fig. 5 $\mathbf{R}^{(2)}$ of the nodes in the line below the complete graph as a function of n_G

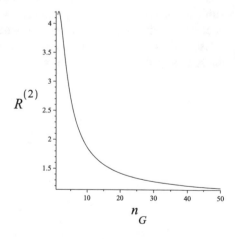

5 A Look at the Normalized PageRank for the Line Connected with a Complete Graph

Looking at the normalized PageRank in our last example with a simple line with one node being part of a complete graph we want to see how the PageRank changes as c or the relation between the size of the line or complete graph changes.

5.1 Dependence on C

Plotting the PageRank with $n_G = 10$, $n_L = 10$, $j = 6$ and $c \in [0.01, 0.99]$ we get the following results. For the nodes in the line above and part of the complete graph $\mathbf{R}^{(1)}_{S_L,i}[S_L \leftrightarrow S_G]$, $i = 7$ we get the result in Fig. 6.

Looking at the results for the nodes above the complete graph we see that the function seems to have a max at about $c = 0.55$ after which it decreases faster the closer to $c = 1$ it gets. Looking at the node part of the complete graph we see the great "winner" as c increases. Do note the difference in the axis for the different images, since this at its lowest point is actually about the same as the highest for the node above the complete graph.

We find the c which maximize the function for some other different parameters n_G, n_L, j, i in Table 1. All the local max/min is calculated using the optimization tool in Maple 15. For the nodes above the complete graph the location of the maximum seems to be moving towards the left as n_G increases and towards the right as n_L increases. In the same manner it moves towards the left as i get closer to n_L. The value of the maximum is only included out of completeness, it is natural that they decrease as either n_G or n_L increases as we in those cases get a larger number of total nodes in the system. It is interesting to note that the max seems to be going towards the right as both n_G, n_L increases as well.

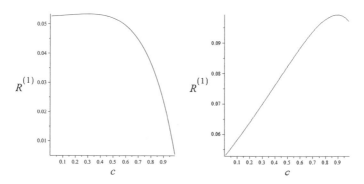

Fig. 6 $\mathbf{R}^{(1)}$ of the node above the complete graph (*left*) and of the node in the line being a part of the complete graph (*right*) as a function of c

Table 1 Maximum PageRank $\mathbf{R}^{(1)}$ of node i "above" the complete graph and node j depending on c for various changes in the graph where one node in a simple line is part of a complete graph

n_G	n_L	j	i	$\arg_c \max \mathbf{R}_i^{(1)}$	$\max \mathbf{R}_i^{(1)}$	$\arg_c \max \mathbf{R}_j^{(1)}$	$\max \mathbf{R}_j^{(1)}$
5	10	6	7	0.349	0.073	1	0.164
10	10	6	7	0.300	0.053	0.894	0.099
20	10	6	7	0.248	0.035	0.776	0.059
10	20	6	7	0.751	0.370	1	0.096
20	20	6	7	0.721	0.027	0.929	0.056
50	50	6	7	0.874	0.010	0.965	0.023
10	10	9	10	0.000	0.053	1	0.091
10	10	3	4	0.515	0.054	0.893	0.107
10	10	6	9	0.300	0.053	-	-

For the node in the line that is part of the complete graph the PageRank of this node is the largest when c is large, sometimes with a local maximum and sometimes not. It seems to be that as the number of nodes in the complete graph increases we are more likely to find a local maximum than not.

For the node just below the complete graph as well as the nodes in the complete graph not part of the line we get the results in Fig. 7.

PageRank of the node below the complete graph decreases as c increases, but compared to the nodes above the complete graph not as fast for large c. This since the PageRank of the nodes in the complete graph increase so fast for large c that even the comparatively small influence it have on the nodes out of it is enough to at least stop the extremely rapid loss of rank as for the nodes above the complete graph.

As with the node in both the line and complete graph, PageRank of the other nodes in the complete graph increases very fast for large c. We once again see a hint to why a to large c could be problematic, it is for large c we get the largest relative

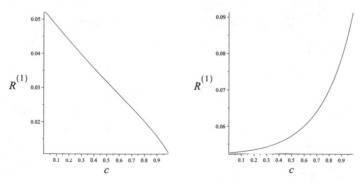

Fig. 7 $\mathbf{R}^{(1)}$ of the node below the complete graph (*left*) and of a node in the complete graph as a function of c

changes in PageRank between nodes. We have no min/max here, instead PageRank increases faster and faster the larger c gets.

We note that these local maximum and minimum are not always present. In these cases we have a PageRank thats decreasing as c increases for the whole interval. If the one exist we can expect the other to as well (since we expect the rank to decrease at the end of the interval). It is hard to say anything conclusive about the location or existence of local maximum or minimum points, but we do note that they exist. There is also a large difference in how PageRank changes for different (especially large) c, we can therefor expect c to have an effect not only in the final rank and the computational aspect, but also the final ranking order of pages.

5.2 A Look at the Partial Derivatives with Respect to C

Since we have the formulas for the normalized PageRank it is also possible to find the partial derivatives. Since the partial derivatives result in very large expressions (multiple pages each) they are not included here. By setting $n_G = n_L = 10, j = 6$ we get the result after taking the partial derivative with respect to c for $0.05 < c < 0.95$ for the node $e_{L,7}$ above the complete graph as well as the node in the line part of the complete graph in Fig. 8.

We see the derivative falling faster as c increases. Here as well we see the more dramatic changes in large c above about 0.8. Apart from seeing the maximum at around $c = 0.3$ in the original function we can also see that the derivative seems to briefly increase in the beginning, reaching a maximum at about $c = 0.1$. For the node part of both the line and the complete graph we get the result in Fig. 8.

We can see a high derivative all the way until we get to very large c where it finally starts going down. We can clearly see the maximum at about $c \approx 0.9$ in the original function. For the node on the line below the complete graph we get the result in Fig. 9.

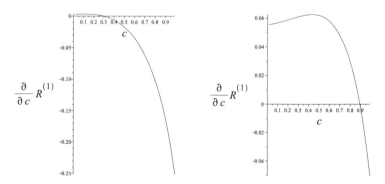

Fig. 8 Partial derivative with respect to c of normalized PageRank of the node in the line above the complete graph (*left*) and of the node in the line part of the complete graph (*right*)

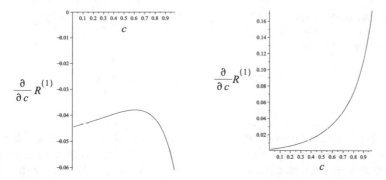

Fig. 9 Partial derivative with respect to c of normalized PageRank of the node in the line below the complete graph (*left*) and of a node in the complete graph not part of the line (*right*)

Although the derivative is decreasing for all c, the derivative have a local maximum at about $c \approx 0.6$. Worth to note is that the axis can be a little misleading, the partial derivative is in fact not that close to 0 at the local maximum. As before the largest changes are at high c. Worth to note that the derivative is decreasing for all c. For the nodes in the complete graph not part of the line we get the result found in Fig. 9.

As before the largest changes are found at large c. Compared to the node part of both the complete graph and the line the derivative for the ones only in the complete graph continue to increase as c increases, however the PageRank itself is not actually ever higher for the ones not part of the line. We have seen that although it is possible to find symbolic expressions for the PageRank and derivative for some simple graphs, as the complexity of the graph increases it becomes very hard to do. Already for these simple examples the partial derivatives a rather large and complicated expressions. Finding more general symbolic expressions for when the derivative is *zero* should be possible although problematic given the constraints and size of the problem.

5.3 A Comparison of Normalized and Non-normalized PageRank

Here we will take a short look at the difference between normalized ($\mathbf{R}^{(1)}$) and non normalized ($\mathbf{R}^{(2)}$) PageRank in order to get a bigger understanding of the differences between them. We already know that $\mathbf{R}^2 \propto \mathbf{R}^{(1)}$ so there will always be the same relation between the PageRank of two nodes. Here we will take a look at how the absolute difference between nodes and the two types of PageRank differ instead.

Since the PageRank is normalized to one in $\mathbf{R}^{(1)}$ we obviously get that the PageRank will decrease as the number of nodes increases, potentially making for problems with number-representation for extremely large graphs unless it is taken into account when making the implementation. This problem is not as large a problem for $\mathbf{R}^{(2)}$ since most nodes will have approximately the same size regardless of the size of the

Fig. 10 A complete graph
(*left*) and a system made of
four dangling nodes (*right*)

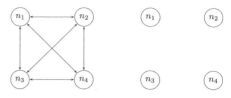

graph. However the possible huge relative difference between nodes is still needed to take into consideration. We note however that with the current way to calculate $\mathbf{R}^{(2)}$ by solving the equation system such large systems that could potentially be a problem in $\mathbf{R}^{(1)}$ is simply to large for us to solve in a timely manner.

We also have one other main difference between the normalized and non-normalized PageRank and that is with dangling nodes and how they effect the global PageRank. In $\mathbf{R}^{(2)}$ a dangling nodes means some of the "probability" escape the graph resulting in a lower total PageRank (but still proportional to $\mathbf{R}^{(1)}$). In $\mathbf{R}^{(1)}$ however dangling nodes can be seen as linking to all nodes and in fact behaves exactly as if they did. We illustrate the difference in a rather extreme example with a graph composed of only four dangling nodes as well as a complete graph composed of four nodes.

An image of the systems can be seen in Fig. 10 below. When computing $\mathbf{R}^{(1)}$ of both systems assuming uniform weight vector $\mathbf{u}$ they are both obviously equal with PageRank $\mathbf{R}^{(1)} = [1/4, 1/4, 1/4, 1/4]$, it does not even matter what c we chose as long as it is between zero and one for convergence. However for the non normalized PageRank we get a large difference between the PageRank of the two systems where we for the complete graph get the PageRank $\mathbf{R}_a^{(2)} = [1/1 - c, 1/1 - c, 1/1 - c, 1/1 - c]$ as seen in [7]. However for the graph made up of only dangling nodes we get the PageRank $\mathbf{R}_b^{(2)} = [1, 1, 1, 1]$ regardless of c. We see that while they might be proportional to each other, the non normalized version behaves differently for dangling nodes making a distinction between dangling nodes and nodes that link to all nodes (including itself which we normally do not allow). While this distinction might seem unnecessary since nodes that link to all nodes do not normally exist or similar nodes such as a node that links to all or most other nodes should either be extremely uncommon or plain do not exist as well, this might not be the case if working with smaller link structures where such a distinction might be useful. It is also this distinction that makes it possible to make comparisons of PageRank between different systems in $\mathbf{R}^{(2)}$ while not generally possible in $\mathbf{R}^{(1)}$.

6 Conclusions

We have seen that we can solve the resulting equation system instead of using the definition directly or using an iterative method such as the Power method. While this method is significantly slower it has made it possible to get a bigger understanding of the different roles of the link matrix $\mathbf{A}$ and how the parameter c influence the ranking. We have seen how PageRank changes when connecting two systems: a line of nodes and a complete graph in various ways. For these systems we also found explicit expressions for the PageRank and showed two different ways to find these. Either by solving the equation system itself or by calculating:

$$
\left(\sum_{e_i \in S, \; e_i \neq e_g} P(e_i \to e_g) + 1 \right) \left(\sum_{k=0}^{\infty} (P(e_g \to e_g))^k \right),
$$

where $P(e_i \to e_g)$ is the sum of probability of all paths from node e_i to node e_g and the weight vector $\mathbf{u}$ is uniform.

Given the expressions for PageRank we looked at the results when changing some parameters. While it is hard to say anything specific, two things seem to be true overall: The most dramatic changes happens as c get large, usually somewhere where $c > 0.8$ some nodes get dramatically larger PageRank compared to the other. We also see that complete graphs, while not gaining a larger rank if the graph is larger, it becomes a lot more reliable (as in not as effected in changes of individual nodes) in keeping its large PageRank as the structure get larger.

We saw that if using uniform $\mathbf{V}$ it is possible to split a large system S into multiple disjoint systems $S_1, S_2, \ldots, S_N$ it is possible to calculate $\mathbf{R}^{(2)}$ for every subsystem itself and they will not differ from $\mathbf{R}^{(1)}$ apart from a normalizing constant that is the same across all subsystems. This is a property we would like to if possible have when using the power method as well. This since it could potentially greatly reduce the work needed primary when doing updates in the system.

Acknowledgements This research was supported in part by the Swedish Research Council (621-2007-6338), Swedish Foundation for International Cooperation in Research and Higher Education (STINT), Royal Swedish Academy of Sciences, Royal Physiographic Society in Lund and Crafoord Foundation.

References

1. Battiston, S., Puliga, M., Kaushik, R., Tasca, P., Caldarelli, G.: DebtRank: Too central to fail? Financial networks, the FED and systemic risk. Sci. Rep. **2**, 541 (2012)
2. Berman, A., Plemmons, R.: Nonnegative Matrices in the Mathematical Sciences. No. del 11 in Classics in Applied Mathematics. Society for Industrial and Applied Mathematics (1994)
3. Bernstein, D.: Matrix Mathematics. Princeton University Press, Princeton (2005)
4. Brin, S., Page, L.: The anatomy of a large-scale hypertextual web search engine. Comput. Network. ISDN Syst. **30**(1–7), 107–117 (1998). Proceedings of the Seventh International World Wide Web Conference
5. Dhyani, D., Bhowmick, S.S., Ng, W.K.: Deriving and verifying statistical distribution of a hyperlink-based web page quality metric. Data Knowl. Eng. **46**(3), 291–315 (2003)
6. Engström, C., Silvestrov, S.: A componentwise pagerank algorithm. In: Applied Stochastic Models and Data Analysis (ASMDA 2015). The 16th Conference of the ASMDA International Society (2015, in press)

7. Engström, C., Silvestrov, S.: Pagerank, a look at small changes in a line of nodes and the complete graph. *Engineering Mathematics II*, Springer Proceedings in Mathematics & Statistics **179** (2016)
8. Gantmacher, F.: The Theory of Matrices. Chelsea, New York (1959)
9. Haveliwala, T., Kamvar, S.: The second eigenvalue of the google matrix. Technical report 2003-20, Stanford InfoLab (2003)
10. Ishii, H., Tempo, R., Bai, E.W., Dabbene, F.: Distributed randomized pagerank computation based on web aggregation. In: Proceedings of the 48th IEEE Conference on Decision and Control, 2009 Held Jointly with the 2009 28th Chinese Control Conference. CDC/CCC 2009, pp. 3026–3031 (2009)
11. Kamvar, S., Haveliwala, T.: The condition number of the pagerank problem. Technical report 2003-36, Stanford InfoLab (2003)
12. Kamvar, S.D., Schlosser, M.T., Garcia-Molina, H.: The eigentrust algorithm for reputation management in p2p networks. In: Proceedings of the 12th International Conference on World Wide Web. WWW 2003, pp. 640–651. ACM, New York (2003)
13. Lancaster, P.: Theory of Matrices. Academic Press, New York (1969)
14. Norris, J.R.: Markov Chains. Cambridge University Press, New York (2009)
15. Tobias, R., Georg, L.: Markovprocesser. Univ., Lund (2000)

Graph Centrality Based Prediction
of Cancer Genes

Holger Weishaupt, Patrik Johansson, Christopher Engström, Sven Nelander,
Sergei Silvestrov and Fredrik J. Swartling

Abstract Current cancer therapies including surgery, radiotherapy and
chemotherapy are often plagued by high failure rates. Designing more targeted
and personalized treatment strategies requires a detailed understanding of druggable
tumor driver genes. As a consequence, the detection of cancer driver genes has
evolved to a critical scientific field integrating both high-throughput experimental
screens as well as computational and statistical strategies. Among such approaches,
network based prediction tools have recently been accentuated and received major
focus due to their potential to model various aspects of the role of cancer genes in
a biological system. In this chapter, we focus on how graph centralities obtained
from biological networks have been used to predict cancer genes. Specifically, we
start by discussing the current problems in cancer therapy and the reasoning behind
using network based cancer gene prediction, followed by an outline of biological

H. Weishaupt (✉) · P. Johansson
Department of Immunology, Genetics and Pathology, Science for Life Laboratory,
Uppsala University, Dag Hammarskjöldsväg 20, 751 85 Uppsala, Sweden
e-mail: holger.weishaupt@igp.uu.se

P. Johansson
e-mail: patrik.johansson@igp.uu.se

C. Engström · S. Silvestrov
Division of Applied Mathematics, School of Education, Culture and Communication,
Mälardalen University, Box 883, 721 23 Västerås, Sweden
e-mail: christopher.engstrom@mdh.se

S. Silvestrov
e-mail: sergei.silvestrov@mdh.se

S. Nelander · F.J. Swartling
Department of Immunology, Genetics and Pathology, Uppsala University, Dag
Hammarskjöldsväg 20, 751 85 Uppsala, Sweden
e-mail: sven.nelander@igp.uu.se

F.J. Swartling
e-mail: fredrik.swartling@igp.uu.se

© Springer International Publishing Switzerland 2016
S. Silvestrov and M. Rančić (eds.), *Engineering Mathematics II*,
Springer Proceedings in Mathematics & Statistics 179,
DOI 10.1007/978-3-319-42105-6_13

networks, their generation and properties. Finally, we review major concepts, recent results as well as future challenges regarding the use of graph centralities in cancer gene prediction.

Keywords Biological networks · Graph centrality · Disease genes · Gene prioritization

1   Introduction

Efforts towards understanding and treating cancer have received major research focus for many decades. However, despite tremendous progress made during this time, mortality rates among cancer patients still remain high [149], implicating cancer as one of the leading causes for human deaths [7].

The recent technological advancements that facilitate high-throughput simultaneous measurements of thousands of biological entities have now given researchers access to an even more detailed insight into the mechanisms underlying cancer. While such data have enabled the identification of numerous genes and pathways mis-regulated in and potentially causing cancer, related analyses at this resolution have also demonstrated that cancer is much more diverse and complex than what was initially expected.

Specifically, transcriptional and epigenetic studies have revealed that individual tumor types can manifest in a multitude of different molecular appearances, also referred to as subtypes or subgroups [61, 153, 159, 165, 174]. While it was often assumed that each patient could be assigned to one unique such subgroup, recent studies in the malignant brain tumor Glioblastoma have further demonstrated that different subgroups of tumors might coexist in different regions [154] or even intervariably change from cell to cell [120] in the same patient. Additionally, the bulk tumor might constitute numerous different cancer cell clones [110, 154], each of which in turn could entail a hierarchy of cancer related progeny cells [109], as well as other cell types from the tumor micro-environment [10]. In summary, cancer has presented itself as a rather heterogeneous disease not only at an inter- but also at an intratumoral level (compare also reviews [109, 197]).

While the presence of different tumor subtypes with varying clinical prognoses suggests a need for more personalized therapies on one hand [32, 36], the aforementioned intratumor heterogeneity on the other hand presents one of the greatest obstacles towards the development of any successful treatment option. Specifically, considering the high degree of cellular diversity in the tumor mass, it is not only difficult to determine the individual tumor driving cells, but also to predict tumor plasticity due to sub-clonal interactions and dynamics upon targeted treatment [92, 105, 108]. A recent study in the malignant childhood brain tumor Medulloblastoma has for instance demonstrated that between diagnosis and relapse there is often a very low agreement between genomic events and thus likely also dominant clones [115]. Coupled to a persisting lack of understanding of what drives the abnormal growth of many such cancer cells, it remains still a challenge to design drug treatments that can effectively eliminate specific let alone all cancer cells in a patient. Ultimately, it is assumed that cancer treatments currently often face the problem that only part of the tumor bulk will be removed, while the remaining cells either inherently or through selective pressure acquiring resistance will survive treatment and ultimately constitute the tumor relapse [105, 108].

Towards overcoming these failures in current therapy, improved strategies will be required, which likely comprise combination drug treatments alone or in connection with other treatment options [39, 92, 115, 200]. Implementing such strategies in turn requires a better knowledge about the drug targetable cancer genes that enable cells to drive the tumor development, metastatic dissemination and to facilitate treatment resistances.

Given the ease of access to high-throughput 'omic's data, cancer related genes can nowadays be predicted in a variety of different ways, the preferred alternative of which is often the direct determination of genomic abnormalities in cancer patients using sequencing or microarray based platforms. In particular, a related systematic approach that has rapidly grown in importance during the last decade is the genome wide association study (GWAS), in which the frequency of genetic variants, often in the form of single-nucleotide polymorphisms (SNPs), are investigated in patient and control cohorts to identify associations between phenotypes and candidate genes [66]. As a result of such efforts, more than 10,000 of such associations have been registered so far [124].

However, the direct detection of candidate genes as exemplified by targeted re-sequencing or GWAS is often dependent on large cohort size for the detection of genetic variants as well as the validation of significant associations [66]. For rarer diseases or traits such a strategy might hence not be possible without extensive pooling of international biobanks. In those cases however, one might fall back on patient derived cell lines or animal models when available, in which one can then exploit several screens for candidate gene identification. Specifically, forward genetic screens attempt to discover the genetic event causing a given phenotype, e.g. a transformation causing tumor initiation, progression or treatment resistance, by introducing a multitude of random sequence variations [114], using for instance transposon based mutagenesis [168] or retroviral mutagenesis [169]. Reverse genetic screens on the other hand are designed to determine the phenotype caused by a given gene alteration through targeted modification of the gene's function [64], for instance trough RNA interference [150] or use of the CRISPR-Cas system [138].

Nevertheless, despite recent advances in the detection of disease specific genetic variants, it has also become clear that oftentimes driver events might not be readily distinguishable. Rather, many diseases have been shown to be polygenic [131, 184], i.e. exhibiting a large number of variations with only modest effects on susceptibility with the disease phenotype being shaped by interactions or complementary effects of the respective loci. For cancer, the identification of dominant driver genes is further complicated by genetic heterogeneity owing to genomic stability, which causes increased rates of mutations and structural aberrations [26, 50]. Specifically, while the determination of genomic landscapes in various tumors has produced a number of recurrently mutated gene loci, they have also demonstrated the genomic diversity within cancer types with the presence of many infrequent events among patients and multiple distinct events within patients [82, 176]. To identify the disease-driving mechanism among the mass of low penetrance variants, it has been suggested to study the over-representation of functional pathways [128, 181] among the affected genes. However, as disease development and subsequent phenotypes are likely being shaped also by other factors, such as signals from the cellular microenvironment or life style and environmental influences, driving mechanisms might need to be studied in an even greater context of entire biological systems and molecular networks [128, 141, 143, 175].

Importantly, biological networks not only present a highly adaptive means of illustrating the various aspects of intrinsic relationships and interactions of biological entities captured in diverse experimental data sets, but they also provide a powerful basis for mathematical and bioinformatic studies of the underlying topological structures (compare reviews by [2, 100, 175, 199, 199, 204]). Specifically, during the last decade a number of graph theory based methods where suggested for the prediction of disease genes from various types of molecular networks, as reviewed in [13, 185, 189].

In this chapter we will review some techniques and aspects of network based gene selection with a focus on graph centrality related methods, how such methods have been employed in cancer and disease gene screens and point out some challenges and future perspectives pertaining to such an approach. For the remainder of this text, depending on the context and source, we will include results and conceptions from both cancer network analysis as well as other disease network analyses. However, for the purpose of cancer gene identification, these two terms can here be considered equivalent.

2 Biological Networks

During the last decades, networks have gradually evolved to one of the most important tools in systems biology. The ubiquitous use of networks in biological science relies on the fact that they can be designed to model a great variety of relationships underlying biological processes, including for instance direct physical interactions, causal dependencies, functional relatedness, as well as the co-regulation or cooperation of molecular entities. As a consequence, networks represent one of the most promising resources for studying and understanding the complex dynamics of biological systems. Specifically, it has been suggested that, while more and more individual genomic variation are being linked to specific diseases, the actual disease phenotypes might be a result of large-scale perturbations of the entire biological system [141, 175]. Hence, the characterization of the underlying biological networks will be a crucial step in unraveling the intricate connections between genetic events, system wide alterations and disease outcomes.

2.1 Network Types

Specific relational data has been gathered from smaller scale studies throughout the world for many decades now. In addition, the recent dawn of modern high-throughput analysis techniques has further facilitated affordable and fast large-scale screens of related biological data. Together, these resources have allowed the establishment of networks in various biological fields and modeling a multitude of different biological relationships. The major types of biological networks spawned by these advances are listed below.

Gene co-expression networks attempt to model how genes are regulated together in certain signaling pathways by displaying genes as nodes and linking them by an edge if they show a correlation of expression values over a set of conditions or tissues [28, 158, 202].

Genetic interaction networks (GIN), which depict the dependency of genomic variations in causing certain phenotypes by modeling genes as nodes and linking them by an edge if a simultaneous alteration of both is required to obtain a given biological outcome, such as lethality [166, 167].

Insertion site interaction networks for forward genetic screens, in which nodes represent genomic loci and/or genes that are targeted by the integration of a transposon or virus into the host genome and edges depict some type of relationship between these sites. Specifically, insertion sites might be linked to each other and to proximal genes, if the genomic distance between these loci is below some threshold thus identifying regions of highly clustered integration sites, which are commonly referred to as common integration sites (CIS) [49]. Alternatively, nodes could also depict the gene targets of a CIS and edges might depict the co-occurrence or mutual exclusivity of these CISs in a given sample of cells [88, 103, 170].

Metabolic networks, which can model biochemical processes as relationships between enzymes, substrates and/or metabolic reactions in various ways based on the specific choice of node and edge representations [74, 76, 179].

Pan-disease networks, in which nodes represent genes and edges connect genes that have been implicated in the same disease [55].

Protein domain networks, with nodes representing protein domains and a link indicating the presence of the two connected domains in the same protein [190, 191].

Protein phosphorilation networks, attempt to model the protein phosphorylation cascades in a cell by modeling proteins as nodes with directed edges indicating a phosphorylation of the target protein (substrate) catalyzed by the source protein (kinase) [125].

Protein-protein interaction networks (PPIN), in which nodes are representing proteins and indirect edges are indicating a direct physical binding of two proteins [52, 75, 95, 157, 196].

Protein-RNA binding networks, which depict the physical interaction of RNA-binding proteins with their target RNAs [8, 93].

RNA-RNA interaction networks aim to capture transcriptional regulatory interactions similar to the TRNs but instead focus on the posttranscriptional regulation of micro RNAs (miRNAs) or non-coding RNAs (ncRNA) and their interaction with each other or with other RNA components of the gene regulatory machinery [71, 93].

Transcriptional/gene regulatory networks (TRN/GRN), with nodes representing genes and directed edges indicating a regulatory relationship, in which the protein product of the first gene, usually a transcription factor (TF), binds to a DNA regulatory region of the second gene to affect a transcriptional activation or repression [9, 60, 130, 139].

Functional Association/Linkage networks (FAN/FLN), in which nodes represent genes or gene products and edges indicate a potential functional similarity, which is determined based on association evidence usually gathered from an integrated collection of other types of biological data [67, 97, 98].

2.2 Generation of Biological Networks

The interactions represented in biological networks should preferably be derived from curated and validated experimental data in order to ensure biological soundness of the modeled relationships. For instance, there exists a plethora of methods for the targeted study of individual protein-protein interactions [111, 121]. However, while such traditional methods have produced valuable collections of high-quality interaction data during the last decades, their low-throughput nature often makes them time consuming and expensive to perform for thousands of biological entities.

On the other hand, the onset of high-throughput technologies has opened up another route to predict biological interactions through large-scale experimental screens. For instance, transcription factor binding sites can be experimentally studied through chromatin immunoprecipitation (ChIP) analyses [42, 173], while transcriptional regulatory interactions can be estimated by combining ChIP and gene expression data alone [126, 183] or in combination with other genomic data [182]. For the detection of PPIs a multitude of different high-throughput methods have been discussed [15]. In addition, with the increasing amount of high-throughput data made available, there have also been substantial advances of methodology for the computational prediction of biological interactions. Specifically, putative TF binding sites (TFBSs) can for instance be computationally predicted using nucleotide position weight matrices of known TF binding motifs [25], as those made

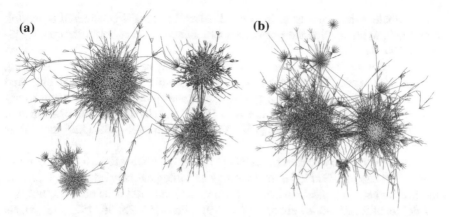

Fig. 1 Structural organization of two biological networks. The figure shows the topology of a BioGrid network including only links of the *'Phenotype Enhancement'* or *'Phenotype Suppression'* type (**a**) and a BioGrid network including only links of the *'PCA'* type (**b**)

available for instance through the TRANSFAC [107] and JASPAR [137] databases, or by related models [106, 148]. For the prediction of Protein-Protein Interactions (PPIs) a plethora of different approaches have been developed, as reviewed in [146, 171]. Finally, the last decade has also seen the development of numerous algorithms for the reverse engineering of TRNs from expression data [4, 44, 65, 68, 69, 104, 151].

Together, the contributions from low-throughput, high-throughput and computational efforts have produced a multitude of resources comprising relationships between biological entities, which are presented as pathways, interaction or reaction data in databases such as the Kyoto Encyclopedia of Genes and Genomes (KEGG) [81], the Reactome pathway knowledge base [43], the Biological General Repository for Interaction Datasets (BioGRID) [156], the STRING database [73], Pathway Commons (PC) [29], the Human Protein Reference Database (HPRD) [123], the Database of Interaction Proteins (DIP) [136], or the IntAct Molecular Interaction Database [85], the majority of which are also integrated in the multi-resource ConsensusPathDB interaction database [80].

2.3 Properties of Biological Networks

Biological networks generated according to the different approaches and depicting different types of interactions will typically look quite different (compare for instance Fig. 1), and do not necessarily exhibit the same edges between any given pair of genes/proteins.

Nevertheless the majority of biological networks are still considered to exhibit certain ubiquitous topological properties, which have been reviewed frequently [2, 12, 13, 100, 175, 199, 204].

Specifically, biological networks have been found to exhibit higher clustering coefficients than random networks [179, 190, 196], which in turn implies a clustered organization with regions with a higher interconnectivity than intra-connectivity to the rest of the network, compare also Fig. 1. A related modular organization has been demonstrated for PPINs [132, 155], metabolic networks [129] and TRNs [37, 60, 130].

As demonstrated by studies on TRNs [112, 144], PPINs [193], as well as composite PPINs and TRNs [195], the topology of biological networks is furthermore enriched for certain typical patterns of connectivity, referred to as network motifs.

Additionally, metabolic networks [74, 179], PPINs [52, 75, 95, 157, 196], PDNs [190], TRNs [60, 99, 139], gene co-expression networks [28] and GINs [167] have also been found to be scale-free, i.e. they are not dominated by nodes with a specific number of connections but the distribution of connections per node follows instead a power law with a large number of nodes having very few connections and a small number of nodes presenting with large number of connections [3], compare Fig. 2.

Scale-free networks are also expected to exhibit the small world characteristic [5, 31], which has for instance been demonstrated for PPINs [52, 95, 196], PDNs [190] and metabolic networks [179]. The small world property implies that the majority of nodes in the network can be reached from any starting node in the network by traversing only a small number of links [186].

Given the nature and properties of these networks, certain topological network analyses present themselves as powerful tools for extracting particular biological features from the respective underlying data, compare for instance the chapters by [2, 100, 199]. Specifically, due to the scale-free property, one can always find nodes

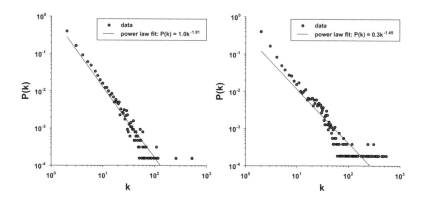

Fig. 2 Scale-free property of two biological networks. The figure shows the estimated probabilities $P(k)$ of a node being connected with k other nodes for a BioGrid network including only links of the *'Phenotype Enhancment'* or *'Phenotype Suppression'* type (**a**) and a BioGrid network including only links of the *'PCA'* type (**b**). *Gray dots* indicate measured probabilities and *red lines* represent power law fits

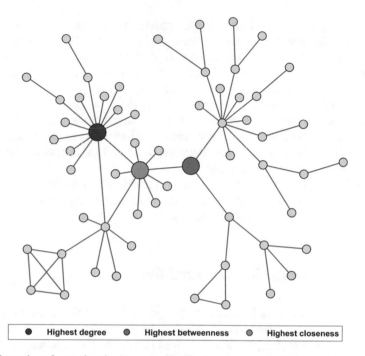

| ● Highest degree | ◉ Highest betweenness | ◉ Highest closeness |

Fig. 3 Illustration of central nodes in a network. The figure shows a scale free network, in which the nodes with highest degree centrality (*red*), highest betweenness centrality (*blue*) and highest closeness centrality (*green*) have been highlighted

in those networks that appear more central with respect to certain parameters as compared to other nodes in the network, compare Fig. 3.

The existence of high degree, i.e. hub, genes in biological systems has early on been recognized a potential avenue for the development of targeted drug treatments [11], but the distinct topological properties of such networks have also been suggested to lead to the discovery of novel disease genes and thus also therapy options. The next section will discuss how such properties might be utilized for cancer gene discovery.

3   Candidate Gene Prediction Using Graph Centralities

During the last decade numerous computational methods have been suggested for the network based prediction of disease and cancer genes. Among those techniques a large number assume a *"guilt-by-association"* [113, 185, 203] or *"guilt-by-proximity"* framework [189] and predict new candidate genes or pathways based on their direct functional linkage or network proximity (e.g. presence within the same network module) to known disease genes (compare for instance reviews and chapters [13, 185, 189]). As a consequence however, these methods rely on prior knowledge about existing disease genes in order to predict novel genes.

In this chapter, we focus on centrality ranking as an alternative network based approach, which has the potential benefit of not requiring prior knowledge about existing disease genes for a specific or similar disorder. Instead, one assumes that disease genes have very characteristic and determinable positions in their respective network. Specifically, considering that highly central nodes, as a more integral component of the particular information flow in the network, also engage in more important roles in the underlying biology, it is of interest to be able to identify such nodes from biological networks. Considering such networks as mathematical graphs, a variety of related centrality equations have been defined and applied to extract nodes with specific connectivity characteristics.

After introducing certain general notations to define mathematical networks and graphs, this section will review some of the more widely used centrality methods and discuss their applicability to biological networks.

3.1 Some Prerequisites to Centralities

In mathematics a network is described by a graph structure $G(V, E)$, where the set $V = \{v_1, v_2, \ldots, v_n\}$ represents the individual nodes, also called vertices, and the set $E = \{e_{ij}\}$, $i, j \in [1, n]$ denotes the edges that connect certain pairs of vertices. Specifically, an edge e_{ij} will imply that node v_i has a link to node v_j. Such a graph with n nodes can then always be described by a $n \times n$ matrix A, with element (i, j) representing the value of edge e_{ij} between nodes i and j, where the specific choice of these edge values dictates the particular type of graph we are modeling. Specifically, we will here distinguish between directed and undirected, weighted and unweighted networks as well as between graph structures allowing or forbidding self loops.

In an undirected network, we will always have $e_{ij} = e_{ji}$ and a symmetric matrix A, i.e. any edge in the network works in both directions in exactly the same fashion, while in a directed network we might encounter pairs of nodes v_i, v_j for which $e_{ij} \neq e_{ji}$ and the matrix A is potentially non-symmetric. The respective undirected graph is usually drawn without any arrow heads, while in the directed graph edges will be replaced by arrows in order to indicate direction.

In an unweighted network, we always have $e_{ij} \in \{0, 1\}$, implying that the edge between nodes v_i and v_j either exists or not, while in a weighted network the respective matrix entries will correspond to so called weights $w_{i,j}$ and can take any value from a predefined interval, e.g. $w_{ij} \in [-1, 1]$. These weights are then meant to indicate the strength of the connection described by the edge or in the case of negative and positive weights can also distinguish between an inhibitory and stimulating meaning of edges. In the case of a unweighted network the matrix A will also be denoted as adjacency matrix, while for a weighted network depending on convention one might refer to A as weighted adjacency matrix.

Self loops designate edges $e_{i,i}$ from node v_i to it self. In some types of graphs these self loops are allowed, while other explicitly prohibit them. In the adjacency matrix A this decision will dictate whether the diagonal is always zero.

Finally, another network property that must be considered here is the so called connectedness of the underlying matrix. Specifically, we say that a network is (strongly) connected, if it is possible to reach any other node $j \neq i$ from node i by traversing the existing edges of the network. If this is true for the directed graph it is strongly connected, while if it is true only for the undirected graph is is merely connected, if neither is true we say that the network is disconnected. This is of importance, because if the network is connected we can define a number of additional network properties, which do not make sense for a disconnected network. Specifically, we can here mention (1) the distance, also referred to as shortest path, between two nodes i and j, which is denoted by $d(i, j)$ or $dist(i, j)$ and is defined as the smallest number of edges that would have to be traversed in order to travel from node i to j and (2) the diameter $diam(G)$ of the network, which is simply the largest value of $d(i, j)$. The diameter and some distances are obviously not computably in disconnected network, in which case they are often set to $d(i, j) = \infty$ or $diam(G) = \infty$.

Above it was mentioned that graphs can be connected or disconnected, weighted or unweighted as well as directed or undirected. Different centrality measures might make specific assumptions about the particular organization of the underlying network and since one might be interested in analyzing more than one type of network configuration with the same centrality measure it will often be necessary to expand the given centrality method to also work on other types of networks that the centrality measure was not perhaps intended for according to its original definition. We have in the following section tried to describe the original definition of centrality measures and its requirements or assumptions regarding underlying network structures, but also attempted to gather potential approaches to modify the given method for other types of network structures.

The majority of centrality measures will produce centrality measures of some absolute magnitude, the specific values of which will depend directly on for instance network size. Comparing these measures between networks of different sizes is therefore not meaningful, but requires a normalization step to transform the absolute centralities to relative centralities. We have attempted to gather proposed normalization schemes for each of the represented centrality measures in the following section.

3.2 Definitions and Visualization of Common Centrality Measures

As of today more than 110 different centralities have been described in the literature.[1] The majority of the centrality measures have been developed in other research fields for other purposes. Still, depending on the choice of interaction data, many of these centrality measures can directly or with slight modifications be applied to biological networks (see also Sects. 3.3 and 6.1 for discussions about applicability and mean-

---

[1]A comprehensive list of centralities can be found in the CentiServer (http://www.centiserver. org/) [72].

Table 1 Centrality measures based on shortest path*

Centrality measure	Equation	Reference
Closeness centrality	$\frac{1}{\sum_{i \neq j} d(v_i, v_j)}$	[47, 135]
Stress centrality	$\sum_{k \neq i} \sum_{l \neq i} \sigma_{kl}(v_i)$	[23, 145]
Betweenness centrality	$\sum_{k \neq i} \sum_{l \neq i} \frac{\sigma_{kl}(v_i)}{\sigma_{kl}}$	[23, 46]
Flow betweenness		[48]
Load centrality		[23, 54]
Eccentricity centrality	$\frac{1}{\max\{d(v_i, v_j): v_j \in V\}}$	[62]
Radiality/integration centrality	$\frac{\sum_{i \neq j}(D+1-d(v_i, v_j))}{N-1}$	[172]

*Where σ is defined as in Sect. 3.3.3, D is the diameter of the graph and $d(v_i, v_j)$ is the distance from vertex v_i to vertex v_j

Table 2 Centrality measures based on powers of the adjacency matrix*

Centrality measure	Equation	Reference
Degree centrality	$\sum_{j=1}^{n} a_{i,j}$	–
Eigenvector centrality	$\frac{1}{\lambda} \sum_{j=1}^{n} a_{i,j} C_{eig}(v_j)$	[19]
Katz status	$\sum_{k=1}^{\infty} \sum_{j=1}^{n} \alpha^k (A^k)_{ij}$	[84]
Page rank	$(I - \alpha M^T)^{-1}\mathbf{v}$	[24]
Cumulative nomination	$(I + A)^n e$	[122]

*Where A is the adjacency matrix M is a scaled and slightly modified adjacency matrix. α, β are scalar parameters chosen appropriately and $\mathbf{e}, \mathbf{v}$ are the one vector and a non-zero weight vector, respectively

Table 3 Other centrality measures*

Centrality measure	Equation	Reference
Centroid value	$\min\{f(v_i, v_j) : v_j \in V/v_i\}$	[152, 192]
Clustering coefficient	$\frac{2|\{e_{jl}: v_j, v_k \in N_i, e_{jk} \in E\}|}{k_i(k_i-1)}$	[186]
Topological coefficient	$C_{TC}(v_i) = \frac{avg(J(v_i, v_j))}{k_i}$	[157]

*Where $f(v_i, v_j)$ is the difference between the number of vertices closer to v_i than v_j and the number of vertices closer to v_j than v_i, k_i is the number of neighbors of vertex v_i and $J(v_i, v_j)$ defined only for vertices which share at least one neighbor is the number of neighbors shared between v_i, v_j plus one if there is an edge between v_i and e_j

ingfulness of centralities in biological networks). Here we list the definitions and show visualizations for some of the centrality measures most frequently applied to candidate gene prediction from biological networks. Specifically, we are separating the centralities here into those methods based on shortest path calculations (Table 1, Fig. 4), based on the calculations of powers of the adjacency matrix (Table 2, Fig. 5) and other centralities (Table 3, Fig. 6).

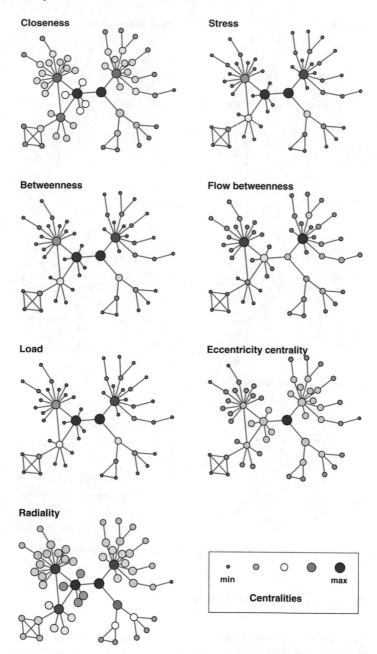

Fig. 4 Illustration of centrality measures based on shortest path

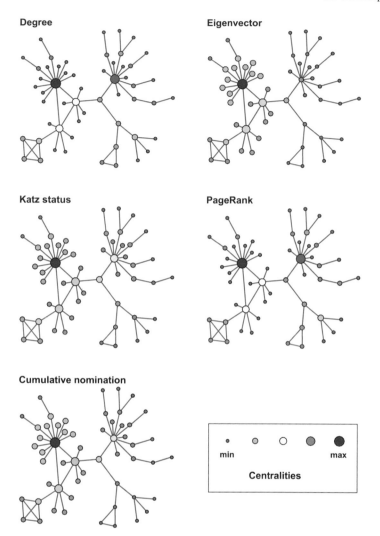

Fig. 5 Illustration of centrality measures based on powers of the adjacency matrix

3.3 Applicability to Biological Networks

As briefly noted above, the implementation and design of individual centralities make them more or less applicable to certain network types. In this section we will discuss some of the factors that impede general applicability of centralities and discuss some modifications and remedies to these problems. Of the centrality measures applied to the identification of biologically important nodes, degree, betweenness as well as closeness centrality are by far the most frequently utilized and studied methods.

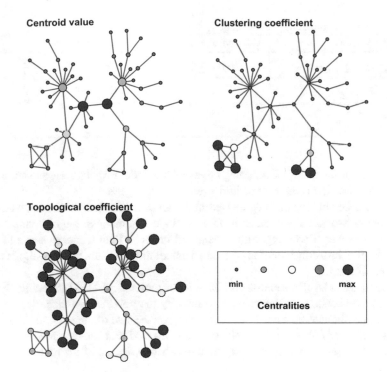

Fig. 6 Illustration of other centralities

Thus, we will start with a more in detail illustration of such considerations about applicability on the example of these three centralities, before concluding this section with a summarization of similar reflections for the other centrality measures.

3.3.1 Degree Centrality

Degree centrality is one of the simplest and most straightforward measures of graph centrality and is based only on the number of edges connected to a specific node. Specifically, for an undirected network with no loops, the degree centrality $deg(v_i)$ of a node v_i is equal to the number of edges connected to the node.

$$C_{deg}(v_i) = \sum_{j=1}^{n} a_{i,j}.$$

If the network has loops, these are typically either ignored (only interested in number of neighbors) or counted twice (once for each end of the loop touching the vertex). For directed, i.e. asymmetric, networks we define in-degree and out-degree to be

Table 4 Modifications of degree centrality

Directed network	Weighted network	Normalization
$C_{deg_{in}}(v_i) = \sum_{j=1}^{n} a_{j,i}$	$C_{deg_w}(v_i) = \sum_{j=1}^{n} w_{i,j}$	$C_{deg_{norm}}(v_i) = \frac{C_{deg}(v_i)}{N-1}$
$C_{deg_{out}}(v_i) = \sum_{j=1}^{n} a_{i,j}$		$C_{deg_{norm}}(v_i) = \frac{\sum_{j=1}^{n} w_{i,j}}{C_{deg_{max}}}$
$C_{deg_{total}}(v_i) =$ $\sum_{j=1}^{n} a_{j,i} + \sum_{j=1}^{n} a_{i,j}$		

the number of incoming or outgoing edges of the node, respectively, as well as the total-degree, which is the sum of in-degree and out-degree.

It should be obvious that degree centrality can be calculated on disconnected and weighted networks as well since it only counts the number of edges connected to a node. However if working with a weighted network then it makes sense to take these weights into consideration by instead calculating the sum of the weights of all connected edges.

Several aspects of normalization have been outlined in [187]. Specifically, we normalize by scaling the centrality measures by the maximal possible degree centrality value obtainable in a network depending on the size of the network, this gives a scaling factor $1/(N-1)$ if no self-loops are allowed. For the weighted network it is common to normalize by dividing by the maximal degree of the non-normalized degree.

A summary of the discussed modifications can be found in Table 4.

3.3.2 Closeness Centrality

Closeness centrality [47, 135] is commonly defined as:

$$C_c(v_i) = \frac{1}{\sum_{v_j \neq v_i} d(v_i, v_j)}.$$

The original definition of closeness centrality as defined above only makes sense for connected undirected networks where the distance $d(v_i, v_j)$ is well defined. If the network is undirected then it is possible that $d(v_i, v_j) \neq d(v_j, v_i)$ and if it is not (strongly) connected then the distance between some nodes will be undefined.

For disconnected networks, a number of potential modifications to the original method have been proposed, some of which have been reviewed in [18]. Throughout this discussion we will assume that the distance between two vertices is infinite if there is no path between them.

A simple solution is achieved by ignoring all unreachable nodes in the computation of closeness [18]. Another solution was proposed by Chavdar Dangalchev [34], by moving the sum out of the quotient and more heavily penalizing long distances by taking powers of two. A third solution goes back to the work of Nan Lin (1976) [96], who

Table 5 Modifications of closeness centrality

Disconnected network	Ref.	Normalization	Ref.
$C_c^{(1)}(v_i) = \frac{1}{\sum_{d(v_i,v_j)<\infty} d(v_i,v_j)}$	[187]	$C_{RC}(v_i) = \frac{n-1}{d(v_i,v_j)}$	[14, 47]
$C_c^{(2)}(v_i) = \sum_{j \neq i} \frac{1}{2^{d(v_i,v_j)}}$	[34]	$C_{NC}(v_i) = \frac{C_c(v_i)}{(n-1)\min(w)}$	[187]
$C_{LI}(v_i) = \frac{|\{y \mid d(v_i,v_i)<\infty\}|^2}{\sum_{d(v_i,v_j)<\infty} d(v_i,v_j)}$	[96]		
$C_H(v_i) = \sum_{i \neq j} \frac{1}{d(v_i,v_j)}$	[133]		

redefined closeness centrality based on so called *"nonempty coreachable sets"* [18], producing a measure also referred to as Lin's index. Finally, the probably most commonly used alternative definition is the so called *"harmonic centrality index"* or simply *"harmonic centrality"* [18, 133]. This latter measure is similar to the one introduced by Chavdar Dangalchev, but the ordinary distance between the nodes is used instead.

A common method to deal with directed graphs is by calculating either in-closeness (using $d(v_j, v_i)$) or out-closeness (using $d(v_i, v_j)$) similarly to how we calculate degree centrality for directed graphs [187].

The question if closeness can be applied to weighted networks depends on the choice of distance function, the most common distance used being the shortest path which can easily be adapted to weighted graphs by regarding edge weights as costs and finding the path with minimum cost.

A scaled version of closeness centrality was proposed by Beauchamp in 1965 [14] and rediscussed by Freeman in 1979 [47] in his definition of closeness centrality, by multiplying the absolute Closeness with $N - 1$ to get the average closeness. Furthermore, [187] extends this normalization to weighted networks by dividing the non-normalized centrality by the maximum possible value.

A summary of the modifications discussed here can be found in Table 5.

3.3.3 Betweenness Centrality

Betweenness centrality [23, 46] can be seen as a measure of how important a node is for the communication between other nodes in the network by estimating how often it is visited when finding shortest paths between other nodes.

If we let σ_{kl} denote the number of shortest paths between two nodes v_k and v_l and let $\sigma_{kl}(v_i)$ denote the number of shortest paths between two nodes v_k and v_l that traverse through node v_i then betweenness centrality for a connected, directed, unweighted network can be formally defined as

$$C_{between}(v_i) = \sum_{k \neq i \neq l} \frac{\sigma_{kl}(v_i)}{\sigma_{kl}}.$$

Since we are calculating shortest paths, disconnected networks pose a problem. While we have not found a documented solution in the literature, a simple solution is to set $\frac{\sigma_{kl}(v_i)}{\sigma_{kl}} \equiv 0$ if $d(v_k, v_l) = \infty$ similar to how we did for closeness centrality.

The above holds for both directed and undirected networks, although for undirected networks where $\frac{\sigma_{kl}(v_i)}{\sigma_{kl}} = \frac{\sigma_{lk}(v_i)}{\sigma_{lk}}$ it makes sense to divide the total with 2 since we would otherwise count each path twice or modify the algorithm to only calculate shortest paths for one symmetric half of the network [187].

Similar to closeness, weighted networks can be handled in a similar way by viewing edge weights as the cost of traversing said edge and finding the path with the smallest cost.

Again, we normalize the raw betweenness centrality values by division with the maximum possible centralities. The respective maximum values have been given in [187], which considered the current node to be the center of a star-network according to [46, 47]. Specifically, for a undirected network the maximum possible centrality value becomes $C_{between}^{max} = \frac{n^2-3n+2}{2}$, while for a directed network the normalization factor is $C_{between}^{max} = n^2 - 3n + 2$.

3.3.4   Summary of Applicability Considerations

Table 6 summarizes for all included centralities, whether they are applicable to directed, weighted and disconnected networks, respectively. A method that can be used on directed but not disconnected networks needs it to be strongly connected in the directed graph unless noted otherwise. Some methods have problems with single unconnected vertices but otherwise work for disconnected networks, in this case centrality is usually set to zero (or some other suitable value) hence we consider these applicable to disconnected networks as well. Similarly most methods relying on shortest paths can easily be used on weighted graphs by instead using the shortest distance on the weighted graph instead.

3.4   Linear Combinations of Centralities

It has been mentioned that different centralities might operate differently or are not defined on certain network types, such as disconnected or directed networks. In addition, when using a set of centralities, one should also consider possible redundancies between certain centralities.

For example on many weighted networks it is reasonable to assume there is only one shortest path between any pair of vertices, which would imply that betweenness, stress and load centrality will in this case all be the same. Another example is Katz Status which can be rewritten as $((I - \alpha A^T)^{-1} - I)e$ using a Neuman series and it is obvious that Katz Status is equivalent to Bonacich Alpha/Beta centrality minus 1. Similarly, if α is small (close to 0), then Katz status and Degree centrality give a

Table 6 Applicability of centrality measures

Centrality measure	Disconnected	Directed	Weighted	Normalization*
Closeness centrality	[18, 34, 96, 133]	[187]	[116]	[14, 47, 187]
Stress centrality	Yes	Yes	[127]	Yes(o)
Betweenness centrality	Yes	[187]	[116]	[23, 187]
Flow betweenness	Yes	Yes	Yes	Yes(t)
Load centrality	Yes	Yes	[23]	[23]
Eccentricity centrality	Yes	Yes	Yes	Yes(t)
Radiality/integration centrality	[172]	Yes	[116]	[172]
Degree centrality	Yes	Yes	Yes	[187]
Eigenvector centrality	No	Yes	[20, 21]	[134, 187]
Katz status	Yes	Yes	Yes	Yes(o)
Page rank	Yes	Yes	Yes	Yes(p)
Cumulative nomination	[122]	Yes	Yes	Yes(o)
Centroid value	No	Yes[†]	Yes	No
Clustering coefficient	Yes	Yes	No	Yes(t)
Topological coefficient	Yes	No	No	Yes(t)

*Where (t) denotes normalization using a theoretical maximum, (o) denotes normalization using the observed maximum using corresponding non-normalized centrality and (p) denotes normalization into a probability density (non-negative ranks with sum one)
[†]Requires a choice of direction for "distance" calculations and the network to be strongly connected instead of simply connected

similar ranking since higher order terms in the sum disappear quickly when calculating Katz Status.

Thus, depending on the network context, centralities might be redundant due to highly similar enrichment profiles or they could act complementary. In order to quantify these ranking similarities on a given network and between a set of centralities one can for instance inspect a correlation matrix [89, 90] (compare Fig. 7), where one would preferably employ a ranking correlation coefficient such as Kendall's tau which allows for ties.

As a consequence of the observed differences in centrality based enrichment patterns, it has been suggested that multiple centralities should be employed when ranking genes in biological networks [89]. Importantly, it can even be argued that linear combinations of certain centrality measures can act complementary and allow for the enrichment of novel, distinct sets of nodes. For instance combining degree and betweenness centrality can lead to the enrichment of specific nodes that are central in terms of connections but also shortest path within the network, compare Fig. 8. Thus, inspections of complementary enrichment patterns and investigations of linear combinations of centrality measures might be beneficial also in the identification of cancer genes.

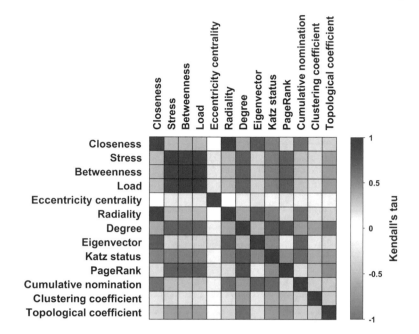

Fig. 7 Rank correlation between centrality rankings. The figure shows a heatmap of the pairwise Kendall's tau rank correlation coefficient between centralities calculated on the BioGrid network including only links of the *'Synthetic Haploinsufficiency'*. For the computation of centralities, only the largest symmetrized adjacency matrix of the largest network component was included

Specifically, a number of studies have employed combined centralities for the prioritization of genes, sometimes even stating combinations as a requirement in order to see a centrality based enrichment of genes with a certain phenotype association. Siddani et al. [147] have used a combination of ten centralities to identify novel candidate genes for the Systemic Lupus Erythematosus disease. Bhattacharyya and Chakrabarti [16] prioritized proteins in PPI networks of *Plasmodium falciparum* and argued that integrating all employed centrality measures was necessary for identifying *"truly central proteins"*. del Rio et al. [35] investigated the centrality based prediction of essential genes from metabolic networks of *Saccharomyces cerevisiae* and found that at least two centrality measures had to be employed together in order to achieve a statistically significant identification of essential genes.

3.5 Implementation

Seeing the general applicability of centrality measures, a wide variety of packages and standalone softwares not specifically tailored for a biological use can be found for the R or MATLAB platforms, such as the R packages *sna* [27], *igraph*, [33]

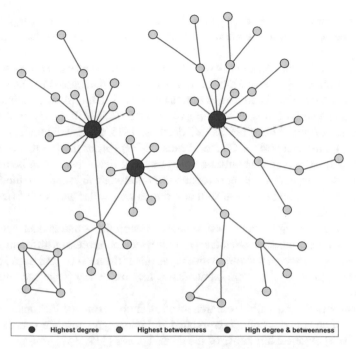

Fig. 8 Illustration of central nodes in a network. The figure shows a scale free network, in which the nodes with highest degree centrality (*red*), highest betweenness centrality (*blue*) and nodes with the a high simultaneous score in both centralities (*violet*) have been highlighted

and *CePa* [58] or the *MatlabBGL* package [53]. In addition, inspired by the growing importance of centrality related questions in biological networks, numerous modules specifically intended for the use on biological networks have been introduced during the last years, including for instance the *CentiBin* [79] and *CentiLib* [57] software tools as well as the *CentiScape* plugin [140] for the widely used biological network illustration tool *CytoScape* [160]. A more comprehensive list of centrality software resources can even be found on the *CentiServer* (http://www.centiserver.org/), a recently published tool for the calculation of a very large collection of network centralities through the use of a web interface or R package [72].

4 Determining Enrichment of Cancer Genes Among High Centrality Nodes

Despite a substantial body of investigations during recent years, the exact relationship between cancer genes and graph centralities remains largely unresolved. Hence, before utilizing centrality measures to nominate gene targets from regulatory networks, we must evaluate to which extent the selection of high centrality nodes will lead to an enrichment of cancer genes. The extent of such a relationship may well vary between different settings and should be considered carefully on a case-by-case basis.

One important consideration is which genes we regard as cancer genes. For this purpose, one can make use of a number of resources and databases in order to classify genes into cancer related and cancer unrelated subsets. Examples of such databases are the Catalogue Of Somatic Mutations In Cancer (COSMIC) [45], the pathways in cancer set of genes from the KEGG database [81], the Candidate Cancer Gene Database [1], the Network of Cancer Genes (NCG) [6], or the IntOGen-mutations platform [56]. However, it should be noted that any list of genes can be used. The gene list used to evaluate the performance of a network inference method or centrality measure should be chosen so that nominating similar genes is of interest for downstream analysis.

When investigating the relationship between centralities and cancer gene status, there are two main questions that can be addressed and represent different forms of enrichment. Specifically, one might investigate (1) if the most central genes are more often cancer genes or (2) if there is a tendency towards cancer genes having a higher centrality.

To answer the first question researchers often simply compare the mean or median centrality value between phenotype-related and phenotype-unrelated genes [63, 70, 77, 117, 194] or select a number of top scored genes [117, 147] among which one could quantify the over-representation of phenotype related genes. The comparison of means or medians can be performed using standard tests. The enrichment of cancer genes among the top central genes can be quantified using a hypergeometric or Fisher's exact test.

The first question is thus straightforward to answer, and it might be informative for nominating gene targets, but it lacks nuance since it does not take into account the distribution of centrality values. The second question may therefore be more useful as a performance benchmark.

The analysis of enrichment of centrality values bears a strong resemblance to gene set enrichment analysis often considered when interpreting gene expression in relation to some measured phenotype. In this setting the question considered is whether a measured phenotype has a significant association with the expression of genes in a certain category (e.g. pathway membership or functional annotation). Many methods exist for this purpose (e.g. [38, 86, 161, 164]). Such methods generally work by first quantifying the association of individual genes with the phenotype, in essence creating a ranking of the genes, and then quantifying the difference in distribution of these associations, comparing genes contained in a category and those not contained in that category. In our case the ranking of genes is the ranking of centralities, and the gene category in question is the set of genes considered to be cancer related.

To measure the significance of centrality enrichment among cancer genes, we here propose a simple method based on the enrichment statistic used in the GSEA method [38]. First, we start by sorting the centralities in decreasing order, then iterate along this ranked list while keeping a running sum that is incremented when we encounter a gene that is cancer related, and decremented when we encounter a gene that is not. From the set of vertices $V = \{v_1, v_2, \ldots, v_i, \ldots, v_n\}$, ordered according to the magnitude of their centralities $C(v_i)$, and a set of nodes $S \in V$, we obtain the size of each term in the sum as:

$$\begin{cases} 1/|S|, & \text{if } v_i \text{ in } S, \\ 1/(|V| - |S|), & \text{if } v_i \text{ not in } S. \end{cases}$$

The test statistic (or, enrichment score (ES)) is defined as the largest absolute value of the running sum obtained throughout this iteration. An empirical p-value for the enrichment is obtained by comparing the observed test statistic to a null-distribution obtained by repeated random permutation of the ranked list and calculation of the ES for each permutation. In Fig. 9 we illustrate one application of this method using the BioGrid network in Fig. 1a and the COSMIC cancer genes. However, this approach can be used with any set of genes, for instance GO or KEGG to determine whether a ranking of centrality enriches for genes with a particular biological function.

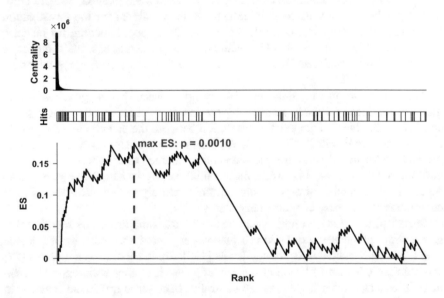

Fig. 9 Test for enrichment of cancer status among high centrality genes. The figure shows the enrichment of COSMIC genes among high page rank scores in the BioGrid network including only links of the *'Phenotype Enhancement'* or *'Phenotype Suppression'* type. *Top panel* distribution of degree values; *middle panel* cancer status of genes; *bottom panel* step function, where at each step the enrichment score (ES) increases if the gene is a cancer gene or decreases if the gene is unrelated. The p values is estimated by calculating the percentage of permutations of cancer gene affiliations with ES scores greater than the one observed. For the computation of the centrality, only the largest symmetrized adjacency matrix of the largest network component was included

5 Recent Results and Progress in Centrality Based Prioritization of Disease and Cancer Genes

Since the first emergence of studies investigating the centrality in biological networks, the further development and application of related methods has grown to an established field of biological research. Advancements in this field have been made in basically two directions, which however progress hand in hand. These directions comprise on one hand the development of novel centrality measures and software facilitating centrality application to biological networks and on the other hand include evermore intricate studies exploring the use of centralities for the ranking of biological entities with certain properties.

Specifically, initial findings relating centralities to important genes or proteins were established using general centrality measures that have previously been defined from other scientific areas such as social sciences. However, recent years have also seen the dawn of many novel centrality measures, inspired by or explicitly defined for ranking problems in biological networks [83, 91, 94, 162, 163]. In addition, a number of software tools or extensions more tailored for the investigation of centralities in biological networks have been developed, including for instance the *CentiBin* [79] and *CentiLib* [57] standalone implementations, the *CentiScape* plugin [140] for *CytoScape* [160], or the web interface and R package provided by the *CentiServer* [72].

The interest in investigating the relationship between disease gene status and graph centralities was likely inspired by the initial observation in model systems suggesting that there might exist a correlation between the essentiality of a protein and its centrality in a PPI [40, 41, 63, 75, 78]. While the identification of essential proteins by the use of centrality measures has continued to draw scientific interest until today [35, 87, 94, 142, 163], these initial findings where subsequently also succeeded by a number of experiments that more closely studied the association of centralities with disease or cancer gene status.

Specifically, a study on lung squamous cell carcinoma tissue has reported that genes with upregulated expression in the cancer tissues showed a higher degree than genes with unaltered expression levels [178]. Similarly, Jonson and Bates [77] showed that in human PPIs consensus cancer genes, i.e. genes with reported mutations in cancer, have a higher degree centrality than genes not found mutated in cancer. Another study investigated the centrality of OMIM derived disease-genes obtained in literature-curated PPIs and found the disease genes to exhibit a higher degree centrality than non-disease genes [194]. Using a small number of prostate cancer genes from the OMIM database as seed genes in a literature-mined interaction network, Özgür et al. [119] found that centrality ranking could be used to enrich for genes with known prostate cancer association. A study of disease genes for primary immunodeficiency (PID) combined network centralities and GO ontologies to rank genes in a human immunome network and was able to identify a number of already known PID genes [117]. Similarly, Sidanni et al. [147] predicted Systemic Lupus Erythematosus (SLE) genes from two different Human immunome networks also using a combination of centralities and GO ontologies and found a large proportion of the predicted genes to represent known SLE genes. Starting with a large dataset of PPIs from the HPRD database, Izudheen and Sheena [70] performed centrality comparisons between cancer, cancer-chance and non-cancer genes in smaller sub-sampled networks and found that cancer and non-cancer genes differed in several centrality measures.

One often cited study that debates the use of particularly degree centrality for the enrichment of disease genes is the work by Goh et al. [55]. The authors established a *"disease gene network"* by connecting any pair of disease genes obtained from the Online Mendelian Inheritance in Man (OMIM) database, which was found associated with the same disease. While disease genes where found to account for high degree nodes in this network, this trend disappeared when excluding disease genes that are also embryonically or postnatally lethal. Particularly, Goh et al. suggest that essential genes in their pan-disease network are likely to form hubs, while the majority of disease genes, being non-essential, are located in the periphery with low degree centrality. However, the authors also report that disease genes caused by somatic mutations actually show a higher degree centrality and tendency to coincide with hubs. In addition, while this study has raised some concerns regarding the separation of essential genes and disease genes and the use of degree centrality to predict disease genes, one should bear in mind that the study investigated only one type of centrality in a pan-disease networks rather than direct molecular interaction network. Thus, the results may not exclude the possibility for associations between centralities and disease genes in other network types.

6 Open Questions and Future Challenges

6.1 Which Network to Choose

When attempting to address a certain biological question, some network types might be more appropriate than others. However, in addition one should also consider how such data analyses might be influenced by the way in which the related networks have been generated.

As mentioned above, the interactions of many biological networks can be derived in a variety of ways, the exact choice of which might bear some influence on the accuracy and completeness of the network. Specifically, networks solely established from low-throughput experimental data might exhibit low false-positive rates, but a large number of false negatives and additionally present with a bias towards interactions of molecules which are of greater scientific interest [51], such as for instance disease proteins [118]. High-throughput methods, as exemplified by protein-protein interaction assays, on the other hand might exhibit larger false-positive rates and could further be influenced by a variety of different biases [17, 51, 177].

Another important factor to consider is the generic nature of many interaction and predominantly PPI databases. Specifically, as such databases often represent aggregations of data from various sources, such as tissues, laboratories or methods, the contained interactions might be considered as a collection of possible interactions in an organism, but often providing insufficient information about when and where a given interaction is present or not. Since such temporal and spatial patterns of interaction might differ substantially between different tissues and diseases, such databases might only be of limited use when attempting to prioritize disease genes for a specific disorder. In order to remedy this lack of tissue-specificity in generic databases, several integrative methods have been suggested during the last years [22, 59, 101, 180]. For other types of interaction data, such as transcriptional regulatory relationships, transcription factor binding or genetic interactions, many tissue- and disease-specific datasets are publicly available and can be utilized to estimate the underlying networks. For instance, as mentioned above, a number of different methods exist for the inference of gene regulatory networks from expression data [4, 44, 65, 68, 69, 104, 151]. Individual techniques and especially community integrations of various techniques achieve ever increasing accuracies for the prediction of individual interactions [102]. However, it is still largely unexplored how well these methods can reconstruct the overall topology and thus also centralities in such estimated networks [188].

6.2 How to Determine Phenotype Specific Candidate Genes?

Above it was discussed that depending on the choice of interaction resource, networks utilized for cancer gene prioritization might lack tissue-specificity. However, even when prioritizing genes from a tissue- and disease-specific network, there still remains the question of whether the high centrality observed for a candidate gene is due to its association with the given phenotype.

Specifically, it can be assumed that genes and proteins with central roles in the normal cell's function also take central positions in respective biological networks, for instance master/global regulators in GRNs [90]. If those genes play crucial roles for cellular function and survival in the healthy tissue, it is reasonable to expect that a proportion of those genes, such as essential proteins and housekeeping genes [55], even has high centrality in the disease network without being actually linked to the disease phenotype. Hence a selection of network nodes with high centrality would naturally also include a number of genes which play a central role in the cells function, regardless of whether it belongs to a cancer or healthy individual. The underlying topological overlap between networks of healthy and disease phenotype creates a marked problem for the prediction of candidate cancer genes.

In order to overcome such a contamination by genes always central in a cell's molecular system, one approach might be to scale or modify centralities observed in a cancer derived network based on the centralities of the equivalent genes in a network derived from the healthy counterpart. Alternative approaches could also make use of the fact that molecular networks are often enriched for small regulatory motifs [37, 112, 144, 195, 198] and that part of cancer development can be understood as a perturbation of the interactions in the healthy network [175, 185]. Thus, identification of cancer specific candidates could involve the identification of changes to central network structures and motifs [175, 185, 201] or the identification of cancer network enriched motifs [30].

6.3 Biological Context of Centralities

As discussed in Sect. 3.3, given a certain network type, it is in most cases straight forward to make a selection of centrality measures that are mathematically applicable and meaningful. However, less is known about the biological meaning associated with individual centralities. Specifically, one has to wonder what principles of distance, neighborhood or information flow as used by centralities would signify in a biological context and if there can actually be some biological property correlating with these centralities. Cases, in which centrality ranking actually leads to the over-representation of cancer or disease genes might provide direct feedback about a potential functional or phenotypical association. However, this particular type of cancer gene prioritization would certainly gain in scientific soundness, if centralities could be shown beforehand to have a biological meaning.

There are some mentions of a further distinction of biologically useful centralities in the literature. For instance, from a exhaustive collection of centralities discussed and implemented in the *CentiServer*, the authors presented a subset of measures more appropriate for biological networks [72], although it is unclear, whether this selection was made due to applicability considerations from a mathematical or biological perspective. On the other hand, the publication introducing the *CentiScaPe* plug-in for *Cytoscape* provides interpretations of the potential biological meaning represented by a number of centrality measures [140]. However, these efforts only cover a small number of the existing centralities. Considering furthermore the vast variety of biological networks and the complex interaction dynamics of the underlying systems, it appears that we have just begun to link the concept of centralities with biological functions. Considering interpretations as such provided by [140], it remains to be shown how one could quantitatively validate a novel interpretation let alone identify such an interpretation for a yet uncharacterized centrality measure.

One potential avenue for associating biological properties and centrality measures could be the exploration of functional annotations such Gene Ontology terms in the context of centrality rankings. It has previously been shown that prioritization based on centrality and GO terms can be combined for the identification of essential proteins [87] or disease genes [117, 147]. Additionally, some studies have investigated the enrichment of functional annotations in centrality prioritized gene signatures. Specifically, Siddani et al. [147] performed GO enrichment analyses on top centrality scored genes in a human immunome network and found an enrichment of important immunology related functional annotations. Wang et al. [180]

performed a GO and KEGG pathway enrichment analysis on different centrality based gene sets obtained from a context-constrained breast cancer network, which was obtained by projecting multiple breast cancer signature genes onto a PPI network. The authors investigated for each centrality the KEGG pathway and GO term with highest significance and interestingly found the *"pathways in cancer"* as the top KEGG category enriched in all centrality derived signatures. Ortutay and Vihinen [117] investigated GO enrichment among the 50 highest ranking genes extracted from a human immunome network. However, the authors only reported a few of the top ranking GO categories and noted the presence of the top scored term in all three centrality selected datasets.

It would be interesting to expand on such investigations, to explore whether and which types of functional annotations could be associated with individual centrality signatures in various types of networks and tissues.

7   Conclusion

Network based ranking methods have emerged as important tools for the prioritization of targetable cancer driver genes. However, many of such techniques rely on *"guilt-by-association"* approaches in order to predict genes or pathways related to known disease genes, which represents with limitations and bias due to the requirement of prior knowledge. Here we review an alternative approach that operates without the requirement of prior knowledge through the use of network centralities. While such topological ranking methods are commonly used, the relationship between centralities in various network types and cancer gene status is still poorly understood. The centrality measure used, and to which network it is applied, impacts how we should interpret the results, and care must be taken when validating each approach. For these purposes it is essential to understand what the network represents, and how different measures of centrality reflect various biological contexts. As always, even though much has been written on the topic, much work remains before we properly understand how network centrality can be used to prioritize targetable cancer driver genes. Two important pieces of this puzzle are the reference gene set used and what measure is employed to benchmark different methods, making them important topics for further study.

Acknowledgements This work was supported by grants from the Swedish Childhood Cancer Foundation.

References

1. Abbott, K.L., Nyre, E.T., Abrahante, J., Ho, Y.Y., Vogel, R.I., Starr, T.K.: The candidate cancer gene database: a database of cancer driver genes from forward genetic screens in mice. Nucleic Acids Res. **43**, D844–D848 (2015)

2. Aittokallio, T., Schwikowski, B.: Graph-based methods for analysing networks in cell biology. Brief. Bioinform. **7**, 243–255 (2006)
3. Albert, R.: Scale-free networks in cell biology. J. Cell Sci. **118**, 4947–4957 (2005)
4. Altay, G., Emmert-Streib, F.: Inferring the conservative causal core of gene regulatory networks. BMC Syst. Biol. **4**, 1–13 (2010)
5. Amaral, L.A., Scala, A., Barthelemy, M., Stanley, H.E.: Classes of small-world networks. In: Proceedings of the National Academy of Sciences of the United States of America vol. 97, pp. 11149–11152 (2000)
6. An, O., Dall'Olio, G.M., Mourikis, T.P., Ciccarelli, F.D.: NCG 5.0: updates of a manually curated repository of cancer genes and associated properties from cancer mutational screenings. Nucleic Acids Res. **44**, D992–D999 (2016)
7. Arias, E., Kochanek, K.D., Anderson, R.N.: How does cause of death contribute to the Hispanic mortality advantage in the United States? NCHS Data Brief **221**, 1–8 (2015)
8. Ascano, M., Hafner, M., Cekan, P., Gerstberger, S., Tuschl, T.: Identification of RNA-protein interaction networks using PAR-CLIP. Wires RNA **3**, 159–177 (2012)
9. Babu, M.M., Luscombe, N.M., Aravind, L., Gerstein, M., Teichmann, S.A.: Structure and evolution of transcriptional regulatory networks. Curr. Opin. Struct. Biol. **14**, 283–291 (2004)
10. Balkwill, F.R., Capasso, M., Hagemann, T.: The tumor microenvironment at a glance. J. Cell Sci. **125**, 5591–5596 (2012)
11. Barabasi, A.L., Bonabeau, E.: Scale-free networks. Sci. Am. **288**, 60–69 (2003)
12. Barabasi, A.L., Oltvai, Z.N.: Network biology: understanding the cell's functional organization. Nat. Rev. Genet. **5**, 101–U115 (2004)
13. Barabasi, A.L., Gulbahce, N., Loscalzo, J.: Network medicine: a network-based approach to human disease. Nat. Rev. Genet. **12**, 56–68 (2011)
14. Beauchamp, M.A.: An improved index of centrality. Behav. Sci. **10**, 161–163 (1965)
15. Berggard, T., Linse, S., James, P.: Methods for the detection and analysis of protein-protein interactions. Proteomics **7**, 2833–2842 (2007)
16. Bhattacharyya, M., Chakrabarti, S.: Identification of important interacting proteins (IIPs) in Plasmodium falciparum using large-scale interaction network analysis and in-silico knock-out studies. Malar. J. **14**, 70 (2015)
17. Björklund, A.K., Light, S., Hedin, L., Elofsson, A.: Quantitative assessment of the structural bias in protein-protein interaction assays. Proteomics **8**, 4657–4667 (2008)
18. Boldi, P., Vigna, S.: Axioms for centrality. Internet Math. **10**, 222–262 (2014)
19. Bonacich, P.: Technique for analyzing overlapping memberships. Sociol. Methodol. **4**, 176–185 (1972)
20. Bonacich, P., Lloyd, P.: Eigenvector-like measures of centrality for asymmetric relations. Soc. Netw. **23**, 191–201 (2001)
21. Borgatti, S.P., Everett, M.G., Johnson, J.C.: Analyzing Social Networks. SAGE Publications Limited, Los Angeles (2013)
22. Bossi, A., Lehner, B.: Tissue specificity and the human protein interaction network. Mol. Syst. Biol. **5**, 260 (2009)
23. Brandes, U.: On variants of shortest-path betweenness centrality and their generic computation. Soc. Netw. **30**, 136–145 (2008)
24. Brin, S., Page, L.: The anatomy of a large-scale hypertextual Web search engine. Comput. Netw. ISDN **30**, 107–117 (1998)
25. Bulyk, M.L.: Computational prediction of transcription-factor binding site locations. Genome Biol. **5**, 201 (2004)
26. Burrell, R.A., McGranahan, N., Bartek, J., Swanton, C.: The causes and consequences of genetic heterogeneity in cancer evolution. Nature **501**, 338–345 (2013)
27. Butts, C.T.: Social network analysis with SNA. J. Stat. Softw. **24**, 1–51 (2008)
28. Carter, S.L., Brechbuhler, C.M., Griffin, M., Bond, A.T.: Gene co-expression network topology provides a framework for molecular characterization of cellular state. Bioinformatics **20**, 2242–2250 (2004)

29. Cerami, E.G., Gross, B.E., Demir, E., Rodchenkov, I., Babur, O., Anwar, N., et al.: Pathway commons, a web resource for biological pathway data. Nucleic Acids Res. **39**, D685–D690 (2011)
30. Chen, L., Qu, X., Cao, M., Zhou, Y., Li, W., Liang, B., et al.: Identification of breast cancer patients based on human signaling network motifs. Sci. Rep. **3**, 3368 (2013)
31. Cohen, R., Havlin, S.: Scale-free networks are ultrasmall. Phys. Rev. Lett. **90**, 058701 (2003)
32. Coombes, R.C.: Drug testing in the patient: toward personalized cancer treatment. Sci. Transl. Med. **7** (2015)
33. Csardi, G., Nepusz, T.: The igraph software package for complex network research. Int. J. Complex Syst. **1695**, 1–9 (2006)
34. Dangalchev, C.: Residual closeness in networks. Phys. A **365**, 556–564 (2006)
35. del Rio, G., Koschützki, D., Coello, G.: How to identify essential genes from molecular networks? BMC Syst. Biol. **3**, 1–12 (2009)
36. Diamandis, M., White, N.M.A., Yousef, G.M.: Personalized medicine: marking a new epoch in cancer patient management. Mol. Cancer Res. **8**, 1175–1187 (2010)
37. Dobrin, R., Beg, Q.K., Barabasi, A.L., Oltvai, Z.N.: Aggregation of topological motifs in the Escherichia coli transcriptional regulatory network. BMC Bioinform. **5**, 1–8 (2004)
38. Efron, B., Tibshirani, R.: On testing the significance of sets of genes. Ann. Appl. Stat. **1**, 107–129 (2007)
39. Eirew, P., Steif, A., Khattra, J., Ha, G., Yap, D., Farahani, H., et al.: Dynamics of genomic clones in breast cancer patient xenografts at single-cell resolution. Nature **518**, 422–426 (2015)
40. Estrada, E.: Virtual identification of essential proteins within the protein interaction network of yeast. Proteomics **6**, 35–40 (2006)
41. Estrada, E.: Protein bipartivity and essentiality in the yeast protein-protein interaction network. J. Proteome Res. **5**, 2177–2184 (2006)
42. Euskirchen, G.M., Rozowsky, J.S., Wei, C.L., Lee, W.H., Zhang, Z.D.D., Hartman, S., et al.: Mapping of transcription factor binding regions in mammalian cells by ChIP: comparison of array- and sequencing-based technologies. Genome Res. **17**, 898–909 (2007)
43. Fabregat, A., Sidiropoulos, K., Garapati, P., Gillespie, M., Hausmann, K., Haw, R., et al.: The reactome pathway knowledgebase. Nucleic Acids Res. **44**, D481–D487 (2016)
44. Faith, J.J., Hayete, B., Thaden, J.T., Mogno, I., Wierzbowski, J., Cottarel, G., et al.: Large- scale mapping and validation of Escherichia coli transcriptional regulation from a compendium of expression profiles. PLoS Biol. **5**, 54–66 (2007)
45. Forbes, S.A., Beare, D., Gunasekaran, P., Leung, K., Bindal, N., Boutselakis, H., et al.: COSMIC: exploring the world's knowledge of somatic mutations in human cancer. Nucleic Acids Res. **43**, D805–D811 (2015)
46. Freeman, L.C.: A set of measures of centrality based on betweenness. Sociometry **40**, 35–41 (1977)
47. Freeman, L.C.: Centrality in social networks conceptual clarification. Soc. Netw. **1**, 215–239 (1979)
48. Freeman, L.C., Borgatti, S.P., White, D.R.: Centrality in valued graphs - a measure of betweenness based on network flow. Soc. Netw. **13**, 141–154 (1991)
49. Fronza, R., Vasciaveo, A., Benso, A., Schmidt, M.: A graph based framework to model virus integration sites. Comput. Struct. Biotechnol. J. **14**, 69–77 (2016)
50. Giam, M., Rancati, G.: Aneuploidy and chromosomal instability in cancer: a jackpot to chaos. Cell Div. **10**, 3 (2015)
51. Gillis, J., Ballouz, S., Pavlidis, P.: Bias tradeoffs in the creation and analysis of protein- protein interaction networks. J. Proteomics **100**, 44–54 (2014)
52. Giot, L., Bader, J.S., Brouwer, C., Chaudhuri, A., Kuang, B., Li, Y., et al.: A protein interaction map of Drosophila melanogaster. Science **302**, 1727–1736 (2003)
53. Gleich, D.F.: Chapter 7 on MatlabBGL. Models and Algorithms for PageRank Sensitivity. Stanford University (2009)
54. Goh, K.I., Kahng, B., Kim, D.: Universal behavior of load distribution in scale-free networks. Phys. Rev. Lett. **87**, 278701 (2001)

55. Goh, K.I., Cusick, M.E., Valle, D., Childs, B., Vidal, M., Barabasi, A.L.: The human disease network. In: Proceedings of the National Academy of Sciences of the United States of America vol. 104, pp. 8685–8690 (2007)
56. Gonzalez-Perez, A., Perez-Llamas, C., Deu-Pons, J., Tamborero, D., Schroeder, M.P., Jene-Sanz, A., et al.: IntOGen-mutations identifies cancer drivers across tumor types. Nat. Methods 10, 1081–1082 (2013)
57. Grassler, J., Koschützki, D., Schreiber, F.: CentiLib: comprehensive analysis and exploration of network centralities. Bioinformatics 28, 1178–1179 (2012)
58. Gu, Z.G., Wang, J.: CePa: an R package for finding significant pathways weighted by multiple network centralities. Bioinformatics 29, 658–660 (2013)
59. Guan, Y.F., Gorenshteyn, D., Burmeister, M., Wong, A.K., Schimenti, J.C., Handel, M.A., et al.: Tissue-specific functional networks for prioritizing phenotype and disease genes. PLoS Comput. Biol. 8, e1002694 (2012)
60. Guelzim, N., Bottani, S., Bourgine, P., Kepes, F.: Topological and causal structure of the yeast transcriptional regulatory network. Nat. Genet. 31, 60–63 (2002)
61. Guinney, J., Dienstmann, R., Wang, X., de Reynies, A., Schlicker, A., Soneson, C., et al.: The consensus molecular subtypes of colorectal cancer. Nat. Med. 21, 1350–1356 (2015)
62. Hage, P., Harary, F.: Eccentricity and centrality in networks. Soc. Netw. 17, 57–63 (1995)
63. Hahn, M.W., Kern, A.D.: Comparative genomics of centrality and essentiality in three eukaryotic protein-interaction networks. Mol. Biol. Evol. 22, 803–806 (2005)
64. Hardy, S., Legagneux, V., Audic, Y., Paillard, L.: Reverse genetics in eukaryotes. Biol. Cell 102, 561–580 (2010)
65. Haury, A.C., Mordelet, F., Vera-Licona, P., Vert, J.P.: TIGRESS: trustful inference of gene regulation using stability selection. BMC Syst. Biol. 6, 1–17 (2012)
66. Hirschhorn, J.N., Daly, M.J.: Genome-wide association studies for common diseases and complex traits. Nat. Rev. Genet. 6, 95–108 (2005)
67. Hu, P.Z., Bader, G., Wigle, D.A., Emili, A.: Computational prediction of cancer-gene function. Nat. Rev. Cancer 7, 23–34 (2007)
68. Huang, X., Zi, Z.K.: Inferring cellular regulatory networks with Bayesian model averaging for linear regression (BMALR). Mol. Biosyst. 10, 2023–2030 (2014)
69. Huynh-Thu, V.A., Irrthum, A., Wehenkel, L., Geurts, P.: Inferring regulatory networks from expression data using tree-based methods. Plos One 5, e12776 (2010)
70. Izudheen, S., Mathew, S.: Cancer gene identification using graph centrality. Curr. Sci. 105, 1143–1148 (2013)
71. Jalali, S., Bhartiya, D., Lalwani, M.K., Sivasubbu, S., Scaria, V.: Systematic transcriptome wide analysis of lncRNA-miRNA interactions. Plos One 8, e53823 (2013)
72. Jalili, M., Salehzadeh-Yazdi, A., Asgari, Y., Arab, S.S., Yaghmaie, M., Ghavamzadeh, A., et al.: CentiServer: a comprehensive resource, web-based application and R package for centrality analysis. Plos One 10, e0143111 (2015)
73. Jensen, L.J., Kuhn, M., Stark, M., Chaffron, S., Creevey, C., Muller, J., et al.: STRING 8-a global view on proteins and their functional interactions in 630 organisms. Nucleic Acids Res. 37, D412–D416 (2009)
74. Jeong, H., Tombor, B., Albert, R., Oltvai, Z.N., Barabasi, A.L.: The large-scale organization of metabolic networks. Nature 407, 651–654 (2000)
75. Jeong, H., Mason, S.P., Barabasi, A.L., Oltvai, Z.N.: Lethality and centrality in protein networks. Nature 411, 41–42 (2001)
76. Jinq, Z., Hong, Y., Jianhua, L., Cao, Z.W., Li, Y.X.: Complex networks theory for analyzing metabolic networks. Chin. Sci. Bull. 51, 1529–1537 (2006)
77. Jonsson, P.F., Bates, P.A.: Global topological features of cancer proteins in the human interactome. Bioinformatics 22, 2291–2297 (2006)
78. Joy, M.P., Brock, A., Ingber, D.E., Huang, S.: High-betweenness proteins in the yeast protein interaction network. J. Biomed. Biotechnol. 2005, 96–103 (2005)
79. Junker, B.H., Koschützki, D., Schreiber, F.: Exploration of biological network centralities with CentiBiN. BMC Bioinform. 7, 1–7 (2006)

80. Kamburov, A., Stelzl, U., Lehrach, H., Herwig, R.: The ConsensusPathDB interaction database: 2013 update. Nucleic Acids Res. **41**, D793–D800 (2013)
81. Kanehisa, M., Sato, Y., Kawashima, M., Furumichi, M., Tanabe, M.: KEGG as a reference resource for gene and protein annotation. Nucleic Acids Res. **44**, D457–D462 (2016)
82. Kandoth, C., McLellan, M.D., Vandin, F., Ye, K., Niu, B., Lu, C., et al.: Mutational landscape and significance across 12 major cancer types. Nature **502**, 333–339 (2013)
83. Karabekmez, M.E., Kirdar, B.: A novel topological centrality measure capturing biologically important proteins. Mol. Biosyst. **12**, 666–673 (2016)
84. Katz, L.: A new status index derived from sociometric analysis. Psychometrika **18**, 39–43 (1953)
85. Kerrien, S., Aranda, B., Breuza, L., Bridge, A., Broackes-Carter, F., Chen, C., et al.: The IntAct molecular interaction database in 2012. Nucleic Acids Res. **40**, D841–D846 (2012)
86. Kim, S.Y., Volsky, D.J.: PAGE: parametric analysis of gene set enrichment. BMC Bioinform. **6**, 144 (2005)
87. Kim, W.: Prediction of essential proteins using topological properties in GO-pruned PPI network based on machine learning methods. Tsinghua Sci. Technol. **17**, 645–658 (2012)
88. Kool, J., Berns, A.: High throughput insertional mutagenesis screens in mice to identify oncogenic networks (vol 9, pg 389, 2009). Nat. Rev. Cancer **9**, 604–604 (2009)
89. Koschützki, D., Schreiber, F.: Comparison of centralities for biological networks. In: German Conference on Bioinformatics (2004)
90. Koschützki, D., Schreiber, F.: Centrality analysis methods for biological networks and their application to gene regulatory networks. Gene Regul. Syst. Biol. **2**, 193–201 (2008)
91. Koschützki, D., Schwobbermeyer, H., Schreiber, F.: Ranking of network elements based on functional substructures. J. Theor. Biol. **248**, 471–479 (2007)
92. Kreso, A., O'Brien, C.A., van Galen, P., Gan, O.I., Notta, F., Brown, A.M.K., et al.: Variable clonal repopulation dynamics influence chemotherapy response in colorectal cancer. Science **339**, 543–548 (2013)
93. Li, J.H., Liu, S., Zhou, H., Qu, L.H., Yang, J.H.: StarBase v2.0: decoding miRNA-ceRNA, miRNA-ncRNA and protein-RNA interaction networks from large-scale CLIP-Seq data. Nucleic Acids Res. **42**, D92–D97 (2014)
94. Li, M., Zhang, H., Wang, J.X., Pan, Y.: A new essential protein discovery method based on the integration of protein-protein interaction and gene expression data. BMC Syst. Biol. **6**, 15 (2012)
95. Li, S.M., Armstrong, C.M., Bertin, N., Ge, H., Milstein, S., Boxem, M., et al.: A map of the interactome network of the metazoan C-elegans. Science **303**, 540–543 (2004)
96. Lin, N.: Foundations of Social Research. McGraw-Hill, New York (1976)
97. Linghu, B., Snitkin, E.S., Hu, Z., Xia, Y., Delisi, C.: Genome-wide prioritization of disease genes and identification of disease-disease associations from an integrated human functional linkage network. Genome Biol. **10**, R91 (2009)
98. Linghu, B., Franzosa, E.A., Xia, Y.: Construction of functional linkage gene networks by data integration. Methods Mol. Biol. **939**, 215–232 (2013)
99. Luscombe, N.M., Babu, M.M., Yu, H.Y., Snyder, M., Teichmann, S.A., Gerstein, M.: Genomic analysis of regulatory network dynamics reveals large topological changes. Nature **431**, 308–312 (2004)
100. Ma, X.K., Gao, L.: Biological network analysis: insights into structure and functions. Brief. Funct. Genomics **11**, 434–442 (2012)
101. Magger, O., Waldman, Y.Y., Ruppin, E., Sharan, R.: Enhancing the prioritization of disease-causing genes through tissue specific protein interaction networks. PLoS Comput. Biol. **8**, e1002690 (2012)
102. Marbach, D., Costello, J.C., Kuffner, R., Vega, N.M., Prill, R.J., Camacho, D.M., et al.: Wisdom of crowds for robust gene network inference. Nat. Methods **9**, 796–804 (2012)
103. March, H.N., Rust, A.G., Wright, N.A., Ten Hoeve, J., de Ridder, J., Eldridge, M., et al.: Insertional mutagenesis identifies multiple networks of cooperating genes driving intestinal tumorigenesis. Nat Genet **43**, 1202–U1255 (2011)

104. Margolin, A.A., Nemenman, I., Basso, K., Wiggins, C., Stolovitzky, G., Dalla Favera, R., et al.: ARACNE: an algorithm for the reconstruction of gene regulatory networks in a mammalian cellular context. BMC Bioinform. **7**, 1–15 (2006)
105. Marusyk, A., Polyak, K.: Tumor heterogeneity: causes and consequences. BBA-Rev. Cancer **1805**, 105–117 (2010)
106. Mathelier, A., Wasserman, W.W.: The next generation of transcription factor binding site prediction. PLoS Comput. Biol. **9**, e1003214 (2013)
107. Matys, V., Kel-Margoulis, O.V., Fricke, E., Liebich, I., Land, S., Barre-Dirrie, A., et al.: TRANSFAC (R) and its module TRANSCompel (R): transcriptional gene regulation in eukaryotes. Nucleic Acids Res. **34**, D108–D110 (2006)
108. McGranahan, N., Swanton, C.: Biological and therapeutic impact of intratumor heterogeneity in cancer evolution (vol 27, pg 15, 2015). Cancer Cell **28**, 141–141 (2015)
109. Meacham, C.E., Morrison, S.J.: Tumour heterogeneity and cancer cell plasticity. Nature **501**, 328–337 (2013)
110. Meyer, M., Reimand, J., Lan, X., Head, R., Zhu, X., Kushida, M., et al.: Single cell-derived clonal analysis of human glioblastoma links functional and genomic heterogeneity. In: Proceedings of the National Academy of Sciences of the United States of America, vol. 112, pp. 851–856 (2015)
111. Miernyk, J.A., Thelen, J.J.: Biochemical approaches for discovering protein-protein interactions. Plant J. **53**, 597–609 (2008)
112. Milo, R., Shen-Orr, S., Itzkovitz, S., Kashtan, N., Chklovskii, D., Alon, U.: Network motifs: simple building blocks of complex networks. Science **298**, 824–827 (2002)
113. Moreau, Y., Tranchevent, L.C.: Computational tools for prioritizing candidate genes: boosting disease gene discovery. Nat. Rev. Genet. **13**, 523–536 (2012)
114. Moresco, E.M.Y., Li, X.H., Beutler, B.: Going forward with genetics recent technological advances and forward genetics in mice. Am. J. Pathol. **182**, 1462–1473 (2013)
115. Morrissy, A.S., Garzia, L., Shih, D.J.H., Zuyderduyn, S., Huang, X., Skowron, P., et al.: Divergent clonal selection dominates medulloblastoma at recurrence. Nature **529**, 351–357 (2016)
116. Newman, M.E.J.: Scientific collaboration networks. II. Shortest paths, weighted networks, and centrality. Phys. Rev. E **64**, 016132 (2001)
117. Ortutay, C., Vihinen, M.: Identification of candidate disease genes by integrating gene ontologies and protein-interaction networks: case study of primary immunodeficiencies. Nucleic Acids Res. **37**, 622–628 (2009)
118. Oti, M., Snel, B., Huynen, M.A., Brunner, H.G.: Predicting disease genes using protein-protein interactions. J. Med. Genet. **43**, 691–698 (2006)
119. Özgür, A., Vu, T., Erkan, G., Radev, D.R.: Identifying gene-disease associations using centrality on a literature mined gene-interaction network. Bioinformatics **24**, 1277–1285 (2008)
120. Patel, A.P., Tirosh, I., Trombetta, J.J., Shalek, A.K., Gillespie, S.M., Wakimoto, H., et al.: Single-cell RNA-seq highlights intratumoral heterogeneity in primary glioblastoma. Science **344**, 1396–1401 (2014)
121. Phizicky, E.M., Fields, S.: Protein-protein interactions - Methods for detection and analysis. Microbiol. Rev. **59**, 94–123 (1995)
122. Poulin, R., Boily, M.C., Masse, B.R.: Dynamical systems to define centrality in social networks. Soc. Netw. **22**, 187–220 (2000)
123. Prasad, T.S.K., Goel, R., Kandasamy, K., Keerthikumar, S., Kumar, S., Mathivanan, S., et al.: Human protein reference database-2009 update. Nucleic Acids Res. **37**, D767–D772 (2009)
124. Price, A.L., Spencer, C.C.A., Donnelly, P.: Progress and promise in understanding the genetic basis of common diseases. Proc. R. Soc. B-Biol. Sci. **282**, 20151684 (2015)
125. Ptacek, J., Devgan, G., Michaud, G., Zhu, H., Zhu, X.W., Fasolo, J., et al.: Global analysis of protein phosphorylation in yeast. Nature **438**, 679–684 (2005)
126. Qin, J., Hu, Y.H., Xu, F., Yalamanchili, H.K., Wang, J.W.: Inferring gene regulatory networks by integrating ChIP-seq/chip and transcriptome data via LASSO-type regularization methods. Methods **67**, 294–303 (2014)

127. Rajasingh, I., Rajan, B., Florence, I.D.: Betweeness-centrality of grid networks. In: Proceedings of the 2009 International Conference on Computer Technology and Development, vol. 1, pp. 407–410 (2009)

128. Ramanan, V.K., Shen, L., Moore, J.H., Saykin, A.J.: Pathway analysis of genomic data: concepts, methods, and prospects for future development. Trends Genet. **28**, 323–332 (2012)

129. Ravasz, E., Somera, A.L., Mongru, D.A., Oltvai, Z.N., Barabasi, A.L.: Hierarchical organization of modularity in metabolic networks. Science **297**, 1551–1555 (2002)

130. Resendis-Antonio, O., Freyre-Gonzalez, J.A., Menchaca-Mendez, R., Gutierrez-Rios, R.M., Martinez- Antonio, A., Avila-Sanchez, C., et al.: Modular analysis of the transcriptional regulatory network of E-coli. Trends Genet. **21**, 16–20 (2005)

131. Risch, N.J.: Searching for genetic determinants in the new millennium. Nature **405**, 847–856 (2000)

132. Rives, A.W., Galitski, T.: Modular organization of cellular networks. In: Proceedings of the National Academy of Sciences of the United States of America, vol. 100, pp. 1128–1133 (2003)

133. Rochat, Y.: Closeness centrality extended to unconnected graphs: the harmonic centrality index. ASNA. No. EPFL-CONF-200525 (2009)

134. Ruhnau, B.: Eigenvector-centrality - a node-centrality? Soc. Netw. **22**, 357–365 (2000)

135. Sabidussi, G.: The centrality index of a graph. Psychometrika **31**, 581–586 (1966)

136. Salwinski, L., Miller, C.S., Smith, A.J., Pettit, F.K., Bowie, J.U., Eisenberg, D.: The database of interacting proteins: 2004 update. Nucleic Acids Res. **32**, D449–D451 (2004)

137. Sandelin, A., Alkema, W., Engstrom, P., Wasserman, W.W., Lenhard, B.: JASPAR: an open-access database for eukaryotic transcription factor binding profiles. Nucleic Acids Res. **32**, D91–D94 (2004)

138. Sander, J.D., Joung, J.K.: CRISPR-Cas systems for editing, regulating and targeting genomes. Nat. Biotechnol. **32**, 347–355 (2014)

139. Sanz, J., Navarro, J., Arbues, A., Martin, C., Marijuan, P.C., Moreno, Y.: The transcriptional regulatory network of Mycobacterium tuberculosis. Plos One **6**, e22178 (2011)

140. Scardoni, G., Petterlini, M., Laudanna, C.: Analyzing biological network parameters with CentiScaPe. Bioinformatics **25**, 2857–2859 (2009)

141. Schadt, E.E.: Molecular networks as sensors and drivers of common human diseases. Nature **461**, 218–223 (2009)

142. Schoch, D., Brandes, U.: Centrality as a predictor of lethal proteins: performance and robustness. In: MMB & DFT (2014)

143. Sharma, A., Gulbahce, N., Pevzner, S.J., Menche, J., Ladenvall, C., Folkersen, L., et al.: Network-based analysis of genome wide association data provides novel candidate genes for lipid and lipoprotein traits. Mol. Cell. Proteomics **12**, 3398–3408 (2013)

144. Shen-Orr, S.S., Milo, R., Mangan, S., Alon, U.: Network motifs in the transcriptional regulation network of Escherichia coli. Nat. Genet. **31**, 64–68 (2002)

145. Shimbel, A.: Structural parameters of communication networks. Bull. Math. Biophys. **15**, 501–507 (1953)

146. Shoemaker, B.A., Panchenko, A.R.: Deciphering protein-protein interactions. Part II. Computational methods to predict protein and domain interaction partners. PLoS Comput. Biol. **3**, 595–601 (2007)

147. Siddani, B.R., Pochineni, L.P., Palanisamy, M.: Candidate gene identification for systemic lupus erythematosus using network centrality measures and gene ontology. Plos One **8**, e81766 (2013)

148. Siddharthan, R.: Dinucleotide weight matrices for predicting transcription factor binding sites: generalizing the position weight matrix. Plos One **5**, e9722 (2010)

149. Siegel, R.L., Miller, K.D., Jemal, A.: Cancer statistics, 2015. CA: Cancer J. Clin. **65**, 5–29 (2015)

150. Silva, J., Chang, K., Hannon, G.J., Rivas, F.V.: RNA-interference-based functional genomics in mammalian cells: reverse genetics coming of age. Oncogene **23**, 8401–8409 (2004)

151. Simoes, R.D., Emmert-Streib, F.: Bagging statistical network inference from large-scale gene expression data. Plos One **7**, e33624 (2012)
152. Slater, P.J.: Maximin facility location. J. Res. NBS B Math. Sci. **79**, 107–115 (1975)
153. Sorlie, T., Tibshirani, R., Parker, J., Hastie, T., Marron, J.S., Nobel, A., et al.: Repeated observation of breast tumor subtypes in independent gene expression data sets. In: Proceedings of the National Academy of Sciences of the United States of America, vol. 100, pp. 8418–8423 (2003)
154. Sottoriva, A., Spiteri, I., Piccirillo, S.G., Touloumis, A., Collins, V.P., Marioni, J.C., et al.: Intratumor heterogeneity in human glioblastoma reflects cancer evolutionary dynamics. In: Proceedings of the National Academy of Sciences of the United States of America, vol. 110, pp. 4009–4014 (2013)
155. Spirin, V., Mirny, L.A.: Protein complexes and functional modules in molecular networks. In: Proceedings of the National Academy of Sciences of the United States of America, vol. 100, pp. 12123–12128 (2003)
156. Stark, C., Breitkreutz, B.J., Reguly, T., Boucher, L., Breitkreutz, A., Tyers, M.: BioGRID: a general repository for interaction datasets. Nucleic Acids Res. **34**, D535–D539 (2006)
157. Stelzl, U., Worm, U., Lalowski, M., Haenig, C., Brembeck, F.H., Goehler, H., et al.: A human protein- protein interaction network: a resource for annotating the proteome. Cell **122**, 957–968 (2005)
158. Stuart, J.M., Segal, E., Koller, D., Kim, S.K.: A gene-coexpression network for global discovery of conserved genetic modules. Science **302**, 249–255 (2003)
159. Sturm, D., Witt, H., Hovestadt, V., Khuong-Quang, D.A., Jones, D.T., Konermann, C., et al.: Hotspot mutations in H3F3A and IDH1 define distinct epigenetic and biological subgroups of glioblastoma. Cancer Cell **22**, 425–437 (2012)
160. Smoot, M.E., Ono, K., Ruscheinski, J., Wang, P.L., Ideker, T.: Cytoscape 2.8: new features for data integration and network visualization. Bioinformatics **27**, 431–432 (2011)
161. Subramanian, A., Tamayo, P., Mootha, V.K., Mukherjee, S., Ebert, B.L., Gillette, M.A., et al.: Gene set enrichment analysis: a knowledge-based approach for interpreting genome-wide expression profiles. In: Proceedings of the National Academy of Sciences of the United States of America, vol. 102, pp. 15545–15550 (2005)
162. Szalay, K.Z., Csermely, P.: Perturbation centrality and turbine: a novel centrality measure obtained using a versatile network dynamics tool. Plos One **8**, e78059 (2013)
163. Tang, X., Wang, J., Zhong, J., Pan, Y.: Predicting essential proteins based on weighted degree centrality. IEEE/ACM Trans. Comput. Biol. Bioinform./IEEE ACM **11**, 407–418 (2014)
164. Tarca, A.L., Draghici, S., Bhatti, G., Romero, R.: Down-weighting overlapping genes improves gene set analysis. BMC Bioinform. **13**, 136 (2012)
165. Taylor, M.D., Northcott, P.A., Korshunov, A., Remke, M., Cho, Y.J., Clifford, S.C., et al.: Molecular subgroups of medulloblastoma: the current consensus. Acta Neuropathol. **123**, 465–472 (2012)
166. Tong, A.H.Y., Evangelista, M., Parsons, A.B., Xu, H., Bader, G.D., Page, N., et al.: Systematic genetic analysis with ordered arrays of yeast deletion mutants. Science **294**, 2364–2368 (2001)
167. Tong, A.H.Y., Lesage, G., Bader, G.D., Ding, H.M., Xu, H., Xin, X.F., et al.: Global mapping of the yeast genetic interaction network. Science **303**, 808–813 (2004)
168. Tschida, B.R., Largaespada, D.A., Keng, V.W.: Mouse models of cancer: sleeping beauty transposons for insertional mutagenesis screens and reverse genetic studies. Semin. Cell Dev. Biol. **27**, 86–95 (2014)
169. Uren, A.G., Kool, J., Berns, A., van Lohuizen, M.: Retroviral insertional mutagenesis: past, present and future. Oncogene **24**, 7656–7672 (2005)
170. Uren, A.G., Kool, J., Matentzoglu, K., de Ridder, J., Mattison, J., van Uitert, M., et al.: Large-scale mutagenesis in p19ARF-and p53-deficient mice identifies cancer genes and their collaborative networks. Cell **133**, 727–741 (2008)
171. Valencia, A., Pazos, F.: Computational methods for the prediction of protein interactions. Curr. Opin. Struct. Biol. **12**, 368–373 (2002)

172. Valente, T.W., Foreman, R.K.: Integration and radiality: measuring the extent of an individual's connectedness and reachability in a network. Soc. Netw. **20**, 89–105 (1998)
173. Valouev, A., Johnson, D.S., Sundquist, A., Medina, C., Anton, E., Batzoglou, S., et al.: Genome-wide analysis of transcription factor binding sites based on ChIP-Seq data. Nat. Methods **5**, 829–834 (2008)
174. Verhaak, R.G., Hoadley, K.A., Purdom, E., Wang, V., Qi, Y., Wilkerson, M.D., et al.: Integrated genomic analysis identifies clinically relevant subtypes of glioblastoma characterized by abnormalities in PDGFRA, IDH1, EGFR, and NF1. Cancer Cell **17**, 98–110 (2010)
175. Vidal, M., Cusick, M.E., Barabasi, A.L.: Interactome networks and human disease. Cell **144**, 986–998 (2011)
176. Vogelstein, B., Papadopoulos, N., Velculescu, V.E., Zhou, S., Diaz Jr., L.A., Kinzler, K.W.: Cancer genome landscapes. Science **339**, 1546–1558 (2013)
177. von Mering, C., Krause, R., Snel, B., Cornell, M., Oliver, S.G., Fields, S., et al.: Comparative assessment of large-scale data sets of protein-protein interactions. Nature **417**, 399–403 (2002)
178. Wachi, S., Yoneda, K., Wu, R.: Interactome-transcriptome analysis reveals the high centrality of genes differentially expressed in lung cancer tissues. Bioinformatics **21**, 4205–4208 (2005)
179. Wagner, A., Fell, D.A.: The small world inside large metabolic networks. Proc. R. Soc. B-Biol. Sci. **268**, 1803–1810 (2001)
180. Wang, J., Chen, G., Li, M., Pan, Y.: Integration of breast cancer gene signatures based on graph centrality. BMC Syst. Biol. **5**(Suppl 3), S10 (2011)
181. Wang, K., Li, M.Y., Hakonarson, H.: Analysing biological pathways in genome-wide association studies. Nat. Rev. Genet. **11**, 843–854 (2010)
182. Wang, P.W., Qin, J., Qin, Y.M., Zhu, Y., Wang, L.L.Y., Li, M.L.J., et al.: ChIP-Array 2: integrating multiple omics data to construct gene regulatory networks. Nucleic Acids Res. **43**, W264–W269 (2015)
183. Wang, S., Sun, H.F., Ma, J., Zang, C.Z., Wang, C.F., Wang, J., et al.: Target analysis by integration of transcriptome and ChIP-seq data with BETA. Nat. Protoc. **8**, 2502–2515 (2013)
184. Wang, W.Y.S., Barratt, B.J., Clayton, D.G., Todd, J.A.: Genome-wide association studies: theoretical and practical concerns. Nat. Rev. Genet. **6**, 109–118 (2005)
185. Wang, X., Gulbahce, N., Yu, H.: Network-based methods for human disease gene prediction. Brief. Funct. Genomics **10**, 280–293 (2011)
186. Watts, D.J., Strogatz, S.H.: Collective dynamics of 'small-world' networks. Nature **393**, 440–442 (1998)
187. Wei, W., Pfeffer, J., Reminga, J., Carley, K.M.: Handling Weighted, Asymmetric, Self-Looped, and Disconnected Networks in ORA (No. CMU-ISR-11-113). Carnegie Mellon University, Pittsburgh (2011)
188. Weishaupt, H., Johansson, P., Engström, C., Nelander, S., Silvestrov, S., Swartling, FJ.: Loss of conservation of graph centralities in reverse-engineered transcriptional regulatory networks. In: 16th Applied Stochastic Models and Data Analysis International Conference (ASMDA2015) with Demographics 2015 Workshop (2015)
189. Wu, X.B., Li, S.: Cancer Gene Prediction Using a Network Approach. Chapman & Hall/CRC Mathematical and Computational Biology, pp. 191–212 (2010)
190. Wuchty, S.: Scale-free behavior in protein domain networks. Mol. Biol. Evol. **18**, 1694–1702 (2001)
191. Wuchty, S., Almaas, E.: Evolutionary cores of domain co-occurrence networks. BMC Evol. Biol. **5**, 1–12 (2005)
192. Wuchty, S., Stadler, P.F.: Centers of complex networks. J. Theor. Biol. **223**, 45–53 (2003)
193. Wuchty, S., Oltvai, Z.N., Barabasi, A.L.: Evolutionary conservation of motif constituents in the yeast protein interaction network. Nat. Genet. **35**, 176–179 (2003)
194. Xu, J.Z., Li, Y.J.: Discovering disease-genes by topological features in human protein-protein interaction network. Bioinformatics **22**, 2800–2805 (2006)
195. Yeger-Lotem, E., Sattath, S., Kashtan, N., Itzkovitz, S., Milo, R., Pinter, R.Y., et al.: Network motifs in integrated cellular networks of transcription-regulation and protein-protein interaction. In: Proceedings of the National Academy of Sciences of the United States of America, vol. 101, pp. 5934–5939 (2004)

196. Yook, S.H., Oltvai, Z.N., Barabasi, A.L.: Functional and topological characterization of protein interaction networks. Proteomics **4**, 928–942 (2004)
197. Zellmer, V.R., Zhang, S.Y.: Evolving concepts of tumor heterogeneity. Cell Biosci. **4**, 1–8 (2014)
198. Zhang, L.V., King, O.D., Wong, S.L., Goldberg, D.S., Tong, A.H., Lesage, G., et al.: Motifs, themes and thematic maps of an integrated Saccharomyces cerevisiae interaction network. J. Biol. **4**, 6 (2005)
199. Zhang, M., Deng, J., Fang, C.V., Zhang, X., Lu, L.J.: Molecular network analysis and applications. In: Alterovitz, G., Ramoni, M. (eds.) Knowledge-Based Bioinformatics: From Analysis to Interpretation, pp. 253. Wiley, Chichester (2011)
200. Zhao, B.Y., Pritchard, J.R., Lauffenburger, D.A., Hemann, M.T.: Addressing genetic tumor heterogeneity through computationally predictive combination therapy. Cancer Discov. **4**, 166–174 (2014)
201. Zhao, H., Liu, T., Liu, L., Zhang, G., Pang, L., Yu, F., et al.: Chromatin states modify network motifs contributing to cell-specific functions. Sci. Rep. **5**, 11938 (2015)
202. Zhao, W., Langfelder, P., Fuller, T., Dong, J., Li, A., Hovarth, S.: Weighted gene coexpression network analysis: state of the art. J. Biopharm. Stat. **20**, 281–300 (2010)
203. Zhu, C., Wu, C., Aronow, B.J., Jegga, A.G.: Computational approaches for human disease gene prediction and ranking. Adv. Exp. Med. Biol. **799**, 69–84 (2014)
204. Zhu, X., Gerstein, M., Snyder, M.: Getting connected: analysis and principles of biological networks. Genes Dev. **21**, 1010–1024 (2007)

Output Rate Variation Problem: Some Heuristic Paradigms and Dynamic Programming

Gyan Bahadur Thapa and Sergei Silvestrov

Abstract The output rate variation problem stands as one of the important research directions in the area of multi-level just-in-time production system. In this short survey, we present the mathematical models of the problem followed by consideration of its NP-hardness. We further carry out the brief review of heuristic approaches that are devised to solve the problem. The dynamic programming approach and pegging assumption are also briefly discussed. The pegging assumption reduces the multi-level problem into weighted single-level problem. A couple of the open problems regarding ORVP are listed at the end.

Keywords Just-in-time · Objectives · Constraints · Heuristics · Dynamic programming

1 Introduction

The output rate variation problem is the multi-level production sequencing problem in just-in-time (JIT) work environment. Toyota company in Japan invented the just-in-time production systems (JITPS) and mostly benefited around the decade of sixties-seventies. The problems in JITPS are categorized in two parts, namely single-level, called production rate variation problem (PRVP) and multi-level, called output rate variation problem (ORVP). The PRVP has been richly studied, for example in [3, 13, 28]. The PRVP deals only with the final assembly line, having polynomial time solutions whereas the multi-level problem deals with overall systems from raw materials to final customers. The ORVP consists of several levels in the production

G.B. Thapa (✉)
Pulchowk Campus, Institute of Engineering, Tribhuvan University, P.O. Box 19758,
Kathmandu, Nepal
e-mail: thapagbt@ioe.edu.np

S. Silvestrov
Division of Applied Mathematics, School of Education,
Culture and Communication, Mälardalen University, Box 883, 721 23 Västerås, Sweden
e-mail: sergei.silvestrov@mdh.se

© Springer International Publishing Switzerland 2016   313
S. Silvestrov and M. Rančić (eds.), *Engineering Mathematics II*,
Springer Proceedings in Mathematics & Statistics 179,
DOI 10.1007/978-3-319-42105-6_14

supply chain, for example, raw materials → components → sub-assemblies → products → distribution centers → retailers → customers. In this supply chain system, the multiple copies of different models are produced at the final assembly level, which is interlinked with several upstream production levels where raw materials are procured, stored and fabricated to produce the final products [2] and with several downstream distribution levels where final products are stored and distributed to the retailers and then to the customers.

The whole body of supply chain consists of inbound logistics along the production levels and outbound logistics along the distribution levels. The synchronized view of seven levels of production and supply chain network has been presented in [27]. There may be several sub-levels in between any two production levels. Therefore, the formulation of the ORVP contains L, $(l = 1, 2, \ldots, L)$ levels.

The rest of the paper is organized as follows: Sect. 2 presents the mathematical formulations of ORVP followed by its NP-hardness. Section 3 describes the Goal chasing heuristics developed by Toyota. The pegging assumption to convert the ORVP in terms of weighted PRVP is exhibited in Sect. 4, whereas the dynamic programming solution is reported in Sect. 5. Finally, Sect. 6 concludes the paper pointing out some of the open problems.

2   Mathematical Formulation of ORVP

Assume that the mixed-model multi-level JITSP (i.e., ORVP) consists of L levels of manufacturing operations, indexed by l, $l = 1, 2, \ldots, L$ with the first product level 1. The number of different part types and the demand of item i in level l are denoted by n_l and d_{il} respectively, where $i = 1, 2, \ldots, n_l$. The number of total units of item i at level l required to produce one unit of product p is denoted by t_{ilp} such that $d_{il} = \sum_{p=1}^{n_1} t_{ilp} d_{p1}$ is the dependent demand for item i at level l determined by the final product demands $d_{p1}, p = 1, 2, \ldots, n_1$ and $l = 1, 2, \ldots, L$. Note that $t_{ilp} = 1$ if $i = p$ and 0 if otherwise. Finally, $D_l = \sum_{i=1}^{n_1} d_{il}$ denotes the total demand at level l, and the ratio $r_{il} = \frac{d_{il}}{D_l}$ gives the demand rate for item i of level l such that $\sum_{i=1}^{n_l} r_{il} = 1$ at each level $l = 1, 2, \ldots, L$. It is noteworthy that the model of ORVP is assumed to be non-preemptive; that is, once commenced production of a product at level 1 must be completed prior to switch into another unit. This creates the concept of various stages or cycles in the production system. The production schedule at level 1 consists of D_1 stages in total and at each stage a single unit of an end-product can be processed. An item is said to be in stage k, $(k = 1, 2, \ldots, D_1)$, if k units of product have been produced at level 1 and there will be k complete units of various products p at level 1 during the first k time units.

Let x_{ilk} be the necessary quantity of item i produced at level l during the time units 1 through k and $y_{lk} = \sum_{i=1}^{n_l} x_{ilk}$ be the cumulative quantity of item i produced at level l during the same time units such that $y_{1k} = \sum_{i=1}^{n_l} x_{i1k} = k$. Due to the pull nature of the JITPS, the particular combination of the highest level products produced during the k time units (the x_{p1k} values) determines the necessary cumulative production at every other level. Thus, the required cumulative production for item i at level l with $l \geq 2$ through k time units is given by $x_{ilk} = \sum_{p=1}^{n_1} t_{ilp} x_{p1k}$. For a unimodal convex penalty function F_i, $i = 1, \ldots, n_l$ with minimum 0 at 0, the maximum deviation and the sum deviation multi-level JIT sequencing problems in mixed-model systems (i.e., ORVP) are mathematically formulated to minimize the objectives Z_{max} and Z_{sum} as the followings [14, 18]:

$$Z_{max} = \min \max_{i,l,k} F_i(x_{ilk} - y_{lk} r_{il}), \tag{1}$$

$$Z_{sum} = \min \sum_{k=1}^{D_1} \sum_{l=1}^{L} \sum_{i=1}^{n_l} F_i(x_{ilk} - y_{lk} r_{il}), \tag{2}$$

subject to

$$x_{ilk} = \sum_{p=1}^{n_1} t_{ilp} x_{p1k}, \quad i = 1, \ldots, n_l, \ l = 1, \ldots, L, \ k = 1, \ldots, D_1, \tag{3}$$

$$y_{lk} = \sum_{i=1}^{n_l} x_{ilk}, \quad l = 2, 3, \ldots, L, \ k = 1, \ldots, D_1, \tag{4}$$

$$y_{1k} = \sum_{p=1}^{n_1} x_{p1k} = k, \quad k = 1, 2, \ldots, D_1, \tag{5}$$

$$x_{p1k} \geq x_{p1(k-1)}, \quad p = 1, 2, \ldots, n_1, \ k = 1, 2, \ldots, D_1, \tag{6}$$

$$x_{p1D_1} = d_{p1}, \quad x_{p10} = 0, \quad p = 1, 2, \ldots, n_1, \tag{7}$$

$$x_{ilk} \geq 0 \ \text{integer}, \quad i = 1, \ldots, n_l, \ l = 1, \ldots, L, \ k = 1, \ldots, D_1. \tag{8}$$

Here the constraint (3) ensures that the necessary cumulative production of part i of level l by the end of time unit k is determined explicitly by the quantity of products produced at level 1. Constraints (4) and (5) show the total cumulative production of level l and level 1 respectively during the time slots 1 through k. Constraint (6) ensures that the total production of every product over k time units is a non-decreasing function of k. Constraint (7) guarantees that the demands for each product are met exactly, and (8) is the integrality constraint. The constraints (5), (6), (8) jointly ensure that exactly one unit of a product is scheduled during one time unit in the product level.

The particular cases of the objectives (1) and (2) are studied in literature [2, 27] as absolute and squared deviation objectives in both cases as follows:

$$Z_{max}^a = \min \max_{i,l,k} |x_{ilk} - y_{lk}r_{il}|, \tag{9}$$

$$Z_{max}^s = \min \max_{i,l,k} (x_{ilk} - y_{lk}r_{il})^2, \tag{10}$$

$$Z_{min}^a = \min \sum_{k=1}^{D_1} \sum_{l=1}^{L} \sum_{i=1}^{n_l} |x_{ilk} - y_{lk}r_{il}|, \tag{11}$$

$$Z_{min}^s = \min \sum_{k=1}^{D_1} \sum_{l=1}^{L} \sum_{i=1}^{n_l} (x_{ilk} - y_{lk}r_{il})^2. \tag{12}$$

The ORVP is a nonlinear integer programming problem, whose objective functions describe the sequence dependent nature of the schedule for lower parts. The required cumulative productions x_{ilk}'s, $l > 1$ are calculated directly from the assembly sequence of the products x_{i1k}'s, and the desired production goal for model i in level l is calculated as the ideal proportion (r_{il}) of the total cumulative production quantity (y_{lk}) of level l. Balanced schedules are generated by keeping the required production of all parts and products as close to this goal as possible.

The min-max objectives of ORVP aim to find a *smooth* schedule in every time period for every output. This is the basic concept underlying Toyota's sequencing algorithm [20]. Moreover, the value of the objective function Z_{max}^a represents an applicable physical application, providing the maximum overproduction or underproduction (the maximum inventory or shortage) from the desired quantity of production that occurs at any time in the schedule. This fact may be used to determine the number of kanbans (or the necessary safety stocks) used [16]. The min-sum objectives of ORVP seek optimal schedules that may have relatively large deviation in a single period or for a certain output while having the lowest possible total deviation.

2.1 The NP-Hardness of ORVP

For an input size n of a problem P, a generally accepted minimum requirement for an algorithm to be considered as efficient is that its running time is polynomial in n, denoted by $O(n^c)$ for some constant c. A decision problem is a problem whose output is a single Boolean value: yes or no, true or false, on or off etc. Based on this definition, there are three classes of decision problems: P (solvable in polynomial time), NP (Non-deterministic polynomial) and Co-NP (complements of problems in NP). To this end, the stunning conjecture is whether P is equal to NP. For a detail literature of computational complexity classes, we recommend [6, 7, 17, 22, 26].

The crux of a combinatorial problem is to develop an algorithm that guarantees identifying an optimal solution for every instance of the problem. Unfortunately as illustrated above, not all combinatorial problems possess an algorithm with small amount of computer time. For example, Steiner tree problem, 3-partition problem, exact 3-dimensional matching problem are some intractable problems.

The ORVP with the sum of the square deviation objective has been shown to be NP-hard in the ordinary sense [13]. This result has been achieved by reducing the scheduling around the shortest job (SASJ) problem to the ORVP. The scheduling around the shortest job problem finds a schedule, on a single machine, of independent jobs i, $i = 1, 2, \ldots, n$, that minimizes the sum $\sum_{i=1}^{n} (C_i - C_1)^2$, where C_i is completion time of job i, $i = 1, 2, \ldots, n$ with processing times $p_1 \leq p_2 \leq \ldots \leq p_n$. The SASJ problem is NP-hard in the ordinary sense [12]. Moreover, the min-sum ORVP problem is computationally more difficult and the results established so far on the completion time variance minimization problem indicate that even special cases of ORVP are NP-hard.

Furthermore, bottleneck ORVP with absolute-deviation objective that considers only two levels of production has been proved to be NP-hard in the strong sense. An instance of the 3-partition problem can be reduced into an instance of ORVP with two levels in pseudo-polynomial time [16]. The 3-partition problem is to decide whether a given multiset of integers can be partitioned into triples that all have the same sum. That is, for $3m$ integers, is there a partition $\{A_1, A_2, \ldots, A_m\}$ of the set $\{1, 2, \ldots, 3m\}$ such that $\sum_{i \in A_i} a_i = B$, $1 \leq i \leq m$, where a_i is a positive integer, $1 \leq i \leq 3m$ and B is a bound such that $\sum_{i=1}^{3m} a_i = mB$, $\frac{B}{4} < a_i < \frac{B}{2}$? The well-known fact is that the 3-partition problem is strongly NP-complete [22].

3 Heuristics Paradigms for ORVP

A number of sequencing methods as heuristics has been developed and reported with comparison in the literature due to the popularity of JITPS evolved during the 1980s [4, 24, 25]. It is noteworthy that the heuristic approach for PRVP has been recently reported in [29]. In this work, we report the heuristics for ORVP. A complex heuristic for selecting the production sequence when the objective is to minimize the chance of stopping the line due to overloading individual stations is proposed in [21]. In this heuristic, the authors suggest a procedure which uses many different initial sequences. For each initial sequence, an improvement routine is applied in which jobs are moved until no improvement occurs, followed by an interchange of jobs until no improvement occurs. The best of the several sequences is the solution. The empirical results are presented for problems with up to 100 jobs, which suggest that the heuristic performs almost as well as a branch and bound procedure with a CPU time trap of 2 seconds (see [9] also).

Monden [20] developed the two greedy heuristics at Toyota, which he referred as goal chasing methods: GCM I and GCM II (see [11] too). The heuristics GCM I and GCM II, designed with product level and sub-assembly level, constructed a sequence filling one position at a time from first slot to the last one, considering the variability at the sub-assembly level. In comparison of GCM I, the GCM II represented a decrease in computational time, since the sum is formed only on the components of a given product [24]. However, the comparative research in [24] and in [25] showed that GCM I performed better than GCM II when compared on the basis of maintaining a constant usage of component parts. These heuristics has been found to yield very good results in the Toyota [10].

Hyundai's heuristic (HH) used an alternative way, which was developed to approximate the result given by GCM I while reducing the steps of computation. Duplaga and Bragg [4] concluded that the reduction in computational effort related to HH may be significant in situations similar to automobile assembly where many options and choices are available for final product configurations.

The GCM has been advanced to the extended goal chasing method (EGCM) to consider all levels in a multi-level production system [18] and introduced another polynomial heuristic to reduce the myopic nature of the previous heuristic. Moreover, the myopic nature of the GCM I has been reduced and an exact procedure based on the bounded dynamic programming is developed in [1]. In the following three Subsections, we briefly formalize the goal chasing heuristics.

3.1   Goal Chasing Method I

The goal chasing method I (GCM I) was developed and used by Toyota to schedule automobile final assembly lines. It constructs a sequence filling one time unit at a time from first slot to the last one. This method is designed with the two levels: the product level and the sub-assembly level, considering the variability at the sub-assembly level only, whereas the variability is ignored at the final level [18].

For a stage k, the objective function used in GCM I to schedule the product i is

$$\text{minimize} \left[\text{GCM I} = \sum_{i=1}^{n_2} \left[x_{i2k} + t_{i2k} - y_{2k} r_{i2} \right]^2 \right]. \tag{13}$$

The GCM I is a myopic heuristic. This heuristic yields infeasible sequence frequently but if it yields a feasible sequence, then the sequence is necessarily optimal too [24]. The time complexity of GCM I is $O(n_l n D)$.

3.2 Goal Chasing Method II

As in the case of GCM I, the GCM II constructs a sequence filling one time unit at a time from first slot to last one. The GCM II is designed to decrease computational time because the sum is formed only on the components of a given model [18]. This indicates that the computational time can be considerably saved if a model encompasses only a small fraction of the total number of parts [25].

For stage k, the objective function used in GCM II to schedule the product i is

$$\text{minimize} \left[\text{GCM II} = \sum_{i \in C} \left[x_{i2k} - y_{2k} r_{i2} \right]^2 \right], \tag{14}$$

where C is the set of components of a given model. If C contains a small fraction of total number of components, the computational time is substantially reduced. This heuristic is also myopic and frequently generates infeasible sequence.

The goal chasing method has been extended to consider all levels in [18], which is called extended goal chasing method (EGCM). It can be said that the GCM I and GCM II are special cases of the EGCM.

3.3 Extended Goal Chasing Method

The extended goal chasing method (EGCM) is also a heuristic for multi-level problem since it includes more levels [18]. For a stage k, the objective function used in EGCM to schedule the product i is

$$\text{minimize} \left[\text{EGCM} = \sum_{k=1}^{D_1} \sum_{l=1}^{L} \sum_{i=1}^{n_l} w_l (x_{ilk} - y_{lk} r_{il})^2 \right], \tag{15}$$

where the weight w_l determines the relative importance of a level l. The heuristic sequences model i at time unit k with minimum

$$\sum_{l=1}^{L} \sum_{i=1}^{n_l} (x_{ilk} + t_{ilk} - y_{lk} r_{il})^2. \tag{16}$$

This is also a myopic polynomial heuristic. There exist two heuristics for the solution of the problem in [18].

4 The Pegging Assumption

The output rate variation problem is NP-hard combinatorial problem. However, this can be solved in pseudo-polynomial time under the pegging assumption, which separates each part at the lower production levels into distinct groups for each product into which that part will be assembled. The pegging process reduces the ORVP into weighted case of the product rate variation problem [8, 23].

In the pegging assumption, parts of output i at level $l, l = 2, 3, \dots, L$, are dedicated or pegged to be assembled into the particular model at level 1. The parts dedicated to be assembled into the different models are distinct in pegging i.e., $h \neq p$ implies $t_{ilh} \neq t_{ilp}$ for each output i at level l. Pegging is useful for high quality model production because high quality parts are required for high quality model and such parts can be used under this assumption. The mathematical formulation of pegging in a JIT production environment has been firstly developed in [8], where some heuristic procedures for the pegged multi-level min-sum model are also presented.

The sequencing model of the pegged ORVP with absolute deviation objective [23] is to minimize the following weighted deviation:

$$\min \ \max_{h,i,l,k} \{w_{h1} \, |x_{h1k} - kr_{h1}| \, , w_{il} \, |x_{h1k}t_{ilh} - kt_{ilh}r_h|\} , \tag{17}$$

where $h = 1, 2, \dots, n_1$; $i = 1, 2, \dots, n_l$; $k = 1, 2, \dots, D_1$; $l = 2, 3, \dots, L$ subject to the constraints in single-level case [27].

For $l = 1, 2, \dots, L$, $t_{ilh} = 1$ if $i = h$ and 0 otherwise, the objective function is reduced to $\min \ \max_{h,i,l,k} \{w_{il}(t_{ilh}) \, |x_{i1k} - kr_{i1}|\}$. With $\tilde{w}_i^* = \max_{i,l} \{w_{il} (t_{ilh})\}$, the pegged ORVP is transformed into the following formulation:

$$\min \ \max_{i,k} \ \tilde{w}_i^* \, |x_{ik} - kr_i| \tag{18}$$

with constraints in single-level case [27].

Clearly, this is the weighted product rate variation problem formulation [23]. The pegged ORVP with total deviation objective can analogously be reduced to a weighted PRVP with total deviation objective [23]. The optimal schedules for the weighted PRVP with total deviation objective can be obtained using the assignment approach [13–15].

5 Dynamic Programming Solution

The efficient algorithms for the solutions of ORVP are unlikely to exist due to the NP-hardness of the problem. Nevertheless, the dynamic programming (DP) procedure gives rise to optimal solutions [16] for small number of products. The DP algorithm has been applied for the problem with the objective that simultaneously minimizes the variability in the usage of parts and smooths the workload in the final assembly process [19].

The DP procedure for ORVP is developed in [16], which is polynomial in D_1 and consequently seems to be effective for small number of products n_1 even when the total product demand D_1 is large. During the enumeration process, an excessive amount of time or space is reduced by using some fast heuristic as a filter which eliminates any states from DP's state space that would lead to no optimality. Two myopic heuristics to generate the filter are proposed in [16]. If the heuristics yield near-optimal sequences, then the state space size can be reduced.

The weighted case of output rate variation problem with the two sequencing objective functions $\min\max_{i,l,k} w_{il} |x_{ilk} - y_{lk} r_{il}|$ and $\min \sum_{k=1}^{D_1} \sum_{l=1}^{L} \sum_{i=1}^{n_l} w_{il}(x_{ilk} - y_{lk} r_{il})^2$ can be concisely transformed into the matrix representation and can be implemented the transformation for the solution of ORVP using DP procedure [16].

First we consider the min-max objective function $\max_{i,l,k} w_{il} |x_{ilk} - y_{lk} r_{il}|$ and denote the deviation matrix $\Gamma = [\gamma_{il}]_{n \times n_1}$ with $n = \sum_{l=1}^{L} n_l$, where γ_{ilp} represents the $\left(\sum_{m=1}^{l-1} n_m + 1 \right)$th row and pth column element.

Now we have,

$$\max_{i,l,k} w_{il} |x_{ilk} - y_{lk} r_{il}|$$

$$= \max_{i,l,k} \left| \sum_{p=1}^{n_1} w_{il}(t_{ilp} x_{p1k} - r_{il} \sum_{i=1}^{n_l} t_{ilp} x_{p1k}) \right|$$

$$= \max_{i,l,k} \left| \sum_{p=1}^{n_1} w_{il}(t_{ilp} - r_{il} \sum_{i=1}^{n_l} t_{ilp}) x_{p1k} \right|$$

$$= \max_{i,l,k} \left| \sum_{p=1}^{n_1} \gamma_{ilp} x_{p1k} \right|$$

where $\gamma_{ilp} = w_{il}(t_{ilp} - r_{il} \sum_{i=1}^{n_l} t_{ilp})$.

Let the column vector $X_k = \left(x_{11k}, x_{21k}, \ldots, x_{n_1 1k} \right)^T$ to be the cumulative production at level 1 during the time period 1 through k. Hence the objective function $\tilde{Z}_{max}^a = \min \max_{i,l,k} w_{il} |x_{ilk} - y_{lk} r_{il}|$ at the time unit k over all parts, is transformed into matrix representation as follows:

$$minimize \quad \max_{i,l,k} \quad w_{il} |x_{ilk} - y_{lk} r_{il}| = \min \max_{k} \| \Gamma X_k \|_1 ,$$

where the norm $\|.\|_1$ is defined as

$$maximum \quad \|a\|_1 = \max_i |a_i|, \quad i = 1, 2, \ldots, n,$$

for a vector $a = (a_1, a_2, \ldots, a_n)$.

Let the demand vector at level 1 be $d = (d_1, d_2, \ldots, d_{n_1})$ and the states in a schedule be $X = (x_1, x_2, \ldots, x_{n_1})$ with cardinality $|X| = \sum_{i=1}^{n_1} x_i$ where x_i is the cumulative production of model i, $x_i \leq d_i$. Let e_i be the unit vector with n_1 entries all of which are zero except for a single 1 in the ith row.

Define $\phi(X)$ to be the minimum of the maximum absolute deviation for all parts and models over all partial schedules of X and $\|\Gamma X\|_1$ is the maximum of the deviation of actual production from the ideal production over all parts and models when X is the amount of model produced.

The DP recursion for $\phi(X)$ is as follows [16]:

$$\phi(\emptyset) = \phi(X : X \equiv 0) = 0,$$

$$\phi(X) = \min_i \{\max \{\phi(X - e_i), \|\Gamma X\|, \} : i = 1, \ldots, n_1, \; x_i \geq 1\}.$$

For any state X, it is observed that $\phi(X) \geq 0$ and $\|\Gamma(X : X = d)\|_1 = 0$.

Now we consider the objective function $\sum_{k=1}^{D_1} \sum_{l=1}^{L} \sum_{i=1}^{n_l} w_{il} (x_{ilk} - y_{lk} r_{il})^2$.

That is,

$$\sum_{k=1}^{D_1} \sum_{l=1}^{L} \sum_{i=1}^{n_l} w_{il} (x_{ilk} - y_{lk} r_{il})^2$$

$$= \sum_{k=1}^{D_1} \sum_{l=1}^{L} \sum_{i=1}^{n_l} w_{il} \left(t_{ilp} x_{p1k} - r_{il} \sum_{i=1}^{n_l} t_{ilp} x_{p1k}\right)^2$$

$$= \sum_{k=1}^{D_1} (\|\Omega X_k\|_2)^2,$$

where $\Omega = w_{il}^{\frac{1}{2}} \delta_{ilp}$, $\delta_{ilp} = t_{ilp} - r_{il} \sum_{i=1}^{n_l} t_{ilp}$.

The euclidean norm $\|.\|_2$ is defined as $\|a\|_2 = \sqrt{\sum_{i=1}^{n} a_i^2}$ for a vector $a = (a_1, \ldots, a_n)$.

Let $\Phi(X)$ to be the minimum of the total square deviation respectively for all parts and models over all partial schedules of X. The term $(\|\Omega X_k\|_2)^2$ is the sum of square deviations of actual production from the ideal production over all parts and models when X is the amount of model produced.

The DP recursion for $\Phi(X)$ is as follows [16]:

$$\Phi(\emptyset) = \Phi(X : X \equiv 0) = 0,$$

$$\Phi(X) = \min_i \left\{ \Phi(X - e_i) + (\|\Omega X\|_2)^2, i = 1, 2, \ldots, n_1, x_i \geq 1 \right\}.$$

It is always true that $\Phi(X) \geq 0$ and $\Phi(X : X = d) = 0$ for any state X. In any state of X, x_i can have any of the values $0, 1, \ldots, d_i$. The number of states in the DP recursion is $\prod_{i=1}^{n_1} (d_i + 1)$.

Any state X can be generated from n_1 states. The computation time for $\|\Gamma X\|_1$ or $(\|\Omega X\|_2)^2$ is $O(n_1 n)$, $n = \sum_{m=1}^{L} n_m$. The space and time complexities of the DP procedures are $O\left(\prod_{i=1}^{n_1} (d_i + 1)\right)$ and $O\left(n_1 n \prod_{i=1}^{n_1} (d_i + 1)\right)$ respectively [19].

The number of feasible schedules for any problem instance is $\frac{D_1!}{d_1! d_2! \ldots d_{n_1}!}$. This is considerably larger than the number of states in the DP recursion. The inequality $\prod_{i=1}^{n_1} (d_i + 1) \leq \left(\frac{D_1 + n_1}{n_1}\right)^{n_1}$ shows that the DP algorithm is effective for small number of products even with large copies.

An excessive amount of time or that of space can be reduced by using some fast heuristics as a filter. The filter eliminates any states from DP's state space that would lead to no optimality. Two myopic heuristics exist for generating the filter [16]. One of the two heuristics shows that model i becomes next model to be scheduled if that minimizes $\|\Gamma(X + e_i)\|_1$ and the other shows to minimize $\max \left\{ \|\Gamma(X + e_i)\|_1, \min_i \|\Gamma(X + e_i + e_i)\|_1 \right\}$.

The DP algorithm progresses through the state space in the forward direction of increasing the cardinality as the procedure generates all states X with $|X| = k$ before $|X| = k + 1$, $k = 1, 2, \ldots, D_1$ [16].

It is noteworthy that the output rate variation problem with a commutative aggregation function that aggregates deviations over all production cycles which is known as the symmetric output rate variation problem has been solved by the dynamic programming procedure [5].

6 Conclusion

The mixed-model just-in-time sequencing problem has been widely studied with various mathematical formulations and solution strategies. However, it is still a challenging area due to its interesting base model of theoretical value and wide real-world applications. The PRVP is solvable in pseudo-polynomial time, but the ORVP is NP-hard. The problem whether cyclic sequences are optimal for ORVP also remains open. The input-output matrix analysis could be another approach to deal the multi-level problem. The simultaneous study of production and logistics is a challenging area having many research issues [30]. Our further work will be focused on synchronized study of production and logistics to balance overall supply chain systems. It is thus, this paper not only provides the review of existing literature but also opens the floor to be worked forward.

References

1. Bautista, J., Companys, R., Corominas, A.: Heuristics and exact algorithms for solving the Monden problem. Eur. J. Oper. Res. **88**, 101–113 (1996)
2. Dhamala, T.N., Kubiak, W.: A brief survey of just-in-time sequencing for mixed-model systems. Int. J. Oper. Res. **2**(2), 38–47 (2005)
3. Dhamala, T.N., Thapa, G.B., Yu, H.: An efficient frontier for sum deviation just-in-time sequencing problem in mixed-model systems via apportionment. Int. J. Autom. Comput. **9**(1), 87–97 (2012)
4. Duplaga, E., Bragg, D.: Mixed-model assembly line sequencing heuristics for smoothing component parts usage: a comparative analysis. Int. J. Prod. Res. **36**(8), 2209–2224 (1998)
5. Fliedner, M., Boysen, N., Scholl, A.: Solving symmetric mixed-model multi-level just-in-time scheduling problems. Discret. Appl. Math. **158**, 222–231 (2010)
6. Garey, M.R., Johnson, D.S.: Computers and Intractability: A Guide to the Theory of NP-Completeness. W. H. Freeman and Co., New York, USA (1979)
7. Garey, M.R., Johnson, D.S.: Complexity results for multiprocessor scheduling under resource constraints. SIAM J. Comput. **4**(4), 397–411 (1975)
8. Goldstein, T., Miltenburg, J.: The effects of pegging in the scheduling of just-in-time production systems, Working paper, 294, Faculty of Business, McMaster University, Hamilton, Ontario (1998)
9. Khashouie, G.M.: Sequencing Mixed-model Assembly Lines in Just-in-time Production Systems, Department of Systems Engineering, Brunel University, UK (2003)
10. Korgaonker, M.G.: Just in Time Manufacturing. Macmillan India Ltd. (1992)
11. Kotani, S., Ito, T., Ohno, K.: Sequencing problem for a mixed-model assembly line in the Toyota production system. Int. J. Prod. Res. **42**(23), 4955–4974 (2004)
12. Kubiak, W.: Completion Time Variance Minimization on Single Machine is Difficult, Working paper, 92-6, Memorial University of Newfoundland (1992)
13. Kubiak, W.: Minimizing variation of production rates in just-in-time systems: a survey. Eur. J. Oper. Res. **66**, 259–271 (1993)
14. Kubiak, W., Sethi, S.: A note on level schedules for mixed-model assembly lines in just-in-time production systems. Manag. Sci. **37**(1), 121–122 (1991)
15. Kubiak, W., Sethi, S.: Optimal just-in-time schedules for flexible transfer lines. Int. J. Flex. Manuf. Syst. **6**, 137–154 (1994)
16. Kubiak, W., Steiner, G., Yeomans, J.: Optimal level schedules for mixed-model multi-level just-in-time assembly systems. Ann. Oper. Res. **69**, 241–259 (1997)
17. Lenstra, J.K., Rinnooy, A.H.G.: Computational complexity of discrete optimization problems. Ann. Discret. Math. **4**, 121–140 (1979)
18. Miltenburg, J., Sinnamon, G.: Scheduling mixed-model multi-level just-in-time production systems. Int. J. Prod. Res. **27**(9), 1487–1509 (1989)
19. Miltenburg, J., Steiner, G., Yeomans, J.: A dynamic programming algorithm for scheduling mixed-model just-in-time production systems. Math. Comput. Model. **13**(3), 57–66 (1990)

20. Monden, Y.: Toyota Production System; Practical Approach to Production Management. Industrial Engineering and Management Press, Norcross, GA (1983)
21. Okamura, K., Yamashina, H.: A heuristic algorithm for the assembly line model-mix sequencing problem to minimize the risk of stopping the conveyor. Int. J. Prod. Res. **17**, 233–241 (1979)
22. Papadimitriou, C., Steiglitz, K.: Combinatorial Optimization: Algorithms and Complexity. Prentice Hall of India (2003)
23. Steiner, G., Yeomans, J.: Optimal level schedules in mixed-model multi-level JIT assembly systems with pegging. Eur. J. Oper. Res. **95**, 38–52 (1996)
24. Sumichrast, R., Russell, R.: Evaluating mixed-model assembly line sequencing heuristics for just-in-time production systems. J. Oper. Manag. **9**(3), 371–390 (1990)
25. Sumichrast, R., Russell, R., Taylor, B.: A comparative analysis of sequencing procedures for mixed-model assembly lines in a just-in-time production system. Int. J. Prod. Res. **30**(1), 199–214 (1992)
26. Thapa, G.B.: Computational complexity and integer programming. Mathematical Sciences and Applications, Kathmandu University, Nepal **60–70** (2006)
27. Thapa, G. B.: Optimization of Just-in-time Sequencing Problems and Supply Chain Logistics, Ph.D. Thesis, Division of Applied Mathematics, Mälardalen University, Sweden (2015)
28. Thapa, G.B., Dhamala, T.N.: Just-in-time sequencing in mixed-model production systems relating with fair representation in apportionment theory. Nepali Math. Sci. Rep. **29**(1&2), 29–68 (2009)
29. Thapa, G.B., Silvestrov, S.: Heuristics for single-level just-in-time sequencing problem. J. Inst. Sci. Technol. **18**(2), 125–131 (2013)
30. Thapa, G.B., Silvestrov, S.: Supply chain logistics in multi-level just-in-time production sequencing problems. J. Inst. Eng. **11**(1), 91–100 (2015)

L^p-Boundedness of Two Singular Integral Operators of Convolution Type

Sten Kaijser and John Musonda

Abstract We investigate boundedness properties of two singular integral operators defined on L^p-spaces ($1 < p < \infty$) on the real line, both as convolution operators on $L^p(\mathbb{R})$ and on the spaces $L^p(\omega)$, where $\omega(x) = 1/(2 \cosh \frac{\pi}{2} x)$. It is proved that both operators are bounded on these spaces and estimates of the norms are obtained. This is achieved by first proving boundedness for $p = 2$ and weak boundedness for $p = 1$, and then using interpolation to obtain boundedness for $1 < p \leq 2$. To obtain boundedness also for $2 \leq p < \infty$, we use duality in the translation invariant case, while the weighted case is partly based on the expositions on the conjugate function operator in (M. Riesz, Mathematische Zeitschrift, 27, 218–244, 1928) [7].

Keywords Convolution operators · Sech (function) · Hilbert transform · Hardy space · Weak type estimates

1 Introduction

In [4, 5], three systems of orthogonal polynomials belonging to the class of Meixner–Pollaczek polynomials were described together with some operators connecting them. The first system was the special case of the Meixner–Pollaczek polynomials [3, 6] with parameter $\lambda = 1/2$, a system that can also be described as the orthogonal polynomials obtained from the weight function $\omega(x) = 1/(2 \cosh \frac{\pi}{2} x)$. The Second system was a limiting case of the Meixner–Pollaczek polynomials with the parameter λ tending to 0. That system could also be described as the polynomials orthogonal in the strip $\mathbb{S} = \{z \in \mathbb{C} : |\mathrm{Im}\, z| < 1\}$ with respect to the Poisson measure for the

S. Kaijser (✉)
Department of Mathematics, Uppsala University, Box 480, 751 06 Uppsala, Sweden
e-mail: sten@math.uu.se

J. Musonda
Division of Applied Mathematics, School of Education,
Culture and Communication, Mälardalen University, Box 883, 721 23 Västerås, Sweden
e-mail: john.musonda@mdh.se

© Springer International Publishing Switzerland 2016
S. Silvestrov and M. Rančić (eds.), *Engineering Mathematics II*,
Springer Proceedings in Mathematics & Statistics 179,
DOI 10.1007/978-3-319-42105-6_15

origin. In the third system, the polynomials were orthogonal with respect to the weight $\omega_2(x) = \omega \star \omega(x) = x/(2\sinh\frac{\pi}{2}x)$.

These systems are connected by two operators R and J (see [1] or [2]) mapping functions in the strip $\mathbb{S}$ to functions on the real line $\mathbb{R}$, defined by

$$Rf(x) = \frac{f(x+i) + f(x-i)}{2} \quad \text{and} \quad Jf(x) = \frac{f(x+i) - f(x-i)}{2i}.$$

Besides these two operators, the operators $B = R^{-1}$ and $S = JR^{-1}$ turn out to have interesting properties with respect to these systems of polynomials. Both operators can be represented as convolution operators

$$Bf(z) = \int_{-\infty}^{\infty} \frac{f(t)dt}{2\cosh\frac{\pi}{2}(z-t)} \quad \text{and} \quad Sf(x) = \lim_{\varepsilon \to 0} \int_{|x-t|>\varepsilon} \frac{f(t)dt}{2\sinh\frac{\pi}{2}(x-t)},$$

leading to the Fourier transforms

$$\widehat{Bf}(t) = \operatorname{sech} t\, \hat{f}(t) \quad \text{and} \quad \widehat{Sf}(t) = i\tanh t\, \hat{f}(t).$$

These two operators can be studied in the context of either real or complex analysis, and in this paper we consider the operator B as an operator from functions on the real line $\mathbb{R}$ to functions in the strip $\mathbb{S}$, while the operator S is studied as an operator on functions on $\mathbb{R}$. Function spaces on $\mathbb{R}$ are denoted by L and those on $\mathbb{S}$ by H. For an arbitrary non-negative and locally integrable function ω on $\mathbb{R}$, $L^p(\omega)$ denotes measurable functions on $\mathbb{R}$ with

$$\|f\|_{L^p(\omega)}^p = \int_{-\infty}^{\infty} |f(x)|^p \omega(x)\, dx < \infty,$$

and $H^p(\omega)$ analytic functions on $\mathbb{S}$ with

$$\|f\|_{H^p(\omega)}^p = \sup_{-1<a<1} \int_{-\infty}^{\infty} |f(x+ia)|^p \omega(x)\, dx < \infty.$$

Furthermore, $L^p(\mathbb{R}) = L^p(1)$ and $H^p(\mathbb{S}) = H^p(1)$. Unless stated otherwise, we assume throughout that $1 < p < \infty$, $F = Bf$ and $\omega(x) = 1/(2\cosh\frac{\pi}{2}x)$.

We investigate boundedness properties of B and S, both as convolution operators in the translation invariant case and for the weight $\omega(x) = 1/(2\cosh\frac{\pi}{2}x)$. Our main results are the following.

Theorem 1 *For $1 < p < \infty$, the operator B is linear and bounded from*

(a) $L^p(\mathbb{R})$ *to* $H^p(\mathbb{R})$,
(b) $L^p(\omega)$ *to* $H^p(\omega)$.

Theorem 2 *For $1 < p < \infty$, the operator S is linear and bounded on the spaces* $L^p(\mathbb{R})$ *and* $L^p(\omega)$.

For $p = 2$, both of these results were proved in [5]. In this paper we first prove weak boundedness for $p = 1$, and then use interpolation to obtain boundedness for $1 < p \le 2$. To obtain boundedness also for $2 \le p < \infty$, we use duality in the translation invariant case and the method of M. Riesz [7] in the weighted case.

2 Weak Boundedness for $p = 1$

We denote the Lebesgue measure of a measurable set E of real numbers by $|E|$ if given by dx, and by $|E|_\omega$ if given by $\omega(x)\,dx$. We further write $A_0(\bar{\mathbb{S}})$ to denote the space of functions f that are analytic in the strip $\mathbb{S}$, continuous on the closed strip $\bar{\mathbb{S}}$ and such that $|f| \to 0$ when $|z| \to \infty$.

We shall prove the following result which, perhaps, is interesting in its own right.

Proposition 1 *Let $\lambda > 0$ and $E_\lambda = \{x : |Bf(x+i)| > \lambda\}$. If $f \in L^1(\mathbb{R})$, then*

$$|E_\lambda| \le \frac{16}{\lambda}\|f\|_{L^1(\mathbb{R})},$$

and if $f \in L^1(\omega)$, then

$$|E_\lambda|_\omega \le \frac{16}{\lambda}\|f\|_{L^1(\omega)}.$$

Corollary 1 *Let $\lambda > 0$ and $E_\lambda^S = \{x : |Sf(x)| > \lambda\}$. If $f \in L^1(\mathbb{R})$, then*

$$|E_\lambda^S| \le \frac{16}{\lambda}\|f\|_{L^1(\mathbb{R})},$$

and if $f \in L^1(\omega)$, then

$$|E_\lambda^S|_\omega \le \frac{16}{\lambda}\|f\|_{L^1(\omega)}.$$

The main idea is to first consider the case when f is positive, and we have then the following.

Lemma 1 *Let f be a continuously differentiable function on $\mathbb{R}$ with compact support, $\lambda > 0$ and $E_\lambda = \{x : |Bf(x \pm i)| > \lambda\}$. If f is such that*

$$\int_{-\infty}^{\infty} f(x)\,dx = 1,$$

then $|E_\lambda| \le 2/\lambda$, and if f is such that

$$\int_{-\infty}^{\infty} f(x)\omega(x)\,dx = 1,$$

then $|E_\lambda|_\omega \le 2/\lambda$.

Proof The first part of the proof is the same for both assertions, and we start by denoting

$$R^{-1}f(z) = Bf(z) = F(z) = \int_{-\infty}^{\infty} \frac{f(t)dt}{2\cosh\frac{\pi}{2}(z-t)}.$$

We observe that F is analytic in $\mathbb{S}$, and using partial integration, we see that F is continuous on the closed strip and that $F \in A_0(\bar{\mathbb{S}})$. It is obvious that $F(\bar{z}) = \overline{F(z)}$ and that F is real-valued on $\mathbb{R}$. Furthermore, we have

$$\frac{1}{\cosh(x+iy)} = \frac{\cosh(x-iy)}{|\cosh(x+iy)|^2} = \frac{\cosh x \cos y - i \sinh x \sin y}{\cosh 2x + \cos 2y},$$

and this implies that $\mathrm{Re}(F(z)) \geq 0$ in $\mathbb{S}$. Let further for a given $\lambda > 0$,

$$\varphi_\lambda(z) = \frac{z}{z+\lambda}.$$

It is easy to see that φ_λ maps the real line to itself, leaves the origin fixed and maps ∞ to 1. This implies that the imaginary axis is mapped to the circle $|z - 1/2| = 1/2$. It is also clear that the circle $|z| = \lambda$ is mapped to the line $\mathrm{Re}(z) = 1/2$ so that the right half-plane is mapped to the interior of the circle $|z - 1/2| = 1/2$ and that $|z| > \lambda$ implies that $\mathrm{Re}(\varphi_\lambda(z)) > 1/2$.

We now consider the first assertion of the lemma, and for this, we use Cauchy's theorem. We observe therefore that for all $-1 \leq a \leq 1$, we have

$$\int_{-\infty}^{\infty} F(x+ia)\,dx = \int_{-\infty}^{\infty} F(x)\,dx = 1.$$

The next step is to see that

$$\int_{-\infty}^{\infty} G(x)\,dx = \int_{-\infty}^{\infty} \varphi_\lambda(F(x))\,dx = \int_{-\infty}^{\infty} \frac{F(x)}{F(x)+\lambda}\,dx \leq \frac{1}{\lambda}\int_{-\infty}^{\infty} F(x)\,dx = \frac{1}{\lambda}.$$

It follows again from Cauchy's theorem that also

$$\int_{-\infty}^{\infty} G(x\pm i)\,dx = \int_{-\infty}^{\infty} G(x)\,dx \leq \frac{1}{\lambda}$$

so that by Chebyshev's inequality,

$$|\{x : \mathrm{Re}(G(x\pm i)) > 1/2\}| \leq \frac{1/\lambda}{1/2} = \frac{2}{\lambda}.$$

However, as observed above,

$$\{x : \operatorname{Re}(G(x \pm i)) > 1/2\} = \{x : |F(x \pm i)| > \lambda\} = E_\lambda,$$

and this proves the first assertion.

To prove the second assertion, we need the fact that the measure $\omega(x)\,dx$ is closely related to the Poisson measure for $\mathbb{S}$; see [8]. In fact, we have for u harmonic in $\mathbb{S}$, continuous on $\bar{\mathbb{S}}$ and such that $|u(x)| < e^{a|x|}$ for some a, $0 < a < \pi/2$ that

$$u(0) = \int_{-\infty}^{\infty} \frac{u(x+i) + u(x-i)}{2} \omega(x)\,dx = \int_{-\infty}^{\infty} Ru(x)\omega(x)\,dx.$$

Applying this to the function F, we see that

$$F(0) = \int_{-\infty}^{\infty} R R^{-1} f(x)\omega(x)\,dx = \int_{-\infty}^{\infty} f(x)\omega(x)\,dx = 1.$$

We next observe that

$$G(0) = \frac{F(0)}{F(0) + \lambda} = \frac{1}{1 + \lambda}.$$

Since $G(x - i) = \overline{G(x + i)}$, it follows that

$$G(0) = \int_{-\infty}^{\infty} RG(x)\omega(x)\,dx = \int_{-\infty}^{\infty} \operatorname{Re}(G(x + i))\omega(x)\,dx = \frac{1}{1 + \lambda}.$$

Therefore, by the same argument as before, it follows that $|E_\lambda|_\omega \leq 2/\lambda$, and this proves the lemma. $\qquad\square$

If f is real-valued, then

$$Bf(x \pm i) = f(x) \pm i Sf(x),$$

so the same conclusion holds *a fortiori* for the set $E_\lambda^S = \{x : |Sf(x)| > \lambda\}$.

We can now prove the proposition.

Proof of Proposition 1 We prove only the first assertion since the proof of the second assertion is exactly the same. We first consider the case when f is real-valued. We write $f^+ = \max(f, 0)$, $f^- = \max(-f, 0)$ and hence $f = f^+ - f^-$. We observe that in order to have $|Bf| > \lambda$, we have to have either $|Bf^+| > \lambda/2$ or $|Bf^-| > \lambda/2$, and since $\|f\| = \|f^+\| + \|f^-\|$, this implies that $|E_\lambda| \leq 4/\lambda$. If f is complex-valued, we write $f = g + ih$ and essentially the same argument as above implies that

$$|E_\lambda| \leq \frac{16}{\lambda} \|f\|_{L^1(\mathbb{R})},$$

and this proves the proposition (under the assumption that f is continuously differentiable with compact support, but since such functions are dense in L^1, the same holds by continuity for all functions in L^1). Again the same result holds for the operator S.   □

3   The Case $1 < p \le 2$

In [5], it was proved that the operator B is bounded from L^2 to H^2 with norm less than or equal to 2 (both in the translation invariant case and for the weight ω), and that S is bounded on L^2 with norm 1 (again in both cases). Using the Marcinkiewicz interpolation theorem, we can now prove the following.

Lemma 2   *The operators B and S are strongly bounded for $1 < p < 2$ with norm at most*

$$\frac{16p}{(p-1)(2-p)}.$$

Proof   This follows immediately from the Marcinkiewicz interpolation theorem.   □

Choosing $p = 4/3$, we see that for $T = B$ or $T = S$, we have

$$\|T\|_{L^{4/3}} \le 96.$$

We can now use the Riesz–Thorin theorem [9] for $4/3 \le p \le 2$ to obtain the following result.

Proposition 2   *Let $1 < p \le 2$.*

(a) The operator S is bounded as an operator from L^p to H^p with norm at most

$$\min\left(\frac{16p}{(p-1)(2-p)}\right)$$

for $1 < p < 4/3$, and at most
$$96^{\theta(p)}$$

where $\theta(p) = 4/p - 2$ for $4/3 \le p \le 2$.

(b) The operator B is bounded as an operator from L^p to H^p with norm at most

$$\min\left(\frac{16p}{(p-1)(2-p)}\right)$$

for $1 < p < 4/3$ and at most

$$96^{\theta(p)} \cdot 2^{1-\theta(p)}$$

where $\theta(p) = 4/p - 2$ for $4/3 \le p \le 2$.

Proof For $1 < p < 4/3$, this was the preceding lemma. For $4/3 \le p \le 2$, it follows from the Riesz–Thorin theorem. □

4 The Case $2 \le P < \infty$

We first observe that in the translation invariant case the operator S is self-adjoint so that by duality we have immediately the following result.

Proposition 3 (a) *The operator S is bounded on the space $L^2(\mathbb{R})$ with norm at most $24p$ for $2 \le p \le 4$, and with norm at most $96^{\theta(p)}$ where $\theta(p) = 2 - 4/p$ for $4 < p < \infty$.*
(b) *The operator B is bounded as an operator from $L^p(\mathbb{R})$ to $H^p(\mathbb{S})$ with norm at most $1 + \|S\|_p$.*

Proof (a) follows from duality while (b) follows from the fact that on the boundary of $\mathbb{S}$, we have $Bf = f + iSf$ so that $\|Bf\| \le \|f\| + \|Sf\|$. □

To prove boundedness also on the space $L^p(\omega)$ for $2 \le p < \infty$, we use the same idea that M. Riesz used when proving boundedness of the conjugate function operator, i.e. by considering even powers.

Proposition 4 *Let $f \in L^{2n}$ be real-valued and such that $\|f\|_{L^{2n}(\omega)} = 1$. Then*

$$\|Sf\|_{L^{2n}(\omega)} \le \frac{2n}{\log 2} \|f\|_{L^{2n}(\omega)}.$$

Proof Let $F = Bf$, then

$$F(0)^{2n} = \int_{-\infty}^{\infty} R(F(x)^{2n})\omega(x)\,dx = \int_{-\infty}^{\infty} \mathrm{Re}(f(x) + iSf(x))^{2n}\omega(x)\,dx.$$

Denoting Sf by g, we have

$$\mathrm{Re}(f + iSf)^{2n} = \mathrm{Re}(f + ig)^{2n} = \sum_{k=0}^{n} \binom{2n}{2k}(-1)^k f(x)^{2n-2k} g(x)^{2k},$$

and this implies that

$$0 \le F(0)^{2n} = \int_{-\infty}^{\infty} \sum_{k=0}^{n} \binom{2n}{2k}(-1)^k f(x)^{2n-2k} g(x)^{2k} \omega(x)\,dx \le 1.$$

Writing X for $\|g\|_{2n}$, we see that

$$X^{2n} \leq \int_{-\infty}^{\infty} \sum_{k=1}^{n-1} \binom{2n}{2k} f(x)^{2k} g(x)^{2n-2k} + 1 - F(0)^{2n}.$$

From Hölder's inequality, It follows that

$$X^{2n} \leq \sum_{k=0}^{2n-1} \binom{2n}{k} X^k = (X+1)^{2n} - X^{2n},$$

and therefore $2X^{2n} \leq (X+1)^{2n}$ so that $2^{1/2n} X \leq X + 1$ or $(2^{1/2n} - 1)X \leq 1$. Since $(2^{1/2n} - 1) \geq \log 2 / 2n$, it follows that $X \leq 2n / \log 2$. $\square$

Remark 1 A more careful analysis at the binomial sum shows that it is actually possible to have at least

$$3X^{2n} \leq (X+1)^{2n},$$

which shows that the denominator $\log 2$ can easily be replaced by $\log 3$.

Remark 2 Using Cauchy's theorem, essentially the same idea can also be used in the translation invariant case. One observes that if f is real-valued and $\|f\|_{2n} = 1$, then $\|\omega \star f\|_{2n} \leq 1$, and since

$$\int_{-\infty}^{\infty} (f(x) + iSf(x))^{2n}\, dx = \int_{-\infty}^{\infty} (\omega \star f(x))^{2n}\, dx \leq 1,$$

we see again that $\|Sf\|_{2n} \leq 2n/\log 2$. If we do not assume f to be real-valued, it follows that

$$\|Sf\|_{2n} \leq \frac{4n}{\log 2} \|f\|_{2n}.$$

Using the Riesz–Thorin theorem, we see that for $2 \leq p < \infty$,

$$\|Sf\|_p \leq \frac{4p}{\log 2} \|f\|_p$$

so that $\|S\|_p \leq 4p/\log 2$.

Remark 3 In the translation invariant case, we can now use duality to move from $2 \leq p < \infty$ to $1 < p \leq 2$ and thus obtain the better estimate that

$$\|S\|_p \leq \frac{p'}{\log 2}$$

where $p' = p/(p-1)$.

Finally as above, we have that for all $1 < p < \infty$,

$$\|B\|_p \leq 1 + \|S\|_p.$$

Acknowledgement This research was supported by the International Science Program (Uppsala University and Sida foundation) and the Research environment MAM (Mathematics and Applied Mathematics), School of Education, Culture and Communication, Mälardalen University.

References

1. Araaya, T.K.: The Meixner-Pollaczek polynomials and a system of orthogonal polynomials in a strip. J. Comput. Appl. Math. **170**, 241–254 (2004)
2. Kaijser, S.: Några nya ortogonala polynom. Normat. **47**(4), 156–165 (1999)
3. Meixner, J.: Orthogonale polynomsysteme mit einer besonderen gestalt der erzeugenden funktion. J. Lond. Math. Soc. **9**, 6–13 (1934)
4. Musonda, J.: Three systems of orthogonal polynomials and associated operators. U.U.D.M. project report. (2012:8)
5. Musonda, J., Kaijser, S.: Three Systems of Orthogonal Polynomials and L^2-Boundedness of Two Associated Operators(2015). http://urn.kb.se/resolve?urn=urn:nbn:se:mdh:diva-29566
6. Pollaczek, F.: Sur une famille de polynomes orthogonaux qui contient les polynomes d'Hermite et de Laguerre comme cas limites. C. R. Acad. Sci. Paris. **230**, 1563–1565 (1950)
7. Riesz, M.: Sur les fonctions conjuguées. Mathematische Zeitschrift. **27**, 218–244 (1928)
8. Stein, E.M., Weiss, G.: Introduction to Fourier Analysis on Euclidean Spaces, pp. 205–209. Princeton University Press, Princeton, N.J. (1971)
9. Thorin, O.: An extension of a convexity theorem due to M. Riesz, Kungl. Fysiogr. Sällsk. i Lund Förh. **8**, 166–170 (1938)

Fractional-Wavelet Analysis of Positive definite Distributions and Wavelets on $\mathscr{D}'(\mathbb{C})$

Emanuel Guariglia and Sergei Silvestrov

Abstract In this chapter we describe a wavelet expansion theory for positive definite distributions over the real line and define a fractional derivative operator for complex functions in the distribution sense. In order to obtain a characterization of the complex fractional derivative through the distribution theory, the Ortigueira-Caputo fractional derivative operator $_{\mathrm{C}}\mathrm{D}^{\alpha}$ [13] is rewritten as a convolution product according to the fractional calculus of real distributions [8]. In particular, the fractional derivative of the Gabor–Morlet wavelet is computed together with its plots and main properties.

Keywords Wavelet basis · Positive definite distribution · Complex fractional derivative · Gabor–Morlet wavelet

1 Introduction

In recent years, wavelet analysis and fractional calculus have shown to be a powerful tool in several areas of mathematics. Indeed, the time-frequency localization property provided by the wavelet approach gives the possibility to use a wavelet basis as a mathematical microscope in order to better investigate the behavior of a function by the well-known Heisenberg box [18] located in the time-frequency plane.

Wavelet expansions are used to characterize different function spaces, such as L^p-spaces, Sobolev spaces, Morrey–Campanato spaces, etc. [19]. In particular, several key concepts of wavelet analysis, such as the wavelet transform, can be extended to the space of tempered distributions $\mathscr{S}'(\mathbb{R})$.

E. Guariglia (✉)
Department of Physics "E. R. Caianiello", University of Salerno,
Via Giovanni Paolo II, 84084 Fisciano, Italy
e-mail: eguariglia@unisa.it

E. Guariglia · S. Silvestrov
Division of Applied Mathematics, School of Education, Culture and Communication,
Mälardalen University, Box 883, 721 23 Västerås, Sweden
e-mail: sergei.silvestrov@mdh.se

© Springer International Publishing Switzerland 2016
S. Silvestrov and M. Rančić (eds.), *Engineering Mathematics II*,
Springer Proceedings in Mathematics & Statistics 179,
DOI 10.1007/978-3-319-42105-6_16

In this chapter, a wavelet expansion for the family of positive definite distributions is presented. It has many applications in different areas of both pure and applied mathematics (Lie groups, maximum entropy methods, etc.) [7, 14] and can be generalized for the class of tempered distributions [16]. Furthermore, an open problem about the reconstruction formula for Shannon wavelets in the distribution sense is proposed.

In [3, 4] a generalization to complex functions of the distribution theory is presented. In particular, these two papers provide a generalization of the classical Dirac delta in the complex plane which gives the possibility to rewrite the complex fractional derivative in the distribution sense.

Indeed, the fractional derivative of complex functions is provided by the Ortigueira-Caputo fractional derivative $_cD^\alpha$ [13], which can be written as a convolution of the given complex function with a suitable function that defines a regular distribution on $\mathbb{C}$ (see (30) and (31)).

The authors have computed the fractional derivative of the Gabor–Morlet wavelet through the Ortigueira-Caputo operator. It represents a wavelet family with several applications in signal theory and geophysics.

This chapter is organized as follows: some preliminaries and notations on function spaces and wavelet analysis are provided in Sect. 2. A wavelet expansion for the positive definite distributions is shown in Sect. 3. The fractional differentiation in the complex plane, together with the complex-variable distribution theory, is given in the first part of the Sect. 4, while in the second part it is widely explained how the Ortigueira-Caputo fractional derivative can be rewritten in the distribution sense. The fractional derivative of the Gabor–Morlet wavelet is computed in Sect. 5.

2   Preliminaries and Notations

In this chapter, n will denote an element of $\mathbb{N}_0 = \mathbb{N} \cup \{0\}$, and i the imaginary unit. The fractional and the integer parts of a real number x will be indicated by $\{x\}$ and $\lfloor x \rfloor$, respectively. The Heaviside step function u and the sinc function are defined, respectively, by

$$u(x) = \begin{cases} 1, & x \geq 0, \\ 0, & x < 0, \end{cases} \tag{1}$$

$$\mathrm{sinc}(x) = \frac{\sin(\pi x)}{\pi x} = \frac{e^{i\pi x} - e^{-i\pi x}}{2i\pi x}. \tag{2}$$

The L^2-inner product for complex-valued functions on an interval $[a, b]$ is given by

$$\langle f, g \rangle = \int_a^b f(x)\overline{g(x)}\mathrm{d}x, \tag{3}$$

where the Dirac bracket notation is adopted (in order to emphasize the action of any tempered distribution on a test function). Suppose that $f, g \in L^1_{\text{loc}}(\mathbb{R})$, the convolution of f and g is defined by

$$(f * g)(x) = \int_{-\infty}^{\infty} f(\tau)g(t - \tau)d\tau, \tag{4}$$

i.e. $(f * g)(x) = \langle f(\tau), \overline{g(-(\tau - t))} \rangle$. The dual of a normed space $V(\mathbb{F})$ will be indicated with $V'(\mathbb{F})$.

2.1 Space Functions, Orthogonal Wavelets and Wavelet Transform of Distributions

The distributions over $\mathbb{R}$ are defined as the dual space of the test functions $\mathscr{D}(\mathbb{R})$, namely $\mathscr{D}'(\mathbb{R})$ is the space of all linear and continuous functionals on the space of the test functions. Similarly, the tempered distribution is the dual of the Schwartz space $\mathscr{S}(\mathbb{R})$. In other words, $\mathscr{S}'(\mathbb{R})$ is the space of all linear and continuous functionals on $\mathscr{S}(\mathbb{R})$, namely the set of all functions

$$f : \mathscr{S}(\mathbb{R}) \to \mathbb{F},$$

where $\mathbb{F}$ is usually $\mathbb{R}$ or $\mathbb{C}$ [17]. In this chapter it will always be taken $\mathbb{F} = \mathbb{C}$.

The space of highly time-frequency localized functions over $\mathbb{R}$ is denoted by $\mathscr{S}_0(\mathbb{R})$ and defined as the space of the functions $\phi \in \mathscr{S}(\mathbb{R})$ such that all the moments vanish, namely

$$\int_{-\infty}^{\infty} x^n \phi(x)dx = 0, \qquad \forall n \in \mathbb{N}_0, \tag{5}$$

where the topology on this function space is defined in the classical way [11].

Let $\psi \in \mathscr{S}_0(\mathbb{R})$ be a wavelet mother and let

$$\psi_{a,b}(t) = \frac{1}{\sqrt{a}} \psi\left(\frac{t - b}{a}\right),$$

be its correspondent daughter wavelet [18]. The wavelet transform of $f \in \mathscr{S}'(\mathbb{R})$ with respect to ψ is given by

$$W_\psi f(a, b) = \langle f(t), \psi_{a,b}(t) \rangle = \frac{1}{\sqrt{a}} \int_{-\infty}^{\infty} f(t) \overline{\psi\left(\frac{t - b}{a}\right)} dt,$$

in which $(a, b) \in \mathbb{R}_{>0} \times \mathbb{R}$ since a and b correspond to the scale factor and time shift, respectively. It is clear that this integral transform represents a $\mathscr{C}^{\infty}$-function on the half-plane $\mathbb{R}_{>0} \times \mathbb{R}$.

In [11] it is shown that $W_{\psi} : \mathscr{S}_0(\mathbb{R}) \mapsto \mathscr{S}(\mathbb{R})$ is a continuous linear map. Naturally, since $\psi \in \mathscr{S}_0(\mathbb{R})$ the wavelet transform above can be also defined for $f \in \mathscr{S}'_0(\mathbb{R})$ [11].

Some concepts about the theory of orthonormal wavelet bases of $L^2(\mathbb{R})$ are briefly recalled below [5]. An orthonormal wavelet on $\mathbb{R}$ is a function $\psi \in L^2(\mathbb{R})$ such that the family $\{\psi_{m,n}\}_{m,n \in \mathbb{Z}}$ represents an orthonormal basis of $L^2(\mathbb{R})$, where $\psi_{m,n}(x) = 2^{m/2}\psi(2^m x - m)$ with $m, n \in \mathbb{Z}$.

The reconstruction formula for orthogonal wavelets claims that every $f \in L^2(\mathbb{R})$ can be written as

$$f = \sum_{m \in \mathbb{Z}} \sum_{n \in \mathbb{Z}} \langle f, \psi_{m,n} \rangle \, \psi_{m,n} \quad \text{in} \quad \|\cdot\|_2 . \tag{6}$$

The series in (6) is often called wavelet series of f. In the literature, the wavelet coefficients of f with respect to ψ are denoted by $c_{m,n}(f)$, i.e.

$$c_{m,n}(f) = \langle f, \psi_{m,n} \rangle_{L^2(\mathbb{R})} = \int_{-\infty}^{\infty} f(x) \, \overline{\psi_{m,n}(x)} \, dx. \tag{7}$$

The notation $c_{m,n}(f)$ does not provide information about the particular family of wavelets chosen. In order to take into account that the coefficients can refer to the wavelet $\overline{\psi}$ (instead of ψ), in this chapter the symbol $c^c_{m,n}(f)$ will be used. The wavelet coefficients $c_{m,n}(f)$ and the wavelet transform of f are linked [16] by

$$c_{m,n}(f) = 2^{-\frac{m}{2}} \, W_{\psi} f \left(n2^{-m}, 2^{-m} \right). \tag{8}$$

In this chapter we are interested in wavelet expansions of positive distributions, i.e. tempered distributions (see Schwartz theorem in the next subsection), hence we need the orthonormal wavelets have to belong to $\mathscr{S}(\mathbb{R})$. In [10] it is shown that an orthonormal wavelet from $\mathscr{S}(\mathbb{R})$ is an element of $\mathscr{S}_0(\mathbb{R})$. The existence of these wavelets, the construction of orthonormal wavelets $\psi \in \mathscr{S}(\mathbb{R})$ such that $\widehat{\psi} \in \mathscr{D}(\mathbb{R})$ and the corresponding multidimensional wavelets can be found in [12].

2.2　Tempered and Positive Definite Distributions

In the distribution theory it is convenient to have the regularizing functions ϕ as positive definite test functions (rarely functions of positive type). In brief, this is realized by passing from the test function ϕ to $\phi * \tilde{\phi}$, where $\widetilde{\phi(x)} = \overline{\phi(-x)}$. It is possible to assume that the regularizing function ϕ is a test function which is both even, positive and definite positive with a never vanishing Fourier transform having the same properties [6]. The classical Bochner's theorem is an important result since it links this family of functions with the Radon measure, thus providing a full characterization of positive definite functions.

A generalization of this mathematical concept can be provided via distribution theory.

Definition 1 (*Positive definite distribution*) A distribution T on $\mathbb{R}$ such that $T\left(\phi * \tilde{\phi}\right)$ ≥ 0, for every test function ϕ, is called positive definite distribution (rarely of positive type).

The definition above generalizes clearly the concept of positive function, i.e. every positive definite function is also a positive definite distribution. The given definition is not constructive and does not provide much information about the position of this family of distributions in the distribution theory. A generalization of Bochner's theorem, as derived by Schwartz, goes in this direction.

Theorem 1 (Schwartz) *A distribution T on $\mathbb{R}$ is definite positive if and only if $T \in \mathscr{S}'(\mathbb{R})$ and its Fourier transform is a positive Radon measure.*

Proof It follows from the Bochner's theorem (see [6]).   □

Three remarkable examples of positive definite distributions are the Dirac impulse δ, the Cantor measure on the Cantor set and the distribution associated to the Poisson summation formula (see [6]).

3 Wavelet Expansions of Positive Definite Distributions

In this section, a theorem concerning the wavelet decomposition in the function space $\mathscr{S}_0(\mathbb{R})$ is presented in order to obtain an its suitable generalization for a tempered distribution. In the last subsection an interesting example is presented and discussed.

3.1 Main Results About the Wavelet Decomposition in $\mathscr{S}_0(\mathbb{R})$

A brief summary of the wavelet expansion theory for the space $\mathscr{S}_0(\mathbb{R})$ is presented. The following statement provides different results with regard to this purpose.

Theorem 2 (Wavelet expansion on $\mathscr{S}_0(\mathbb{R})$) *Let $\phi \in \mathscr{S}_0(\mathbb{R})$, $f \in \mathscr{S}'_0(\mathbb{R})$ and let $\psi \in \mathscr{S}_0(\mathbb{R})$ be an orthonormal wavelet. Then*

1. ϕ can be expressed as

$$\phi = \sum_{m \in \mathbb{Z}} \sum_{n \in \mathbb{Z}} c_{m,n}(\phi) \, \psi_{m,n}, \tag{9}$$

with convergence in $\mathscr{S}_0(\mathbb{R})$;

2. *the wavelet expansion series of f, namely*

$$f = \sum_{m \in \mathbb{Z}} \sum_{n \in \mathbb{Z}} c_{m,n}(f) \, \psi_{m,n}, \tag{10}$$

converges in $\mathscr{S}'_0(\mathbb{R})$;

3.

$$\langle f, \phi \rangle = \sum_{m \in \mathbb{Z}} \sum_{n \in \mathbb{Z}} c_{m,n}(f) \, c^c_{m,n}(\phi).$$

Proof The first property follows from the definition of wavelet coefficient provided in (8). The second and third properties are direct consequences of (9) (for more details see [16]).   □

The previous theorem shows the convergence of the wavelet series on the function spaces $\mathscr{S}_0(\mathbb{R})$ and $\mathscr{S}'_0(\mathbb{R})$. In the next subsection we will try to extend these results to the case of a tempered distribution f.

3.2 Wavelet Decomposition for Positive Definite Distributions

In Sect. 2.1 the functional space $\mathscr{S}_0(\mathbb{R})$ has been defined as the space of the functions $\phi \in \mathscr{S}(\mathbb{R})$ such that all the moments vanish, namely (5) holds. Therefore $\mathscr{S}_0(\mathbb{R})$ is a closed subspace of $\mathscr{S}(\mathbb{R})$. From the Theorem 2 the following convergence result for the wavelet expansions of positive definite distributions follows.

Theorem 3 *Let $\phi \in \mathscr{S}_0(\mathbb{R})$, $f \in \mathscr{S}'(\mathbb{R})$ and let $\psi \in \mathscr{S}_0(\mathbb{R})$ be an orthonormal wavelet. Under these hypotheses it is*

$$\langle f, \phi \rangle = \sum_{m \in \mathbb{Z}} \sum_{n \in \mathbb{Z}} c_{m,n}(f) \, c^c_{m,n}(\phi). \tag{11}$$

Proof Since $\phi \in \mathscr{S}_0(\mathbb{R})$, it follows that the product $\langle f, \phi \rangle$ is an element of $\mathscr{S}_0(\mathbb{R})$. This means that even if the property 2 of the Theorem 2 does not hold, the left-hand side of (11) belonging to $\mathscr{S}_0(\mathbb{R})$ exists in the sense of the Theorem 2. By using (9), and taking into account that $\overline{\phi} \in \mathscr{S}_0(\mathbb{R})$, we get

$$\overline{\phi} = \sum_{m \in \mathbb{Z}} \sum_{n \in \mathbb{Z}} c_{m,n}\left(\overline{\phi}\right) \psi_{m,n},$$

therefore

$$\langle f, \phi \rangle = \sum_{m \in \mathbb{Z}} \sum_{n \in \mathbb{Z}} c_{m,n}\left(\overline{\phi}\right) \langle f, \psi_{m,n} \rangle = \sum_{m \in \mathbb{Z}} \sum_{n \in \mathbb{Z}} c_{m,n}(f) \, \langle \overline{\psi_{m,n}}, \phi \rangle.$$

The proof follows directly by recalling the meaning of $c^c_{m,n}$ (see Sect. 2.1).   □

If all the hypotheses of the previous theorem are satisfied and the Fourier transform of f is a positive Radon measure, according to the Schwartz theorem (see Sect. 2.2) f is nothing but a positive definite distribution. Therefore the thesis of the Theorem 3 holds for every positive definite distribution.

3.3 Example and Open Problem

Let us consider the distribution associated to the Poisson summation formula [6], namely the distribution T given by

$$T = f(x) = \sum_{k=-\infty}^{\infty} c_k \frac{e^{ixk}}{\sqrt{2\pi}}, \tag{12}$$

where $f \in L^1_{loc}(\mathbb{R})$ is a 2π-periodic function, c_k are its Fourier coefficients and the Fourier transform of f is defined by

$$\widehat{f}(\omega) = \frac{1}{\sqrt{2\pi}} \int_{-\infty}^{\infty} f(x) e^{-i\omega x} \, dx. \tag{13}$$

In [6] it is shown that T is a positive definite distribution. In order to apply the Theorem 3 to this distribution, a wavelet family belonging to the function space $\mathscr{S}_0(\mathbb{R})$ has to be chosen. The Shannon wavelet is defined [2] by

$$\psi_{SHA}(x) = \text{sinc}(x - 1/2) - 2\,\text{sinc}(2x - 1), \tag{14}$$

Since its scaling function is simply given by $\phi(x) = \text{sinc}(x)$ [2], this wavelet family satisfies the hypotheses of the Theorem 3. Hence (11) holds, namely

$$\langle f, \phi \rangle = \sum_{m \in \mathbb{Z}} \sum_{n \in \mathbb{Z}} c_{m,n}(f) \, c^c_{m,n}(\phi). \tag{15}$$

where the coefficient $c^c_{m,n}(\phi)$ can be easily computed being $\phi(x) = \text{sinc}(x)$. A remarkable result [2] is that the reconstruction formula (6) for Shannon wavelets enables to compute the derivatives of f in terms of the wavelet decomposition, i.e.

$$\frac{d^l}{dx^l} f(x) = \sum_{h=-\infty}^{\infty} \alpha_h \frac{d^l}{dx^l} \phi^0_h(x) + \sum_{n=0}^{\infty} \sum_{k=-\infty}^{\infty} \beta^n_k \frac{d^l}{dx^l} \psi^n_k(x), \tag{16}$$

if $f \in L^2(\mathbb{R})$ and $f \in C^q$ with q sufficiently high. In order to generalize the previous equation, given that in our case f is a tempered distribution, it should be written in the distribution sense. At the state of the art, this is an open problem. Since (16) enables us to approximate a function and its derivatives, the proposed problem might provide interesting results.

4 Fractional Calculus of Complex Functions in the distribution sense

In the first part of this section, the generalization of the distribution theory to any complex-valued function of a complex variable is presented in order to lay the foundations for a fractional-wavelet analysis of complex functions in the distribution sense. In the second part, a definition of fractional derivative in the complex plane, which can be reinterpreted in terms of the distribution theory, is presented.

4.1 A Complex-Variable Distribution Theory

A generalization of the distribution theory to $\mathbb{C}$ now becomes necessary. Luckily, this topic is widely discussed in [4] and the basic properties of the complex generalized functions are summarized in the following statement below.

Theorem 4 *Let f, g, ϕ, ψ be complex-valued functions of a complex variable s, $a, b \in \mathbb{C}$, $\gamma \in \mathbb{R}_{\neq 0}$, and suppose that the inner products of (1)–(5) are defined and convergent over $(0, \infty)$. If $\psi \in \mathscr{D}'(\mathbb{C})$, $\phi \in \mathscr{D}(\mathbb{C})$ and $s_0 \in \mathbb{C} : \mathfrak{R}(s_0) = \sigma_0$, the following properties hold true.*

1.

$$\begin{cases} \langle f(s), a\,\phi(s) + b\,\psi(s)\rangle = a\langle f(s), \phi(s)\rangle + b\langle f(s), \psi(s)\rangle, \\ \langle f(s) + g(s), \phi(s)\rangle = \langle f(s), \phi(s)\rangle + \langle g(s), \phi(s)\rangle, \end{cases} \quad \text{(linearity)}$$

2.

$$\langle f(s - s_0), \phi(s)\rangle_{\mathfrak{R}(s)=\sigma} = \langle f(y), \phi(y + s_0)\rangle_{\mathfrak{R}(y)=\sigma - \sigma_0}. \quad \text{(shift in } \mathbb{C})$$

3.

$$\langle f(\gamma s), \phi(s)\rangle_{\mathfrak{R}(s)=\sigma} = \frac{1}{|\gamma|}\left\langle f(y), \phi\left(\frac{y}{\gamma}\right)\right\rangle_{\mathfrak{R}(y)=\gamma\sigma}. \quad \text{(scaling)}$$

4.

$$\langle f(s) * g(s), \phi(s) \rangle_{\Re(s)=\sigma} = \left\langle f(y), \int_{\sigma-i\infty}^{\sigma+i\infty} g(s-y)\phi(s)\, ds \right\rangle_{\Re(y)=\Sigma},$$

$$\text{with } y = \Sigma + i\Omega. \quad (convolution)$$

5.

$$\begin{cases} \langle f^{(n)}(s), \phi(s) \rangle = (-1)^n \langle f(s), \phi^{(n)}(s) \rangle, \\ \langle f'(s)g(s), \phi(s) \rangle = -\langle f(s), g'(s)\phi(s) \rangle. \end{cases} \quad (derivation)$$

Proof These properties immediately follow from the definition of distribution (for further details, see [4]).   □

The Dirac impulse δ can be extended in the complex plane without any difficulties introducing the generalized Dirac impulse ξ by a definition based on its integral about the origin [3]. The ξ impulse (or generalized impulse) is a complex-valued function of the complex variable $s = \sigma + iw$ defined by

$$\langle \xi(s), \phi(s) \rangle_{\Re(s)=\sigma} = \begin{cases} i\phi(0), & \sigma = 0, \\ 0, & \sigma \neq 0, \end{cases} \quad (17)$$

i.e.

$$\begin{cases} \xi(s) = 0, & s \neq 0, \\ \\ \int_{-\infty}^{\infty} \xi(iw)\, dw = 1, & \text{otherwise,} \end{cases}$$

hence

$$\xi(iw) = \delta(w). \quad (18)$$

Its main properties (linearity, scaling, convolution, etc.) immediately follow from the Theorem 4. Another important result is that the ξ impulse can be viewed as the limit of a progressively narrowing and increasing height sequence of functions as $\varepsilon \to 0$, namely the same property which holds for δ [3, 4]. In the classical theory, this sequence is represented by a family of rectangular or Gaussian windows, while it is provided by a cylindrical sequence in the complex domain (see Fig. 1). The Gaussian sequence used to approximate the Dirac impulse is

$$g(x) = \frac{1}{\sqrt{\pi\varepsilon}} e^{-x^2/\varepsilon},$$

while its complex generalization is simply given by

$$w_\varepsilon(s) = \frac{1}{\sqrt{\pi\varepsilon}} e^{-|s|^2/\varepsilon}. \quad (19)$$

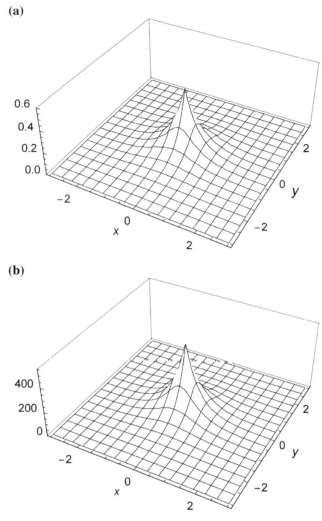

Fig. 1 3D plots of the $\left|w_\varepsilon(s = x + iy)\right|$ for $\varepsilon = 1$ (**a**) and $\varepsilon = 10^{-2}$ (**b**). They provide a geometrical interpretation of (20)

It is not hard to show [4] that

$$\xi(s) = \lim_{\varepsilon \to 0} w_\varepsilon(s). \tag{20}$$

Figure 1 illustrates how w_ε approximates ξ for ε close to 0 according to the previous equation.

Let

$$I = \int_l \xi(s)\phi(s)\mathrm{d}s, \tag{21}$$

be the integral along a straight line l in the complex plane. If the origin does not belong to the line l, it can be described by $s = s_0 + r\,e^{i\theta}$, where $s_0 \neq 0$ and $s \neq 0$ for every s_0, r and θ. Both in this case and if the impulse is a Gaussian sequence, $I = 0$ [4]. Hence, we have to consider only the case when the origin belongs to l.

If w_ε is a Gaussian or cylinder sequence, since $s = r\,e^{i\theta}$, it is [4]

$$I = e^{i\theta} \int_{-\infty}^{\infty} \delta(r)\,\phi\left(r\,e^{i\theta}\right)\mathrm{d}r,$$

therefore

$$\xi\left(r\,e^{i\theta}\right) = \delta(r) \Rightarrow \xi(i\omega) = \delta(\omega). \tag{22}$$

The previous equation justifies the definition (18). This generalization of the distribution theory to $\mathbb{C}$ is suitable to several applications in Laplace, z transforms, as well as in differential and difference equations [3, 4].

4.2   The Complex Fractional Derivative Operator in the Distribution Sense

The well-know convolution method, due to Schwartz [17], is based on the possibility to write the Riemann–Liouville fractional integral I^α [15] with lower boundary $a = 0$ as a convolution, i.e.

$$I^\alpha f(x) \overset{det}{=} \frac{1}{\Gamma(\alpha)} \int_{a=0}^{x} f(t)(x - t)^{\alpha - 1}\mathrm{d}t = f(x) * \frac{x_+^{\alpha - 1}}{\Gamma(\alpha)}, \tag{23}$$

where the function

$$x_+^\alpha \overset{det}{=} x^\alpha\, u(x), \tag{24}$$

defines the regular distribution

$$\langle x_+^\alpha, \phi \rangle = \int_0^\infty x^\alpha \phi(x)\mathrm{d}x, \qquad \forall \phi \in \mathscr{D}(\mathbb{R}), \tag{25}$$

for $\alpha > -1$ (indeed, the integral above makes sense if $\Re(\alpha) > -1$) [8]. Formula (23) can be written as a convolution without the restriction $a = 0$ defining $x_+^\alpha = x^\alpha u(x - a)$ instead of (24). Clearly (23) makes sense if the associated convolution is valid.

Following the same approach provided for the real fractional calculus (distribution sense) [8], we can denote

$$\phi_\alpha(x) = \frac{x_+^{\alpha-1}}{\Gamma(\alpha)}, \tag{26}$$

and hence (23) becomes

$$\mathrm{I}^\alpha f(x) = f(x) * \phi_\alpha(x). \tag{27}$$

Formula (27) shows that I^α is suitable to define real fractional derivatives in the distribution sense. Moreover, $\phi_\alpha(x)$ satisfies the semigroup property [8], i.e. $\phi_\alpha(x) * \phi_\eta(x) = \phi_{\alpha+\eta}(x)$ if α and η have real parts greater than zero.

Indeed, under these hypotheses it is

$$\Gamma(\alpha)\, \Gamma(\eta)\, \phi_\alpha(x) * \phi_\eta(x) = x_+^\alpha * x_+^\eta = \int_0^x y^{\alpha-1}(x - y)^{\eta-1}\, \mathrm{d}y.$$

By a change of variables $y = xt$, the right-hand side (RHS) becomes

$$\int_0^x (xt)^{\alpha-1}(x - xt)^{\eta-1}\, x\, \mathrm{d}t = x_+^{\alpha+\eta-1} \int_0^1 t^{\alpha-1}(1 - t)^{\eta-1}\, x\mathrm{d}t = x_+^{\alpha+\eta-1}\, \mathrm{B}(\alpha, \eta)$$

$$= x_+^{\alpha+\eta-1}\, \frac{\Gamma(\alpha)\, \Gamma(\eta)}{\Gamma(\alpha + \eta)} = \Gamma(\alpha)\, \Gamma(\eta)\, \phi_{\alpha+\eta}(x),$$

where B is the Euler beta function [1].

The close link between the Riemann–Liouville fractional integral and the Caputo fractional derivative [15] justifies the reason for which the distribution theory can be applied to the latter.

Ortigueira's generalization of the α-order Caputo fractional derivative $_c\mathrm{D}^\alpha$ to the complex plane is defined [13] by

$$_c\mathrm{D}^\alpha f(s) \overset{def}{=} \frac{e^{i(\pi-\theta)(\alpha-m)}}{\Gamma(m - \alpha)} \int_0^\infty \frac{f^{(m)}(xe^{i\theta} + s)}{x^{\alpha-m+1}}\mathrm{d}x, \tag{28}$$

where f is a complex-valued function of the complex variable s, $m - 1 < \alpha < m \in \mathbb{Z}^+$ and $\theta \in [0, 2\pi)$. This fractional operator is defined such that the integral and derivative signs can be exchanged [13], hence

$$f^{(\alpha)}(s) = \frac{e^{i(\pi - \theta)(\alpha - m)}}{\Gamma(m - \alpha)} \frac{d^m}{ds^m} \int_0^\infty \frac{f(xe^{i\theta} + s)}{x^{\alpha - m + 1}} dx$$

$$= \frac{e^{i(\pi - \theta)(\alpha - m)}}{\Gamma(m - \alpha)} \frac{d^m}{ds^m} \int_0^\infty f(z + s) z^{m - \alpha - 1} e^{-i\theta(m - \alpha - 1)} e^{-i\theta} dz$$

$$= \frac{e^{i\pi(\alpha - m)}}{\Gamma(m - \alpha)} \frac{d^m}{ds^m} \int_0^\infty f(s - z')(-z')^{m - \alpha - 1}(-1) dz' = \frac{e^{i\pi(\alpha - m)}}{\Gamma(m - \alpha)} e^{i\pi(m - \alpha)}$$

$$\cdot \frac{d^m}{ds^m}\left(f(s) * s_+^{m - \alpha - 1}\right) = \frac{d^m}{ds^m}\left(f(s) * \frac{s_+^{m - \alpha - 1}}{\Gamma(m - \alpha)}\right),$$

where the function $s_+^{m - \alpha - 1} = s^{m - \alpha - 1} u(\Re(s)) u(\Im(s))$ is the complex counterpart of (24). Indeed, the function $s_+^{m - \alpha - 1}$ defines the regular distribution

$$\langle s_+^{m - \alpha}, \phi(s) \rangle = \int_0^\infty s^{m - \alpha} \phi(s) ds, \qquad \forall \phi \in \mathscr{D}(\mathbb{C}), \tag{29}$$

in the sense of distribution theory for complex functions (see Theorem 4 in Sect. 4.1). This definition holds because $m - \alpha > 0$ and makes sense if the associated convolution is valid.

As in the real case, we can introduce the function

$$\phi_{m - \alpha}(s) = \frac{s_+^{m - \alpha - 1}}{\Gamma(m - \alpha)}, \tag{30}$$

hence

$$f^{(\alpha)}(s) = \frac{d^m}{ds^m}\left(f(s) * \phi_{m - \alpha}(s)\right). \tag{31}$$

Therefore (31) is the complex counterpart of (27), namely it provides the fractional derivative on the function space $\mathscr{D}'(\mathbb{C})$.

In this case the semigroup property is not satisfied, i.e. $\phi_{m - \alpha}(s) * \phi_{n - \beta}(s) \neq \phi_{m - \alpha + n - \beta}(s)$ if $m - 1 < \alpha < m \in \mathbb{Z}^+$ and $n - 1 < \beta < n \in \mathbb{Z}^+$.

Indeed,

$$\Gamma(m - \alpha)\,\Gamma(n - \beta)\,\phi_{m-\alpha}(s) * \phi_{n-\beta} = s_+^{m-\alpha-1} * s_+^{n-\beta-1}$$

$$= \int_0^\infty z^{m-\alpha-1}(s - z)^{n-\beta-1}\,dz. \qquad (32)$$

By a change of variables $z = sw$, the RHS becomes

$$\int_0^\infty (sw)^{m-\alpha-1}(s - sw)^{n-\beta-1}\,s\,dw = s_+^{m-\alpha+n-\beta-1} \cdot \int_0^\infty w^{m-\alpha-1}(1 - w)^{n-\beta-1}\,dw$$

$$\neq s_+^{m-\alpha+n-\beta-1} \int_0^1 w^{m-\alpha-1}(1 - w)^{n-\beta-1}\,dw$$

$$= s_+^{m-\alpha+n-\beta-1}\,\mathrm{B}(m - \alpha, n - \beta) = \Gamma(m - \alpha)\,\Gamma(n - \beta)\,\phi_{m-\alpha+n-\beta}.$$

The semigroup property does not hold since in (28) the upper limit of integration is infinity, hence the last computation does not provide the Euler beta function.

5 Fractional Derivative of Complex Wavelets

The aim of this section is to show the power of the Ortigueira-Caputo fractional derivative by computing the fractional derivative of the Gabor–Morlet wavelet.

5.1 Gabor–Morlet Wavelet

The Gabor–Morlet wavelet (sometimes Gabor wavelet or complex Morlet wavelet) is a complex wavelet widely used in geophysical applications. It is given [18] by

$$\psi_{GM}(x) = \frac{1}{\sqrt{\pi f_b}}\,e^{-x^2/f_b}\,e^{i2\pi f_c x}. \qquad (33)$$

In the current literature, a common choice is taking $f_b = 2$ and $f_c = \dfrac{5}{2\pi}$ since the value $\omega_c = 2\pi f_c = 5$ is often used in the applications. Hence, we get

$$\psi_{GM}(x) = \frac{1}{\sqrt{2\pi}}e^{-x^2/2}\,e^{ix\omega_c}. \qquad (34)$$

Its fractional derivative is provided by the following statement.

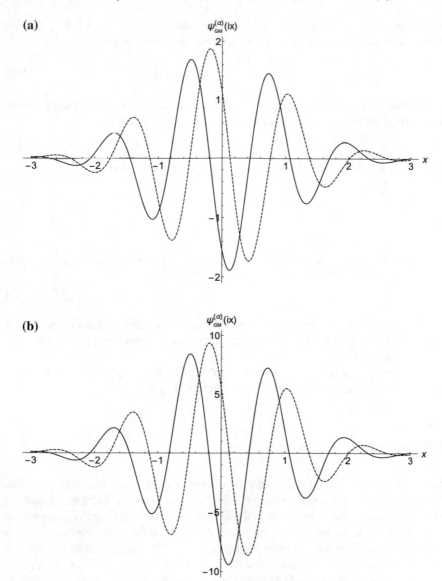

Fig. 2 Real part (*solid curve*) and imaginary part (*dashed curve*) for the α-order fractional derivative of the Gabor–Morlet wavelet with $\alpha = 0.4$ (**a**) and $\alpha = 1.4$ (**b**)

Theorem 5 *Let $s \in \mathbb{C}$ such that $\Re(s) = 0$ and let $m - 1 < \alpha < m \in \mathbb{Z}^+$. The α-order fractional derivative of the Gabor–Morlet wavelet is given by*

$$\psi_{GM}^{(\alpha)}(s) = e^{i2\pi\{\alpha\}} \, \omega_c^\alpha \, \psi_{GM}(s). \tag{35}$$

Proof It is easy to show this property with a direct computation following the same approach given in [9] for the Riemann ζ function. Indeed, since

$$\int_0^\infty e^{(ye^{i\theta} + ix)\omega_c} \frac{1}{y^{\alpha-m+1}} \mathrm{d}y = e^{-i\theta(m-\alpha)} \, e^{ix\omega_c} (-\omega_c)^{\alpha-m} \, \Gamma(m - \alpha),$$

we get

$$\psi_{GM}^{(\alpha)}(s = ix) = \frac{1}{\sqrt{2\pi}} e^{i2\pi\alpha} \, e^{-x^2/2} \, \omega_c^{\alpha-m} \frac{\mathrm{d}^m}{\mathrm{d}s^m} \left(e^{s\omega_c}\right) = e^{i2\pi\{\alpha\}} \, \omega_c^\alpha \, \psi_{GM}(s). \tag{36}$$

$\square$

This theorem shows that the α-order fractional derivative of the Gabor–Morlet wavelet is nothing more than the product of the same wavelet $\psi_{GM}(s = ix)$ and the complex factor $e^{i2\pi\{\alpha\}} \, \omega_c^\alpha$.

Therefore the real and imaginary parts of this fractional derivative still belong to the family of Gabor–Morlet wavelets, as shown in Fig. 2.

6 Conclusion

In this chapter, a wavelet expansion of positive definite distributions is provided. The Ortigueira-Caputo fractional operator, which provides the fractional derivative of a complex function, is rewritten in the distribution sense. The fractional derivative of the Gabor–Morlet wavelet is computed. An open problem, concerning the possibility to generalize the reconstruction formula for Shannon wavelets through the distribution theory, is proposed. This fractional derivative operator has already given some results in analytic number theory (see for instance [9]) and could be able to describe different physical phenomena, while the proposed wavelet expansion could be applied to other distribution families in order to obtain another generalization of the Theorem 2.

Acknowledgements Emanuel Guariglia would like to thank the Division of Applied Mathematics, School of Education, Culture and Communication, Mälardalens University for giving him the opportunity to work in an extremely favourable research environment.

References

1. Abramowitz, M., Stegun, I.A.: Handbook of Mathematical Functions with Formulas, Graphs, and Mathematical Tables. Martino Fine Books, New York (2014)
2. Cattani, C.: Shannon wavelets theory. Math. Probl. Eng. **2008**, 1–24 (2008)
3. Corinthios, M.J.: Generalisation of the Dirac-delta impulse extending Laplace and z transform domains. IEE Proc. Vis. Image Sig. Process. **150**, 69–81 (2003)
4. Corinthios, M.J.: Complex-variable distribution theory for Laplace and z transforms. IEE Proc. Vis. Image Sig. Process. **152**, 97–106 (2005)
5. Daubechies, I.: Ten Lectures on Wavelets. SIAM, Philadelphia (1992)
6. Donoghue Jr., W.F.: Distributions and Fourier Transform. Academic Press, New York (1969)
7. Gabardo, J.-P.: Extension of Positive-Definite Distributions and Maximum Entropy. Memoirs of the American Mathematical Society, vol. 102. American Mathematical Society, Providence (1993)
8. Gel'fand, I.M., Shilov, G.E.: Generalized Functions. Academic Press, New York (1964)
9. Guariglia, E.: Fractional derivative of the Riemann zeta function. In: Cattani C., Srivastava H., Yang X.J. (eds.) Fractional Dynamics, De Gruyter Open, Chap. 21 (2015)
10. Hernández, E., Weiss, G.: A First Course of Wavelets. CRC Press, Boca Raton (1996)
11. Hölschneider, M.: Wavelets: An Analysis Tool. Clarendon Press, New York (1999)
12. Lemarié, P.G., Meyer, Y.: Ondeletles et bases hilbertiennes. Revista Matemática Iberoamericana **2**, 1–18 (1986)
13. Li, C., Dao, X., Guo, P.: Fractional derivatives in complex planes. Nonlinear Anal. Theory Methods Appl. **71**, 1857–1869 (2009)
14. Neeb, K.-H., Ólafsson, G.: Reflection positivity and conformal symmetry. J. Funct. Anal. Elsevier **266**, 2174–2224 (2014)
15. Ortigueira, M.D.: Fractional Calculus for Scientists and Engineers, pp. 35–37. Springer, London (2011)
16. Saneva, K., Vindas, J.: Wavelet expansions and asymptotic behaviour of distributions. J. Math. Anal. Appl. **370**, 543–554 (2010)
17. Schwartz, L.: Théorie des Distributions, vol. 1-2. Hermann, Paris (1951)
18. Teolis, A.: Computational Signal Processing with Wavelets. Birkhäuser, Boston (1998)
19. Triebel, H.: Function Spaces and Wavelets on Domains. European Mathematical Society, Berlin (2008)

Linear Classification of Data with Support Vector Machines and Generalized Support Vector Machines

Talat Nazir, Xiaomin Qi and Sergei Silvestrov

Abstract In this paper, we study the support vector machine and introduced the notion of generalized support vector machine for classification of data. We show that the problem of generalized support vector machine is equivalent to the problem of generalized variational inequality and establish various results for the existence of solutions. Moreover, we provide various examples to support our results.

Keywords Linear classification · Support vector machine · Generalized support vector machine · Kernel function

1 Support Vector Machine

Over the last decade, support vector machines (SVMs) [2, 3, 13, 14, 18] have been revealed as a powerful and important tool for pattern classification and regression. It has been used in various applications such as text classification [5], facial expression recognition [9], gene analysis [4] and many others [1, 6–8, 10–12, 15, 19–22]. Recently, Wang et al. [16] presented SVM based fault classifier design for a water level control system. They also studied the SVM classifier based fault diagnosis for a water level process [17].

T. Nazir (✉) · X. Qi · S. Silvestrov
Division of Applied Mathematics, School of Education,
Culture and Communication, Mälardalen University, Box 883, 721 23 Västerås, Sweden
e-mail: talat@ciit.net.pk

X. Qi
e-mail: xiaomin.qi@mdh.se

S. Silvestrov
e-mail: sergei.silvestrov@mdh.se

T. Nazir
Department of Mathematics, COMSATS Institute of Information Technology,
Abbottabad 22060, Pakistan

© Springer International Publishing Switzerland 2016
S. Silvestrov and M. Rančić (eds.), *Engineering Mathematics II*,
Springer Proceedings in Mathematics & Statistics 179,
DOI 10.1007/978-3-319-42105-6_17

For the standard support vector classification (SVC), the basic idea is to find the optimal separating hyperplane between the positive and negative examples. The optimal hyperplane may be obtained by maximizing the margin between two parallel hyperplanes, which involves the minimization of a quadratic programming problem.

Support Vector Machines are based on the concept of decision planes that define decision boundaries. A decision plane is one that separates between a set of objects having different class memberships.

Support Vector Machines can be thought of as a method for constructing a special kind of rule, called a linear classifier, in a way that produces classifiers with theoretical guarantees of good predictive performance (the quality of classification on unseen data).

In this paper, we study the problems of support vector machine and define generalized support vector machine. We also show the sufficient conditions for the existence of solutions for problems of generalized support vector machine. We also support our results with various examples.

Thought this paper, by $\mathbb{N}$, $\mathbb{R}$, $\mathbb{R}^n$ and $\mathbb{R}_n^+$ we denote the set of all natural numbers, the set of all real numbers, the set of all n-tuples real numbers, the set of all n-tuples of nonnegative real numbers, respectively.

Also, we consider $\|\cdot\|$ and $< \cdot, \cdot >$ as Euclidean norm and usual inner product on $\mathbb{R}^n$, respectively.

Furthermore, for two vectors $\mathbf{x}, \mathbf{y} \in \mathbb{R}^n$, we say that $\mathbf{x} \leq \mathbf{y}$ if and only if $x_i \leq y_i$ for all $i \in \{1, 2, \ldots, n\}$, where x_i and y_i are the components of $\mathbf{x}$ and $\mathbf{y}$, respectively.

2 Linear Classifiers

Binary classification is frequently performed by using a function $f : \mathbb{R}^n \to \mathbb{R}$ in the following way: the input $\mathbf{x} = (x_1, \ldots, x_n)$ is assigned to the positive class if, $f(\mathbf{x}) \geq 0$ and otherwise to the negative class. We consider the case where $f(\mathbf{x})$ is a linear function of $\mathbf{x}$, so that it can be written as

$$f(\mathbf{x}) = \langle \mathbf{w}, \mathbf{x} \rangle + b =$$
$$= \sum_{i=1}^{n} w_i x_i + b,$$

where $\mathbf{w} \in \mathbb{R}^n$, $b \in \mathbb{R}$ are the parameters that control the function and the decision rule is given by $sgn(f(\mathbf{x}))$. The learning methodology implies that these parameters must be learned from the data.

Definition 1 We define the functional margin of an example $(\mathbf{x}_k, y_k)$ with respect to a hyperplane $(\mathbf{w}, b)$ to be the quantity

$$\gamma_k = y_k (\langle \mathbf{w}, \mathbf{x}_k \rangle + b),$$

where $y_k \in \{-1, 1\}$. Note that $\gamma_k > 0$ implies correct classification of $(\mathbf{x}_k, y_k)$. If we replace functional margin by geometric margin we obtain the equivalent quantity for the normalized linear function $\left(\frac{1}{\|\mathbf{w}\|} \mathbf{w}, \frac{1}{\|\mathbf{w}\|} b \right)$, which therefore measures the Euclidean distances of the points from the decision boundary in the input space.

Actually geometric margin can be written as

$$\tilde{\gamma} = \frac{1}{\|\mathbf{w}\|} \gamma.$$

To find the hyperplane which has maximal geometric margin for a training set S means to find maximal $\tilde{\gamma}$. For convenience, we let $\gamma = 1$, the objective function can be written as

$$\max \frac{1}{\|\mathbf{w}\|}.$$

Of course, there are some constraints for the optimization problem. According to the definition of margin, we have $y_k (\langle \mathbf{w}, \mathbf{x}_k \rangle + b) \geq 1, k = 1, \ldots, l$. We rewrite the equivalent formation of the objective function with the constraints as

$$\min \frac{1}{2} \|\mathbf{w}\|^2 \quad \text{s.t.} \quad y_k (\langle \mathbf{w}, \mathbf{x}_k \rangle + b) \geq 1, k = 1, \ldots, l.$$

We denote this problem by SVM.

3 Generalized Support Vector Machines

We replace $\mathbf{w}, b$ by W, B respectively. The control function $F : \mathbb{R}^n \to \mathbb{R}^n$ is defined as

$$F(\mathbf{x}) = W.\mathbf{x} + B, \tag{1}$$

where $W \in \mathbb{R}^{n \times n}$, $B \in \mathbb{R}^n$ are the parameters of control function.
Define

$$\tilde{\gamma}_k^* = \mathbf{y}_k (W\mathbf{x}_k + B) > 1 \quad \text{for} \quad k = 1, 2, \ldots, l, \tag{2}$$

where $\mathbf{y}_k \in \{(-1, -1, \ldots, -1), (1, 1, \ldots, 1)\}$ is n-dimensional vector.

Definition 2 We define a map $G : \mathbb{R}^n \to \mathbb{R}_+^n$ by

$$G(\mathbf{w}_i) = (\|\mathbf{w}_i\|, \|\mathbf{w}_i\|, \ldots, \|\mathbf{w}_i\|) \quad \text{for} \quad i = 1, 2, \ldots, n, \tag{3}$$

where $\mathbf{w}_i$ are the rows of $W_{n \times n}$ for $i = 1, 2, \ldots, n$.

The problem is to find $\mathbf{w}_i \in \mathbb{R}^n$ that satisfy

$$\min_{\mathbf{w}_i \in W} G(\mathbf{w}_i) \quad \text{s.t.} \quad \eta > 0, \tag{4}$$

where $\eta = \mathbf{y}_k (W\mathbf{x}_k + B) - 1$.

We call this problem a Generalized Support Vector Machine (GSVM).
The GSVM is equivalent to

$$\text{find} \quad \mathbf{w}_i \in W : \; \langle G'(\mathbf{w}_i), \mathbf{v} - \mathbf{w}_i \rangle \geq 0 \quad \text{for all} \quad \mathbf{v} \in \mathbb{R}^n \quad \text{with} \quad \eta > 0,$$

or more specifically

$$\text{find} \quad \mathbf{w}_i \in W : \; \langle \eta G'(\mathbf{w}_i), \mathbf{v} - \mathbf{w}_i \rangle \geq 0 \quad \text{for all} \quad \mathbf{v} \in \mathbb{R}^n. \tag{5}$$

Hence the problem of GSVM becomes the problem of generalized variational inequality.

Example 1 Let us take the group of points of positive class $(1, 0)$, $(0, 1)$ and negative class $(-1, 0)$, $(0, -1)$.

First we use <u>SVM</u> to solve this problem to find the hyperplane $< \mathbf{w}, \mathbf{x} > +b = 0$ that separates these two kinds of points. Obviously, we know that the hyperplane is H which is shown in the Fig. 1.

Fig. 1 Example 1

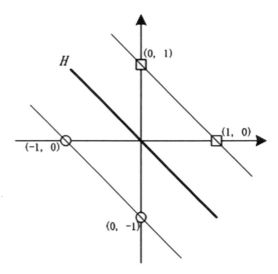

For two positive points, we have

$$(w_1, w_2) \begin{bmatrix} 1 \\ 0 \end{bmatrix} + b = 1,$$

$$(w_1, w_2) \begin{bmatrix} 0 \\ 1 \end{bmatrix} + b = 1,$$

which implies

$$w_1 + b = 1,$$
$$w_2 + b = 1.$$

For two negative points, we have

$$(w_1, w_2) \begin{bmatrix} -1 \\ 0 \end{bmatrix} + b = -1,$$

$$(w_1, w_2) \begin{bmatrix} 0 \\ -1 \end{bmatrix} + b = -1,$$

which implies that

$$-w_1 + b = -1,$$
$$-w_2 + b = -1.$$

From the equations, we get $\mathbf{w} = (1, 1)$ and $b = 0$. The result is $\|\mathbf{w}\| = \sqrt{2}$. Now we apply GSVM for this data. For two positive points, we have

$$\begin{bmatrix} w_{11} & w_{12} \\ w_{21} & w_{22} \end{bmatrix} \begin{bmatrix} 1 \\ 0 \end{bmatrix} + \begin{bmatrix} b_1 \\ b_2 \end{bmatrix} = \begin{bmatrix} 1 \\ 1 \end{bmatrix},$$

and

$$\begin{bmatrix} w_{11} & w_{12} \\ w_{21} & w_{22} \end{bmatrix} \begin{bmatrix} 0 \\ 1 \end{bmatrix} + \begin{bmatrix} b_1 \\ b_2 \end{bmatrix} = \begin{bmatrix} 1 \\ 1 \end{bmatrix},$$

which gives

$$\begin{bmatrix} w_{11} \\ w_{21} \end{bmatrix} + \begin{bmatrix} b_1 \\ b_2 \end{bmatrix} = \begin{bmatrix} 1 \\ 1 \end{bmatrix} \quad \text{and} \quad \begin{bmatrix} w_{12} \\ w_{22} \end{bmatrix} + \begin{bmatrix} b_1 \\ b_2 \end{bmatrix} = \begin{bmatrix} 1 \\ 1 \end{bmatrix}. \tag{6}$$

For two negative points, we have

$$\begin{bmatrix} w_{11} & w_{12} \\ w_{21} & w_{22} \end{bmatrix} \begin{bmatrix} -1 \\ 0 \end{bmatrix} + \begin{bmatrix} b_1 \\ b_2 \end{bmatrix} = \begin{bmatrix} -1 \\ -1 \end{bmatrix},$$

and

$$\begin{bmatrix} w_{11} & w_{12} \\ w_{21} & w_{22} \end{bmatrix} \begin{bmatrix} 0 \\ -1 \end{bmatrix} + \begin{bmatrix} b_1 \\ b_2 \end{bmatrix} = \begin{bmatrix} -1 \\ -1 \end{bmatrix},$$

which provides

$$\begin{bmatrix} -w_{11} \\ -w_{21} \end{bmatrix} + \begin{bmatrix} b_1 \\ b_2 \end{bmatrix} = \begin{bmatrix} -1 \\ -1 \end{bmatrix} \quad \text{and} \quad \begin{bmatrix} -w_{12} \\ -w_{22} \end{bmatrix} + \begin{bmatrix} b_1 \\ b_2 \end{bmatrix} = \begin{bmatrix} -1 \\ -1 \end{bmatrix}. \tag{7}$$

From (6) and (7), we get

$$W = \begin{bmatrix} 1 & 1 \\ 1 & 1 \end{bmatrix}, \quad B = \begin{bmatrix} b_1 \\ b_2 \end{bmatrix} = \begin{bmatrix} 0 \\ 0 \end{bmatrix}.$$

Thus
$$\min G\left(\mathbf{w}_i\right) = \min\left\{G\left(\mathbf{w}_1\right), G\left(\mathbf{w}_2\right)\right\} = (\sqrt{2}, \sqrt{2}).$$

Hence we get $\mathbf{w} = (1, 1)$ that minimizes $G\left(\mathbf{w}_i\right)$ for $i = 1, 2$.

Remark 1 The above example shows that we get the same result by applying any method SVM and GSVM.

In the next example, we consider the two distinct groups of data, first solve both data for separate cases and then solve it for combined case for both methods SVM and GSVM.

Example 2 Let us consider the three categories of data.
   Situation 1 Suppose that, we have data $(1, 0)$, $(0, 1)$ as positive class and data $(-1/2, 0)$, $(0, -1/2)$ as negative class shown in Fig. 2.
   Using <u>SVM</u> to solve this problem, we have

$$(w_1, w_2) \begin{bmatrix} 1 \\ 0 \end{bmatrix} + b = 1,$$

and

$$(w_1, w_2) \begin{bmatrix} 0 \\ 1 \end{bmatrix} + b = 1,$$

which implies

$$w_1 + b = 1 \quad \text{and} \quad w_2 + b = 1. \tag{8}$$

For two negative points, we have

$$(w_1, w_2) \begin{bmatrix} -1/2 \\ 0 \end{bmatrix} + b = -1,$$

Fig. 2 Example 2,
situation 1

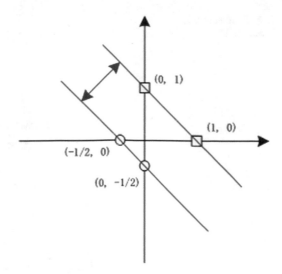

and

$$(w_1, w_2) \begin{bmatrix} 0 \\ -1/2 \end{bmatrix} + b = -1,$$

which gives

$$-\frac{w_1}{2} + b = -1 \quad \text{and} \quad -\frac{w_2}{2} + b = -1. \tag{9}$$

From (8) and (9), we get $\mathbf{w} = (\frac{4}{3}, \frac{4}{3})$ with $b = \frac{-1}{3}$, where $\|\mathbf{w}\| = \frac{\sqrt{32}}{3}$.

For **situation 2**, we consider the data $(\frac{1}{2}, 0)$ and $(0, \frac{1}{2})$ as positive class, data $(-2, 0)$ and $(0, -2)$ as negative class shown in Fig. 3.

Using SVM to solve this problem, we have

$$(w_1, w_2) \begin{bmatrix} 1/2 \\ 0 \end{bmatrix} + b = 1,$$

and

$$(w_1, w_2) \begin{bmatrix} 0 \\ 1/2 \end{bmatrix} + b = 1,$$

which implies

$$\frac{1}{2}w_1 + b = 1 \quad \text{and} \quad \frac{1}{2}w_2 + b = 1. \tag{10}$$

From the negative points, we have

$$(w_1, w_2) \begin{bmatrix} -2 \\ 0 \end{bmatrix} + b = -1,$$

Fig. 3 Example 2, situation 2

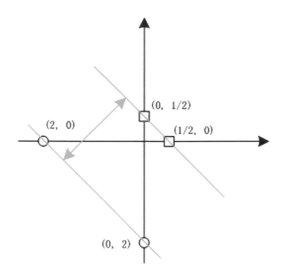

and

$$(w_1, w_2) \begin{bmatrix} 0 \\ -2 \end{bmatrix} + b = -1,$$

implies that

$$-2w_1 + b = -1 \quad \text{and} \quad -2w_2 + b = -1. \tag{11}$$

From (10) and (11), we get $\mathbf{w} = (\frac{4}{5}, \frac{4}{5})$ and $b = \frac{3}{5}$ with $\|\mathbf{w}\| = \frac{\sqrt{32}}{5}$.

In the next **situation 3,** we combine these two groups of data. Now, we have data $(1/2, 0), (0, 1/2), (1, 0), (0, 1)$ as positive class and $(-1/2, 0), (0, -1/2), (-2, 0), (0, -2)$ as negative class shown in Fig. 4.

Using <u>SVM</u> to solve this problem, we have

$$(w_1, w_2) \begin{bmatrix} 1/2 \\ 0 \end{bmatrix} + b = 1,$$

and

$$(w_1, w_2) \begin{bmatrix} 0 \\ 1/2 \end{bmatrix} + b = 1,$$

which implies

$$w_1/2 + b = 1 \quad \text{and} \quad w_2/2 + b = 1. \tag{12}$$

For two negative points, we have

$$(w_1, w_2) \begin{bmatrix} -1/2 \\ 0 \end{bmatrix} + b = -1,$$

Fig. 4 Example 2, situation 3

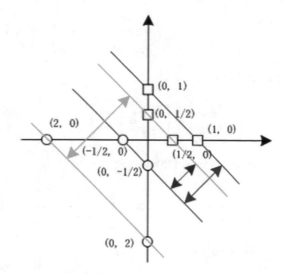

and

$$(w_1, w_2) \begin{bmatrix} 0 \\ -1/2 \end{bmatrix} + b = -1,$$

implies that

$$-\frac{1}{2}w_1 + b = -1 \quad \text{and} \quad -\frac{1}{2}w_2 + b = -1. \tag{13}$$

From (12) and (13), we obtain $\mathbf{w} = (2, 2)$ and $b = 0$, where $\|\mathbf{w}\| = 2\sqrt{2}$. Now we solve the same problem for all three situations by using <u>GSVM</u>. For two positive points of **situation 1**, we have

$$\begin{bmatrix} w_{11} & w_{12} \\ w_{21} & w_{22} \end{bmatrix} \begin{bmatrix} 1 \\ 0 \end{bmatrix} + \begin{bmatrix} b_1 \\ b_2 \end{bmatrix} = \begin{bmatrix} 1 \\ 1 \end{bmatrix},$$

and

$$\begin{bmatrix} w_{11} & w_{12} \\ w_{21} & w_{22} \end{bmatrix} \begin{bmatrix} 0 \\ 1 \end{bmatrix} + \begin{bmatrix} b_1 \\ b_2 \end{bmatrix} = \begin{bmatrix} 1 \\ 1 \end{bmatrix},$$

which implies

$$\begin{bmatrix} w_{11} \\ w_{21} \end{bmatrix} + \begin{bmatrix} b_1 \\ b_2 \end{bmatrix} = \begin{bmatrix} 1 \\ 1 \end{bmatrix} \quad \text{and} \quad \begin{bmatrix} w_{12} \\ w_{22} \end{bmatrix} + \begin{bmatrix} b_1 \\ b_2 \end{bmatrix} = \begin{bmatrix} 1 \\ 1 \end{bmatrix}. \tag{14}$$

Again, for the negative points, we have

$$\begin{bmatrix} w_{11} & w_{12} \\ w_{21} & w_{22} \end{bmatrix} \begin{bmatrix} -1/2 \\ 0 \end{bmatrix} + \begin{bmatrix} b_1 \\ b_2 \end{bmatrix} = \begin{bmatrix} -1 \\ -1 \end{bmatrix},$$

and

$$\begin{bmatrix} w_{11} & w_{12} \\ w_{21} & w_{22} \end{bmatrix} \begin{bmatrix} 0 \\ -1/2 \end{bmatrix} + \begin{bmatrix} b_1 \\ b_2 \end{bmatrix} = \begin{bmatrix} -1 \\ -1 \end{bmatrix},$$

which gives

$$\begin{bmatrix} -\frac{1}{2}w_{11} \\ -\frac{1}{2}w_{21} \end{bmatrix} + \begin{bmatrix} b_1 \\ b_2 \end{bmatrix} = \begin{bmatrix} -1 \\ -1 \end{bmatrix} \quad \text{and} \quad \begin{bmatrix} -\frac{1}{2}w_{12} \\ -\frac{1}{2}w_{22} \end{bmatrix} + \begin{bmatrix} b_1 \\ b_2 \end{bmatrix} = \begin{bmatrix} -1 \\ -1 \end{bmatrix}. \tag{15}$$

From (14) and (15), we get

$$W = \begin{bmatrix} \frac{4}{3} & \frac{4}{3} \\ \frac{4}{3} & \frac{4}{3} \end{bmatrix} \quad \text{and} \quad B = \begin{bmatrix} -\frac{1}{3} \\ -\frac{1}{3} \end{bmatrix}.$$

Thus we get

$$\min_{\mathbf{w}_i \in W} G\left(\mathbf{w}_i\right) = (\frac{4\sqrt{2}}{3}, \frac{4\sqrt{2}}{3}).$$

Hence we get $\mathbf{w} = (\frac{4}{3}, \frac{4}{3})$ that minimize $G\left(\mathbf{w}_i\right)$ for $i = 1, 2$. Now, for positive points of **situation 2**, we have

$$\begin{bmatrix} w_{11} & w_{12} \\ w_{21} & w_{22} \end{bmatrix} \begin{bmatrix} 1/2 \\ 0 \end{bmatrix} + \begin{bmatrix} b_1 \\ b_2 \end{bmatrix} = \begin{bmatrix} 1 \\ 1 \end{bmatrix},$$

and

$$\begin{bmatrix} w_{11} & w_{12} \\ w_{21} & w_{22} \end{bmatrix} \begin{bmatrix} 0 \\ 1/2 \end{bmatrix} + \begin{bmatrix} b_1 \\ b_2 \end{bmatrix} = \begin{bmatrix} 1 \\ 1 \end{bmatrix},$$

which gives

$$\begin{bmatrix} \frac{1}{2}w_{11} \\ \frac{1}{2}w_{21} \end{bmatrix} + \begin{bmatrix} b_1 \\ b_2 \end{bmatrix} = \begin{bmatrix} 1 \\ 1 \end{bmatrix} \quad \text{and} \quad \begin{bmatrix} \frac{1}{2}w_{12} \\ \frac{1}{2}w_{22} \end{bmatrix} + \begin{bmatrix} b_1 \\ b_2 \end{bmatrix} = \begin{bmatrix} 1 \\ 1 \end{bmatrix}. \tag{16}$$

For two negative points for this case, we have

$$\begin{bmatrix} w_{11} & w_{12} \\ w_{21} & w_{22} \end{bmatrix} \begin{bmatrix} -2 \\ 0 \end{bmatrix} + \begin{bmatrix} b_1 \\ b_2 \end{bmatrix} = \begin{bmatrix} -1 \\ -1 \end{bmatrix},$$

and

$$\begin{bmatrix} w_{11} & w_{12} \\ w_{21} & w_{22} \end{bmatrix} \begin{bmatrix} 0 \\ -2 \end{bmatrix} + \begin{bmatrix} b_1 \\ b_2 \end{bmatrix} = \begin{bmatrix} -1 \\ -1 \end{bmatrix},$$

which gives

$$\begin{bmatrix} -2w_{11} \\ -2w_{21} \end{bmatrix} + \begin{bmatrix} b_1 \\ b_2 \end{bmatrix} = \begin{bmatrix} -1 \\ -1 \end{bmatrix} \quad \text{and} \quad \begin{bmatrix} -2w_{12} \\ -2w_{22} \end{bmatrix} + \begin{bmatrix} b_1 \\ b_2 \end{bmatrix} = \begin{bmatrix} -1 \\ -1 \end{bmatrix}. \tag{17}$$

Thus, we obtain that

$$W = \begin{bmatrix} \frac{4}{5} & \frac{4}{5} \\ \frac{4}{5} & \frac{4}{5} \end{bmatrix} \quad \text{and} \quad B = \begin{bmatrix} \frac{3}{5} \\ \frac{3}{5} \end{bmatrix}.$$

Thus we get

$$\min_{i \in \{1,2\}} G\left(\mathbf{w}_i\right) = \left(\frac{4\sqrt{2}}{5}, \frac{4\sqrt{2}}{5} \right).$$

Hence we get $\mathbf{w} = \left(\frac{4}{5}, \frac{4}{5} \right)$ that minimize $G\left(\mathbf{w}_i\right)$ for $i = 1, 2$.
For the positive points of the combined data for **situation 3**, we have

$$\begin{bmatrix} w_{11} & w_{12} \\ w_{21} & w_{22} \end{bmatrix} \begin{bmatrix} 1/2 \\ 0 \end{bmatrix} + \begin{bmatrix} b_1 \\ b_2 \end{bmatrix} = \begin{bmatrix} 1 \\ 1 \end{bmatrix},$$

and

$$\begin{bmatrix} w_{11} & w_{12} \\ w_{21} & w_{22} \end{bmatrix} \begin{bmatrix} 0 \\ 1/2 \end{bmatrix} + \begin{bmatrix} b_1 \\ b_2 \end{bmatrix} = \begin{bmatrix} 1 \\ 1 \end{bmatrix},$$

which gives

$$\begin{bmatrix} \frac{1}{2}w_{11} \\ \frac{1}{2}w_{21} \end{bmatrix} + \begin{bmatrix} b_1 \\ b_2 \end{bmatrix} = \begin{bmatrix} 1 \\ 1 \end{bmatrix} \quad \text{and} \quad \begin{bmatrix} \frac{1}{2}w_{12} \\ \frac{1}{2}w_{22} \end{bmatrix} + \begin{bmatrix} b_1 \\ b_2 \end{bmatrix} = \begin{bmatrix} 1 \\ 1 \end{bmatrix}. \tag{18}$$

For two negative points for this case, we have

$$\begin{bmatrix} w_{11} & w_{12} \\ w_{21} & w_{22} \end{bmatrix} \begin{bmatrix} -\frac{1}{2} \\ 0 \end{bmatrix} + \begin{bmatrix} b_1 \\ b_2 \end{bmatrix} = \begin{bmatrix} -1 \\ -1 \end{bmatrix},$$

and

$$\begin{bmatrix} w_{11} & w_{12} \\ w_{21} & w_{22} \end{bmatrix} \begin{bmatrix} 0 \\ -\frac{1}{2} \end{bmatrix} + \begin{bmatrix} b_1 \\ b_2 \end{bmatrix} = \begin{bmatrix} -1 \\ -1 \end{bmatrix},$$

which gives

$$\begin{bmatrix} -\frac{1}{2}w_{11} \\ -\frac{1}{2}w_{21} \end{bmatrix} + \begin{bmatrix} b_1 \\ b_2 \end{bmatrix} = \begin{bmatrix} -1 \\ -1 \end{bmatrix} \quad \text{and} \quad \begin{bmatrix} -\frac{1}{2}w_{12} \\ -\frac{1}{2}w_{22} \end{bmatrix} + \begin{bmatrix} b_1 \\ b_2 \end{bmatrix} = \begin{bmatrix} -1 \\ -1 \end{bmatrix}. \tag{19}$$

From this, we obtain that

$$W = \begin{bmatrix} 2 & 2 \\ 2 & 2 \end{bmatrix} \quad \text{and} \quad B = \begin{bmatrix} 0 \\ 0 \end{bmatrix}.$$

Thus we get

$$\min_{i \in \{1,2\}} G(\mathbf{w}_i) = (2\sqrt{2}, 2\sqrt{2}).$$

Hence we get $\mathbf{w} = (2, 2)$ that minimizes $G(\mathbf{w}_i)$ for $i = 1, 2$.

Proposition 1 *Let $G : \mathbb{R}^n \to \mathbb{R}_+^n$ be a differentiable operator. An element $\mathbf{w}^* \in \mathbb{R}^n$ minimize G if and only if $G'(\mathbf{w}^*) = 0$, that is, $\mathbf{w}^* \in \mathbb{R}^n$ solves GSVM if and only if $G'(\mathbf{w}^*) = 0$.*

Proof Let $G'(\mathbf{w}^*) = 0$, then for all $\mathbf{v} \in \mathbb{R}^n$,

$$< \eta G'(\mathbf{w}^*), \mathbf{v} - \mathbf{w}^* > \ = \ < 0, \mathbf{v} - \mathbf{w}^* > \ = \ 0.$$

Consequently, the inequality

$$< \eta G'(\mathbf{w}^*), \mathbf{v} - \mathbf{w}^* > \ = \ < 0, \mathbf{v} - \mathbf{w}^* > \ \geq 0$$

holds for all $\mathbf{v} \in \mathbb{R}^n$. Hence $\mathbf{w}^* \in \mathbb{R}^n$ solves the problem of GSVM.

Conversely, assume that $\mathbf{w}^* \in \mathbb{R}^n$ satisfies

$$< \eta G'(\mathbf{w}^*), \mathbf{v} - \mathbf{w}^* > \ \geq 0 \ \forall \ \mathbf{v} \in \mathbb{R}^n.$$

Taking $\mathbf{v} = \mathbf{w}^* - G'(\mathbf{w}^*)$ in the above inequality implies that

$$< \eta G'(\mathbf{w}^*), -G'(\mathbf{w}^*) > \ \geq \ 0,$$

which further implies

$$-\eta \|G'(\mathbf{w}^*)\|^2 \ \geq \ 0.$$

Since $\eta > 0$, we get $G'(\mathbf{w}^*) = 0$. $\qquad\square$

Definition 3 Let K be a closed and convex subset of $\mathbb{R}^n$. Then, for every point $\mathbf{x} \in \mathbb{R}^n$, there exists a unique nearest point in K, denoted by $P_K(\mathbf{x})$, such that $\|\mathbf{x} - P_K(\mathbf{x})\| \leq \|\mathbf{x} - \mathbf{y}\|$ for all $\mathbf{y} \in K$ and also note that $P_K(\mathbf{x}) = \mathbf{x}$ if $\mathbf{x} \in K$. The mapping P_K is called the metric projection of $\mathbb{R}^n$ onto K. It is well known that $P_K : \mathbb{R}^n \to K$ is characterized by the properties:

(i) $P_K(\mathbf{x}) = \mathbf{z}$ for $\mathbf{x} \in \mathbb{R}^n$ if and only if $< \mathbf{z}, \mathbf{y} - \mathbf{z} > \ \geq \ < \mathbf{x}, \mathbf{y} - \mathbf{z} >$ for all $\mathbf{y} \in \mathbb{R}^n$;

(ii) For every $\mathbf{x}, \mathbf{y} \in \mathbb{R}^n$, $\|P_K(\mathbf{x}) - P_K(\mathbf{y})\|^2 \leq \ < \mathbf{x} - \mathbf{y}, P_K(\mathbf{x}) - P_K(\mathbf{y}) >$;

(iii) $\|P_K(\mathbf{x}) - P_K(\mathbf{y})\| \leq \|\mathbf{x} - \mathbf{y}\|$, for every $\mathbf{x}, \mathbf{y} \in \mathbb{R}^n$, that is, P_K is a nonexpansive map.

Proposition 2 *Let $G : \mathbb{R}^n \to \mathbb{R}^n_+$ be a differentiable operator. An element $\mathbf{w}^* \in \mathbb{R}^n$ minimizes mapping G defined in (3) if and only if $\mathbf{w}^*$ is the fixed point of map*

$$P_{\mathbb{R}^n_+}(I - \rho G') : \mathbb{R}^n \to \mathbb{R}^n_+ \ \text{ for any } \ \rho > 0.$$

that is,

$$\begin{aligned}\mathbf{w}^* &= P_{\mathbb{R}^n_+}(I - \rho G')(\mathbf{w}^*) \\ &= P_{\mathbb{R}^n_+}(\mathbf{w}^* - \rho G'(\mathbf{w}^*)),\end{aligned}$$

where $P_{\mathbb{R}^n_+}$ is a projection map from $\mathbb{R}^n$ to $\mathbb{R}^n_+$.

Proof Suppose $\mathbf{w}^* \in \mathbb{R}^n_+$ is solution of $GSVM$. Then for $\eta > 0$, we have

$$< \eta G'(\mathbf{w}^*), \mathbf{w} - \mathbf{w}^* > \ \geq \ 0 \ \text{ for all } \ \mathbf{w} \in \mathbb{R}^n.$$

Adding $< \mathbf{w}^*, \mathbf{w} - \mathbf{w}^* >$ on both sides, we get

$$< \mathbf{w}^*, \mathbf{w} - \mathbf{w}^* > \ + < \eta G'(\mathbf{w}^*), \mathbf{w} - \mathbf{w}^* > \ \geq \ < \mathbf{w}^*, \mathbf{w} - \mathbf{w}^* > \ \text{ for all } \ \mathbf{w} \in \mathbb{R}^n,$$

which further implies that

$$< \mathbf{w}^*, \mathbf{w} - \mathbf{w}^* > \ \geq \ < \mathbf{w}^* - \eta G'(\mathbf{w}^*), \mathbf{w} - \mathbf{w}^* > \ \text{ for all } \ \mathbf{w} \in \mathbb{R}^n,$$

which is possible only if $\mathbf{w}^* = P_{\mathbb{R}^n_+}(\mathbf{w}^* - \rho G'(\mathbf{w}^*))$, that is, $\mathbf{w}^*$ is the fixed point of G'.

Conversely, let $\mathbf{w}^* = P_{\mathbb{R}^n_+}(\mathbf{w}^* - \rho G'(\mathbf{w}^*))$. Then we have

$$< \mathbf{w}^*, \mathbf{w} - \mathbf{w}^* > \ \geq \ < \mathbf{w}^* - \eta G'(\mathbf{w}^*), \mathbf{w} - \mathbf{w}^* > \ \text{ for all } \ \mathbf{w} \in \mathbb{R}^n,$$

which implies

$$< \mathbf{w}^*, \mathbf{w} - \mathbf{w}^* > - < \mathbf{w}^* - \eta G'(\mathbf{w}^*), \mathbf{w} - \mathbf{w}^* > \ \geq \ 0 \ \text{ for all } \ \mathbf{w} \in \mathbb{R}^n,$$

and so

$$< \eta G'(\mathbf{w}^*), \mathbf{w} - \mathbf{w}^* > \ \geq \ 0 \ \text{ for all } \ \mathbf{w} \in \mathbb{R}^n.$$

Thus $\mathbf{w}^* \in \mathbb{R}^n_+$ is the solution of GSVM. $\qquad\square$

Definition 4 A map $G : \mathbb{R}^n \to \mathbb{R}^n$ is said to be
(I) L-Lipschitz if for every $L > 0$,

$$\|G(\mathbf{x}) - G(\mathbf{y})\| \leq L \|\mathbf{x} - \mathbf{y}\| \ \text{ for all } \ \mathbf{x}, \mathbf{y} \in \mathbb{R}^n.$$

(II) monotone if

$$< G\left(\mathbf{x}\right) - G\left(\mathbf{y}\right), \mathbf{x} - \mathbf{y} > \ \geq \ 0 \quad \text{for all} \quad \mathbf{x}, \mathbf{y} \in \mathbb{R}^n.$$

(III) strictly monotone if

$$< G\left(\mathbf{x}\right) - G\left(\mathbf{y}\right), \mathbf{x} - \mathbf{y} > \ > \ 0 \quad \text{for all} \quad \mathbf{x}, \mathbf{y} \in \mathbb{R}^n \quad \text{with} \quad \mathbf{x} \neq \mathbf{y}.$$

(IV) α-strongly monotone if

$$< G\left(\mathbf{x}\right) - G\left(\mathbf{y}\right), \mathbf{x} - \mathbf{y} > \ \geq \ \alpha \left\|\mathbf{x} - \mathbf{y}\right\|^2 \quad \text{for all} \quad \mathbf{x}, \mathbf{y} \in \mathbb{R}^n.$$

Note that, every α-strongly monotone map $G : \mathbb{R}^n \to \mathbb{R}^n$ is strictly monotone and every strictly monotone map is monotone.

Example 3 Let $G : \mathbb{R}^n \to \mathbb{R}^n$ be a mapping defined as

$$G\left(\mathbf{w}_i\right) = \alpha \mathbf{w}_i + \beta,$$

where α is any non negative scalar and β is any real number. Then G is Lipschitz continuous with Lipschitz constant $L = \alpha$.

Also, for any $\mathbf{x}, \mathbf{y} \in \mathbb{R}^n$,

$$< G\left(\mathbf{x}\right) - G\left(\mathbf{y}\right), \mathbf{x} - \mathbf{y} > \ = \ \alpha \left\|\mathbf{x} - \mathbf{y}\right\|^2,$$

which shows that G is α-strongly monotone.

Theorem 1 *Let $K \subseteq \mathbb{R}^n$ be closed and convex and $G' : \mathbb{R}^n \to K$ is strictly monotone. If there exists a $\mathbf{w}^* \in K$ which is the solution of $GSVM$, then $\mathbf{w}^*$ is unique in K.*

Proof Suppose that $\mathbf{w}_1^*, \mathbf{w}_2^* \in K$ with $\mathbf{w}_1^* \neq \mathbf{w}_2^*$ be the two solutions of $GSVM$. Then we have

$$< \eta G'\left(\mathbf{w}_1^*\right), \mathbf{w} - \mathbf{w}_1^* > \ \geq \ 0 \quad \text{for all} \quad \mathbf{w} \in \mathbb{R}^n, \tag{20}$$

and

$$< \eta G'\left(\mathbf{w}_2^*\right), \mathbf{w} - \mathbf{w}_2^* > \ \geq \ 0 \quad \text{for all} \quad \mathbf{w} \in \mathbb{R}^n, \tag{21}$$

where $\eta > 0$. Putting $\mathbf{w} = \mathbf{w}_2^*$ in (20) and $\mathbf{w} = \mathbf{w}_1^*$ in (21), we get

$$< \eta G'\left(\mathbf{w}_1^*\right), \mathbf{w}_2^* - \mathbf{w}_1^* > \ \geq \ 0, \tag{22}$$

and

$$< \eta G'\left(\mathbf{w}_2^*\right), \mathbf{w}_1^* - \mathbf{w}_2^* > \ \geq \ 0. \tag{23}$$

Equation (22) can be further written as

$$< -\eta G'\left(\mathbf{w}_1^*\right), \mathbf{w}_1^* - \mathbf{w}_2^* > \geq 0. \tag{24}$$

Adding (23) and (24) implies that

$$< \eta G'\left(\mathbf{w}_2^*\right) - \eta G'\left(\mathbf{w}_1^*\right), \mathbf{w}_1^* - \mathbf{w}_2^* > \geq 0,$$

which implies

$$\eta < G'\left(\mathbf{w}_1^*\right) - G'\left(\mathbf{w}_2^*\right), \mathbf{w}_1^* - \mathbf{w}_2^* > \leq 0,$$

or equivalently,

$$< G'\left(\mathbf{w}_1^*\right) - G'\left(\mathbf{w}_2^*\right), \mathbf{w}_1^* - \mathbf{w}_2^* > \leq 0. \tag{25}$$

Since G' is strictly monotone, we must have

$$< G'\left(\mathbf{w}_1^*\right) - G'\left(\mathbf{w}_2^*\right), \mathbf{w}_1^* - \mathbf{w}_2^* > > 0, \tag{26}$$

which contradicts (25). Thus $\mathbf{w}_1^* = \mathbf{w}_2^*$. $\qquad\square$

Theorem 2 *Let $K \subseteq \mathbb{R}^n$ be closed and convex. If the map $G' : \mathbb{R}^n \to K$ is L-Lipchitz and α-strongly monotone, then there exists a unique $\mathbf{w}^* \in K$ which is the solution of $GSVM$.*

Proof **Uniqueness**: Suppose that $\mathbf{w}_1^*, \mathbf{w}_2^* \in K$ be the two solutions of $GSVM$, then for $\eta > 0$, we have

$$< \eta G'\left(\mathbf{w}_1^*\right), \mathbf{w} - \mathbf{w}_1^* > \geq 0 \quad \text{for all} \quad \mathbf{w} \in \mathbb{R}^n, \tag{27}$$

and

$$< \eta G'\left(\mathbf{w}_2^*\right), \mathbf{w} - \mathbf{w}_2^* > \geq 0 \quad \text{for all} \quad \mathbf{w} \in \mathbb{R}^n. \tag{28}$$

Putting $\mathbf{w} = \mathbf{w}_2^*$ in (27) and $\mathbf{w} = \mathbf{w}_1^*$ in (28), we get

$$< \eta G'\left(\mathbf{w}_1^*\right), \mathbf{w}_2^* - \mathbf{w}_1^* > \geq 0 \tag{29}$$

and

$$< \eta G'\left(\mathbf{w}_2^*\right), \mathbf{w}_1^* - \mathbf{w}_2^* > \geq 0. \tag{30}$$

Equation (29) can be further written as

$$< -\eta G'\left(\mathbf{w}_1^*\right), \mathbf{w}_1^* - \mathbf{w}_2^* > \geq 0. \tag{31}$$

Adding (30) and (31) implies that

$$< \eta G'\left(\mathbf{w}_2^*\right) - \eta G'\left(\mathbf{w}_1^*\right), \mathbf{w}_1^* - \mathbf{w}_2^* > \geq 0, \tag{32}$$

which implies

$$\eta < G'\left(\mathbf{w}_1^*\right) - G'\left(\mathbf{w}_2^*\right), \mathbf{w}_1^* - \mathbf{w}_2^* > \ \leq \ 0. \tag{33}$$

Since G' is α-strongly monotone, we have

$$\alpha\eta \left\| \mathbf{w}_1^* - \mathbf{w}_2^* \right\|^2 \leq \eta < G'\left(\mathbf{w}_1^*\right) - G'\left(\mathbf{w}_2^*\right), \mathbf{w}_1^* - \mathbf{w}_2^* >$$
$$\leq \ 0,$$

which implies that

$$\alpha\eta \left\| \mathbf{w}_1^* - \mathbf{w}_2^* \right\|^2 \ \leq \ 0.$$

As $\alpha\eta > 0$, we conclude $\left\| \mathbf{w}_1^* - \mathbf{w}_2^* \right\| = 0$ and hence $\mathbf{w}_1^* = \mathbf{w}_2^*$.

Existence: As we know, if $\mathbf{w}^* \in \mathbb{R}_+^n$ is solution of $GSVM$, then for $\eta > 0$ we have

$$< \eta G'\left(\mathbf{w}^*\right), \mathbf{w} - \mathbf{w}^* > \ \geq \ 0 \quad \text{for all} \quad \mathbf{w} \in \mathbb{R}^n,$$

if and only if

$$\mathbf{w}^* = P_{\mathbb{R}_+^n}\left(\mathbf{w}^* - \rho G'(\mathbf{w}^*)\right) \tag{34}$$
$$\equiv F\left(\mathbf{w}^*\right) \quad (say).$$

Now for any $\mathbf{w}_1^*, \mathbf{w}_2^* \in \mathbb{R}_+^n$, we have

$$\left\| F\left(\mathbf{w}_1^*\right) - F\left(\mathbf{w}_2^*\right) \right\|^2 \tag{35}$$
$$= \left\| P_{\mathbb{R}_+^n}(\mathbf{w}_1^* - \rho G'(\mathbf{w}_1^*)) - P_{\mathbb{R}_+^n}(\mathbf{w}_2^* - \rho G'(\mathbf{w}_2^*)) \right\|^2$$
$$\leq \left\| (\mathbf{w}_1^* - \rho G'(\mathbf{w}_1^*)) - (\mathbf{w}_2^* - \rho G'(\mathbf{w}_2^*)) \right\|^2$$
$$= \left\| (\mathbf{w}_1^* - \mathbf{w}_2^*) - \rho [G'(\mathbf{w}_1^*) - G'(\mathbf{w}_2^*)] \right\|^2$$
$$= < (\mathbf{w}_1^* - \mathbf{w}_2^*) - \rho [G'(\mathbf{w}_1^*) - G'(\mathbf{w}_2^*)], (\mathbf{w}_1^* - \mathbf{w}_2^*) - \rho [G'(\mathbf{w}_1^*) - G'(\mathbf{w}_2^*)] >$$
$$= \left\| \mathbf{w}_1^* - \mathbf{w}_2^* \right\|^2 - 2\rho < \mathbf{w}_1^* - \mathbf{w}_2^*, G'(\mathbf{w}_1^*) - G'(\mathbf{w}_2^*) >$$
$$+ \rho^2 \left\| G'(\mathbf{w}_1^*) - G'(\mathbf{w}_2^*) \right\|^2.$$

Since G' is L-Lipchitz and α-strongly monotone, we get

$$\left\| F\left(\mathbf{w}_1^*\right) - F\left(\mathbf{w}_2^*\right) \right\|^2 \leq \left\| \mathbf{w}_1^* - \mathbf{w}_2^* \right\|^2 - 2\alpha\rho \left\| \mathbf{w}_1^* - \mathbf{w}_2^* \right\|^2$$
$$+ \rho^2 L^2 \left\| \mathbf{w}_1^* - \mathbf{w}_2^* \right\|^2$$
$$= (1 + \rho^2 L^2 - 2\rho\alpha) \left\| \mathbf{w}_1^* - \mathbf{w}_2^* \right\|^2,$$

that is

$$\left\| F\left(\mathbf{w}_1^*\right) - F\left(\mathbf{w}_2^*\right) \right\| \le \theta \left\| \mathbf{w}_1^* - \mathbf{w}_2^* \right\|, \tag{36}$$

where $\theta = \sqrt{1 + \rho^2 L^2 - 2\rho\alpha}$. Since $\rho > 0$, so that when $\rho \in (0, \frac{2\alpha}{L^2})$, then we get $\theta \in [0, 1)$. Now, by using Banach contraction principle,, we obtain the fixed point of map F, that is, there exists a unique $\mathbf{w}^* \in \mathbb{R}_+^n$ such that

$$F\left(\mathbf{w}^*\right) = P_{\mathbb{R}_+^n}(\mathbf{w}^* - \rho G'(\mathbf{w}^*))$$
$$= \mathbf{w}^*.$$

Hence $\mathbf{w}^* \in \mathbb{R}_+^n$ is the solution of GSVM. $\square$

Example 4 Let us take the group of data of positive class $(\alpha_1, \alpha_2, \ldots, \alpha_{n-1}, 0)$, $(\alpha_1, \alpha_2, \ldots, \alpha_{n-2}, 0, \alpha_n), \ldots, (0, \alpha_2, \alpha_3, \ldots, \alpha_n)$ and negative class $(k\alpha_1, k\alpha_2, \ldots, k\alpha_{n-1}, 0)$, $(k\alpha_1, k\alpha_2, \ldots, k\alpha_{n-2}, 0, k\alpha_n), \ldots, (0, k\alpha_2, k\alpha_3, \ldots, k\alpha_n)$ for $n \ge 2$, where each $\alpha_i \ne 0$ for $i \in \mathbb{R}$ and $k \ne 1$.

A map $G : \mathbb{R}^n \to \mathbb{R}_+^n$ is given by

$$G\left(\mathbf{w}_i\right) = (\|\mathbf{w}_i\|, \|\mathbf{w}_i\|, \ldots, \|\mathbf{w}_i\|) \quad \text{for} \quad i = 1, 2, \ldots, n,$$

where $\mathbf{w}_i$ are the rows of $W_{n \times n}$ for $i = 1, 2, \ldots, n$. Then we have

$$G'\left(\mathbf{w}_i\right) = \frac{1}{\|\mathbf{w}_i\|} \mathbf{w}_i \quad \text{for} \quad i = 1, 2, \ldots, n.$$

Now from the given data, we get

$$W = \frac{2}{(n-1)(1-k)} \begin{bmatrix} \frac{1}{\alpha_1} & \frac{1}{\alpha_2} & \cdots & \frac{1}{\alpha_n} \\ \frac{1}{\alpha_1} & \frac{1}{\alpha_2} & \cdots & \frac{1}{\alpha_n} \\ \cdot & \cdot & \cdot & \cdot \\ \cdot & \cdot & \cdot & \cdot \\ \cdot & \cdot & \cdot & \cdot \\ \frac{1}{\alpha_1} & \frac{1}{\alpha_2} & \cdots & \frac{1}{\alpha_n} \end{bmatrix}$$

and so we have

$$G\left(\mathbf{w}_i\right) = \frac{2}{(n-1)(1-k)} \sqrt{\frac{1}{\alpha_1^2} + \frac{1}{\alpha_2^2} + \cdots + \frac{1}{\alpha_n^2}}(1, 1, \ldots, 1) \quad \text{for} \quad i = 1, 2, \ldots, n,$$

and

$$G'\left(\mathbf{w}_i\right) = \frac{1}{\sqrt{\frac{1}{\alpha_1^2} + \frac{1}{\alpha_2^2} + \cdots + \frac{1}{\alpha_n^2}}} \left(\frac{1}{\alpha_1}, \frac{1}{\alpha_2}, \ldots, \frac{1}{\alpha_n}\right) \quad \text{for} \quad i = 1, 2, \ldots, n.$$

Note that, for any $\mathbf{w}_1, \mathbf{w}_2 \in W$,

$$\left\| G'\left(\mathbf{w}_1\right) - G'\left(\mathbf{w}_2\right) \right\| = 0 = L\left\|\mathbf{w}_1 - \mathbf{w}_2\right\|$$

is satisfied, where L is any nonnegative real number. Also

$$< G'\left(\mathbf{w}_1\right) - G'\left(\mathbf{w}_2\right), \mathbf{w}_1 - \mathbf{w}_2 > \ \geq \ 0,$$

which shows that G' is monotone operator. Moreover, $\mathbf{w} = \frac{2}{(n-1)(1-k)}\left(\frac{1}{\alpha_1}, \frac{1}{\alpha_2}, \ldots, \frac{1}{\alpha_n}\right)$ is the solution of GSVM with $\|\mathbf{w}\| = \frac{2}{(n-1)(1-k)}\sqrt{\frac{1}{\alpha_1^2} + \frac{1}{\alpha_2^2} + \cdots + \frac{1}{\alpha_n^2}}$.

Example 5 Let us take the group of data of positive class $(\alpha_1, \alpha_2, \ldots, \alpha_m, 0, 0, \ldots,$ $0)$, $(0, \alpha_2, \alpha_3, \ldots, \alpha_{m+1}, 0, 0, \ldots, 0)$, $\ldots$, $(\alpha_1, \alpha_2, \ldots, \alpha_{m-1}, 0, 0, \ldots, 0, \alpha_n)$ and negative class $(\kappa\alpha_1, \kappa\alpha_2, \ldots, \kappa\alpha_m, 0, 0, \ldots, 0)$, $(0, \kappa\alpha_2, \kappa\alpha_3, \ldots, \kappa\alpha_{m+1}, 0,$ $0, \ldots, 0)$, $\ldots$, $(\kappa\alpha_1, \kappa\alpha_2, \ldots, \kappa\alpha_{m-1}, 0, 0, \ldots, 0, \kappa\alpha_n)$ for $n > m \geq 1$, where each $\alpha_i \neq 0$ for $i \in \mathbb{N}$ and $\kappa \neq 1$.

A map $G : \mathbb{R}^n \to \mathbb{R}_+^n$ is given by

$$G\left(\mathbf{w}_i\right) = \left(\|\mathbf{w}_i\|, \|\mathbf{w}_i\|, \ldots, \|\mathbf{w}_i\|\right) \quad \text{for} \quad i = 1, 2, \ldots, n,$$

where $\mathbf{w}_i$ are the rows of $W_{n \times n}$ for $i = 1, 2, \ldots, n$. Then we have

$$G'\left(\mathbf{w}_i\right) = \frac{1}{\|\mathbf{w}_i\|}\mathbf{w}_i \quad \text{for} \quad i = 1, 2, \ldots, n.$$

Now from the given data, we get

$$W = \frac{2}{m\left(1-k\right)}\begin{bmatrix} \frac{1}{\alpha_1} & \frac{1}{\alpha_2} & \cdots & \frac{1}{\alpha_n} \\ \frac{1}{\alpha_1} & \frac{1}{\alpha_2} & \cdots & \frac{1}{\alpha_n} \\ \cdot & \cdot & & \cdot \\ \cdot & \cdot & & \cdot \\ \cdot & \cdot & & \cdot \\ \frac{1}{\alpha_1} & \frac{1}{\alpha_2} & \cdots & \frac{1}{\alpha_n} \end{bmatrix}$$

and so we have

$$G\left(\mathbf{w}_i\right) = \frac{2}{m\left(1-k\right)}\sqrt{\frac{1}{\alpha_1^2} + \frac{1}{\alpha_2^2} + \cdots + \frac{1}{\alpha_n^2}}\left(1, 1, \ldots, 1\right) \quad \text{for} \quad i = 1, 2, \ldots, n,$$

and

$$G'\left(\mathbf{w}_i\right) = \frac{1}{\sqrt{\frac{1}{\alpha_1^2} + \frac{1}{\alpha_2^2} + \cdots + \frac{1}{\alpha_n^2}}}\left(\frac{1}{\alpha_1}, \frac{1}{\alpha_2}, \ldots, \frac{1}{\alpha_n}\right) \quad \text{for} \quad i = 1, 2, \ldots, n.$$

It is easy to verify that G' is monotone and Lipchitz continuous operator. The vector $\mathbf{w} = \frac{2}{m(1-k)}\left(\frac{1}{\alpha_1}, \frac{1}{\alpha_2}, \ldots, \frac{1}{\alpha_n}\right)$ solves GSVM and $\|\mathbf{w}\| = \frac{2}{m(1-k)}\sqrt{\frac{1}{\alpha_1^2} + \frac{1}{\alpha_2^2} + \cdots + \frac{1}{\alpha_n^2}}$.

Example 6 Consider $(\alpha_1, 0, 0)$, $(0, \alpha_2, 0)$, $(0, 0, \alpha_3)$, $(\beta_1, 0, 0)$, $(0, \beta_2, 0)$, $(0, 0, \beta_3)$ as data of positive class and $(k\alpha_1, 0, 0)$, $(0, k\alpha_2, 0)$, $(0, 0, k\alpha_3)$, $(k\beta_1, 0, 0)$, $(0, k\beta_2, 0)$, $(0, 0, k\beta_3)$ as negative class of data, where α_i, β_i and k are positive real numbers with each $\alpha_i \leq \beta_i$ for $i = 1, 2, 3$ and $k \neq 1$.

The map $G : \mathbb{R}^n \to \mathbb{R}^n_+$ is given as

$$G(\mathbf{w}_i) = (\|\mathbf{w}_i\|, \|\mathbf{w}_i\|, \ldots, \|\mathbf{w}_i\|) \quad \text{for} \quad i = 1, 2, \ldots, n,$$

where $\mathbf{w}_i$ are the rows of $W_{3 \times 3}$ for $i = 1, 2, 3$. Then we have

$$G'(\mathbf{w}_i) = \frac{1}{\|\mathbf{w}_i\|} \mathbf{w}_i \quad \text{for} \quad i = 1, 2, 3.$$

Now from the given data, we get

$$W = \frac{2}{(1-k)} \begin{bmatrix} \frac{1}{\alpha_1} & \frac{1}{\alpha_2} & \frac{1}{\alpha_3} \\ \frac{1}{\alpha_1} & \frac{1}{\alpha 2} & \frac{1}{\alpha_3} \\ \frac{1}{\alpha_1} & \frac{1}{\alpha_2} & \frac{1}{\alpha_3} \end{bmatrix}$$

and so we have

$$G(\mathbf{w}_i) = \frac{2}{(1-k)} \left(\sqrt{\frac{1}{\alpha_1^2} + \frac{1}{\alpha_2^2} + \frac{1}{\alpha_3^2}}, \sqrt{\frac{1}{\alpha_1^2} + \frac{1}{\alpha_2^2} + \frac{1}{\alpha_3^2}}, \sqrt{\frac{1}{\alpha_1^2} + \frac{1}{\alpha_2^2} + \frac{1}{\alpha_3^2}} \right),$$

and

$$G'(\mathbf{w}_i) = \frac{1}{\sqrt{\frac{1}{\alpha_1^2} + \frac{1}{\alpha_2^2} + \frac{1}{\alpha_3^2}}} \left(\frac{1}{\alpha_1}, \frac{1}{\alpha_2}, \frac{1}{\alpha_3} \right).$$

Note that, for any $\mathbf{w}_1, \mathbf{w}_2 \in W$,

$$\left\| G'(\mathbf{w}_1) - G'(\mathbf{w}_2) \right\| = 0 = L \|\mathbf{w}_1 - \mathbf{w}_2\|$$

is satisfied for $L > 0$. Also

$$< G'(\mathbf{w}_1) - G'(\mathbf{w}_2), \mathbf{w}_1 - \mathbf{w}_2 > \geq 0$$

is satisfied, which shows that G' is monotone operator. Moreover, $\mathbf{w} = \frac{2}{(1-k)}$ $(\frac{1}{\alpha_1}, \frac{1}{\alpha_2}, \frac{1}{\alpha_3})$ is the solution of GSVM with $\|\mathbf{w}\| = \frac{2}{(1-k)} \sqrt{\frac{1}{\alpha_1^2} + \frac{1}{\alpha_2^2} + \frac{1}{\alpha_3^2}}$.

Example 7 Let us take the group of data of positive class $(1, 0, 0)$, $(1, 1, 0)$, $(0, 1, 1)$ and negative class $(-\frac{1}{2}, 0, 0)$, $(-\frac{1}{2}, -\frac{1}{2}, 0)$, $(0, -\frac{1}{2}, -\frac{1}{2})$. Now from the given data, we have

$$W = \begin{bmatrix} \frac{4}{3} & 0 & \frac{4}{3} \\ \frac{4}{3} & 0 & \frac{4}{3} \\ \frac{4}{3} & 0 & \frac{4}{3} \end{bmatrix}$$

with

$$G(\mathbf{w}_i) = \frac{4}{3}(\sqrt{2}, \sqrt{2}, \sqrt{2}) \quad \text{for} \quad i = 1, 2, 3,$$

and

$$G'(\mathbf{w}_i) = \frac{1}{\sqrt{2}}(1, 0, 1) \quad \text{for} \quad i = 1, 2, 3.$$

It is easy to verify that G' is monotone operator and Lipchitz continuous. Moreover, $\mathbf{w} = (\frac{4}{3}, 0, \frac{4}{3})$ is the solution of GSVM with $\|\mathbf{w}\| = \frac{4}{3}\sqrt{2}$.

4 Conclusion

Recently many results appeared in the literature addressing the problems related to the support vector machine and its applications. In this paper, we initiated the study of generalized support vector machine and presented linear classification of data by using support vector machine and generalized support vector machine. We also provided sufficient conditions under which the solution of generalized support vector machine exists. Various examples are also presented to show the validity of these results. Furthermore, one can study the results of generalized support vector machine for nonlinear classification of data.

Acknowledgements Talat Nazir and Xiaomin Qi are grateful to the Erasmus Mundus project FUSION for supporting the research visit to Mälardalen University, Sweden, and to the Research environment MAM in Mathematics and Applied Mathematics, Division of Applied Mathematics, the School of Education, Culture and Communication of Mälardalen University for creating excellent research environment.

References

1. Adankon, M.M., Cheriet, M.: Model selection for the LS-SVM. Application to handwriting recognition. Pattern Recognit. **42**(12), 3264–3270 (2009)
2. Cortes, C., Vapnik, V.N.: Support-vector networks. Mach. Learn. **20**(3), 273–297 (1995)
3. Cristianini, N., Shawe-Taylor, J.: An Introduction to Support Vector Machines and other Kernel Based Learning Methods. Cambridge University Press, Cambridge (2000)
4. Guyon, I., Weston, J., Barnhill, S., Vapnik, V.: Gene selection for cancer classification using support vector machines. Mach. Learn. **46**(1–3), 389–422 (2002)
5. Joachims, T.: Text categorization with support vector machines: learning with many relevant features. In: Proceedings of the European Conference on Machine Learning. Springer (1998)

6. Khan, N., Ksantini, R., Ahmad, I., Boufama, B.: A novel SVM+NDA model for classification with an application to face recognition. Pattern Recognit. **45**(1), 66–79 (2012)
7. Li, S., Kwok, J.T., Zhu, H., Wang, Y.: Texture classification using the support vector machines. Pattern Recognit. **36**(12), 2883–2893 (2003)
8. Liu, R., Wang, Y., Baba, T., Masumoto, D., Nagata, S.: SVM-based active feedback in image retrieval using clustering and unlabeled data. Pattern Recognit. **41**(8), 2645–2655 (2008)
9. Michel, P., Kaliouby, R. E.: Real time facial expresion recognition in video using support vector machines. In: Proceedings of ICMI 2003, pp. 258–264 (2003)
10. Noble, W.S.: Support Vector Machine Applications in Computational Biology. MIT Press, Cambridge (2004)
11. Shao, Y., Lunetta, R.S.: Comparison of support vector machine, neural network, and CART algorithms for the land-cover classification using limited training data points. ISPRS J. Photogramm. Remote Sens. **70**, 78–87 (2012)
12. Shao, Y.H., Chen, W.J., Deng, N.Y.: Nonparallel hyperplane support vector machine for binary classification problems. Inf. Sci. **263**, 22–35 (2014)
13. Vapnik, V.N.: The Nature of Statistical Learning Theory. Springer, New York (1996)
14. Vapnik, V.N.: Statistical Learning Theory. Wiley, New York (1998)
15. Wang, X.Y., Wang, T., Bu, J.: Color image segmentation using pixel wise support vector machine classification. Pattern Recognit. **44**(4), 777–787 (2011)
16. Wang, D., Qi, X., Wen, S., Deng., M.: SVM based fault classifier design for a water level control system. In: Proceedings of 2013 International Conference on Advanced Mechatronic Systems, Luoyang, China, pp. 152–157 (2013)
17. Wang, D., Qi, X., Wen, S., Dan, Y., Ouyang, L., Deng, M.: Robust nonlinear control and SVM classifier based fault diagnosis for a water level process. ICIC Express Lett. **5**(1), 767–774 (2014)
18. Weston, J., Watkins, C.: Multi-class Support Vector Machines. Technical report CSD-TR-98-04, Department of Computer Science, Royal Holloway, University of London (1998)
19. Wu, Y.C., Lee, Y.-S., Yang, J.-C.: Robust and efficient multiclass SVM models for phrase pattern recognition. Pattern Recognit. **41**(9), 2874–2889 (2008)
20. Xue, Z., Ming, D., Song, W., Wan, B., Jin, S.: Infrared gait recognition based on wavelet transform and support vector machine. Pattern Recognit. **43**(8), 2904–2910 (2010)
21. Zhao, Z., Liu, J., Cox, J.: Safe and efficient screening for sparse support vector machine. In: Proceedings of the 20th ACM SIGKDD International Conference on Knowledge Discovery and Data Mining, KDD, New York, NY, USA, vol. 14, pp. 542–551 (2014)
22. Zuo, R., Carranza, E.J.M.: Support vector machine: a tool for mapping mineral prospectivity. Comput. Geosci. **37**(12), 1967–1975 (2011)

Linear and Nonlinear Classifiers of Data with Support Vector Machines and Generalized Support Vector Machines

Talat Nazir, Xiaomin Qi and Sergei Silvestrov

Abstract The support vector machine for linear and nonlinear classification of data is studied. The notion of generalized support vector machine for data classifications is used. The problem of generalized support vector machine is shown to be equivalent to the problem of generalized variational inequality and various results for the existence of solutions are established. Moreover, examples supporting the results are provided.

Keywords Linear and nonlinear classification · Support vector machine · Generalized support vector machine · Kernel function

1 Support Vector Machine

Support vector machines (SVM) [2, 3, 13, 14, 18] were developed by Vapnik et al. (1995) and are gaining popularity due to many attractive features. As a very powerful tool for data classification and regression, it has been used in many fields, such as text classification [5], facial expression recognition [9], gene analysis [4] and many others [1, 6–8, 10–12, 17, 19–22]. Recently, it has been used for faults classification in a water level control system [15]. And a faults classifier based SVM is used to diagnose the faults for a water level control process [16].

T. Nazir (✉) · X. Qi · S. Silvestrov
Division of Applied Mathematics, School of Education, Culture and Communication,
Mälardalen University, Box 883, 721 23 Västerås, Sweden
e-mail: talat@ciit.net.pk

X. Qi
e-mail: xiaomin.qi@mdh.se

S. Silvestrov
e-mail: sergei.silvestrov@mdh.se

T. Nazir
Department of Mathematics, COMSATS Institute of Information Technology,
Abbottabad 22060, Pakistan

© Springer International Publishing Switzerland 2016
S. Silvestrov and M. Rančić (eds.), *Engineering Mathematics II*,
Springer Proceedings in Mathematics & Statistics 179,
DOI 10.1007/978-3-319-42105-6_18

The classification problems can be restricted to consideration of the two-class problems without loss of generality. The goal of support vector classification (SVC) is to separate the two classes by a hyperplane which can also work well on unseen examples. The method is to find the optimal hyperplane that maximizes the margin between two classes of data. The set of data is said to be optimally separated by the hyperplane if it is separated without error and the distance between the closest data is maximal. Support vector classification can be thought of a process using given data to find the decision plane which can guarantee good predictive performance on unseen data. And the process of finding the decision plane is a quadratic programming process.

In this paper, we study the problems of support vector machine and generalized support vector machine. We also show the sufficient conditions for the existence of solutions for problems of generalized support vector machine. We also present various examples to support these results.

Throughout this paper, by $\mathbb{N}$, $\mathbb{R}$, $\mathbb{R}^n$ and $\mathbb{R}_n^+$ we denote the set of all natural numbers, the set of all real numbers, the set of all n-tuples real numbers, the set of all n-tuples of nonnegative real numbers, respectively.

Also, we consider $\|\cdot\|$ and $<\cdot,\cdot>$ as Euclidean norm and usual inner product on $\mathbb{R}^n$, respectively, such as, $<\mathbf{x}, \mathbf{y}>= \mathbf{x}.\mathbf{y} = x_1 y_1 + x_2 y_2 + \cdots + x_n y_n$ for all $\mathbf{x} = (x_1, x_2, \ldots, x_n)$, $\mathbf{y} = (y_1, y_2, \ldots, y_n)$ in $\mathbb{R}^n$. Furthermore, for any two vectors $\mathbf{x}, \mathbf{y} \in \mathbb{R}^n$, we say that $\mathbf{x} \leq \mathbf{y}$ if and only if $x_i \leq y_i$ for all $i \in \{1, 2, \ldots, n\}$, where x_i and y_i are the components of $\mathbf{x}$ and $\mathbf{y}$, respectively.

1.1 Data Classification

Actually, complex real-world applications are always not linearly separable. Kernel representations offer an alternative solution by projecting the data into a higher dimensional feature space to increase the computational power of the linear learning machine.

In order to learn linear or non-linear relations with a linear machine, a set of non-linear features is selected. This is equivalent to applying a fixed non-linear mapping function Φ that transforms data in input space X to data in feature space F, in which the linear machine can be used. For this classification, both spaces X and F need to be vector spaces, where dimension of these two spaces may or may not be same. When the given data is linearly separable, we consider Φ as identity operator. For binary classification of data, we consider the decision function $f : \mathbb{R}^n \to \mathbb{R}$, where the input $\mathbf{x} = (x_1, \ldots, x_n)$ is assigned to the positive class if, $f(\mathbf{x}) \geq 0$ and otherwise to the negative class. The decision function is defined as

$$f(\mathbf{x}) =< \mathbf{w}, \Phi(\mathbf{x}) > +b. \tag{1}$$

This means two steps will be built for non-linear machine: first a fixed non-linear mapping of the data to a feature space, and then a linear machine is used to classify them in the feature space.

In addition, the vector $\mathbf{w}$ is a linear combination of the support vectors in the training data and can be written as

$$\mathbf{w} = \sum_i \alpha_i \Phi(\mathbf{x}_i), \tag{2}$$

where each α_i is Lagrange multiplier of the support vectors.

So the decision function can be rewritten as

$$f(\mathbf{x}) = \sigma\left(\sum_i \alpha_i(\Phi(\mathbf{x}_i) \cdot \Phi(\mathbf{x})) + b\right), \tag{3}$$

where σ is a sign function.

The Kernel K has an associated feature with mapping Φ, and it takes two inputs and give their similarity in feature space F, that is, $K : F \times F \to \mathbb{R}$ is defined as

$$K(\mathbf{x}_i, \mathbf{x}) = \Phi(\mathbf{x}_i) \cdot \Phi(\mathbf{x}). \tag{4}$$

Thus, the decision function from (3) becomes

$$f(\mathbf{x}) = \sigma(\sum_i \alpha_i K(\mathbf{x}_i, \mathbf{x}) + b). \tag{5}$$

Some useful kernels for real valued vectors are defined below:

(I) Linear kernel
$$K(\mathbf{x}_i, \mathbf{x}) = \mathbf{x}_i \cdot \mathbf{x}.$$

(II) Polynomial kernel (of degree p)

$$K(\mathbf{x}_i, \mathbf{x}) = (\mathbf{x}_i \cdot \mathbf{x})^p \ \text{ or } \ (\mathbf{x}_i \cdot \mathbf{x} + 1)^p,$$

where p is a tunable parameter.
(III) Radial Basis Function (RBF) kernel

$$K(\mathbf{x}_i, \mathbf{x}) = \exp[-\gamma||\mathbf{x}_i - \mathbf{x}||^2],$$

where γ is a hyperparameter (also called kernel bandwidth). The RBF kernel corresponds to an infinite feature space.
(IV) Sigmoid Kernel
$$K(\mathbf{x}_i, \mathbf{x}) = \tanh(k\mathbf{x}_i \cdot \mathbf{x} + \theta),$$

where k is a scalar and θ is the displacement.

(V) Inverse multi-quadratic kernel

$$K(\mathbf{x}_i, \mathbf{x}) = \left(\|\mathbf{x}_i - \mathbf{x}\|^2 + \gamma^{-2} \right)^{-1/2},$$

where γ is a hyperparameter (also called kernel bandwidth).

Now, from (1), we define the functional margin of an example $(\Phi(\mathbf{x}_i), y_i)$ with respect to a hyperplane $(\mathbf{w}, b)$ to be the quantity

$$\gamma_i = y_i \left(\langle \mathbf{w}, \Phi(\mathbf{x}_i) \rangle + b \right),$$

where $y_i \in \{-1, 1\}$. Note that $\gamma_i > 0$ implies correct classification of $(\mathbf{x}_i, y_i)$. If we replace functional margin by geometric margin we obtain the equivalent quantity for the normalized linear function $(\frac{1}{\|\mathbf{w}\|}\mathbf{w}, \frac{1}{\|\mathbf{w}\|}b)$, which therefore measures the Euclidean distances of the points from the decision boundary in the input space.

Actually geometric margin can be written as

$$\tilde{\gamma} = \frac{1}{\|\mathbf{w}\|} \gamma.$$

To find the hyperplane which has maximal geometric margin for a training set S means to find maximal $\tilde{\gamma}$. For convenience, we let $\gamma = 1$, the objective function can be written as

$$\max \frac{1}{\|\mathbf{w}\|}.$$

Of course, there are some constraints for the optimization problem. According to the definition of margin, we have $y_i \left(\langle \mathbf{w}, \Phi(\mathbf{x}_i) \rangle + b \right) \geq 1, i = 1, \ldots, l$. We rewrite in the equivalent form the objective function with the constraints as

$$\min \frac{1}{2} \|\mathbf{w}\|^2 \quad \text{such that} \quad y_i \left(\langle \mathbf{w}, \Phi(\mathbf{x}_i) \rangle + b \right) \geq 1, \ i = 1, \ldots, l. \tag{6}$$

We denote this problem by SVM for data classification.

Example 1 Let's take the group of points $(0, 2), (0, -2), (1, 1), (1, -1),$ $(-1, 1), (-1, -1)$ as positive class and the group of points $(2, 0), (-2, 0), (2, 1),$ $(2, -1), (-2, 1), (-2, -1)$ as negative class shown in Fig. 1.

By using the mapping function

$$\Phi(\mathbf{x}) = \left(x_1^2, \sqrt{2}x_1x_2, x_2^2 \right),$$

which transforms data from two-dimensional input space to three-dimensional feature space, that is $(1, \sqrt{2}, 1), (1, -\sqrt{2}, 1)$ and $(0, 0, 4)$ as positive class and $(4, 2\sqrt{2}, 1), (4, -2\sqrt{2}, 1)$ and $(4, 0, 0)$ as negative data shown in Fig. 2.

Fig. 1 The data points given in Example 1

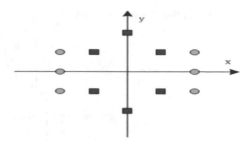

Fig. 2 The data separation in three dimensional feature space

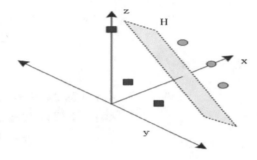

Now by using this data in three dimensional feature space, we consider the following: For positive points, we have

$$(w_1, w_2, w_3) \begin{bmatrix} 1 \\ \sqrt{2} \\ 1 \end{bmatrix} + b \geq 1,$$

$$(w_1, w_2, w_3) \begin{bmatrix} 1 \\ -\sqrt{2} \\ 1 \end{bmatrix} + b \geq 1,$$

$$(w_1, w_2, w_3) \begin{bmatrix} 0 \\ 0 \\ 4 \end{bmatrix} + b \geq 1,$$

which implies

$$w_1 + \sqrt{2}w_2 + w_3 + b \geq 1,$$
$$w_1 - \sqrt{2}w_2 + w_3 + b \geq 1,$$
$$4w_3 + b \geq 1.$$

For negative points, we have

$$(w_1, w_2, w_3) \begin{bmatrix} 4 \\ 2\sqrt{2} \\ 1 \end{bmatrix} + b \le -1,$$

$$(w_1, w_2, w_3) \begin{bmatrix} 4 \\ -2\sqrt{2} \\ 1 \end{bmatrix} + b \le -1,$$

$$(w_1, w_2, w_3) \begin{bmatrix} 4 \\ 0 \\ 0 \end{bmatrix} + b \le -1,$$

implying that

$$4w_1 + 2\sqrt{2}w_2 + w_3 + b \le -1,$$
$$4w_1 - 2\sqrt{2}w_2 + w_3 + b \le -1,$$
$$4w_1 + b \le -1.$$

From the equations, we get $\mathbf{w} = (-0.6667, 0, 0)$ with $\|\mathbf{w}\| = 0.6667$ and shown in Fig. 3.

Further, if we use Radial Basis Function (RBF) Kernel $K(\mathbf{x}_i, \mathbf{x}) = \exp[-\gamma \|\mathbf{x}_i - \mathbf{x}\|^2]$, with $\gamma = 1/3$, we get $w = (0.0031, 0.0012)$ which is shown in Fig. 4.

Also if we use Sigmoid Kernel $K(\mathbf{x}_i, \mathbf{x}) = \tanh(k\mathbf{x}_i \cdot \mathbf{x} + \theta)$ with $k = 1/3$ and $\theta = 2.85$, we get $w = (0, 0)$ shown in Fig. 5.

Example 2 Let us look at another example. The positive data be shown as red square and the negative data be shown as blue circle respectively as shown in Fig. 6.

It is also a non-linear separable problem. Now, if we transfer the original data into the feature space by using the mapping function $\Phi(\mathbf{x})$, we can see that the data in the feature space is linear separable see Fig. 7.

Fig. 3 The data separation using Polynomial Kernel of degree 2

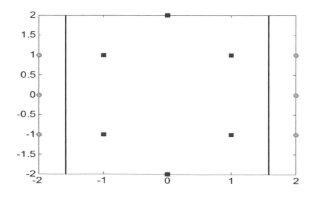

Fig. 4 The data separation using Radial Basis Function (RBF)

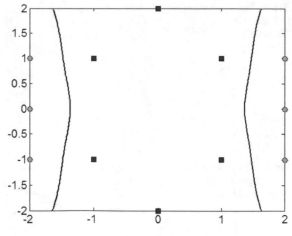

Fig. 5 The data separation using Sigmoid Kernel

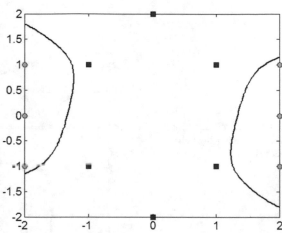

Fig. 6 The data points given in Example 2

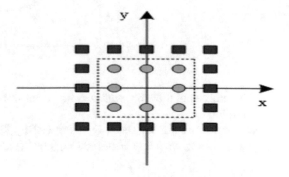

Fig. 7 The data separation in feature space of Example 2

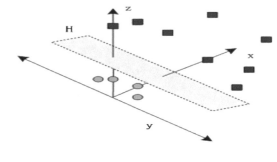

Fig. 8 The data separation of Example 2 using Polynomial Kernel

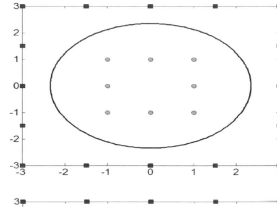

Fig. 9 The data separation of Example 2 using RBF Kernel

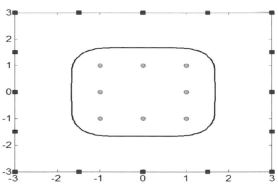

Using Polynomial Kernel with $p = 2$, we get $\mathbf{w} = (-0.4898, -0.1633)$ which is shown in Fig. 8.

Next if we use Radial Basis Function (RBF) Kernel $K(\mathbf{x}_i, \mathbf{x}) = \exp[-\gamma \|\mathbf{x}_i - \mathbf{x}\|^2]$, with $\gamma = 2$, we get $w = (-0.0016, 0.0014)$ as shown in Fig. 9.

Example 3 Consider the points $(0, 0)$, $(1, 0)$, $(-1, 0)$ as positive class and points $(2, 0)$, $(3, 0)$, $(-2, 0)$, $(-3, 0)$ as negative class see in Fig. 10.

Fig. 10 The data points given in Example 3

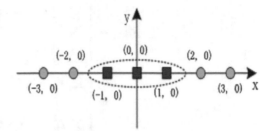

Fig. 11 Data separation of Example 3 by using Polynomial Kernel of degree 2

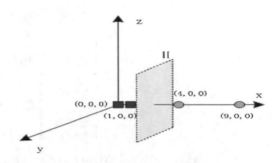

Note that, no linear separator exists for this data in the input space. Now, if we use $\Phi(\mathbf{x}) = (x_1^2, \sqrt{2}x_1x_2, x_2^2)$, then it transforms two-dimensional data into three-dimensional feature space, which can be separated by hyperplane H as shown in the Fig. 11.

2 Generalized Support Vector Machines

Consider a new control function $F : \mathbb{R}^p \to \mathbb{R}^p$ defined as

$$F(\mathbf{x}) = W\Phi(\mathbf{x}) + B,\tag{7}$$

where $W \in \mathbb{R}^{p \times p}$, $B \in \mathbb{R}^p$ are parameters and p is the dimension of feature space. In addition, W contains the $\mathbf{w}_i$ as a row, where each $\mathbf{w}_i$ is the linear combination of the support vectors in the feature space and can be written as

$$\mathbf{w}_i = \sum_j \alpha_j^{(i)} \Phi(\mathbf{x}_j),\tag{8}$$

where Φ is a mapping that transforms data in input space X to data in feature space F. From (7), we obtain

$$F(\mathbf{x}) = \begin{bmatrix} \sum_j \alpha_j^{(1)} \Phi(\mathbf{x}_j) \\ \vdots \\ \sum_j \alpha_j^{(p)} \Phi(\mathbf{x}_j) \end{bmatrix} \Phi(\mathbf{x}) + B$$

$$= \begin{bmatrix} \sum_j \alpha_j^{(1)} \Phi(\mathbf{x}_j) \Phi(\mathbf{x}) \\ \vdots \\ \sum_j \alpha_j^{(p)} \Phi(\mathbf{x}_j) \Phi(\mathbf{x}) \end{bmatrix} + B$$

$$= \begin{bmatrix} \sum_j \alpha_j^{(1)} K(\mathbf{x}_j, \mathbf{x}) \\ \vdots \\ \sum_j \alpha_j^{(p)} K(\mathbf{x}_j, \mathbf{x}) \end{bmatrix} + B$$

$$= \begin{bmatrix} \sum_j \alpha_j^{(1)} \\ \vdots \\ \sum_j \alpha_j^{(p)} \end{bmatrix} K(\mathbf{x}_j, \mathbf{x}) + B,$$

where $K(\mathbf{x}_j, \mathbf{x})$ is the kernel having associated feature with mapping Φ.
Define

$$\tilde{\gamma}_k^* = \mathbf{y}_k (W\Phi(\mathbf{x}_k) + B)$$

$$= \mathbf{y}_k \left(\begin{bmatrix} \sum_j \alpha_j^{(1)} \\ \vdots \\ \sum_j \alpha_j^{(p)} \end{bmatrix} K(\mathbf{x}_j, \mathbf{x}) + B \right)$$

$$= \mathbf{y}_k (\zeta K(\mathbf{x}_j, \mathbf{x}) + B) \geq 1 \quad \text{for} \quad k = 1, 2, \ldots, l,$$

where $\mathbf{y}_k \in \{(-1, -1, \ldots, -1), (1, 1, \ldots, 1)\}$ is a p-dimensional vector, $K(\mathbf{x}_j, \mathbf{x}) = \Phi(\mathbf{x}) \Phi(\mathbf{x}_k)$ and $\zeta = \begin{bmatrix} \sum_j \alpha_j^{(1)} \\ \vdots \\ \sum_j \alpha_j^{(p)} \end{bmatrix}$.

Definition 1 We define a map $G : \mathbb{R}^p \to \mathbb{R}_+^p$ by

$$G(\mathbf{w}_i) = (\|\mathbf{w}_i\|, \|\mathbf{w}_i\|, \ldots, \|\mathbf{w}_i\|) \quad \text{for} \quad i = 1, 2, \ldots, p, \tag{9}$$

where $\mathbf{w}_i$ are the rows of $W_{p \times p}$ for $i = 1, 2, \ldots, p$.

Now, the problem is to find $\mathbf{w}_i \in \mathbb{R}^p$ that satisfy

$$\min_{\mathbf{w}_i \in W} G(\mathbf{w}_i) \quad \text{such that} \quad \eta \geq 0, \tag{10}$$

where $\eta = \mathbf{y}_k (\zeta K(\mathbf{x}_j, \mathbf{x}) + B) - 1$.

We call this problem as the Generalized Support Vector Machine (GSVM).

Note that, if $\begin{bmatrix} \sum_j \alpha_j^{(1)} \\ \vdots \\ \sum_j \alpha_j^{(p)} \end{bmatrix} K(\mathbf{x}_j, \mathbf{x}) = -B$, then $\eta = -1$ and we obtain no solution of GSVM problem.

Example 4 Consider the data of points for positive and negative class as given in Example 1. Then by using polynomial Kernel of degree two, we obtain $(1, \sqrt{2}, 1)$, $(1, -\sqrt{2}, 1)$, $(0, 0, 4)$ the vectors of positive data and $(4, 2\sqrt{2}, 1)$, $(4, -2\sqrt{2}, 1)$, $(4, 0, 0)$ the vector negative data in feature space. From positive data points, we have

$$\begin{bmatrix} w_{11} & w_{12} & w_{13} \\ w_{21} & w_{22} & w_{23} \\ w_{31} & w_{32} & w_{33} \end{bmatrix} \begin{bmatrix} 1 \\ \sqrt{2} \\ 1 \end{bmatrix} + \begin{bmatrix} b_1 \\ b_2 \\ b_3 \end{bmatrix} \geq \begin{bmatrix} 1 \\ 1 \\ 1 \end{bmatrix},$$

$$\begin{bmatrix} w_{11} & w_{12} & w_{13} \\ w_{21} & w_{22} & w_{23} \\ w_{31} & w_{32} & w_{33} \end{bmatrix} \begin{bmatrix} 1 \\ -\sqrt{2} \\ 1 \end{bmatrix} + \begin{bmatrix} b_1 \\ b_2 \\ b_3 \end{bmatrix} \geq \begin{bmatrix} 1 \\ 1 \\ 1 \end{bmatrix},$$

$$\begin{bmatrix} w_{11} & w_{12} & w_{13} \\ w_{21} & w_{22} & w_{23} \\ w_{31} & w_{32} & w_{33} \end{bmatrix} \begin{bmatrix} 0 \\ 0 \\ 4 \end{bmatrix} + \begin{bmatrix} b_1 \\ b_2 \\ b_3 \end{bmatrix} \geq \begin{bmatrix} 1 \\ 1 \\ 1 \end{bmatrix},$$

which gives

$$w_{11} + \sqrt{2}w_{12} + w_{13} + b_1 \geq 1,$$
$$w_{21} + \sqrt{2}w_{22} + w_{23} + b_2 \geq 1,$$
$$w_{31} + \sqrt{2}w_{32} + w_{33} + b_3 \geq 1,$$

$$w_{11} - \sqrt{2}w_{12} + w_{13} + b_1 \geq 1,$$
$$w_{21} - \sqrt{2}w_{22} + w_{23} + b_2 \geq 1,$$
$$w_{31} - \sqrt{2}w_{32} + w_{33} + b_3 \geq 1,$$

$$4w_{13} + b_1 \geq 1,$$
$$4w_{23} + b_2 \geq 1,$$
$$4w_{33} + b_3 \geq 1.$$

Also from negative data points,

$$\begin{bmatrix} w_{11} & w_{12} & w_{13} \\ w_{21} & w_{22} & w_{23} \\ w_{31} & w_{32} & w_{33} \end{bmatrix} \begin{bmatrix} 4 \\ 2\sqrt{2} \\ 1 \end{bmatrix} + \begin{bmatrix} b_1 \\ b_2 \\ b_3 \end{bmatrix} \leq \begin{bmatrix} -1 \\ -1 \\ -1 \end{bmatrix},$$

$$\begin{bmatrix} w_{11} & w_{12} & w_{13} \\ w_{21} & w_{22} & w_{23} \\ w_{31} & w_{32} & w_{33} \end{bmatrix} \begin{bmatrix} 4 \\ -2\sqrt{2} \\ 1 \end{bmatrix} + \begin{bmatrix} b_1 \\ b_2 \\ b_3 \end{bmatrix} \leq \begin{bmatrix} -1 \\ -1 \\ -1 \end{bmatrix},$$

$$\begin{bmatrix} w_{11} & w_{12} & w_{13} \\ w_{21} & w_{22} & w_{23} \\ w_{31} & w_{32} & w_{33} \end{bmatrix} \begin{bmatrix} 4 \\ 0 \\ 0 \end{bmatrix} + \begin{bmatrix} b_1 \\ b_2 \\ b_3 \end{bmatrix} \leq \begin{bmatrix} -1 \\ -1 \\ -1 \end{bmatrix},$$

which gives

$$4w_{11} + 2\sqrt{2}w_{12} + w_{13} + b_1 \leq -1,$$
$$4w_{21} + 2\sqrt{2}w_{22} + w_{23} + b_2 \leq -1,$$
$$4w_{31} + 2\sqrt{2}w_{32} + w_{33} + b_3 \leq -1,$$

$$4w_{11} - 2\sqrt{2}w_{12} + w_{13} + b_1 \leq -1,$$
$$4w_{21} - 2\sqrt{2}w_{22} + w_{23} + b_2 \leq -1,$$
$$4w_{31} - 2\sqrt{2}w_{32} + w_{33} + b_3 \leq -1,$$

$$4w_{11} + b_1 \leq -1,$$
$$4w_{12} + b_2 \leq -1,$$
$$4w_{13} + b_3 \leq -1.$$

By solving these equations, we get

$$W = \begin{bmatrix} -1.39 & -0.512 & -0.627 \\ 0.667 & 0 & -0.667 \\ 0.667 & 0 & 0 \end{bmatrix} \quad \text{and} \quad B = \begin{bmatrix} 3.742 \\ 1.047 \\ 1.51 \end{bmatrix},$$

with smallest norm of $\mathbf{w}_i$

$$\min_{\mathbf{w}_i \in W} G(\mathbf{w}_i) = (0.667, 0.667, 0.667).$$

Hence we get $\mathbf{w} = (0.667, 0, 0)$ that minimize $G(\mathbf{w}_i)$ for $i = 1, 2, 3$.

Fig. 12 Data for situation 1
in Example 5

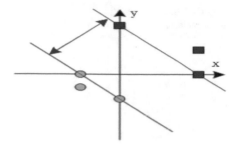

If we are dealing with the data that can linearly separable, then in the process of
GSVM, map Φ deals as identity operator. The next example we show the situations
for this case.

Example 5 Let us consider the three categories of data:

Situation 1 Suppose that we have data $(2, 0)$, $(0, 2)$, $(2, 1)$ as positive class and
data $(-1, 0)$, $(0, -1)$, $(-1, -1/2)$ as negative class shown in Fig. 12.

For positive points, we have $(2, 0)$, $(0, 2)$, $(2, 1)$, so

$$\begin{bmatrix} w_{11} & w_{12} \\ w_{21} & w_{22} \end{bmatrix} \begin{bmatrix} 2 \\ 0 \end{bmatrix} + \begin{bmatrix} b_1 \\ b_2 \end{bmatrix} \geq \begin{bmatrix} 1 \\ 1 \end{bmatrix},$$

$$\begin{bmatrix} w_{11} & w_{12} \\ w_{21} & w_{22} \end{bmatrix} \begin{bmatrix} 0 \\ 2 \end{bmatrix} + \begin{bmatrix} b_1 \\ b_2 \end{bmatrix} \geq \begin{bmatrix} 1 \\ 1 \end{bmatrix},$$

$$\begin{bmatrix} w_{11} & w_{12} \\ w_{21} & w_{22} \end{bmatrix} \begin{bmatrix} 2 \\ 1 \end{bmatrix} + \begin{bmatrix} b_1 \\ b_2 \end{bmatrix} \geq \begin{bmatrix} 1 \\ 1 \end{bmatrix},$$

which implies

$$\begin{bmatrix} 2w_{11} \\ 2w_{21} \end{bmatrix} + \begin{bmatrix} b_1 \\ b_2 \end{bmatrix} \geq \begin{bmatrix} 1 \\ 1 \end{bmatrix},$$

$$\begin{bmatrix} 2w_{12} \\ 2w_{22} \end{bmatrix} + \begin{bmatrix} b_1 \\ b_2 \end{bmatrix} \geq \begin{bmatrix} 1 \\ 1 \end{bmatrix},$$

$$\begin{bmatrix} 2w_{11} + w_{12} \\ 2w_{21} + w_{22} \end{bmatrix} + \begin{bmatrix} b_1 \\ b_2 \end{bmatrix} \geq \begin{bmatrix} 1 \\ 1 \end{bmatrix}.$$

Again, for the negative points, we have $(-1, 0)$, $(0, -1)$, $(-1, -1/2)$ and

$$\begin{bmatrix} w_{11} & w_{12} \\ w_{21} & w_{22} \end{bmatrix} \begin{bmatrix} -1 \\ 0 \end{bmatrix} + \begin{bmatrix} b_1 \\ b_2 \end{bmatrix} \leq \begin{bmatrix} -1 \\ -1 \end{bmatrix},$$

$$\begin{bmatrix} w_{11} & w_{12} \\ w_{21} & w_{22} \end{bmatrix} \begin{bmatrix} 0 \\ -1 \end{bmatrix} + \begin{bmatrix} b_1 \\ b_2 \end{bmatrix} \leq \begin{bmatrix} -1 \\ -1 \end{bmatrix},$$

$$\begin{bmatrix} w_{11} & w_{12} \\ w_{21} & w_{22} \end{bmatrix} \begin{bmatrix} -1 \\ -1/2 \end{bmatrix} + \begin{bmatrix} b_1 \\ b_2 \end{bmatrix} \leq \begin{bmatrix} -1 \\ -1 \end{bmatrix},$$

Fig. 13 The data separation
for situation 2

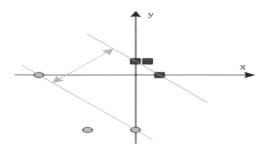

which gives

$$\begin{bmatrix} -w_{11} \\ -w_{21} \end{bmatrix} + \begin{bmatrix} b_1 \\ b_2 \end{bmatrix} \leq \begin{bmatrix} -1 \\ -1 \end{bmatrix},$$

$$\begin{bmatrix} -w_{12} \\ -w_{22} \end{bmatrix} + \begin{bmatrix} b_1 \\ b_2 \end{bmatrix} \leq \begin{bmatrix} -1 \\ -1 \end{bmatrix},$$

$$\begin{bmatrix} -w_{11} - \frac{1}{2}w_{12} \\ -w_{21} - \frac{1}{2}w_{22} \end{bmatrix} + \begin{bmatrix} b_1 \\ b_2 \end{bmatrix} \leq \begin{bmatrix} 1 \\ 1 \end{bmatrix}.$$

From above equations, we get

$$W = \begin{bmatrix} \frac{2}{3} & \frac{2}{3} \\ \frac{2}{3} & \frac{2}{3} \end{bmatrix} \quad \text{and} \quad B = \begin{bmatrix} -\frac{1}{3} \\ -\frac{1}{3} \end{bmatrix}.$$

Thus we get

$$\min_{\mathbf{w}_i \in W} G(\mathbf{w}_i) = \left(\frac{2\sqrt{2}}{3}, \frac{2\sqrt{2}}{3} \right).$$

Hence we get $\mathbf{w} = (\frac{2}{3}, \frac{2}{3})$ that minimizes $G(\mathbf{w}_i)$ for $i = 1, 2$.

Situation 2 We consider the data $(1, 0)$, $(0, 1)$, $(1/2, 1)$ as positive class, data $(-4, 0)$, $(0, -4)$, $(-2, -4)$ as negative class which is shown in Fig. 13.

Now, for positive points of **Situation 2**, we have $(1, 0)$, $(0, 1)$, $(1/2, 1)$ and

$$\begin{bmatrix} w_{11} & w_{12} \\ w_{21} & w_{22} \end{bmatrix} \begin{bmatrix} 1 \\ 0 \end{bmatrix} + \begin{bmatrix} b_1 \\ b_2 \end{bmatrix} \geq \begin{bmatrix} 1 \\ 1 \end{bmatrix},$$

$$\begin{bmatrix} w_{11} & w_{12} \\ w_{21} & w_{22} \end{bmatrix} \begin{bmatrix} 0 \\ 1 \end{bmatrix} + \begin{bmatrix} b_1 \\ b_2 \end{bmatrix} \geq \begin{bmatrix} 1 \\ 1 \end{bmatrix},$$

$$\begin{bmatrix} w_{11} & w_{12} \\ w_{21} & w_{22} \end{bmatrix} \begin{bmatrix} \frac{1}{2} \\ 1 \end{bmatrix} + \begin{bmatrix} b_1 \\ b_2 \end{bmatrix} \geq \begin{bmatrix} 1 \\ 1 \end{bmatrix},$$

which gives

$$\begin{bmatrix} w_{11} \\ w_{21} \end{bmatrix} + \begin{bmatrix} b_1 \\ b_2 \end{bmatrix} \geq \begin{bmatrix} 1 \\ 1 \end{bmatrix},$$

$$\begin{bmatrix} w_{12} \\ w_{22} \end{bmatrix} + \begin{bmatrix} b_1 \\ b_2 \end{bmatrix} \geq \begin{bmatrix} 1 \\ 1 \end{bmatrix},$$

$$\begin{bmatrix} \frac{1}{2}w_{11} + w_{12} \\ \frac{1}{2}w_{21} + w_{22} \end{bmatrix} + \begin{bmatrix} b_1 \\ b_2 \end{bmatrix} \geq \begin{bmatrix} 1 \\ 1 \end{bmatrix}.$$

For negative points for this case, we have

$$\begin{bmatrix} w_{11} & w_{12} \\ w_{21} & w_{22} \end{bmatrix} \begin{bmatrix} -4 \\ 0 \end{bmatrix} + \begin{bmatrix} b_1 \\ b_2 \end{bmatrix} \leq \begin{bmatrix} -1 \\ -1 \end{bmatrix},$$

$$\begin{bmatrix} w_{11} & w_{12} \\ w_{21} & w_{22} \end{bmatrix} \begin{bmatrix} 0 \\ -4 \end{bmatrix} + \begin{bmatrix} b_1 \\ b_2 \end{bmatrix} \leq \begin{bmatrix} -1 \\ -1 \end{bmatrix},$$

$$\begin{bmatrix} w_{11} & w_{12} \\ w_{21} & w_{22} \end{bmatrix} \begin{bmatrix} -2 \\ -4 \end{bmatrix} + \begin{bmatrix} b_1 \\ b_2 \end{bmatrix} \leq \begin{bmatrix} -1 \\ -1 \end{bmatrix},$$

which gives

$$\begin{bmatrix} -4w_{11} \\ -4w_{21} \end{bmatrix} + \begin{bmatrix} b_1 \\ b_2 \end{bmatrix} \leq \begin{bmatrix} -1 \\ -1 \end{bmatrix},$$

$$\begin{bmatrix} -4w_{12} \\ 4w_{22} \end{bmatrix} + \begin{bmatrix} b_1 \\ b_2 \end{bmatrix} < \begin{bmatrix} -1 \\ -1 \end{bmatrix},$$

$$\begin{bmatrix} -2w_{11} - 4w_{12} \\ -2w_{21} - 4w_{22} \end{bmatrix} + \begin{bmatrix} b_1 \\ b_2 \end{bmatrix} \leq \begin{bmatrix} -1 \\ -1 \end{bmatrix}.$$

Thus, we obtain that

$$W = \begin{bmatrix} \frac{2}{5} & \frac{2}{5} \\ \frac{2}{5} & \frac{2}{5} \end{bmatrix} \quad \text{and} \quad B = \begin{bmatrix} \frac{3}{5} \\ \frac{3}{5} \end{bmatrix}.$$

Thus we get

$$\min_{i \in \{1,2\}} G(\mathbf{w}_i) = \left(\frac{2\sqrt{2}}{5}, \frac{2\sqrt{2}}{5} \right).$$

Hence we get $\mathbf{w} = (\frac{2}{5}, \frac{2}{5})$ that minimize $G(\mathbf{w}_i)$ for $i = 1, 2$.

Fig. 14 The data separation
for situation 3

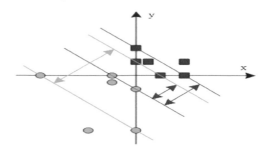

In the next **Situation 3**, we combine of this two groups of data. Now, we have data $(2, 0)$, $(0, 2)$, $(2, 1)$, $(1, 0)$, $(0, 1)$, $(1/2, 1)$ as positive class and $(-1, 0)$, $(0, -1)$, $(-1, -1/2)$, $(-4, 0)$, $(0, -4)$, $(-2, -4)$ as negative class see Fig. 14.

For the positive points of the combination, we have

$$\begin{bmatrix} w_{11} & w_{12} \\ w_{21} & w_{22} \end{bmatrix} \begin{bmatrix} 1 \\ 0 \end{bmatrix} + \begin{bmatrix} b_1 \\ b_2 \end{bmatrix} = \begin{bmatrix} 1 \\ 1 \end{bmatrix},$$

and

$$\begin{bmatrix} w_{11} & w_{12} \\ w_{21} & w_{22} \end{bmatrix} \begin{bmatrix} 0 \\ 1 \end{bmatrix} + \begin{bmatrix} b_1 \\ b_2 \end{bmatrix} = \begin{bmatrix} 1 \\ 1 \end{bmatrix},$$

which gives

$$\begin{bmatrix} w_{11} \\ w_{21} \end{bmatrix} + \begin{bmatrix} b_1 \\ b_2 \end{bmatrix} = \begin{bmatrix} 1 \\ 1 \end{bmatrix} \quad \text{and} \quad \begin{bmatrix} w_{12} \\ w_{22} \end{bmatrix} + \begin{bmatrix} b_1 \\ b_2 \end{bmatrix} = \begin{bmatrix} 1 \\ 1 \end{bmatrix}.$$

For negative points for this case, we have

$$\begin{bmatrix} w_{11} & w_{12} \\ w_{21} & w_{22} \end{bmatrix} \begin{bmatrix} -1 \\ 0 \end{bmatrix} + \begin{bmatrix} b_1 \\ b_2 \end{bmatrix} = \begin{bmatrix} -1 \\ -1 \end{bmatrix},$$

and

$$\begin{bmatrix} w_{11} & w_{12} \\ w_{21} & w_{22} \end{bmatrix} \begin{bmatrix} 0 \\ -1 \end{bmatrix} + \begin{bmatrix} b_1 \\ b_2 \end{bmatrix} = \begin{bmatrix} -1 \\ -1 \end{bmatrix},$$

which gives

$$\begin{bmatrix} -w_{11} \\ -w_{21} \end{bmatrix} + \begin{bmatrix} b_1 \\ b_2 \end{bmatrix} = \begin{bmatrix} -1 \\ -1 \end{bmatrix} \quad \text{and} \quad \begin{bmatrix} -w_{12} \\ -w_{22} \end{bmatrix} + \begin{bmatrix} b_1 \\ b_2 \end{bmatrix} = \begin{bmatrix} -1 \\ -1 \end{bmatrix}.$$

From this, we obtain that

$$W = \begin{bmatrix} 1 & 1 \\ 1 & 1 \end{bmatrix} \quad \text{and} \quad B = \begin{bmatrix} 0 \\ 0 \end{bmatrix}.$$

Thus we get

$$\min_{i \in \{1,2\}} G(\mathbf{w}_i) = (\sqrt{2}, \sqrt{2}).$$

Hence we get $\mathbf{w} = (1, 1)$ that minimize $G(\mathbf{w}_i)$ for $i = 1, 2$.

The problem of GSVM defined in (10) is equivalent to

$$\text{find} \quad \mathbf{w}_i \in W : \langle G'(\mathbf{w}_i), \mathbf{v} - \mathbf{w}_i \rangle \geq 0 \quad \text{for all} \quad \mathbf{v} \in \mathbb{R}^p \quad \text{with} \quad \eta \geq 0. \tag{11}$$

Hence the problem of GSVM becomes to the problem of generalized variational inequality.

Note that it we take $G'(\mathbf{w}_i) = \frac{\mathbf{w}_i}{\|\mathbf{w}_i\|}$, then from (11), we obtain

$$\text{find} \quad \mathbf{w}_i \in W : \langle \mathbf{w}_i, \mathbf{v} - \mathbf{w}_i \rangle \geq 0 \quad \text{for all} \quad \mathbf{v} \in \mathbb{R}^p \quad \text{with} \quad \eta \geq 0, \tag{12}$$

or

$$\text{find} \quad \mathbf{w}_i \in W : \langle \mathbf{w}_i, \mathbf{v} \rangle \geq \|\mathbf{w}_i\|^2 \quad \text{for all} \quad \mathbf{v} \in \mathbb{R}^p \quad \text{with} \quad \eta \geq 0. \tag{13}$$

We study the sufficient conditions for the existence of solutions for GSVM problems.

Proposition 1 *Let $G : \mathbb{R}^p \to \mathbb{R}^p_+$ be a differentiable operator. An element $\mathbf{w}^* \in \mathbb{R}^p$ minimizes G if and only if $G'(\mathbf{w}^*) = \mathbf{0}$, that is, $\mathbf{w}^* \in \mathbb{R}^p$ solves GSVM if and only if $G'(\mathbf{w}^*) = \mathbf{0}$.*

Proof Let $G'(\mathbf{w}^*) = \mathbf{0}$, then for all $\mathbf{v} \in \mathbb{R}^p$ with $\eta = y_k (\zeta K(\mathbf{x}_j, \mathbf{x}) + B) - 1 \geq 0$,

$$< G'(\mathbf{w}^*), \mathbf{v} - \mathbf{w}^* > = < 0, \mathbf{v} - \mathbf{w}^* > = 0,$$

and consequently, the inequality

$$< G'(\mathbf{w}^*), \mathbf{v} - \mathbf{w}^* > \geq 0$$

holds for all $\mathbf{v} \in \mathbb{R}^p$. Hence $\mathbf{w}^* \in \mathbb{R}^p$ solves problem of GSVM.

Conversely, assume that $\mathbf{w}^* \in \mathbb{R}^p$ satisfies

$$< G'(\mathbf{w}^*), \mathbf{v} - \mathbf{w}^* > \geq 0 \ \forall \mathbf{v} \in \mathbb{R}^n \quad \text{such that} \quad \eta \geq 0.$$

Taking $\mathbf{v} = \mathbf{w}^* - G'(\mathbf{w}^*)$ in the above inequality implies that

$$< G'(\mathbf{w}^*), -G'(\mathbf{w}^*) > \geq 0,$$

which further implies

$$-\|G'(\mathbf{w}^*)\|^2 \geq 0,$$

and we get $G'(\mathbf{w}^*) = \mathbf{0}$. $\square$

Remark 1 Note that if $G'(\mathbf{w}^*) = \mathbf{0}$ at some $\mathbf{w}^* \in \mathbb{R}^p$, then we obtain $\frac{\mathbf{w}^*}{\|\mathbf{w}^*\|} = \mathbf{0}$ which implies $\mathbf{w}^* = \mathbf{0}$. Thus it follows from Proposition 2.4 that if $G'(\mathbf{w}^*) = \mathbf{0}$ at some $\mathbf{w}^* \in \mathbb{R}^p$, then $\mathbf{w}^* = \mathbf{0}$ solves GSVM problem.

Remark 2 If $\mathbf{w}^* = \mathbf{0}$, then from (8), we obtain

$$\sum_j \alpha_j^{(*)} \Phi(\mathbf{x}_j) = \mathbf{0},$$

which implies

$$\sum_j \alpha_j^{(*)} \Phi(\mathbf{x}_j) \Phi(\mathbf{x}) = 0,$$

that is

$$\sum_j \alpha_j^{(*)} K(\mathbf{x}_j, \mathbf{x}) = 0. \tag{14}$$

Since $\alpha_j^{(*)} > 0$ for all j, so we have

$$K(\mathbf{x}_j, \mathbf{x}) = 0.$$

Definition 2 Let K be a closed and convex subset of $\mathbb{R}^n$. Then, for every point $\mathbf{x} \in \mathbb{R}^n$, there exists a unique nearest point in K, denoted by $P_K(\mathbf{x})$, such that $\|\mathbf{x} - P_K(\mathbf{x})\| \leq \|\mathbf{x} - \mathbf{y}\|$ for all $\mathbf{y} \in K$ and also note that $P_K(\mathbf{x}) = \mathbf{x}$ if $\mathbf{x} \in K$. P_K is called the metric projection of $\mathbb{R}^n$ onto K. It is well known that $P_K : \mathbb{R}^n \to K$ is characterized by the properties:

(i) $P_K(\mathbf{x}) = \mathbf{z}$ for $\mathbf{x} \in \mathbb{R}^n$ if and only if $< \mathbf{z} - \mathbf{x}, \mathbf{y} - \mathbf{z} > \geq 0$ for all $\mathbf{y} \in \mathbb{R}^n$;
(ii) For every $\mathbf{x}, \mathbf{y} \in \mathbb{R}^n$, $\|P_K(\mathbf{x}) - P_K(\mathbf{y})\|^2 \leq < \mathbf{x} - \mathbf{y}, P_K(\mathbf{x}) - P_K(\mathbf{y}) >$;
(iii) $\|P_K(\mathbf{x}) - P_K(\mathbf{y})\| \leq \|\mathbf{x} - \mathbf{y}\|$, for every $\mathbf{x}, \mathbf{y} \in \mathbb{R}^n$, that is, P_K is nonexpansive map.

Proposition 2 *Let $G : \mathbb{R}^p \to \mathbb{R}^p_+$ be a differentiable operator. An element $\mathbf{w}^* \in \mathbb{R}^p$ minimize mapping G defined in (11) if and only if $\mathbf{w}^*$ is the fixed point of map*

$$P_{\mathbb{R}^n_+}(I - \rho G') : \mathbb{R}^p \to \mathbb{R}^p_+ \text{ for any } \rho > 0,$$

that is,

$$\mathbf{w}^* = P_{\mathbb{R}_+^p} \left(I - \rho G' \right) (\mathbf{w}^*)$$
$$= P_{\mathbb{R}_+^p} \left(\mathbf{w}^* - \rho G' \left(\mathbf{w}^* \right) \right),$$

where $P_{\mathbb{R}_+^p}$ is a projection map from $\mathbb{R}^p$ to $\mathbb{R}_+^p$ and $\eta = \mathbf{y}_k \left(\zeta K(\mathbf{x}_j, \mathbf{x}) + B \right) - 1 \geq 0$.

Proof Suppose $\mathbf{w}^* \in \mathbb{R}_+^p$ is solution of GSVM. Then for $\eta = \mathbf{y}_k \left(\zeta K(\mathbf{x}_j, \mathbf{x}) + B \right) - 1 \geq 0$, we have

$$< G' \left(\mathbf{w}^* \right), \mathbf{w} - \mathbf{w}^* > \geq 0 \quad \text{for all} \quad \mathbf{w} \in \mathbb{R}^p.$$

Adding $< \mathbf{w}^*, \mathbf{w} - \mathbf{w}^* >$ on both sides, we get

$$< \mathbf{w}^*, \mathbf{w} - \mathbf{w}^* > + < G' \left(\mathbf{w}^* \right), \mathbf{w} - \mathbf{w}^* > \geq \; < \mathbf{w}^*, \mathbf{w} - \mathbf{w}^* > \quad \text{for all} \quad \mathbf{w} \in \mathbb{R}^p,$$

which further implies that

$$< \mathbf{w}^* - \left(\mathbf{w}^* - G' \left(\mathbf{w}^* \right) \right), \mathbf{w} - \mathbf{w}^* > \geq 0 \quad \text{for all} \quad \mathbf{w} \in \mathbb{R}^p,$$

which is possible only if $\mathbf{w}^* = P_{\mathbb{R}_+^p} \left(\mathbf{w}^* - \rho G' (\mathbf{w}^*) \right)$, that is, $\mathbf{w}^*$ is the fixed point of G'.

Conversely, let $\mathbf{w}^* = P_{\mathbb{R}_+^p} \left(\mathbf{w}^* - \rho G' (\mathbf{w}^*) \right)$ with $\eta = \mathbf{y}_k \left(\zeta K(\mathbf{x}_j, \mathbf{x}) + B \right) - 1 \geq 0$, then we have

$$< \mathbf{w}^* - \left(\mathbf{w}^* - G' \left(\mathbf{w}^* \right) \right), \mathbf{w} - \mathbf{w}^* > \geq 0 \quad \text{for all} \quad \mathbf{w} \in \mathbb{R}^p,$$

which implies

$$< G' \left(\mathbf{w}^* \right), \mathbf{w} - \mathbf{w}^* > \geq 0 \quad \text{for all} \quad \mathbf{w} \in \mathbb{R}^p,$$

and so $\mathbf{w}^* \in \mathbb{R}_+^p$ is the solution of GSVM. $\qquad \square$

3 Conclusion

The linear and nonlinear data classifications by using support vector machine and generalized support vector machine have been studied. We also studied the sufficient conditions for existence of the solution of generalized support vector machine. Some examples are shown for supporting these results.

Acknowledgements Talat Nazir and Xiaomin Qi are grateful to the Erasmus Mundus project FUSION for supporting the research visit to Mälardalen University, Sweden, and to the Research environment MAM in Mathematics and Applied Mathematics, Division of Applied Mathematics, the School of Education, Culture and Communication of Mälardalen University for creating excellent research environment.

References

1. Adankon, M.M., Cheriet, M.: Model selection for the LS-SVM. Application to handwriting recognition. Pattern Recognit. **42**(12), 3264–3270 (2009)
2. Cortes, C., Vapnik, V.N.: Support-vector networks. Mach. Learn. **20**(3), 273–297 (1995)
3. Cristianini, N., Shawe-Taylor, J.: An Introduction to Support Vector Machines and other Kernel Based Learning Methods. Cambridge University Press, Cambridge (2000)
4. Guyon, I., Weston, J., Barnhill, S., Vapnik, V.: Gene selection for cancer classification using support vector machines. Mach. Learn. **46**(1–3), 389–422 (2002)
5. Joachims, T.: Text categorization with support vector machines: learning with many relevant features. In: Proceedings of the European Conference on Machine Learning. Springer, Heidelberg (1998)
6. Khan, N., Ksantini, R., Ahmad, I., Boufama, B.: A novel SVM+NDA model for classification with an application to face recognition. Pattern Recognit. **45**(1), 66–79 (2012)
7. Li, S., Kwok, J.T., Zhu, H., Wang, Y.: Texture classification using the support vector machines. Pattern Recognit. **36**(12), 2883–2893 (2003)
8. Liu, R., Wang, Y., Baba, T., Masumoto, D., Nagata, S.: SVM-based active feedback in image retrieval using clustering and unlabeled data. Pattern Recognit. **41**(8), 2645–2655 (2008)
9. Michel, P., Kaliouby, R.E.: Real time facial expresion recognition in video using support vector machines. In: Proceedings of ICMI'03, pp. 258–264 (2003)
10. Noble, W.S.: Support Vector Machine Applications in Computational Biology. MIT Press, Cambridge (2004)
11. Shao, Y., Lunetta, R.S.: Comparison of support vector machine, neural network, and CART algorithms for the land-cover classification using limited training data points. ISPRS J. Photogramm. Remote Sens. **70**, 78–87 (2012)
12. Shao, Y.H., Chen, W.J., Deng, N.Y.: Nonparallel hyperplane support vector machine for binary classification problems. Inf. Sci. **263**, 22–35 (2014)
13. Vapnik, V.N.: The Nature of Statistical Learning Theory. Springer, New York (1996)
14. Vapnik, V.N.: Statistical Learning Theory. Wiley, New York (1998)
15. Wang, D., Qi, X., Wen, S., Deng, M.: SVM based fault classifier design for a water level control system. In: Proceedings of 2013 International Conference on Advanced Mechatronic Systems, pp. 152–157. Luoyang, China (2013)
16. Wang, D., Qi, X., Wen, S., Dan, Y., Ouyang, L., Deng, M.: Robust nonlinear control and SVM classifier based fault diagnosis for a water level process. ICIC Express Lett. **5**(1), 767–774 (2014)
17. Wang, X.Y., Wang, T., Bu, J.: Color image segmentation using pixel wise support vector machine classification. Pattern Recognit. **44**(4), 777–787 (2011)
18. Weston, J., Watkins, C.: Multi-class support vector machines. Technical report CSD-TR-98-04, Department of Computer Science, Royal Holloway, University of London (1998)
19. Wu, Y.C., Lee, Y.-S., Yang, J.-C.: Robust and efficient multiclass SVM models for phrase pattern recognition. Pattern Recognit. **41**(9), 2874–2889 (2008)
20. Xue, Z., Ming, D., Song, W., Wan, B., Jin, S.: Infrared gait recognition based on wavelet transform and support vector machine. Pattern Recognit. **43**(8), 2904–2910 (2010)
21. Zhao, Z., Liu, J., Cox, J.: Safe and efficient screening for sparse support vector machine. In: Proceedings of the 20th ACM SIGKDD International Conference on Knowledge Discovery and Data Mining, KDD 14, pp. 542–551, New York, NY, USA (2014)
22. Zuo, R., Carranza, E.J.M.: Support vector machine: a tool for mapping mineral prospectivity. Comput. Geosci. **37**(12), 1967–1975 (2011)

Common Fixed Points of Weakly Commuting Multivalued Mappings on a Domain of Sets Endowed with Directed Graph

Talat Nazir and Sergei Silvestrov

Abstract In this paper, the existence of coincidence points and common fixed points for multivalued mappings satisfying certain graphic ψ-contraction contractive conditions with set-valued domain endowed with a graph, without appealing to continuity, is established. Some examples are presented to support the results proved herein. Our results unify, generalize and extend various results in the existing literature.

Keywords Common fixed point · Multivalued mapping · Graphic contraction · Directed graph

1 Introduction and Preliminaries

Order oriented fixed point theory is studied in an environment created by a class of partially ordered sets with appropriate mappings satisfying certain order condition like monotonicity, expansivity or order continuity. Existence of fixed points in partially ordered metric spaces has been studied by Ran and Reurings [26]. Recently, many researchers have obtained fixed point results for single and multivalued mappings defined on partially ordered metrics spaces (see, e.g., [6, 8, 18, 25]). Jachymski and Jozwik [20] introduced a new approach in metric fixed point theory by replacing the order structure with a graph structure on a metric space. In this way, the results proved in ordered metric spaces are generalized (see also [19] and the reference therein); in fact, in 2010, Gwodzdz-Lukawska and Jachymski [17], developed the Hutchinson-Barnsley theory for finite families of mappings on a metric space endowed with a directed graph. Abbas and Nazir [2] obtained some fixed point

T. Nazir (✉) · S. Silvestrov
Division of Applied Mathematics, School of Education, Culture and Communication,
Mälardalen University, Box 883, 721 23 Västerås, Sweden
e-mail: talat@ciit.net.pk

S. Silvestrov
e-mail: sergei.silvestrov@mdh.se

T. Nazir
Department of Mathematics, COMSATS Institute of Information Technology,
Abbottabad 22060, Pakistan

© Springer International Publishing Switzerland 2016
S. Silvestrov and M. Rančić (eds.), *Engineering Mathematics II*,
Springer Proceedings in Mathematics & Statistics 179,
DOI 10.1007/978-3-319-42105-6_19

397

results for power graph contraction pair endowed with a graph. Bojor [13] proved fixed point theorems for Reich type contractions on metric spaces with a graph. For more results in this direction, we refer to [4, 5, 12, 14, 15, 24] and reference mentioned therein.

Beg and Butt [9] proved the existence of fixed points of multivalued mapping in metric spaces endowed with a graph G. Recently, Abbas et al. [3] obtained fixed points of multivalued mappings satisfying certain graphic contraction conditions with set-valued domain endowed with a graph. Nicolae et al. [24] established some fixed points of multivalued generalized contractions in metric spaces endowed with a graph.

The aim of this paper is to prove some coincidence point and common fixed point results for multivalued graphic ψ-contractive mappings defined on the family of closed and bounded subsets of a metric space endowed with a graph G. These results extend and strengthen various comparable results in the existing literature [3, 9, 12, 19, 20, 23].

We denote, the letters $\mathbb{R}$, $\mathbb{R}^+$ and $\mathbb{N}$ denote the set of all real numbers, the set of all positive real numbers and the set of all natural numbers, respectively.

Consistent with Jachymski [19], let (X, d) be a metric space and Δ denotes the diagonal of $X \times X$. Let G be a directed graph, such that the set $V(G)$ of its vertices coincides with X and $E(G)$ be the set of edges of the graph which contains all loops, that is, $\Delta \subseteq E(G)$. Also assume that the graph G has no parallel edges and, thus, one can identify G with the pair $(V(G), E(G))$.

Definition 1 [19] An operator $f : X \to X$ is called a Banach G-contraction or simply G-contraction if

(a) f preserves edges of G; for each $x, y \in X$ with $(x, y) \in E(G)$, we have $(f(x), f(y)) \in E(G)$,
(b) f decreases weights of edges of G; there exists $\alpha \in (0, 1)$ such that for all $x, y \in X$ with $(x, y) \in E(G)$, we have $d(f(x), f(y)) \leq \alpha d(x, y)$.

If x and y are vertices of G, then a path in G from x to y of length $k \in \mathbb{N}$ is a finite sequence $\{x_n\}$ ($n \in \{0, 1, 2, \ldots, k\}$) of vertices such that $x_0 = x$, $x_k = y$ and $(x_{i-1}, x_i) \in E(G)$ for $i \in \{1, 2, \ldots, k\}$.

Notice that a graph G is connected if there is a directed path between any two vertices and it is weakly connected if $\widetilde{G}$ is connected, where $\widetilde{G}$ denotes the undirected graph obtained from G by ignoring the direction of edges. Denote by G^{-1} the graph obtained from G by reversing the direction of edges. Thus,

$$E\left(G^{-1}\right) = \{(x, y) \in X \times X : (y, x) \in E(G)\}.$$

It is more convenient to treat $\widetilde{G}$ as a directed graph for which the set of its edges is symmetric, under this convention; we have that

$$E(\widetilde{G}) = E(G) \cup E(G^{-1}).$$

In $V(G)$, we define the relation R in the following way:

For $x, y \in V(G)$, we have $x R y$ if and only if, there is a path in G from x to y. If G is such that $E(G)$ is symmetric, then for $x \in V(G)$, the equivalence class $[x]_{\tilde{G}}$ in $V(G)$ defined by the relation R is $V(G_x)$.

Recall that if $f : X \to X$ is an operator, then by F_f we denote the set of all fixed points of f. Set

$$X_f := \{x \in X : (x, f(x)) \in E(G)\}.$$

Jachymski [20] used the following property:

(P): for any sequence $\{x_n\}$ in X, if $x_n \to x$ as $n \to \infty$ and $(x_n, x_{n+1}) \in E(G)$, then $(x_n, x) \in E(G)$.

Theorem 1 [20] *Let (X, d) be a complete metric space and G a directed graph such that $V(G) = X$ and $f : X \to X$ a G-contraction. Suppose that $E(G)$ and the triplet (X, d, G) have property (P). Then the following statements hold:*

(i) *$F_f \neq \emptyset$ if and only if $X_f \neq \emptyset$;*
(ii) *if $X_f \neq \emptyset$ and G is weakly connected, then f is a Picard operator, i.e., $F_f = \{x^*\}$ and sequence $\{f^n(x)\} \to x^*$ as $n \to \infty$, for all $x \in X$;*
(iii) *for any $x \in X_f$, $f \mid_{[x]_{\tilde{G}}}$ is a Picard operator;*
(iv) *if $X_f \subseteq E(G)$, then f is a weakly Picard operator, i.e., $F_f \neq \emptyset$ and, for each $x \in X$, we have sequence $\{f^n(x)\} \to x^* \in F_f$ as $n \to \infty$.*

For detailed discussion on Picard operators, we refer to Berinde [10, 11].

Let (X, d) be a metric space and $CB(X)$ a class of all nonempty closed and bounded subsets of X. For $A, B \in CB(X)$, let

$$H(A, B) = \max\{\sup_{b \in B} d(b, A), \sup_{a \in A} d(a, B)\},$$

where $d(x, B) = \inf\{d(x, b) : b \in B\}$ is the distance of a point x to the set B. The mapping H is said to be the Pompeiu–Hausdorff metric induced by d.

Throughout this paper, we assume that a directed graph G has no parallel edges and G is a weighted graph in the sense that each vertex x is assigned the weight $d(x, x) = 0$ and each edge (x, y) is assigned the weight $d(x, y)$. Since d is a metric on X, the weight assigned to each vertex x to vertex y need not be zero and, whenever a zero weight is assigned to some edge (x, y), it reduces to a loop (x, x) having weight 0. Further, in Pompeiu–Hausdorff metric induced by metric d, the Pompeiu–Hausdorff weight assigned to each $U, V \in CB(X)$ need not be zero (that is, $H(U, V) \neq 0$) and, whenever a zero Pompeiu–Hausdorff weight is assigned to some $U, V \in CB(X)$, then it reduces to $U = V$.

Definition 2 [3] Let A and B be two nonempty subsets of X. Then by:

(a) 'there is an edge between A and B', we mean there is an edge between some $a \in A$ and $b \in B$ which we denote by $(A, B) \subset E(G)$.

(b)   'there is a path between A and B', we mean that there is a path between some
       $a \in A$ and $b \in B$.

In $CB(X)$, we define a relation R in the following way: For $A, B \in CB(X)$, we
have ARB if and only if, there is a path between A and B. We say that the relation
R on $CB(X)$ is transitive if there is a path between A and B, and there is a path
between B and C, then there is a path between A and C.

Consider the mapping $T : CB(X) \rightarrow CB(X)$ instead of a mapping T from X to
X or from X to $CB(X)$.

For mappings $T : CB(X) \rightarrow CB(X)$, the set X_T is defined as

$$X_T := \{U \in CB(X) : (U, T(U)) \subseteq E(G)\}.$$

Recently, Abbas et al. [3] gave the following definition.

Definition 3 Let $T : CB(X) \rightarrow CB(X)$ be a multivalued mapping. The mapping T
is said to be a graph ϕ-contraction if the following conditions hold:

 (i)   There is an edge between A and B implies there is an edge between $T(A)$ and
       $T(B)$ for all $A, B \in CB(X)$.
 (ii) There is a path between A and B implies there is a path between $T(A)$ and
       $T(B)$ for all $A, B \in CB(X)$.
 (iii) There exists an upper semi-continuous and nondecreasing function $\phi : \mathbb{R}^+ \rightarrow$
       $\mathbb{R}^+$ with $\phi(t) < t$ for each $t > 0$ such that there is an edge between A and B
       implies that

$$H(T(A), T(B)) \leq \phi(H(A, B)) \quad \text{for all} \quad A, B \quad \text{in} \quad CB(X).$$

Definition 4 Let $S, T : CB(X) \rightarrow CB(X)$ be two multivalued mappings. The set
$U \in CB(X)$ is said to be a coincidence point of S and T, if $S(U) = T(U)$. Also, a
set $A \in CB(X)$ is said to be a fixed point of S if $S(A) = A$. The set of all coincidence
points of S and T is denoted by $CP(S, T)$ and the set of all fixed points of S is denoted
by $Fix(S)$.

Definition 5 Two maps $S, T : CB(X) \rightarrow CB(X)$ are said to be weakly compatible
if they commute at their coincidence point.

For more details to the weakly compatible maps, we refer the reader to
[1, 21, 22].

A subset Γ of $CB(X)$ is said to be complete if for any set $X, Y \in \Gamma$, there is an
edge between X and Y.

Abbas et al. [3] used the property P^* stated as follows: A graph G is said to have
property

(P*): if for any sequence $\{X_n\}$ in $CB(X)$ with $X_n \rightarrow X$ as $n \rightarrow \infty$, there exists
edge between X_{n+1} and X_n for $n \in \mathbb{N}$, implies that there is a subsequence $\{X_{n_k}\}$ of
$\{X_n\}$ with an edge between X and X_{n_k} for $n \in \mathbb{N}$.

Theorem 2 [3] *Let (X, d) be a complete metric space endowed with a directed graph G such that $V(G) = X$ and $E(G) \supseteq \Delta$. If $T : CB(X) \to CB(X)$ is a graph ϕ-contraction mapping such that the relation R on $CB(X)$ is transitive, then following statements hold:*

(a) *if $Fix(T)$ is complete, then the Pompeiu–Hausdorff weight assigned to the $U, V \in Fix(T)$ is 0.*

(b) *$X_T \neq \emptyset$ provided that $Fix(T) \neq \emptyset$.*

(c) *If $X_T \neq \emptyset$ and the weakly connected graph G satisfies the property (P^*), then T has a fixed point.*

(d) *$Fix(T)$ is complete if and only if $Fix(T)$ is a singleton.*

We denote Ψ the set of all functions $\psi : \mathbb{R}^+ \to \mathbb{R}^+$, where ψ is nondecreasing function with $\sum_{i=1}^{\infty} \psi^n(t)$ convergent. It is easy to show that if $\psi \in \Psi$, then $\psi(t) < t$ for any $t > 0$. We give the following definition.

Definition 6 Let (X, d) be a metric space endowed with a directed graph G such that $V(G) = X$, $E(G) \supseteq \Delta$ and for every U in $CB(X)$, $(S(U), U) \subseteq E(G)$ and $(U, T(U)) \subseteq E(G)$. Let $S, T : CB(X) \to CB(X)$ be two multivalued mappings. The pair (S, T) of maps is said to be

(I) graph ψ_1-contraction pair if there exists a $\psi \in \Psi$ and there is an edge between A and B such that

$$H(S(A), S(B)) \leq \psi(M_1(A, B))$$

holds, where

$$M_1(A, B) = \max\{H(T(A), T(B)), H(S(A), T(A)), H(S(B), T(B)),$$
$$\frac{H(S(A), T(B)) + H(S(B), T(A))}{2}\}.$$

(II) graph ψ_2-contraction pair if there exists a $\psi \in \Psi$ and there is an edge between A and B such that

$$H(S(A), S(B)) \leq \psi(M_2(A, B))$$

holds, where

$$M_2(A, B) = \alpha H(T(A), T(B)) + \beta H(S(A), T(A)) + \gamma H(S(B), T(B))$$
$$+ \delta_1 H(S(A), T(B)) + \delta_2 H(S(B), T(A)),$$

and $\alpha, \beta, \gamma, \delta_1, \delta_2 \geq 0$, $\delta_1 \leq \delta_2$ with $\alpha + \beta + \gamma + \delta_1 + \delta_2 \leq 1$.

It is obvious that if a pair (S, T) of multivalued mappings on $CB(X)$ is a graph ψ_1-contraction or graph ψ_2-contraction for graph G, then pair (S, T) is also graph

ψ_1-contraction or graph ψ_2-contraction respectively, for the graphs G^{-1}, $\widetilde{G}$ and G_0, here the graph G_0 is defined by $E(G_0) = X \times X$.

Definition 7 A metric space (X, d) is called an ε-chainable metric space for some $\varepsilon > 0$ if for given $x, y \in X$, there is $n \in \mathbb{N}$ and a sequence $\{x_n\}$ such that

$$x_0 = x, \quad x_n = y \quad \text{and} \quad d(x_{i-1}, x_i) < \varepsilon \quad \text{for} \quad i = 1, \ldots, n.$$

For fixed point result of mappings defined on ε-chainable metric space, we refer to [9] and references mentioned therein. We also need of the following lemma of Nadler [23] (see also [7]).

Lemma 1 *Let (X, d) be a metric space. If $U, V \in CB(X)$ with $H(U, V) < \varepsilon$, then for each $u \in U$ there exists an element $v \in V$ such that $d(u, v) < \varepsilon$.*

2   Common Fixed Points

In this section, we obtain coincidence point and common fixed point results for multivalued selfmaps on $CB(X)$ satisfying graph ψ-contraction conditions endowed with a directed graph.

Theorem 3 *Let (X, d) be a metric space endowed with a directed graph G such that $V(G) = X$, $E(G) \supseteq \Delta$ and $S, T : CB(X) \rightarrow CB(X)$ a graph ψ_1-contraction pair such that the range of T contains the range of S. Then the following statements hold:*

(i)   *$CP(S, T) \neq \emptyset$ provided that G is weakly connected with satisfies the property (P^*) and $T(X)$ is complete subspace of $CB(X)$.*

(ii) *if $CP(S, T)$ is complete, then the Pompeiu–Hausdorff weight assigned to the $S(U)$ and $S(V)$ is 0 for all $U, V \in CP(S, T)$.*

(iii) *if $CP(S, T)$ is complete and S and T are weakly compatible, then $Fix(S) \cap Fix(T)$ is a singleton.*

(iv) *$Fix(S) \cap Fix(T)$ is complete if and only if $Fix(S) \cap Fix(T)$ is a singleton.*

Proof To prove (i), let A_0 be an arbitrary element in $CB(X)$. Since range of T contains the range of S, chosen $A_1 \in CB(X)$ such that $S(A_0) = T(A_1)$. Continuing this process, having chosen A_n in $CB(X)$, we obtain an A_{n+1} in $CB(X)$ such that $S(x_n) = T(x_{n+1})$ for $n \in \mathbb{N}$. The inclusion $(A_{n+1}, T(A_{n+1})) \subseteq E(G)$ and $(T(A_{n+1}), A_n) = (S(A_n), A_n) \subseteq E(G)$ implies that $(A_{n+1}, A_n) \subseteq E(G)$.

We may assume that $S(A_n) \neq S(A_{n+1})$ for all $n \in \mathbb{N}$. If not, then $S(A_{2k}) = S(A_{2k+1})$ for some k, implies $T(A_{2k+1}) = S(A_{2k+1})$, and thus $A_{2k+1} \in CP(S, T)$. Now, since $(A_{n+1}, A_n) \subseteq E(G)$ for all $n \in \mathbb{N}$, and pair (S, T) form a graph ψ_1-contraction, so we have

$$H(T\,(A_{n+1})\,,T\,(A_{n+2})) = H(S\,(A_n)\,,S\,(A_{n+1}))$$
$$\leq \psi\,(M_1\,(A_n,A_{n+1}))\,,$$

where

$$M_1\,(A_n,A_{n+1}) = \max\{H(T\,(A_n)\,,T\,(A_{n+1})),\,H\,(S\,(A_n)\,,T\,(A_n)),\,H\,(S\,(A_{n+1})\,,T\,(A_{n+1}))\,,$$
$$\frac{H\,(S\,(A_n)\,,T\,(A_{n+1})) + H\,(S\,(A_{n+1})\,,T\,(A_n))}{2}\}$$
$$= \max\{H(T\,(A_n)\,,T\,(A_{n+1})),\,H\,(T\,(A_{n+1})\,,T\,(A_n)),\,H\,(T\,(A_{n+2})\,,T\,(A_{n+1}))\,,$$
$$\frac{H\,(T\,(A_{n+1})\,,T\,(A_{n+1})) + H\,(T\,(A_{n+2})\,,T\,(A_n))}{2}\}$$
$$\leq \max\{H\,(T\,(A_n)\,,T\,(A_{n+1}))\,,\,H\,(T\,(A_{n+1})\,,T\,(A_{n+2}))\,,$$
$$\frac{H\,(T\,(A_{n+2})\,,T\,(A_{n+1})) + H\,(T\,(A_{n+1})\,,T\,(A_n))}{2}\}$$
$$= \max\{H\,(T\,(A_n)\,,T\,(A_{n+1}))\,,\,H\,(T\,(A_{n+1})\,,T\,(A_{n+2}))\}.$$

Thus, we have

$$H(T\,(A_{n+1})\,,T\,(A_{n+2}) \leq \psi\,(\max\{H\,(T\,(A_n)\,,T\,(A_{n+1}))\,,\,H\,(T\,(A_{n+1})\,,T\,(A_{n+2}))\})$$
$$= \psi(H\,(T\,(A_n)\,,T\,(A_{n+1})))$$

for all $n \in \mathbb{N}$. Therefore for $i = 1, 2, \ldots, n$, we have

$$H(T\,(A_{i-1})\,,T\,(A_i)) \leq \psi(H(A_{i-1},A_i)),$$
$$H(T\,(A_{i-2})\,,T\,(A_{i-1})) \leq \psi(H(A_{i-2},A_{i-1})),$$
$$\cdots$$
$$H(T\,(A_0)\,,T\,(A_1)) \leq \psi(H(A_0,A_1)),$$

and so we obtain

$$H(T\,(A_n)\,,T\,(A_{n+1}) \leq \psi^n(H(A_0,T\,(A_1)))$$

for all $n \in \mathbb{N}$. Now for $m, n \in \mathbb{N}$ with $m > n \geq 1$, we have

$$H\,(T\,(A_n)\,,T\,(A_m)) \leq H\,(T\,(A_n)\,,T\,(A_{n+1})) + H\,(T\,(A_{n+1})\,,T\,(A_{n+2}))$$
$$+ \cdots + H\,(T\,(A_{m-1})\,,T\,(A_m))$$
$$\leq \psi^n(H(A_0,T\,(A_1))) + \psi^{n+1}(H(A_0,T\,(A_1)))$$
$$+ \cdots + \psi^{m-1}(H(A_0,T\,(A_1))).$$

By the convergence of the series $\sum_{i=1}^{\infty} \psi^i(H(A_0,T\,(A_1)))$, we get $H\,(T\,(A_n)\,,$ $T\,(A_m)) \to 0$ as $n, m \to \infty$. Therefore $\{T\,(A_n)\}$ is a Cauchy sequence in $T\,(X)$. Since $(T\,(X)\,,d)$ is complete in $CB\,(X)$, we have $T\,(A_n) \to V$ as $n \to \infty$ for some $V \in CB\,(X)$. Also, we can find U in $CB\,(X)$ such that $T(U) = V$.

We claim that $S(U) = T(U)$. If not, then since $(T(A_{n+1}), T(A_n)) \subseteq E(G)$ so by property (P*), there exists a subsequence $\{T(A_{n_k+1})\}$ of $\{T(A_{n+1})\}$ such that $(T(U), T(A_{n_k+1})) \subseteq E(G)$ for every $n \in \mathbb{N}$. As $(U, T(U)) \subseteq E(G)$ and $(T(A_{n_k+1}), A_{n_k}) = (S(A_{n_k}), A_{n_k}) \subseteq E(G)$ implies that $(U, A_{n_k}) \subseteq E(G)$. Now

$$H(S(U), T(A_{n_k+1})) = H(S(U), S(A_{n_k})) \leq \psi(M_1(U, A_{n_k})), \tag{1}$$

where

$$M_1(U, A_{n_k}) = \max\left\{ H(T(U), T(A_{n_k})), H(S(U), T(U)), H(S(A_{n_k}), T(A_{n_k})), \right.$$
$$\left. \frac{H(S(U), T(A_{n_k})) + H(S(A_{n_k}), T(U))}{2} \right\}$$

$$= \max\left\{ H(T(U), T(A_{n_k})), H(S(U), T(U)), H(T(A_{n_k+1}), T(A_{n_k})), \right.$$
$$\left. \frac{H(S(U), T(A_{n_k})) + H(T(A_{n_k+1}), T(U))}{2} \right\}.$$

Now we consider the following cases:

If $M_1(U, A_{n_k}) = H(T(U), T(A_{n_k}))$, then on taking limit as $k \to \infty$ in (1), we have

$$H(S(U), T(U)) \leq \psi(H(T(U), T(U))),$$

a contradiction.

When $M_1(U, A_{n_k}) = H(S(U), T(U))$, then

$$H(S(U), T(U)) \leq \psi(H(S(U), T(U))),$$

gives a contradiction.

In case $M_1(U, A_{n_k}) = H(T(A_{n_k+1}), T(A_{n_k}))$, then on taking limit as $k \to \infty$ in (1), we get

$$H(S(U), T(U)) \leq \psi(H(T(U), T(U))),$$

a contradiction.

Finally, if $M_1(U, A_{n_k}) = \frac{H(S(U), T(A_{n_k})) + H(T(A_{n_k+1}), T(U))}{2}$, then on taking limit as $k \to \infty$, we have

$$H(S(U), T(U)) \leq \psi\left(\frac{H(S(U), T(U)) + H(T(U), T(U))}{2} \right)$$
$$= \psi\left(\frac{H(S(U), T(U))}{2} \right),$$

a contradiction.

Hence $S(U) = T(U)$, that is, $U \in CP(S, T)$.

To prove (ii), suppose that $CP\,(S,\,T)$ is complete set in G. Let $U,\,V \in CP\,(S,\,T)$ and suppose that the Pompeiu–Hausdorff weight assign to the $S\,(U)$ and $S\,(V)$ is not zero. Since pair $(S,\,T)$ is a graph ψ_1 -contraction, we obtain that

$$
\begin{aligned}
H(S\,(U),\,S\,(V)) \leq\, & \psi(M_1(U,\,V)) \\
\leq\, & \psi(\max\{H(T\,(U),\,T\,(V)),\,H(S\,(U),\,T\,(U)),\,H(S\,(V),\,T\,(V)), \\
& \frac{H(S\,(U),\,T\,(V)) + H(S\,(V),\,T\,(U))}{2}\}) \\
=\, & \psi(\max\{H\,(S\,(U),\,S\,(V)),\,H(S\,(U),\,S\,(U)),\,H(S\,(V),\,T\,(V)), \\
& \frac{H(S\,(U),\,S\,(V)) + H(S\,(V),\,S\,(U))}{2}\}) \\
=\, & \psi\,(H(S\,(U),\,S\,(V))),
\end{aligned}
$$

a contradiction as $\psi\,(t) < t$ for all $t > 0$. Hence (ii) is proved.

To prove (iii), suppose the set $CP\,(S,\,T)$ is weakly compatible. First we are to show that $Fix\,(T) \cap Fix(S)$ is nonempty. Let $W = S\,(U) = T\,(U)$, then we have $T\,(W) = TS\,(U) = ST\,(U) = S\,(W)$, which shows that $W \in CP\,(S,\,T)$. Thus the Pompeiu–Hausdorff weight assign to the $S\,(U)$ and $S\,(W)$ is zero (by ii). Hence $W = S\,(W) = T\,(W)$, that is, $W \in Fix\,(S) \cap Fix\,(T)$. Since $CP\,(S,\,T)$ is singleton set, implies $Fix\,(S) \cap Fix\,(T)$ is singleton.

Finally to prove (iv), suppose the set $Fix\,(S) \cap Fix\,(T)$ is complete. We are to show that $Fix\,(T) \cap Fix(S)$ is singleton. Assume on contrary that there exist $U,V \in CB\,(X)$ such that $U,\,V \in Fix\,(S) \cap Fix\,(T)$ and $U \neq V$. By completeness of $Fix\,(S) \cap Fix\,(T)$, there exists an edge between U and V. As pair $(S,\,T)$ is a graph ψ_1-contraction, so we have

$$
\begin{aligned}
H(U,\,V) =\, & H(S\,(U),\,S\,(V)) \\
\leq\, & \psi(M_1(U,\,V)) \\
=\, & \psi(\max\{H(T\,(U),\,T\,(V)),\,H(S\,(U),\,T\,(U)),\,H(S\,(V),\,T\,(V)), \\
& \frac{H(S\,(U),\,T\,(V)) + H(S\,(V),\,T\,(U))}{2}\}) \\
=\, & \psi(\max\{H(U,\,V),\,H(U,\,U),\,H(V,\,V), \\
& \frac{H(U,\,V) + H(V,\,U)}{2}\}) \\
=\, & \psi\,(H\,(U,\,V)),
\end{aligned}
$$

a contradiction. Hence $U = V$. Conversely, if $Fix(S) \cap Fix(T)$ is singleton, then since $E(G) \supseteq \Delta$, so it is obvious that $F(S) \cap F(T)$ is complete set.   □

Example 1 Let $X = \{1, 2, \ldots, n\} = V\,(G)$, $n > 2$ and $E\,(G) = \{(i,\,j) \in X \times X :\ i \leq j\}$. Let $V\,(G)$ be endowed with metric $d : X \times X \to \mathbb{R}^+$ defined by

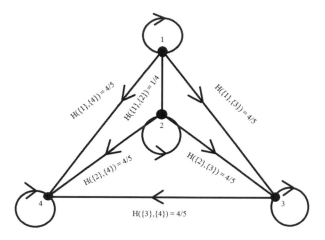

Fig. 1 Graph G defined in Example 1

$$d\,(x,\,y) = \begin{cases} 0 & \text{if } x = y, \\[2mm] \frac{1}{n} & \text{if } x \in \{1, 2\} \text{ with } x \neq y, \\[2mm] \frac{n}{n+1} & \text{otherwise.} \end{cases}$$

Furthermore, the Pompeiu–Hausdorff metric is given by

$$H(A,\,B) = \begin{cases} \frac{1}{n} & \text{if } A, B \subseteq \{1, 2\} \text{ with } A \neq B, \\[2mm] \frac{n}{n+1} & \text{if } A \text{ or } B \text{ or both } \nsubseteq \{1, 2\} \text{ with } A \neq B, \\[2mm] 0 & \text{if } A = B. \end{cases}$$

The Pompeiu–Hausdorff weights (for $n = 4$) assigned to $A, B \in CB(X)$ are shown in the Fig. 1.

Define $S, T : CB(X) \to CB(X)$ as follows:

$$S(U) = \begin{cases} \{1\}, & \text{if } U \subseteq \{1, 2\}, \\ \{1, 2\}, & \text{if } U \nsubseteq \{1, 2\} \end{cases}$$

$$T(U) = \begin{cases} \{1\}, & \text{if } U = \{1\}, \\ \{1, 2, 3\}, & \text{if } U \subseteq \{2, 3\}. \\ \{1, 2, \ldots, n\}, & \text{otherwise.} \end{cases}$$

Note that, for all $V \in CB(X)$, $(V, S\,(V)) \subseteq E\,(G)$ and $(V, T\,(V)) \subseteq E\,(G)$.
Let $\psi : \mathbb{R}_+ \to \mathbb{R}_+$ be defined by

$$\psi\left(\alpha\right) = \begin{cases} \frac{1}{2}\alpha^2 & 0 \leq \alpha < \frac{1}{2} \\ \\ \frac{\alpha}{\alpha+1}, & \frac{1}{2} \leq \alpha. \end{cases}$$

It is easy to verify that $\psi \in \Psi$. Now for all $A, B \in CB\left(X\right)$ with $S\left(A\right) \neq S\left(B\right)$, we consider the following cases:

(i) If $A \subseteq \{1, 2\}$ and $B = \{3\}$ with $\left(A, B\right) \subseteq E\left(G\right)$, then we have

$$
\begin{aligned}
H(S\left(A\right), S\left(B\right)) &= H\left(\{1\}, \{1, 2\}\right) \\
&= \frac{1}{n} \\
&< \frac{n}{2n+1} \\
&= \psi\left(\frac{n}{n+1}\right) \\
&= \psi\left(H\left(\{1, 2\}, \{1, 2, 3\}\right)\right) \\
&= \psi\left(H(S\left(B\right), T\left(B\right))\right) \leq \psi(M_1(A, B)).
\end{aligned}
$$

(ii) When $A \subseteq \{1, 2\}$ and $B \subsetneq \{1, 2, 3\}$ with $\left(A, B\right) \subseteq E\left(G\right)$, implies that

$$
\begin{aligned}
H(S\left(A\right), S\left(B\right)) &= H\left(\{1\}, \{1, 2\}\right) \\
&= \frac{1}{n} \\
&< \frac{n}{2n+1} \\
&= \psi\left(\frac{n}{n+1}\right) \\
&= \psi\left(H\left(\{1, 2\}, \{1, 2, ..., n\}\right)\right) \\
&= \psi\left(H(S\left(B\right), T\left(B\right))\right) \leq \psi(M_1(A, B)).
\end{aligned}
$$

(iii) In case $A = \{3\}$ and $B \subseteq \{1, 2\}$ and with $\left(A, B\right) \subseteq E\left(G\right)$, we have

$$
\begin{aligned}
H(S\left(A\right), S\left(B\right)) &= H\left(\{1, 2\}, \{1\}\right) \\
&= \frac{1}{n} \\
&< \frac{n}{2n+1} \\
&= \psi\left(\frac{n}{n+1}\right) \\
&= \psi\left(H\left(\{1, 2\}, \{1, 2, 3\}\right)\right) \\
&= \psi\left(H(S\left(A\right), T\left(A\right))\right) \leq \psi(M_1(A, B)).
\end{aligned}
$$

(iv) When $A \subsetneq \{1, 2, 3\}$ and $B \subseteq \{1, 2\}$ with $(A, B) \subseteq E(G)$, implies that

$$
\begin{aligned}
H(S(A), S(B)) &= H(\{1, 2\}, \{1\}) \\
&= \frac{1}{n} \\
&< \frac{n}{2n + 1} \\
&= \psi\left(\frac{n}{n + 1}\right) \\
&= \psi(H(\{1, 2\}, \{1, 2, ..., n\})) \\
&= \psi(H(S(A), T(A))) \leq \psi(M_1(A, B)).
\end{aligned}
$$

Hence pair (S, T) is graph ψ_1-contraction. Thus the conditions of Theorem 3 hold. Moreover, $\{1\}$ is the common fixed point of S and T, and $Fix(S) \cap Fix(T)$ is complete.

In the next example we show that it is not necessary that given graph $(V(G), E(G))$ will always be a complete graph.

Example 2 Let $X = \{1, 2, \ldots, n\} = V(G)$, $n > 2$ and

$$
\begin{aligned}
E(G) = \{&(1, 1), (2, 2), \ldots, (n, n), \\
&(1, 2), \ldots, (1, n)\}.
\end{aligned}
$$

On $V(G)$, the metric $d : X \times X \to \mathbb{R}^+$ and Pompeiu–Hausdorff metric $H : CB(X) \to \mathbb{R}^+$ are defined as in Example 1. The Pompeiu–Hausdorff weights (for $n = 4$) assigned to $A, B \in CB(X)$ are shown in the Fig. 2.

Fig. 2 Graph G defined in Example 2

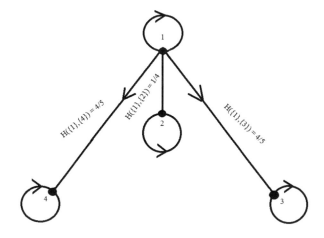

Define $S, T : CB(X) \to CB(X)$ as follows:

$$S(U) = \begin{cases} \{1\}, & \text{if } U = \{1\}, \\ \{1, 2\}, & \text{if } U \neq \{1\} \end{cases}$$

$$T(U) = \begin{cases} \{1\}, & \text{if } U = \{1\}, \\ \{1, \ldots, n\}, & \text{if } U \neq \{1\}. \end{cases}$$

Note that, $(S(A), A) \subseteq E(G)$ and $(A, T(A)) \subseteq E(G)$ for all $A \in CB(X)$.

Take $\psi(\alpha) = \begin{cases} \frac{1}{8}\alpha, & \alpha \in [0, \frac{1}{4}] \\ \\ \frac{\alpha+1}{\alpha+2}, & \alpha \geq \frac{1}{4}. \end{cases}$

Note that $\psi \in \Psi$. Now, for all $A, B \in CB(X)$ with $S(A) \neq S(B)$, we consider the following cases:

(I) If $A = \{1\}$ and $B \neq \{1\}$, then we have

$$H(S(A), S(B)) = \frac{1}{n}$$
$$< \frac{2n+1}{3n+1}$$
$$= \psi\left(\frac{n}{n+1}\right)$$
$$= \psi(H(S(B), T(B))) \leq \psi(M_1(A, B)).$$

(II) If $A \neq \{1\}$ and $B = \{1\}$, then we have

$$H(S(A), S(B)) = \frac{1}{n}$$
$$< \frac{2n+1}{3n+1}$$
$$= \psi\left(\frac{n}{n+1}\right)$$
$$= \psi(H(S(A), T(A))) \leq \psi(M_1(A, B)).$$

Hence pair (S, T) is graph ψ_1-contraction. Thus all the conditions of Theorem 3 are satisfied. Moreover, S and T have a common fixed point and $Fix(S) \cap Fix(T)$ is complete in $CB(X)$.

Theorem 4 *Let (X, d) be a ε-chainable complete metric space for some $\varepsilon > 0$ and $S, T : CB(X) \to CB(X)$ be multivalued mappings. Suppose that for all $A, B \in CB(X)$,*

$$0 < H(S(A), S(B)) < \varepsilon$$

and there exists a $\psi \in \Psi$ such

$$H\left(S\left(A\right),S\left(B\right)\right) \leq \psi(M(A,B)),$$

hold where

$$M(A,B) = \max\left\{ H(T\left(A\right),T\left(B\right)),H(S\left(A\right),T\left(A\right)),H(S\left(B\right),T\left(B\right)),\right.$$

$$\left. \frac{H(S\left(A\right),T\left(B\right)) + H(S\left(B\right),T\left(A\right))}{2}\right\}.$$

Then S and T have a common fixed point provided that S and T are weakly compatible.

Proof By Lemma 1, from $H\left(A,B\right) < \varepsilon$, we have for each $a \in A$, an element $b \in B$ such that $d(a,b) < \varepsilon$. Consider the graph G as $V(G) = X$ and

$$E(G) = \{(a,b) \in X \times X : 0 < d(a,b) < \varepsilon\}.$$

Then the ε-chainability of (X,d) implies that G is connected. For $(A,B) \subset E(G)$, we have from the hypothesis

$$H\left(S\left(A\right),S\left(B\right)\right) \leq \psi(M(A,B)),$$

where $M(A,B) = \max\left\{ H(S\left(A\right),T\left(B\right)),H(S\left(A\right),T\left(A\right)),H(S\left(B\right),T\left(B\right)),\right.$

$$\left. \frac{H(S\left(A\right),T\left(B\right)) + H(S\left(B\right),T\left(A\right))}{2}\right\}$$

implies that pair (S,T) is graph ψ_1−contraction.

Also, G has property (P*). Indeed, if $\{X_n\}$ in $CB(X)$ with $X_n \to X$ as $n \to \infty$ and $(X_n, X_{n+1}) \subset E(G)$ for $n \in \mathbb{N}$, implies that there is a subsequence $\{X_{n_k}\}$ of $\{X_n\}$ such that $\left(X_{n_k}, X\right) \subset E(G)$ for $n \in \mathbb{N}$. So by Theorem 3 (iii), S and T have a common fixed point.   □

Corollary 1 *Let (X,d) be a complete metric space endowed with a directed graph G such that $V(G) = X$ and $E(G) \supseteq \Delta$. Suppose that the mapping $S : CB(X) \to CB(X)$ satisfies the following:*

(a)   *for every V in $CB(X)$, $(V, S(V)) \subset E(G)$.*
(b)   *There exists $a > 0$ such that for $\psi \in F$ there is an edge between A and B implies that*

$$H\left(S\left(A\right),S\left(B\right)\right) \leq \psi(M_1(A,B)),$$

where

$$M_1(A, B) = \max \left\{ H(A, B), H(A, S(A)), H(B, S(B)), \right.$$
$$\left. \frac{H(A, S(B)), H(B, S(A))}{2} \right\}).$$

Then following statements hold:

(i) *if Fix (S) is complete, then the Pompeiu–Hausdorff weight assigned to the U, V ∈ Fix (S) is 0.*

(ii) *If the weakly connected graph G satisfies the property (P*), then S has a fixed point.*

(iii) *Fix (S) is complete if and only if Fix (S) is a singleton.*

Proof Take $T = I$ (identity map) in (I), then Corollary 1 follows from Theorem 3.   □

Theorem 5 *Let (X, d) be a metric space endowed with a directed graph G such that $V(G) = X$, $E(G) \supseteq \Delta$ and $S, T : CB(X) \to CB(X)$ a graph ψ_2 -contraction pair such that the range of T contains the range of S. Then the following statements hold:*

(i) *$CP(S, T) \neq \emptyset$ provided that G is weakly connected with satisfies the property (P*) and $T(X)$ is complete subspace of $CB(X)$.*

(ii) *if $CP(S, T)$ is complete, then the Pompeiu–Hausdorff weight assigned to the $S(U)$ and $S(V)$ is 0 for all $U, V \in CP(S, T)$.*

(iii) *if $CP(S, T)$ is complete and S and T are weakly compatible, then $Fix(S) \cap Fix(T)$ is a singleton.*

(iv) *$Fix(S) \cap Fix(T)$ is complete if and only if $Fix(S) \cap Fix(T)$ is a singleton.*

Proof To prove (i), let A_0 be an arbitrary element in $CB(X)$. Since range of T contains the range of S, chosen $A_1 \in CB(X)$ such that $S(A_0) = T(A_1)$. Continuing this process, having chosen A_n in $CB(X)$, we obtain an A_{n+1} in $CB(X)$ such that $S(x_n) = T(x_{n+1})$ for $n \in \mathbb{N}$. The inclusion $(A_{n+1}, T(A_{n+1})) \subseteq E(G)$ and $(T(A_{n+1}), A_n) = (S(A_n), A_n) \subseteq E(G)$ implies that $(A_{n+1}, A_n) \subseteq E(G)$.

We may assume that $S(A_n) \neq S(A_{n+1})$ for all $n \in \mathbb{N}$. If not, then $S(A_{2k}) = S(A_{2k+1})$ for some k, implies $T(A_{2k+1}) = S(A_{2k+1})$, and thus $A_{2k+1} \in CP(S, T)$. Now, since $(A_{n+1}, A_n) \subseteq E(G)$ for all $n \in \mathbb{N}$, and pair (S, T) form a graph ψ_2-contraction, so we have

$$H(T(A_{n+1}), T(A_{n+2})) = H(S(A_n), S(A_{n+1}))$$
$$\leq \psi(M_2(A_n, A_{n+1})),$$

where

$$M_2 (A_n, A_{n+1})$$
$$= \alpha H(T (A_n), T (A_{n+1})) + \beta H (S (A_n), T (A_n)) + \gamma H (S (A_{n+1}), T (A_{n+1}))$$
$$\quad \delta_1 H (S (A_n), T (A_{n+1})) + \delta_2 H (S (A_{n+1}), T (A_n))$$
$$= \alpha H(T (A_n), T (A_{n+1})) + \beta H (T (A_{n+1}), T (A_n)) + \gamma H (T (A_{n+2}), T (A_{n+1}))$$
$$\quad \delta_1 H (T (A_{n+1}), T (A_{n+1})) + \delta_2 H (T (A_{n+2}), T (A_n))$$
$$\leq (\alpha + \beta) H (T (A_n), T (A_{n+1})) + \gamma H (T (A_{n+1}), T (A_{n+2})),$$
$$\quad \delta_2 [H (T (A_{n+2}), T (A_{n+1})) + H (T (A_{n+1}), T (A_n))]$$
$$= (\alpha + \beta + \delta_2) H (T (A_n), T (A_{n+1})) + (\gamma + \delta_2) H (T (A_{n+1}), T (A_{n+2})).$$

Now, if $H (T (A_n), T (A_{n+1})) \leq H (T (A_{n+1}), T (A_{n+2}))$, we have

$$H(T (A_{n+1}), T (A_{n+2}) \leq \psi \left(\max \{ H (T (A_n), T (A_{n+1})), H (T (A_{n+1}), T (A_{n+2})) \} \right)$$
$$= \psi (H (T (A_n), T (A_{n+1})))$$

for all $n \in \mathbb{N}$. Therefore for $i = 1, 2, \ldots, n$, we have

$$H(T (A_{i-1}), T (A_i)) \leq \psi (H(A_{i-1}, A_i)),$$
$$H(T (A_{i-2}), T (A_{i-1})) \leq \psi (H(A_{i-2}, A_{i-1})),$$
$$\cdots,$$
$$H(T (A_0), T (A_1)) \leq \psi (H(A_0, A_1)),$$

and so we obtain

$$H(T (A_n), T (A_{n+1}) \leq \psi^n (H(A_0, T (A_1)))$$

for all $n \in \mathbb{N}$. Follows the similar argument to those in the proof of Theorem 3, we get $H (T (A_n), T (A_m)) \to 0$ as $n, m \to \infty$. Therefore $\{T (A_n)\}$ is a Cauchy sequence in $T (X)$. Since $(T (X), d)$ is complete in $CB (X)$, we have $T (A_n) \to V$ as $n \to \infty$ for some $V \in CB (X)$. Also, we can find U in $CB (X)$ such that $T(U) = V$.

We claim that $S(U) = T(U)$. If not, then since $(T (A_{n+1}), T (A_n)) \subseteq E (G)$ so by property (P*), there exists a subsequence $\{T (A_{n_k+1})\}$ of $\{T (A_{n+1})\}$ such that $(T (U), T (A_{n_k+1})) \subseteq E (G)$ for every $n \in \mathbb{N}$. As $(U, T (U)) \subseteq E (G)$ and $(T (A_{n_k+1}), A_{n_k}) = (S (A_{n_k}), A_{n_k}) \subseteq E (G)$ implies that $(U, A_{n_k}) \subseteq E (G)$. Now

$$H(S (U), T (A_{n_k+1})) = H(S (U), S (A_{n_k})) \leq \psi (M_2 (U, A_{n_k})), \qquad (2)$$

where

$$M_2\left(U, A_{n_k}\right) = \alpha H(T\left(U\right), T\left(A_{n_k}\right)) + \beta H(S\left(U\right), T\left(U\right)) + \gamma H(S\left(A_{n_k}\right), T\left(A_{n_k}\right))$$
$$+ \delta_1 H(S\left(U\right), T\left(A_{n_k}\right)) + \delta_2 H(S\left(A_{n_k}\right), T\left(U\right))$$
$$= \alpha H(T\left(U\right), T\left(A_{n_k}\right)) + \beta H(S\left(U\right) + T\left(U\right)) + \gamma H(T\left(A_{n_k+1}\right), T\left(A_{n_k}\right))$$
$$+ \delta_1 H(S\left(U\right), T\left(A_{n_k}\right)) + \delta_2 H(T\left(A_{n_k+1}\right), T\left(U\right)).$$

On taking limit as $k \to \infty$ in (2), we have

$$H(S\left(U\right), T\left(U\right)) \leq \psi\left((\beta + \delta_1)H\left(T\left(U\right), T\left(U\right)\right)\right)$$
$$< H(S\left(U\right), T\left(U\right)),$$

a contradiction. Hence $S\left(U\right) = T\left(U\right)$, that is, $U \in CP\left(S, T\right)$.

To prove (ii), suppose that $CP\left(S, T\right)$ is complete set in G. Let $U, V \in CP\left(S, T\right)$ and suppose that the Pompeiu–Hausdorff weight assign to the $S\left(U\right)$ and $S\left(V\right)$ is not zero. Since pair (S, T) is a graph ψ_2 -contraction, we obtain that

$$H(S\left(U\right), S\left(V\right)) \leq \psi(M_2(U, V)), \tag{3}$$

where

$$M_2(U, V)) = \alpha H(T\left(U\right), T\left(V\right)) + \beta H(S\left(U\right), T\left(U\right)) + \gamma H(S\left(V\right), T\left(V\right))$$
$$\delta_1 H(S\left(U\right), T\left(V\right)) + \delta_2 H(S\left(V\right), T\left(U\right))$$
$$= \alpha H\left(S\left(U\right), S\left(V\right)\right) + \beta H(S\left(U\right), S\left(U\right)) + \gamma H(S\left(V\right), T\left(V\right))$$
$$= (\alpha + \delta_1 + \delta_2)H(S\left(U\right), S\left(V\right)),$$

thus

$$H(S\left(U\right), S\left(V\right)) \leq \psi((\alpha + \delta_1 + \delta_2)H(S\left(U\right), S\left(V\right)))$$
$$< \psi\left(H(S\left(U\right), S\left(V\right))\right),$$

a contradiction as $\psi\left(t\right) < t$ for all $t > 0$. Hence (ii) is proved.

To prove (iii), suppose the set $CP\left(S, T\right)$ is weakly compatible. First we are to show that $Fix\left(T\right) \cap Fix(S)$ is nonempty. Let $W = S\left(U\right) = T\left(U\right)$, then we have $T\left(W\right) = TS\left(U\right) = ST\left(U\right) = S\left(W\right)$, which shows that $W \in CP\left(S, T\right)$. Thus the Pompeiu–Hausdorff weight assign to the $S\left(U\right)$ and $S\left(W\right)$ is zero (by ii). Hence $W = S\left(W\right) = T\left(W\right)$, that is, $W \in Fix\left(S\right) \cap Fix\left(T\right)$. Since $CP\left(S, T\right)$ is singleton set, implies $Fix\left(S\right) \cap Fix\left(T\right)$ is singleton.

Finally to prove (iv), suppose the set $Fix\left(S\right) \cap Fix\left(T\right)$ is complete. We are to show that $Fix\left(T\right) \cap Fix(S)$ is singleton. Assume on contrary that there exist $U, V \in CB\left(X\right)$ such that $U, V \in Fix\left(S\right) \cap Fix\left(T\right)$ and $U \neq V$. By completeness of $Fix\left(S\right) \cap Fix\left(T\right)$, there exists an edge between U and V. As pair (S, T) is a graph ψ_2-contraction, so we have

$$H(U, V) = H(S(U), S(V))$$
$$\le \psi(M_2(U, V))$$
$$= \psi(\alpha H(T(U), T(V)) + \beta H(S(U), T(U)) + \gamma H(S(V), T(V))$$
$$+ \delta_1 H(S(U), T(V)) + \delta_2 H(S(V), T(U)))$$
$$= \psi(\alpha H(U, V) + \beta H(U, U) + \gamma H(V, V) + \delta_1 H(U, V) + \delta_2 H(V, U))$$
$$\le \psi(H(U, V)),$$

a contradiction. Hence $U = V$. Conversely, if $Fix(S) \cap Fix(T)$ is singleton, then since $E(G) \supseteq \Delta$, so it is obvious that $F(S) \cap F(T)$ is complete set.   □

Example 3 Let $X = \mathbb{R}_+ = V(G)$ be endowed with Euclidean metric d. Let $f : X \to X$ be defined as $f(x) = \begin{cases} 10, & if \ x \in [0, 10] \\ 20, & otherwise \end{cases}$ and $(a, b) \in E(G)$ for some $a \in A, b \in B$ if $b = f(a)$. Define $S, T : CB(X) \to CB(X)$ as follows:

$$S(U) = \begin{cases} [0, 10], & if \ U \subseteq [0, 10] \\ [10, 20], & otherwise, \end{cases}$$

$$T(U) = \begin{cases} [0, 10], & if \ U \subseteq [0, 10] \\ [5, 25], & otherwise. \end{cases}$$

Note that, for all $V \in CB(X)$, $(S(V), V) \subseteq E(G)$ and $(V, T(V)) \subseteq E(G)$. Let $\psi : \mathbb{R}_+ \dashrightarrow \mathbb{R}_+$ be defined by

$$\psi(\alpha) = \begin{cases} \frac{3}{4}t & 0 \le t < 1 \\ \\ \frac{5}{6}t, & 1 \le t. \end{cases}$$

It is easy to verify that $\psi \in \Psi$. Now for all $A, B \in CB(X)$ with $S(A) \ne S(B)$, we consider $A \subseteq [0, 10]$ and $B \not\subseteq [0, 10]$ with $(A, B) \subseteq E(G)$, implies

$$H(S(A), S(B)) = H([0, 10], [10, 20])$$
$$= 10 < \frac{100}{9}$$
$$= \psi(15\alpha + 5\beta)$$
$$= \psi(\alpha H([0, 10], [5, 25]) + \gamma H([10, 20], [5, 25]))$$
$$= \psi(\alpha H(T(A), T(B)) + \gamma H(S(B), T(B))) \le \psi(M_2(A, B)),$$

where $\alpha = \frac{5}{6}, \gamma = \frac{1}{6}, \beta = \delta_1 = \delta_2 = 0$ and

$$M_2(A, B) = \alpha H(T(A), T(B)) + \beta H(S(A), T(A)) + \gamma H(S(B), T(B))$$
$$+ \delta_1 H(S(A), T(B)) + \delta_2 H(S(B), T(A))$$

Hence pair (S, T) is graph ψ_2-contraction. Thus all the conditions of Theorem 5 are satisfied. Moreover, the set $[0, 10]$ is the common fixed point of S and T, and $Fix(S) \cap Fix(T)$ is complete.

The following corollary generalizes and extends [3, Theorem 2.1].

Corollary 2 *Let (X, d) be a complete metric space endowed with a directed graph G such that $V(G) = X$ and $E(G) \supseteq \Delta$. Suppose that the mappings $S, T : CB(X) \to CB(X)$ satisfies the following:*

(a) *for every V in $CB(X)$, $(S(V), V) \subset E(G)$ and $(V, T(V)) \subseteq E(G)$.*
(b) *There exists $\psi \in \Psi$ such that for all $A, B \in CB(X)$ with there is an edge between A and B implies*

$$H(S(A), S(B)) \le \psi(\alpha H(T(A), T(B)) + \beta H(S(A), T(A)) + \gamma H(S(B), T(B)))$$

hold, where α, β, γ are nonnegative real numbers with $\alpha + \beta + \gamma \le 1$. If the range of T contains the range of S, then the following statements hold:

(i) *$CP(S, T) \ne \emptyset$ provided that G is weakly connected with satisfies the property (P^*) and $T(X)$ is complete subspace of $CB(X)$.*
(ii) *if $CP(S, T)$ is complete, then the Pompeiu–Hausdorff weight assigned to the $S(U)$ and $S(V)$ is 0 for all $U, V \in CP(S, T)$.*
(iii) *if $CP(S, T)$ is complete and S and T are weakly compatible, then $Fix(S) \cap Fix(T)$ is a singleton.*
(iv) *$Fix(S) \cap Fix(T)$ is complete if and only if $Fix(S) \cap Fix(T)$ is a singleton.*

Corollary 3 *Let (X, d) be a complete metric space endowed with a directed graph G such that $V(G) = X$ and $E(G) \supseteq \Delta$. Suppose that the mappings $S : CB(X) \to CB(X)$ satisfies the following:*

(a) *for every V in $CB(X)$, $(S(V), V) \subset E(G)$.*
(b) *There exists $\psi \in \Psi$ such that for all $A, B \in CB(X)$ with there is an edge between A and B implies*

$$H(S(A), S(B)) \le \psi(\alpha H(A, B) + \beta H(S(A), A) + \gamma H(B, S(B)))$$

hold, where α, β, γ are nonnegative real numbers with $\alpha + \beta + \gamma \le 1$. Then the following statements hold:

(i) *if $Fix(S)$ is complete, then the Pompeiu–Hausdorff weight assigned to the $U, V \in Fix(S)$ is 0.*
(ii) *If the weakly connected graph G satisfies the property (P^*), then S has a fixed point.*
(iii) *$Fix(S)$ is complete if and only if $Fix(S)$ is a singleton.*

Proof If we take $T = I$ (identity map) in Corollary 2, the result follows.   □

Remark 1 (1) If $E(G) := X \times X$, then clearly G is connected and our Theorem 2.1 improves and generalizes Theorem 2.1 in [3], Theorem 2.1 in [9], and Theorem 3.1 in [20].

(2) If $E(G) := X \times X$, then clearly G is connected and our Theorem 2.4 extends and generalizes Theorem 2.5 in [9], Theorem 3.2 in [23], Theorem 5.1 in [16] and Theorem 3.1 in [20].

(3) If $E(G) := X \times X$, then clearly G is connected and our Corollary 2.5 improves and generalizes Theorem 2.1 in [9], Theorem 3.2 in [23] and Theorem 3.1 in [20].

3 Conclusion

Jachymski and Jozwik initiated the study of ordered structured metric fixed point theory by using the ordered structured with a graph structure on a metric space. Recently many results appeared in the literature giving the fixed point problems of mappings endow with graph. We presented the common fixed points of a class of multivalued maps with set-valued domain that are only commuting at their coincidence points endow with a directed graph. We presented some examples to show the validity of obtained results.

Acknowledgements Talat Nazır and Xiaomin Qi are grateful to the Erasmus Mundus project FUSION for supporting the research visit to Mälardalen University, Sweden, and to the Research environment MAM in Mathematics and Applied Mathematics, Division of Applied Mathematics, the School of Education, Culture and Communication of Mälardalen University for creating excellent research environment.

References

1. Abbas, M., Jungck, G.: Common fixed point results for noncommuting mappings without continuity in cone metric spaces. J. Math. Anal. Appl. **341**, 416–420 (2008)
2. Abbas, M., Nazir, T.: Common fixed point of a power graphic contraction pair in partial metric spaces endowed with a graph. Fixed Point Theory Appl. **2013**(20) (2013). doi:10.1186/1687-1812-2013-20
3. Abbas, M., Alfuraidan, M.R., Khan, A.R., Nazir, T.: Fixed point results for set-contractions on metric spaces with a directed graph. Fixed Point Theory Appl. **2015**(14) (2015). doi:10.1186/s13663-015-0263-z
4. Aleomraninejad, S.M.A., Rezapoura, Sh, Shahzad, N.: Some fixed point results on a metric space with a graph. Topol. Appl. **159**, 659–663 (2012)
5. Alfuraidan, M.R., Khamsi, M.A.: Caristi fixed point theorem in metric spaces with a graph. Abstract Appl. Anal. **2014**, Article ID 303484 (2014). http://dx.doi.org/10.1155/2014/303484
6. Amini-Harandi, A., Emami, H.: A fixed point theorem for contraction type maps in partially ordered metric spaces and application to ordinary differential equations. Nonlinear Anal. **72**, 2238–2242 (2010)
7. Assad, N.A., Kirk, W.A.: Fixed point theorems for setvalued mappings of contractive type. Pacific. J. Math. **43**, 533–562 (1972)

8. Beg, I., Butt, A.R.: Fixed point theorems for set valued mappings in partially ordered metric spaces. Int. J. Math. Sci. **7**(2), 66–68 (2013)
9. Beg, I., Butt, A.R.: Fixed point of set-valued graph contractive mappings. J. Inequ. Appl. **2013**(252) (2013). doi:10.1186/1029-242X-2013-252
10. Berinde, V.: Iterative Approximation of Fixed Points. Springer, Heidelberg (2007)
11. Berinde, M., Berinde, V.: On a general class of multivalued weakly Picard mappings. J. Math. Anal. Appl. **326**, 772–782 (2007)
12. Bojor, F.: Fixed point of φ-contraction in metric spaces endowed with a graph. Ann. Uni. Craiova Math. Comp. Sci. Ser. **37**(4), 85–92 (2010)
13. Bojor, F.: Fixed point theorems for Reich type contractions on metric spaces with a graph. Nonlinear Anal. **75**, 3895–3901 (2012)
14. Bojor, F.: On Jachymski's theorem. Ann. Univ. Craiova. Math. Comp. Sci. Ser. **40**(1), 23–28 (2012)
15. Chifu, C.I., Petrusel, G.R.: Generalized contractions in metric spaces endowed with a graph. Fixed Point Theory Appl. **2012**(161) (2012). doi:10.1186/1687-1812-2012-161
16. Edelstein, M.: An extension of Banach's contraction principle. Proc. Am. Math. Soc. **12**, 07–10 (1961)
17. Gwozdz-Lukawska, G., Jachymski, J.: IFS on a metric space with a graph structure and extensions of the Kelisky–Rivlin theorem. J. Math. Anal. Appl. **356**, 453–463 (2009)
18. Harjani, J., Sadarangani, K.: Fixed point theorems for weakly contractive mappings in partially ordered sets. Nonlinear Anal. **71**, 3403–3410 (2009)
19. Jachymski, J.: The contraction principle for mappings on a metric space with a graph. Proc. Am. Math. Soc. **136**, 1359–1373 (2008)
20. Jachymski, J., Jozwik, I.: Nonlinear contractive conditions: a comparison and related problems. Banach Center Publ. **77**, 123–146 (2007)
21. Jungck, G.: Common fixed points for commuting and compatible maps on compacta. Proc. Am. Math. Soc. **103**, 977–983 (1988)
22. Jungck, G.: Common fixed points for noncontinuous nonself maps on nonmetric spaces. Far East J. Math. Sci. **4**, 199–215 (1996)
23. Nadler, S.B.: Multivalued contraction mappings. Pacific J. Math. **30**, 475–488 (1969)
24. Nicolae, A., O'Regan, D., Petrusel, A.: Fixed point theorems for singlevalued and multivalued generalized contractions in metric spaces endowed with a graph. J. Georgian Math. Soc. **18**, 307–327 (2011)
25. Nieto, J.J., López, R.R.: Contractive mapping theorems in partially ordered sets and applications to ordinary differential equations. Order **22**(3), 223–239 (2005)
26. Ran, A.C.M., Reurings, M.C.B.: A fixed point theorem in partially ordered sets and some application to matrix equations. Proc. Am. Math. Soc. **132**, 1435–1443 (2004)

Common Fixed Point Results for Family of Generalized Multivalued F-Contraction Mappings in Ordered Metric Spaces

Talat Nazir and Sergei Silvestrov

Abstract In this paper, we study the existence of common fixed points of family of multivalued mappings satisfying generalized F-contractive conditions in ordered metric spaces. These results establish some of the general common fixed point theorems for family of multivalued maps.

Keywords Common fixed point · Multivalued mapping · F-contraction · Ordered metric space

1 Introduction and Preliminaries

Markin [16] initiated the study of fixed points for multivalued nonexpansive and contractive maps. Later, a useful and interesting fixed point theory for such maps was developed. Later, a rich and interesting fixed point theory for such multivalued maps was developed; see, for instance [6, 7, 9–11, 14, 15, 18–20, 23]. The theory of multivalued maps has various applications in convex optimization, dynamical systems, commutative algebra, differential equations and economics. Recently, Wardowski [25] introduced a new contraction called F-contraction and proved a fixed point result as a generalization of the Banach contraction principle. Abbas et al. [3] obtained common fixed point results by employed the F-contraction condition. Further in this direction, Abbas et al. [4] introduced a notion of generalized F-contraction mapping and employed there results to obtain a fixed point of a generalized nonexpansive mappings on star shaped subsets of normed linear spaces. Minak et al. [17] proved some fixed point results for Ciric type generalized F-contractions

T. Nazir (✉) · S. Silvestrov
Division of Applied Mathematics, School of Education, Culture and Communication,
Mälardalen University, Box 883, 721 23 Västerås, Sweden
e-mail: talat@ciit.net.pk

T. Nazir
Department of Mathematics, COMSATS Institute of Information Technology,
Abbottabad 22060, Pakistan
e-mail: sergei.silvestrov@mdh.se

© Springer International Publishing Switzerland 2016
S. Silvestrov and M. Rančić (eds.), *Engineering Mathematics II*,
Springer Proceedings in Mathematics & Statistics 179,
DOI 10.1007/978-3-319-42105-6_20

419

on complete metric spaces. Recently, [5] established some fixed point results for multivalued F-contraction maps on complete metrics spaces.

The aim of this paper is to prove common fixed points theorems for family of multivalued generalized F-contraction mappings without using any commutativity condition in partially ordered metric space. These results extend and unify various comparable results in the literature [12, 13, 21, 22].

We begin with some basic known definitions and results which will be used in the sequel. Throughout this article, $\mathbb{N}$, $\mathbb{R}^+$, $\mathbb{R}$ denote the set of natural numbers, the set of positive real numbers and the set of real numbers, respectively.

Let F be the collection of all mappings $F : \mathbb{R}^+ \to \mathbb{R}$ that satisfy the following conditions:

(F_1) F is strictly increasing, that is, for all $a, b \in \mathbb{R}^+$ such that $a < b$ implies that $F(a) < F(b)$.
(F_2) For every sequence $\{a_n\}$ of positive real numbers, $\lim_{n \to \infty} a_n = 0$ and $\lim_{n \to \infty} F(a_n) = -\infty$ are equivalent.
(F_3) There exists $\lambda \in (0, 1)$ such that $\lim_{a \to 0^+} a^\lambda F(\lambda) = 0$.

Definition 20.1 ([25]) Let (X, d) be a metric space and $F \in F$. A mapping $f : X \to X$ is said to be an F-contraction on X if there exists $\tau > 0$ such that $d(fx, fy) > 0$ implies that

$$\tau + F(d(fx, fy)) \le F(d(x, y))$$

for all $x, y \in X$.

Wardowski [25] gave the following result.

Theorem 20.1 *Let (X, d) be a complete metric space and mapping $f : X \to X$ be and F−contraction. Then there exists a unique x in X such that $x = fx$. Moreover, for any $x_0 \in X$, the iterative sequence $x_{n+1} = f(x_n)$ converges to x.*

Kannan [12] has proved a fixed point theorem for a single valued self mapping T of a metric space X satisfying the property

$$d(Tx, Ty) \le h\{d(x, Tx) + d(y, Ty)\}$$

for all x, y in X and for a fixed where $h \in [0, \frac{1}{2})$.

Ciric [8] considered a mapping $T : X \to X$ satisfying the following contractive condition:

$$d(Tx, Ty) \le q \max\{d(x, y), d(x, Tx), d(y, Ty), d(x, Ty), d(y, Tx)\},$$

where $q \in [0, 1)$. He proved the existence of a fixed point when X is a T-orbitally complete metric space.

Latif and Beg [13] extended mappings considered by Kannan to multivalued mappings and introduced the notion of a K-multivalued mapping. Rus [21] coined the term R-multivalued mapping, which is a generalization of a K-multivalued mapping (see also, [2]). Abbas and Rhoades [1] studied common fixed point problems for multivalued mappings and introduced the notion of generalized R-multivalued mappings which in turn generalizes R-multivalued mappings.

Let (X, d) be a metric space. Denote by $P(X)$ be the family of all nonempty subsets of X, and by $P_{cl}(X)$ the family of all nonempty closed subsets of X.

A point x in X is called fixed point of a multivalued mapping $T : X \rightarrow P_{cl}(X)$ provided $x \in Tx$. The collection of all fixed point of T is denoted by $Fix(T)$.

Recall that, a map $T : X \rightarrow P_{cl}(X)$ is said to be upper semicontinuous, if for $x_n \in X$ and $y_n \in Tx_n$ with $x_n \rightarrow x_0$ and $y_n \rightarrow y_0$, implies $y_0 \in Tx_0$ (see [24]).

Definition 20.2 Let X be a nonempty set. Then $(X, d, \preceq)$ is called partially ordered metric space if and only if d is a metric on a partially ordered set $(X, \preceq)$.

We define $\Delta_1, \Delta_2 \subseteq X \times X$ as follows:

$$\Delta_1 = \{(x, y) \in X \times X \mid x \preceq y\},$$
$$\Delta_2 = \{(x, y) \in X \times X \mid x \prec y\}.$$

Definition 20.3 A subset Γ of a partially ordered set X is said to be well-ordered if every two elements of Γ are comparable.

2   Common Fixed Point Theorems

In this section, we obtain common fixed point theorems for family of multivalued mappings. We begin with the following result.

Theorem 20.2 *Let* $(X, d, \preceq)$ *be an ordered complete metric space and* $\{T_i\}_{i=1}^{m}$: $X \rightarrow P_{cl}(X)$ *be family of multivalued mappings. Suppose that for every* $(x, y) \in \Delta_1$ *and* $u_x \in T_i(x)$, *there exists* $u_y \in T_{i+1}(y)$ *for* $i \in \{1, 2, \ldots, m\}$ *(with* $T_{m+1} = T_1$ *by convention) such that,* $(u_x, u_y) \in \Delta_2$ *implies*

$$\tau + F\left(d(u_x, u_y)\right) \leq F(M(x, y; u_x, u_y)), \tag{1}$$

where τ *is a positive real number and*

$$M(x, y; u_x, u_y) = \max\left\{d(x, y), d(x, u_x), d(y, u_y), \frac{d\left(x, u_y\right) + d\left(y, u_x\right)}{2}\right\}.$$

Then the following statements hold:

(i) *$Fix(T_i) \neq \emptyset$ for any $i \in \{1, 2, \ldots, m\}$ if and only if $Fix(T_1) = Fix(T_2) = \cdots = Fix(T_m) \neq \emptyset$.*

(ii) $Fix(T_1) = Fix(T_2) = \cdots = Fix(T_m) \neq \emptyset$ *provided that any one* T_i *for* $i \in \{1, 2, \ldots, m\}$ *is upper semicontinuous.*

(iii) $\bigcap_{i=1}^{m} Fix(T_i)$ *is well-ordered if and only if* $\bigcap_{i=1}^{m} Fix(T_i)$ *is singleton set.*

Proof To prove (i), let $x^* \in T_k(x^*)$ for any $k \in \{1, 2, \ldots, m\}$. Assume that $x^* \notin T_{k+1}(x^*)$, then there exists an $x \in T_{k+1}(x^*)$ with $(x^*, x) \in \Delta_2$ such that

$$\tau + F\left(d(x^*, x)\right) \leq F(M(x^*, x^*; x^*, x)),$$

where

$$M(x^*, x^*; x^*, x) = \max\left\{d(x^*, x^*), d(x^*, x^*), d(x, x^*), \frac{d(x^*, x) + d(x^*, x^*)}{2}\right\}$$
$$= d(x, x^*),$$

implies that

$$\tau + F\left(d(x^*, x)\right) \leq F(d(x^*, x)),$$

a contradiction as $\tau > 0$. Thus $x^* = x$. Thus $x^* \in T_{k+1}(x^*)$ and so $Fix(T_k) \subseteq Fix(T_{k+1})$. Similarly, we obtain that $Fix(T_{k+1}) \subseteq Fix(T_{k+2})$ and continuing this way, we get $Fix(T_1) = Fix(T_2) = \cdots = Fix(T_k)$. The converse is straightforward.

To prove (ii), suppose that x_0 is an arbitrary point of X. If $x_0 \in T_{k_0}(r_0)$ for any $k_0 \in \{1, 2, \ldots, m\}$, then by using (i), the proof is finished. So we assume that $x_0 \notin T_{k_0}(x_0)$ for any $k_0 \in \{1, 2, \ldots, m\}$. Now for $i \in \{1, 2, \ldots, m\}$, if $x_1 \in T_i(x_0)$, then there exists $x_2 \in T_{i+1}(x_1)$ with $(x_1, x_2) \in \Delta_2$ such that

$$\tau + F\left(d(x_1, x_2)\right) \leq F(M(x_0, x_1; x_1, x_2)),$$

where

$$M(x_0, x_1; x_1, x_2) = \max\left\{d(x_0, x_1), d(x_0, x_1), d(x_1, x_2), \frac{d(x_0, x_2) + d(x_1, x_1)}{2}\right\}$$
$$= \max\left\{d(x_0, x_1), d(x_1, x_2), \frac{d(x_0, x_2)}{2}\right\}$$
$$= \max\{d(x_0, x_1), d(x_1, x_2)\}.$$

Now, if $M(x_0, x_1; x_1, x_2) = d(x_1, x_2)$ then

$$\tau + F\left(d(x_1, x_2)\right) \leq F(d(x_1, x_2)),$$

a contradiction as $\tau > 0$. Therefore $M(x_0, x_1; x_1, x_2) = d(x_0, x_1)$ and we have

$$\tau + F\left(d(x_1, x_2)\right) \leq F\left(d(x_0, x_1)\right).$$

Next for this $x_2 \in T_{i+1}(x_1)$, there exists $x_3 \in T_{i+2}(x_2)$ with $(x_2, x_3) \in \Delta_2$ such that

$$\tau + F\left(d(x_2, x_3)\right) \le F(M(x_1, x_2; x_2, x_3)),$$

where

$$M(x_1, x_2; x_2, x_3) = \max\left\{d(x_1, x_2), d(x_1, x_2), d(x_2, x_3), \frac{d(x_1, x_3) + d(x_2, x_2)}{2}\right\}$$

$$= \max\{d(x_1, x_2), d(x_2, x_3)\}.$$

Now, if $M(x_1, x_2; x_2, x_3) = d(x_2, x_3)$ then

$$\tau + F\left(d(x_2, x_3)\right) \le F(d(x_2, x_3)),$$

a contradiction as $\tau > 0$. Therefore $M(x_1, x_2; x_2, x_3) = d(x_1, x_2)$ and we have

$$\tau + F\left(d(x_2, x_3)\right) \le F\left(d(x_1, x_2)\right).$$

Continuing this process, for $x_{2n} \in T_i(x_{2n-1})$, there exist $x_{2n+1} \in T_{i+1}(x_{2n})$ with $(x_{2n}, x_{2n+1}) \in \Delta_2$ such that

$$\tau + F\left(d(x_{2n}, x_{2n+1})\right) \le F\left(M(x_{2n-1}, x_{2n}; x_{2n}, x_{2n+1})\right),$$

where

$$M(x_{2n-1}, x_{2n}; x_{2n}, x_{2n+1}) = \max\left\{d(x_{2n-1}, x_{2n}), d(x_{2n-1}, x_{2n}), d(x_{2n}, x_{2n+1}),\right.$$

$$\left.\frac{d(x_{2n-1}, x_{2n+1}) + d(x_{2n}, x_{2n})}{2}\right\}$$

$$= \left\{d(x_{2n-1}, x_{2n}), d(x_{2n}, x_{2n+1}), \frac{d(x_{2n-1}, x_{2n+1})}{2}\right\}$$

$$\le d(x_{2n-1}, x_{2n}),$$

that is,

$$\tau + F\left(d(x_{2n}, x_{2n+1})\right) \le F\left(d(x_{2n-1}, x_{2n})\right).$$

Similarly, for $x_{2n+1} \in T_{i+1}(x_{2n})$, there exist $x_{2n+2} \in T_{i+2}(x_{2n+1})$ such that for $(x_{2n+1}, x_{2n+2}) \in \Delta_2$ implies

$$\tau + F\left(d(x_{2n+1}, x_{2n+2})\right) \le F\left(d(x_{2n}, x_{2n+1})\right).$$

Hence, we obtain a sequence $\{x_n\}$ in X such that for $x_n \in T_i(x_{n-1})$, there exist $x_{n+1} \in T_{i+1}(x_n)$ with $(x_n, x_{n+1}) \in \Delta_2$ such that

$$\tau + F\left(d(x_n, x_{n+1})\right) \le F\left(d(x_{n-1}, x_n)\right).$$

Therefore

$$F\left(d(x_n, x_{n+1})\right) \le F\left(d(x_{n-1}, x_n)\right) - \tau \le F\left(d(x_{n-2}, x_{n-1})\right) - 2\tau$$
$$\le \cdots \le F\left(d(x_0, x_1)\right) - n\tau. \tag{2}$$

From (2), we obtain $\lim_{n \to \infty} F\left(d(x_n, x_{n+1})\right) = -\infty$ that together with (F_2) gives

$$\lim_{n \to \infty} d(x_n, x_{n+1}) = 0.$$

From (F_3), there exists $\lambda \in (0, 1)$ such that

$$\lim_{n \to \infty} [d(x_n, x_{n+1})]^\lambda F\left(d(x_n, x_{n+1})\right) = 0.$$

From (2), we have

$$[d(x_n, x_{n+1})]^\lambda F\left(d(x_n, x_{n+1})\right) - [d(x_n, x_{n+1})]^\lambda F\left(d(x_0, x_{n+1})\right)$$
$$\le -n\tau [d(x_n, x_{n+1})]^\lambda \le 0.$$

On taking limit as $n \to \infty$ we obtain

$$\lim_{n \to \infty} n[d(x_n, x_{n+1})]^\lambda - 0.$$

Hence $\lim_{n \to \infty} n^{\frac{1}{\lambda}} d(x_n, x_{n+1}) = 0$ and there exists $n_1 \in \mathbb{N}$ such that $n^{\frac{1}{\lambda}} d(x_n, x_{n+1}) \le 1$ for all $n \ge n_1$. So we have

$$d(x_n, x_{n+1}) \le \frac{1}{n^{1/\lambda}}$$

for all $n \ge n_1$. Now consider $m, n \in \mathbb{N}$ such that $m > n \ge n_1$, we have

$$d(x_n, x_m) \le d(x_n, x_{n+1}) + d(x_{n+1}, x_{n+2}) + \cdots + d(x_{m-1}, x_m)$$
$$\le \sum_{i=n}^{\infty} \frac{1}{i^{1/\lambda}}.$$

By the convergence of the series $\sum_{i=1}^{\infty} \frac{1}{i^{1/\lambda}}$, we get $d(x_n, x_m) \to 0$ as $n, m \to \infty$. Therefore $\{x_n\}$ is a Cauchy sequence in X. Since X is complete, there exists an element $x^* \in X$ such that $x_n \to x^*$ as $n \to \infty$.

Now, if T_i is upper semicontinuous for any $i \in \{1, 2, \ldots, m\}$, then as $x_{2n} \in X$, $x_{2n+1} \in T_i(x_{2n})$ with $x_{2n} \to x^*$ and $x_{2n+1} \to x^*$ as $n \to \infty$ implies that $x^* \in T_i(x^*)$. Thus from (i), we get $x^* \in T_1(x^*) = T_2(x^*) = \cdots = T_m(x^*)$.

Finally to prove (iii), suppose the set $\cap_{i=1}^{m} Fix(T_i)$ is a well-ordered. We are to show that $\cap_{i=1}^{m} Fix(T_i)$ is singleton. Assume on contrary that there exist u and v such that $u, v \in \cap_{i=1}^{m} Fix (T_i)$ but $u \neq v$. As $(u, v) \in \Delta_2$, so for $(u_x, v_y) \in \Delta_2$ implies

$$\tau + F(d(u, v)) \leq F(M(u, v; u, v))$$
$$= F\left(\max \left\{ d(u, v), d(u, u), d(v, v), \frac{d(u, v) + d(v, u)}{2} \right\} \right)$$
$$= F(d(u, v)),$$

a contradiction as $\tau > 0$. Hence $u = v$. Conversely, if $\cap_{i=1}^{m} Fix (T_i)$ is singleton, then it follows that $\cap_{i=1}^{m} Fix (T_i)$ is a well-ordered. $\qquad \square$

The following corollary extends and generalizes Theorem 4.1 of [13] and Theorem 3.4 of [21] for two maps in ordered metric spaces.

Corollary 20.1 *Let* $(X, d, \preceq)$ *be an ordered complete metric space and* $T_1, T_2 :$ $X \to P_{cl}(X)$ *be two multivalued mappings. Suppose that for every* $(x, y) \in \Delta_1$ *and* $u_x \in T_i(x)$*, there exists* $u_y \in T_j(y)$ *for* $i, j \in \{1, 2\}$ *with* $i \neq j$ *such that,* $(u_x, u_y) \in$ Δ_2 *implies*

$$\tau + F\left(d(u_x, u_y) \right) \leq F(M(x, y; u_x, u_y)), \qquad (3)$$

where τ *is a positive real number and*

$$M(x, y; u_x, u_y) = \max \left\{ d(x, y), d(x, u_x), d(y, u_y), \frac{d(x, u_y) + d(y, u_x)}{2} \right\}.$$

Then the following statements hold:

(1) $Fix(T_i) \neq \emptyset$ for any $i \in \{1, 2\}$ if and only if $Fix(T_1) = Fix(T_2) \neq \emptyset$.
(2) $Fix(T_1) = Fix(T_2) \neq \emptyset$ provided that T_1 or T_2 is upper semicontinuous.
(3) $Fix(T_1) \cap Fix(T_2)$ is well-ordered if and only if $Fix(T_1) \cap Fix(T_2)$ is singleton set.

Example 20.1 Let $X = \{x_n = \frac{n(n+1)}{2} : n \in \{1, 2, 3, \ldots\}\}$ endow with usual order $\leq$. Let

$$\Delta_1 = \{(x, y) : x \leq y \text{ where } x, y \in X\} \text{ and}$$
$$\Delta_2 = \{(x, y) : x < y \text{ where } x, y \in X\}.$$

Define $T_1, T_2 : X \to P_{cl}(X)$ as follows:

$$T_1(x) = \{x_1\} \quad for \quad x \in X,$$
$$T_2(x) = \begin{cases} \{x_1\} & , \quad x = x_1 \\ \{x_1, x_{n-1}\}, & x = x_n, \quad for \quad n > 1. \end{cases}$$

Take $F(\alpha) = \ln \alpha + \alpha$, $\alpha > 0$ and $\tau = 1$. For a Euclidean metric d on X, and $(u_x, u_y) \in \Delta_2$, we consider the following cases:

(i) If $x = x_1$, $y = x_m$, for $m > 1$, then for $u_x = x_1 \in T_1(x)$, there exists $u_y = x_{m-1} \in T_2(y)$, such that

$$d(u_x, u_y)e^{d(u_x,u_y)-M(x,y;u_x,u_y)} \leq d(u_x, u_y)e^{d(u_x,u_y)-d(x,y)}$$
$$= \frac{m^2 - m - 2}{2}e^{-m}$$
$$< \frac{m^2 + m - 2}{2}e^{-1}$$
$$= e^{-1}d(x, y)$$
$$\leq e^{-1}M(x, y; u_x, u_y).$$

(ii) If $x = x_n$, $y = x_{n+1}$ with $n > 1$, then for $u_x = x_1 \in T_1(x)$, there exists $u_y = x_{n-1} \in T_2(y)$, such that

$$d(u_x, u_y)e^{d(u_x,u_y)-M(x,y;u_x,u_y)} \leq d(u_x, u_y)e^{d(u_x,u_y)-[\frac{d(x,u_y)+d(y,u_x)}{2}]}$$
$$= \frac{n^2 - n - 2}{2}e^{\frac{-3n-2}{2}}$$
$$< \frac{n^2 + 4n}{2}e^{-1}$$
$$= e^{-1}\left[\frac{d(x, u_y) + d(y, u_x)}{2}\right]$$
$$\leq e^{-1}M(x, y; u_x, u_y).$$

(iii) When $x = x_n$, $y = x_m$ with $m > n > 1$, then for $u_x = x_1 \in T_1(x)$, there exists $u_y = x_{n-1} \in T_2(y)$, such that

$$d(u_x, u_y)e^{d(u_x,u_y)-M(x,y;u_x,u_y)} \leq d(u_x, u_y)e^{d(u_x,u_y)-d(x,u_x)}$$
$$= \frac{n^2 - n - 2}{2}e^{-n}$$
$$< \frac{n^2 + n - 2}{2}e^{-1}$$
$$= e^{-1}d(x, u_x)$$
$$\leq e^{-1}M(x, y; u_x, u_y).$$

Now we show that for $x, y \in X$, $u_x \in T_2(x)$; there exists $u_y \in T_1(y)$ such that $(u_x, u_y) \in \Delta_2$ and (3) of Corollary 20.1 is satisfied. For this, we consider the following cases:

(i) If $x = x_n$, $y = x_1$ with $n > 1$, we have for $u_x = x_{n-1} \in T_2(x)$, there exists $u_y = x_1 \in T_1(y)$, such that

$$d(u_x, u_y)e^{d(u_x, u_y) - M(x, y; u_x, u_y)} \leq d(u_x, u_y)e^{d(u_x, u_y) - d(x, y)}$$
$$= \frac{n^2 - n - 2}{2}e^{-n}$$
$$< \frac{n^2 + n - 2}{2}e^{-1}$$
$$= e^{-1}d(x, y)$$
$$\leq e^{-1}M(x, y; u_x, u_y).$$

(ii) In case $x = x_n$, $y = x_m$ with $m > n > 1$, then for $u_x = x_{n-1} \in T_2(x)$, there exists $u_y = x_1 \in T_2(y)$, such that

$$d(u_x, u_y)e^{d(u_x, u_y) - M(x, y; u_x, u_y)} \leq d(u_x, u_y)e^{d(u_x, u_y) - d(y, u_y)}$$
$$= \frac{n^2 - n - 2}{2}e^{n^2 - n - m^2 - m}$$
$$< \frac{m^2 + m - 2}{2}e^{-1}$$
$$= e^{-1}d(y, u_y)$$
$$\leq e^{-1}M(x, y; u_x, u_y).$$

Hence all the conditions of Corollary 20.1 are satisfied. Moreover, $x_1 = 1$ is the unique common fixed point of T_1 and T_2 with $Fix(T_1) = Fix(T_2)$.

The following result generalizes Theorem 3.4 of [21] and Theorem 3.4 of [22].

Theorem 20.3 *Let $(X, d, \preceq)$ be an ordered complete metric space and $\{T_i\}_{i=1}^m$: $X \to P_{cl}(X)$ be family of multivalued mappings. Suppose that for every $(x, y) \in \Delta_1$ and $u_x \in T_i(x)$, there exists $u_y \in T_{i+1}(y)$ for $i \in \{1, 2, \ldots, m\}$ (with $T_{m+1} = T_1$ by convention) such that, $(u_x, u_y) \in \Delta_2$ implies*

$$\tau + F\left(d(u_x, u_y)\right) \leq F(M_2(x, y; u_x, u_y)), \qquad (4)$$

where τ is a positive real number and

$$M_2(x, y; u_x, u_y) = \alpha d(x, y) + \beta d(x, u_x) + \gamma d(y, u_y) + \delta_1 d(x, u_y) + \delta_2 d(y, u_x),$$

and $\alpha, \beta, \gamma, \delta_1, \delta_2 \geq 0$, $\delta_1 \leq \delta_2$ with $\alpha + \beta + \gamma + \delta_1 + \delta_2 \leq 1$. Then the following statements hold:

(I) *$Fix(T_i) \neq \emptyset$ for any $i \in \{1, 2, \ldots, m\}$ if and only if $Fix(T_1) = Fix(T_2) = \cdots = Fix(T_m) \neq \emptyset$.*

(II) $Fix(T_1) = Fix(T_2) = \cdots = Fix(T_m) \neq \emptyset$ provided that any one T_i for $i \in \{1, 2, \ldots, m\}$ is upper semicontinuous.

(III) $\cap_{i=1}^m Fix(T_i)$ is well-ordered if and only if $\cap_{i=1}^m Fix(T_i)$ is singleton set.

Proof To prove (I), let $x^* \in T_k(x^*)$ for any $k \in \{1, 2, \ldots, m\}$. Assume that $x^* \notin T_{k+1}(x^*)$, then there exists an $x \in T_{k+1}(x^*)$ with $(x^*, x) \in \Delta_2$ such that

$$\tau + F\left(d(x^*, x)\right) \leq F(M_2(x^*, x^*; x^*, x)),$$

where

$$M_2(x^*, x^*; x^*, x) = \alpha d(x^*, x^*) + \beta d(x^*, x^*) + \gamma d(x, x^*)$$
$$+ \delta_1 d(x^*, x) + \delta_2 d(x^*, x^*)$$
$$= (\gamma + \delta_1)d(x, x^*),$$

implies that

$$\tau + F\left(d(x^*, x)\right) \leq F((\gamma + \delta_1)d(x^*, x))$$
$$\leq F(d(x^*, x)),$$

a contradiction as $\tau > 0$. Thus $x^* = x$. Thus $x^* \in T_{k+1}(x^*)$ and so $Fix(T_k) \subseteq Fix(T_{k+1})$. Similarly, we obtain that $Fix(T_{k+1}) \subseteq Fix(T_{k+2})$ and continuing this way, we get $Fix(T_1) = Fix(T_2) = \cdots = Fix(T_k)$. The converse is straightforward.

To prove (II), suppose that x_0 is an arbitrary point of X. If $x_0 \in T_{k_0}(x_0)$ for any $k_0 \in \{1, 2, \ldots, m\}$, then by using (I), the proof is finishes. So we assume that $x_0 \notin T_{k_0}(x_0)$ for any $k_0 \in \{1, 2, \ldots, m\}$. Now for $i \in \{1, 2, \ldots, m\}$, if $x_1 \in T_i(x_0)$, then there exists $x_2 \in T_{i+1}(x_1)$ with $(x_1, x_2) \in \Delta_2$ such that

$$\tau + F\left(d(x_1, x_2)\right) \leq F(M_2(x_0, x_1; x_1, x_2)),$$

where

$$M_2(x_0, x_1; x_1, x_2) = \alpha d(x_0, x_1) + \beta d(x_0, x_1) + \gamma d(x_1, x_2)$$
$$+ \delta_1 d(x_0, x_2) + \delta_2 d(x_1, x_1)$$
$$\leq (\alpha + \beta + \delta_1)d(x_0, x_1) + (\gamma + \delta_1)d(x_1, x_2).$$

Now, if $d(x_0, x_1) \leq d(x_1, x_2)$, then we have

$$\tau + F\left(d(x_1, x_2)\right) \leq F((\alpha + \beta + \gamma + 2\delta_1)d(x_1, x_2))$$
$$\leq F(d(x_1, x_2)),$$

a contradiction. Therefore

$$\tau + F\left(d(x_1, x_2)\right) \leq F\left(d(x_0, x_1)\right).$$

Next for this $x_2 \in T_{i+1}(x_1)$, there exists $x_3 \in T_{i+2}(x_2)$ with $(x_2, x_3) \in \Delta_2$ such that

$$\tau + F\left(d(x_2, x_3)\right) \leq F(M_2(x_1, x_2; x_2, x_3)),$$

where

$$\begin{aligned} M_2(x_1, x_2; x_2, x_3) &= \alpha d(x_1, x_2) + \beta d(x_1, x_2) + \gamma d(x_2, x_3) \\ &\quad + \delta_1 d(x_1, x_3) + \delta_2 d(x_2, x_2) \\ &\leq (\alpha + \beta + \delta_1) d(x_1, x_2) + (\gamma + \delta_1) d(x_2, x_3). \end{aligned}$$

Now, if $d(x_1, x_2) \leq d(x_2, x_3)$ then

$$\begin{aligned} \tau + F\left(d(x_2, x_3)\right) &\leq F((\alpha + \beta + \gamma + 2\delta_1) d(x_2, x_3)) \\ &\leq F\left(d(x_2, x_3)\right), \end{aligned}$$

a contradiction as $\tau > 0$. Therefore

$$\tau + F\left(d(x_2, x_3)\right) \leq F\left(d(x_1, x_2)\right).$$

Continuing this process, for $x_{2n} \in T_i(x_{2n-1})$, there exist $x_{2n+1} \in T_{i+1}(x_{2n})$ with $(x_{2n}, x_{2n+1}) \in \Delta_2$ such that

$$\tau + F\left(d(x_{2n}, x_{2n+1})\right) \leq F\left(M_2(x_{2n-1}, x_{2n}; x_{2n}, x_{2n+1})\right),$$

where

$$\begin{aligned} M_2(x_{2n-1}, x_{2n}; x_{2n}, x_{2n+1}) &= \alpha d(x_{2n-1}, x_{2n}) + \beta d(x_{2n-1}, x_{2n}) + \gamma d(x_{2n}, x_{2n+1}) \\ &\quad + \delta_1 d(x_{2n-1}, x_{2n+1}) + \delta_2 d(x_{2n}, x_{2n}) \\ &\leq (\alpha + \beta + \delta_1) d(x_{2n-1}, x_{2n}) + (\gamma + \delta_1) d(x_{2n}, x_{2n+1}) \\ &\leq d(x_{2n-1}, x_{2n}), \end{aligned}$$

that is,

$$\tau + F\left(d(x_{2n}, x_{2n+1})\right) \leq F\left(d(x_{2n-1}, x_{2n})\right).$$

Similarly, for $x_{2n+1} \in T_{i+1}(x_{2n})$, there exist $x_{2n+2} \in T_{i+2}(x_{2n+1})$ such that for $(x_{2n+1}, x_{2n+2}) \in \Delta_2$ implies

$$\tau + F\left(d(x_{2n+1}, x_{2n+2})\right) \leq F\left(d(x_{2n}, x_{2n+1})\right).$$

Hence, we obtain a sequence $\{x_n\}$ in X such that for $x_n \in T_i(x_{n-1})$, there exist $x_{n+1} \in T_{i+1}(x_n)$ with $(x_n, x_{n+1}) \in \Delta_2$ such that

$$\tau + F\left(d(x_n, x_{n+1})\right) \leq F\left(d(x_{n-1}, x_n)\right).$$

Therefore

$$F\left(d(x_n, x_{n+1})\right) \le F\left(d(x_{n-1}, x_n)\right) - \tau \le F\left(d(x_{n-2}, x_{n-1})\right) - 2\tau$$
$$\le \cdots \le F\left(d(x_0, x_1)\right) - n\tau. \tag{5}$$

From (4), we obtain $\lim_{n \to \infty} F\left(d(x_n, x_{n+1})\right) = -\infty$ that together with (F_2) gives

$$\lim_{n \to \infty} d(x_n, x_{n+1}) = 0.$$

Follows the arguments those in proof of Theorem 20.2, $\{x_n\}$ is a Cauchy sequence in X. Since X is complete, there exists an element $x^* \in X$ such that $x_n \to x^*$ as $n \to \infty$.

Now, if T_i is upper semicontinuous for any $i \in \{1, 2, \ldots, m\}$, then as $x_{2n} \in X$, $x_{2n+1} \in T_i(x_{2n})$ with $x_{2n} \to x^*$ and $x_{2n+1} \to x^*$ as $n \to \infty$ implies that $x^* \in T_i(x^*)$. Thus from (I), we get $x^* \in T_1(x^*) = T_2(x^*) = \cdots = T_m(x^*)$.

Finally to prove (III), suppose the set $\cap_{i=1}^m Fix(T_i)$ is a well-ordered. We are to show that $\cap_{i=1}^m Fix(T_i)$ is singleton. Assume on contrary that there exist u and v such that $u, v \in \cap_{i=1}^m Fix(T_i)$ but $u \ne v$. As $(u, v) \in \Delta_2$, so for $(u_x, v_y) \in \Delta_2$ implies

$$\tau + F\left(d(u, v)\right) \le F(M_2(u, v; u, v)),$$

where

$$M_2(u, v; u, v) = \alpha d(u, v) + \beta d(u, u) + \gamma d(v, v)$$
$$+ \delta_1 d(u, v) + \delta_2 d(v, u)$$
$$= (\alpha + \delta_1 + \delta_2)\, d(x, y),$$

that is,

$$\tau + F\left(d(u, v)\right) \le F((\alpha + \delta_1 + \delta_2)\, d(x, y))$$
$$\le F(d(u, v)),$$

a contradiction as $\tau > 0$. Hence $u = v$. Conversely, if $\cap_{i=1}^m Fix(T_i)$ is singleton, then it follows that $\cap_{i=1}^m Fix(T_i)$ is a well-ordered.   □

The following corollary extends Theorem 3.1 of [21], in the case of family of mappings in ordered metric space.

Corollary 20.2 *Let* $(X, d, \preceq)$ *be an ordered complete metric space and* $\{T_i\}_{i=1}^m :$ $X \to P_{cl}(X)$ *be family of multivalued mappings. Suppose that for every* $(x, y) \in \Delta_1$ *and* $u_x \in T_i(x)$, *there exists* $u_y \in T_{i+1}(y)$ *for* $i \in \{1, 2, \ldots, m\}$ *(with* $T_{m+1} = T_1$ *by convention) such that,* $(u_x, u_y) \in \Delta_2$ *implies*

$$\tau + F\left(d(u_x, u_y)\right) \le F(\alpha d(x, y) + \beta d(x, u_x) + \gamma d(y, u_y)]), \tag{6}$$

where τ is a positive real number and $\alpha, \beta, \gamma \geq 0$ with $\alpha, \beta, \gamma \leq 1$. Then the conclusions obtained in Theorem 20.3 remains true.

The following corollary extends Theorem 4.1 of [13].

Corollary 20.3 *Let $(X, d, \preceq)$ be an ordered complete metric space and $\{T_i\}_{i=1}^m :$ $X \to P_{cl}(X)$ be family of multivalued mappings. Suppose that for every $(x, y) \in \Delta_1$ and $u_x \in T_i(x)$, there exists $u_y \in T_{i+1}(y)$ for $i \in \{1, 2, \dots, m\}$ (with $T_{m+1} = T_1$ by convention) such that, $(u_x, u_y) \in \Delta_2$ implies*

$$\tau + F\left(d(u_x, u_y)\right) \leq F(h[d(x, u_x) + d(y, u_y)]), \tag{7}$$

where τ is a positive real number and $h \in [0, \frac{1}{2}]$. Then the conclusions obtained in Theorem 20.3 remain true.

Corollary 20.4 *Let $(X, d, \preceq)$ be an ordered complete metric space and $\{T_i\}_{i=1}^m :$ $X \to P_{cl}(X)$ be family of multivalued mappings. Suppose that for every $(x, y) \in \Delta_1$ and $u_x \in T_i(x)$, there exists $u_y \in T_{i+1}(y)$ for $i \in \{1, 2, \dots, m\}$ (with $T_{m+1} = T_1$ by convention) such that, $(u_x, u_y) \in \Delta_2$ implies*

$$\tau + F\left(d(u_x, u_y)\right) \leq F(d(x, y)), \tag{8}$$

where τ is a positive real number. Then the conclusions obtained in Theorem 20.3 remain true.

The above corollary extends Theorem 4.1 of [13].

3 Conclusion

Recently many results appeared in the literature giving the problems related to the common fixed point for multivalued maps. In this paper we obtained the results for existence of common fixed points of family of maps that satisfying generalized F-contractions in ordered structured metric spaces. We presented some examples to show the validity of established results.

Acknowledgements Talat Nazir and Xiaomin Qi are grateful to the Erasmus Mundus project FUSION for supporting the research visit to Mälardalen University, Sweden, and to the Research environment MAM in Mathematics and Applied Mathematics, Division of Applied Mathematics, the School of Education, Culture and Communication of Mälardalen University for creating excellent research environment.

References

1. Abbas, M., Rhoades, B.E.: Fixed point theorems for two new classes of multivalued mappings. Appl. Math. Lett. **22**, 1364–1368 (2009)
2. Abbas, M., Khan, A.R., Nazir, T.: Common fixed point of multivalued mappings in ordered generalized metric spaces. Filomat. **26**(5), 1045–1053 (2012)
3. Abbas, M., Ali, B., Romaguera, S.: Fixed and periodic points of generalized contractions in metric spaces. Fixed Point Theory Appl. **243**, 1–11 (2013)
4. Abbas, M., Ali, B., Romaguera, S.: Generalized contraction and invariant approximation results on nonconvex subsets of normed spaces. Abstr. Appl. Anal. **2014** Article ID 391952, 1–5 (2014)
5. Acar, O., Durmaz, G., Minak, G.: Generalized multivalued F−contraction on complete metric spaces. Bull. Iranian Math. Soc. **40**(6), 1469–1478 (2014)
6. Al-Thagafi, M.A., Shahzad, N.: Coincidence points, generalized I− nonexpansive multimaps and applications. Nonlinear Anal. **67**, 2180–2188 (2007)
7. Berinde, M., Berinde, V.: On general class of multivalued weakly Picard mappings. J. Math. Anal. Appl. **326**, 772–782 (2007)
8. Cirić, L.: Generalization of Banach's contraction principle. Proc. Amer. Math. Soc. **2**(45), 267–273 (1974)
9. Cirić, L.: Multi-valued nonlinear contraction mappings. Nonlinear Anal. Theory Method Appl. **71**(7–8), 2716–2723 (2009)
10. Jungck, G., Rhoades, B.E.: Fixed points for set valued functions without continuity. Indian J. Pure Appl. Math. **29**, 227–238 (1998)
11. Kamran, T.: Multivalued f−weakly Picard mappings. Nonlinear Anal. **67**, 2289–2296 (2007)
12. Kannan, R.: Some results on fixed points. Bull. Calcutta. Math. Soc. **60**, 71–76 (1968)
13. Latif, A., Beg, I.: Geometric fixed points for single and multivalued mappings. Demonstratio Math. **30**(4), 791–800 (1997)
14. Latif, A., Tweddle, I.: On multivalued nonexpansive maps. Demonstratio. Math. **32**, 565–574 (1999)
15. Lazar, T., O'Regan, D., Petrusel, A.: Fixed points and homotopy results for Ciric-type multi-valued operators on a set with two metrices. Bull. Korean Math. Soc. **45**(1), 67–73 (2008)
16. Markin, J.T.: Continuous dependence of fixed point sets. Proc. Amer. Math. Soc. **38**, 545–547 (1973)
17. Minak, G., Helvasi, A., Altun, I.: Ciric type generalized F−contractions on complete metric spaces and fixed point results. Filomat. **28**(6), 1143–1151 (2014)
18. Mot, G., Petrusel, A.: Fixed point theory for a new type of contractive multivalued operators. Nonlinear Anal. Theory Method Appl. **70**(9), 3371–3377 (2009)
19. Nadler, S.B.: Multivalued contraction mappings. Pacific J. Math. **30**, 475–488 (1969)
20. Rhaodes, B.E.: On multivalued f−nonexpansive maps. Fixed Point Theory Appl. **2**, 89–92 (2001)
21. Rus, I.A., Petrusel, A., Sintamarian, A.: Data dependence of fixed point set of some multivalued weakly Picard operators. Nonlinear Anal. **52**, 1944–1959 (2003)
22. Sgroi, M., Vetro, C.: Multi-valued F−contractions and the Solution of certain functional and integral equations. Filomat. **27**(7), 1259–1268 (2013)
23. Sintunavarat, W., Kumam, P.: Common fixed point theorem for cyclic generalized multi-valued contraction mappings. Appl. Math. Lett. **25**(11), 1849–1855 (2012)
24. Turkoglu, D., Binbasioglu, D.: Some fixed-point theorems for multivalued monotone mappings in ordered uniform space. Fixed Point Theory Appl. **2011** Article ID 186237, 1–12 (2011)
25. Wardowski, D.: Fixed points of new type of contractive mappings in complete metric spaces. Fixed Point Theory Appl. **94**, 1–6 (2012)

Index

Printed in the United States
By Bookmasters